Antifreeze Proteins Volume 2

Hans Ramløv • Dennis Steven Friis
Editors

Antifreeze Proteins Volume 2

Biochemistry, Molecular Biology and Applications

Springer

Editors
Hans Ramløv
Department of Natural Sciences
Roskilde University
Roskilde, Denmark

Dennis Steven Friis
Copenhagen, Denmark

ISBN 978-3-030-41947-9 ISBN 978-3-030-41948-6 (eBook)
https://doi.org/10.1007/978-3-030-41948-6

This Springer imprint is published by the registered company Springer Nature Switzerland AG.
The registered company address is: Gewerbestrasse 11, 6330 Cham, Switzerland

This book is dedicated to the two great minds and most creative researchers within the field of antifreeze proteins and physiology of cold tolerance: Professor Arthur L. DeVries and the late Professor Karl Erik Zachariassen.

Preface

When I (Hans Ramløv) was asked by Springer Verlag if I would consider editing a book on antifreeze proteins, I was much honored and gladly took on the project. I asked my good friend and previous PhD student (Dennis Steven Friis) if he would join me as editor, as we could complement each other with my background in comparative physiology and his in molecular biology. He was also excited by the idea and gladly teamed up with me.

Two books have been published focusing on antifreeze proteins, the last one in 2010. Since then considerable progress has been made on various aspects of the subject. There are an increasing number of laboratories around the world taking part in the research on antifreeze proteins, and there is a series of conferences dedicated specifically to this subject. Thus, our understanding of the diversity, the structures, the mechanisms by which the antifreeze proteins interact with ice, the evolution, and the adaptive value as well as the ideas for applications and the problems involved in this has increased considerably during the past decade. We therefore thought it timely to edit a book where all aspects of antifreeze proteins were included showing the "state of the art" of the subject at this time. Achieving this could not be possible without the help from leading experts within the different branches of the antifreeze protein research. We therefore contacted the respective scientists and could luckily gather a strong team of contributing authors within a short time. We would like to extend our sincerest gratitude to all of them.

We would like to dedicate this book to the two brilliant scientists within the field of antifreeze proteins and physiological adaptations to cold in ectotherms: Professor Arthur L. DeVries and the late Professor Karl Erik Zachariassen.

Art DeVries is the key figure in the discovery of antifreeze proteins in the late 1960s, and he has ever since worked relentlessly to shed light on all aspects of the biology, role, and mechanism of these proteins, especially in polar fishes. His discoveries have opened a whole new field of research, which is now investigated by a large number of scientists and which has opened the eyes of the public to the exciting adaptations harbored by cold-tolerant ectothermic organisms.

Karl Erik Zachariassen was one of the most nondogmatic, unorthodox, and creative scientists one could ever meet. Together with Ted Hammel he discovered the ice-nucleating agents in insects, and during his career he made several large contributions to the understanding of drought and cold tolerance in insects. Karl Erik Zachariassen worked for many years on cold tolerance in Norwegian beetles. He always tried to put this work into a wider and more generalized perspective often making unexpected observations and sometimes bold statements, which would create heated discussions in the scientific community.

We thank them both for all their efforts, contributions, and enlightenment.

Roskilde, Denmark — Hans Ramløv
Copenhagen, Denmark — Dennis Steven Friis

Contents

Abbreviations

AA	Anti-agglomeration
ABA	Abscisic acid
ABP	Aluminum-binding peptide
ACC	Antarctic circumpolar current
ADIS	Anti-icing and de-icing systems
AFGL	Antifreeze glycolipid
AFGP	Antifreeze glycoprotein
AF(G)P	Antifreeze protein or antifreeze glycoprotein
AFP/AP	Antifreeze protein (xAFP, x = species or type identifier)
AFPP	Antifreeze potentiating protein
ALT	Alanine aminotransferase
AQP	Aquaporin
ASW	Amorphous solid water
BAC	Bacterial chromosome arm vector
BLAST(P)	Basic local alignment search tools (protein)
BSA	Bovine serum albumin
CCP	Carrot concentrate protein
CD	Circular dichroism
CDS	Coding sequence
CF	Carboxyfluorescein
CNT	Classic nucleation theory
CPD	Cryoprotective dehydration
CRD	Carbohydrate recognition domain
CTLD(cps)	C-Type lectin-like domain (containing proteins)
CT_{min}	Critical thermal minimum
DDC	Double siamond cages
DEPC	Dielaidoylphosphatidylcholine
DEPE	Dielaidoylphosphatidylethanolamine
DEPG	Dielaidoylphosphatidylglycerol
DGDG	Digalactosyldiacylglycerol

DMPC	Dimyristoylphosphatidylcholine
DMSO	Di-methyl sulfoxide
DPC	Dodecylphosphocholine
DPPC	Dipalmitoylphosphatidylcholine
DSC	Differential scanning microcalorimeter/microcalorimetry
DT	Delay time
EAC	Escape from adaptive conflict
EPC	Egg phophatidylcholine
EST	Expressed sequence tag
FFS	Forward-flux sampling
F_i	Ice fraction
FIPA	Fluorescent-based ice plane affinity
FP	Freezing point
FTIR	Fourier transform infrared
GC-MS	Gas chromatography mass spectrometry
GFP	Green fluorescent protein
GH	Growth hormone
GI	Gastrointestinal
HC	Hexagonal cages
HDA	High-density amorphous ice
HDL	High-density liquid
HES	Hydroxyethyl starch
hFP	hysteresis freezing point
HGW	Hyperquenched glassy water
hMP	hysteresis melting point
HP-DSC	High-pressure differential scanning microcalorimeter/microcalorimetry
HRP	Horse radish peroxidase
IBD	Ice-binding domain
IBF	Ice-binding face
IBP	Ice-binding protein
IBS	Ice-binding site
I_c	Cubic ice
IDTA	Infrared differential thermal analysis
I_h	Hexagonal ice
INA	Ice-nucleating agent
INLP	Ice-nucleator lipoprotein
INP	Ice-nucleating protein
IR	Infrared
IRI	Ice recrystallization inhibition
IRRINA	Ice recrystallization rate inhibition assay
ISP	Ice structuring protein
ITC	Isothermal calorimetry or calorimetric
JH	Juvenile hormone
KHI	Kinetic hydrate inhibition

LDA	Low-density amorphous ice
LDH	Lactate dehydrogenase
LDHI	Low-dosage hydrate inhibitor
LDL	Low-density liquid
LGT	Lateral gene transfer
LL	Ladderlectin
LLCP	Liquid-liquid critical point
LLT	Lower lethal temperature
LPIN	Lipoprotein ice nucleator
LT_{50}	Temperature with 50% mortality (survival studies)
LUV	Large unicellular vesicles
MD	Molecular dynamics
MGDG	Monogalactosyldiacylglycerol
MP	Melting point
MW	Monoatomic water
Mya	Million years ago
Myr	Million years
NANS	N-Acetylneuraminic acid synthase
ND	Nuclear diffraction
NIBF	Non-ice-binding face
NMR	Nuclear magnetic resonance
Nt	Nucleotide
ORF	Open reading frame
PBS	Phosphate-buffered saline
PCR	Polymerase chain reaction
PDB (XXXX)	Protein Data Bank (ID no.)
PDMS	Polydimethylsiloxane
PEG	Poly(Ethylene) glycol
PI	Phosphatidylinositol
PNAG	Poly-N-Acetylglucosamine
PNIPAAm	Poly(N-Isopropyl acrylamide)
PPII	Polyproline type II
PR	Pathogenesis-related
PSP	Pancreatic stone protein
PTFE	Polytetrafluoroethylene (Teflon)
PVA	Poly(Vinyl alcohol)
PVCap	Polyvinylcaprolactam
PVP	Polyvinylpyrrolidone
PVPip	Poly(N-Vinylpiperidone)
QAE	Quaternary aminoethyl
RCH	Rapid cold hardening
REG protein	Regenerating protein
RFU	Relative fluorescent unit
RIP	Recrystallization inhibition protein

ROS	Reactive oxygen species
SAS	Sialic acid synthase, eukaryotes (NeuB in prokaryotes). Also called NANS
SAXS	Small angle X-ray scattering
SCP	Supercooling point
SDM	Site-directed mutagenesis
SDS	Sodium dodecyl sulfate
SEM	Scanning electron microscopy
SHS	Superhydrophobic surface
sI, sII	Structures of ice types, I, II
SO	Southern Ocean
SP	Sulfopropyl
SPI	Soy protein isolate
SSR	Simple sequence repeats
STM	Scanning tunneling microscopy
TCA	Trichloroacetic acid
TEM	Transmitted electron microscopy
Tf	Freezing temperature
Tg	Glass transition temperature
TH/THA	Thermal hysteresis/thermal hysteresis activity
THF	Tetrahydrofuran/thermal hysteresis factor
THP	Thermal hysteresis protein
TL	Thaumatin-like
TLP	Trypsinogen-like protease
T_m	Melting temperature or transition temperature
TMB	Tetramethylbenzidine
TSS	Transcription start site
ULT	Upper lethal temperature
UTR	Untranslated region
VHDA	Very high-density amorphous ice
VSFG	Vibrational sum-frequency generation
WAP	West Antarctic Peninsula
XDM	X-ray diffraction
ZnO	Zinc oxide

Chapter 1
Contents of Volume 2—Antifreeze Proteins: Biochemistry, Molecular Biology, and Application

Hans Ramløv and Dennis Steven Friis

In the first volume of the book, we present the history of the antifreeze proteins (AFPs), their physiological role in freeze-avoiding or freeze-tolerating species, as well as the evolution and grouping of the antifreeze proteins. In the present volume we delve into the functions and interactions of antifreeze proteins at the molecular level. The volume is divided in three parts (Part I, II, and III of this work) followed by a final summary of both volumes. Part I focuses on the biochemistry of the antifreeze proteins and their interaction with both ice and water. Part II describes the mode of actions for obtaining thermal hysteresis and inhibition of recrystallization and how to measure it. Part III concerns the fields in which the applications of antifreeze proteins have come the furthest, as well as the possibilities of mutating and tailoring the proteins. In this chapter we give a short summary of the chapters that constitutes this second and final volume of the book.

1.1 Part I Biochemistry and Molecular Biology of Antifreeze Proteins

In **Chap. 2**, Dr. Erlend Kristiansen presents the characteristics of antifreeze proteins found in the different taxonomic phyla, focusing on arthropods and polar fish as several different AFPs have been found and characterized in these groups. As the chapter is the first of the second volume, it also contains some introductive information to the AFPs also found in the first volume. The characterization of the AFPs

H. Ramløv (✉)
Department of Natural Sciences, Roskilde University, Roskilde, Denmark
e-mail: hr@ruc.dk

D. S. Friis
Copenhagen, Denmark

H. Ramløv, D. S. Friis (eds.), *Antifreeze Proteins Volume 2*,
https://doi.org/10.1007/978-3-030-41948-6_1

covers their structure, isoform diversity, synthesis, tissue distribution, and the ice-binding sites. Much difference is seen among the AFPs, even between relatively closely related species such as the teleost fish, proposedly due the convergent evolution of the proteins.

In **Chap. 3**, Dr. Dennis Steven Friis and Prof. Dr. Hans Ramløv present current knowledge of the physicochemical properties of the AFPs, also with focus on the arthropods and polar fish. The properties this chapter concerns are protein sizes/weights put in relation to the antifreeze activity, solubility/hydrophobicity, and stability in regard to temperature and pH. The sizes of AFPs vary greatly, and in general, there seems to be a positive correlation between AFP size and the hysteresis the proteins can evoke. However, the different ways of quantifying the activity among research groups make the results difficult to compare. Though the investigated AFPs are pH stable, the thermostability varies; however, the stability experiments are quite sparse.

Both Chaps. 2 and 3 concludes that the AFPs are a quite diverse group of proteins, which essentially does not share other properties than their ice-binding ability which also defines them. Several species have developed AFPs convergently, explaining the large differences. The fact that different taxonomic phyla have AFPs also implies different demands of the proteins. For example, the temperatures the organisms are exposed to, and the way to prevent intruding ice varies in fish and insect, and plants for that matter, giving rise to AFPs with different properties.

In **Chap. 4**, Prof. Dr. Peter L. Davies et al. dive deep into the molecular level and describe the ice-binding sites of AFPs and other ice-binding proteins and their interaction with ice surfaces. The interaction between protein and ice is in itself an unusual ligand/substrate pair. The key features of protein ice-binding sites which are important for the ice interaction are described, as well as the variations of the proteins in aspect of sizes, ice planes they interact with and the degree of thermal hysteresis they can evoke. The difficulties in unraveling the ice recognition issue and the previous and current theories of the ice–protein interaction are discussed. The most plausible theory is the hydration shell model, where the interaction between the protein and free water molecules is the key to the subsequent binding to ice.

In **Chap. 5**, Prof. Dr. Ilja Karina Voets and Dr. Konrad Meister describe the mechanism and importance of water for protein function in general, as well as for AFPs in particular, including the formation of a hydration shell. Molecular dynamics models have been a valuable tool when investigating proteins' interaction with water molecules, and are supporting the theories of the hydration shell model. However, vibrational sum-frequency generation spectroscopy studies did not show the same ordered ice-like water for all types of AFPs. The different methods for studying protein–water interactions and their results on AFPs are discussed in this chapter.

Chapters 4 and 5 thus give a thorough insight to the interactions between antifreeze proteins, and their "substrate" H_2O, being in either its liquid or solid state. The fact that the substrate is H_2O and that the reaction occurs just around the phase transition from water to ice has made the studies quite difficult, and thus several theories of the binding mechanism have emerged during time. The prevailing theory of a hydration shell is confirmed by several studies; however, the degree of this hydration shell might differ among the different AFP types, and might thus only explain the mechanism partly. The studies continue on this complex and crucial interaction.

1.2 Part II Molecular Mechanisms Affected by Antifreeze Proteins

In **Chap. 6**, Dr. Erlend Kristiansen elaborates on the concept of thermal hysteresis and its connection to the Kelvin effect and vapor pressures, including the determinants of the magnitude of the hysteresis, as well as the mechanisms behind the "failure point," where the system returns to its equilibrium state. It is described how AFPs would cause thermal hysteresis when they bind to the ice surface, and what determines the limit of thermal hysteresis, or antifreeze activity, they cause. The differences between AFPs are discussed in relation to their antifreeze activity, including sizes and ice plane recognition, as are the general factors shown to affect antifreeze activity, such as protein solubility and effect of "salting-out."

The basic knowledge of the thermal hysteresis is well founded, and the expected binding of antifreeze proteins to ice surfaces aligns with the theory. Furthermore, the factors that affect the AFP potency seem to rely on both protein properties related to the ice-binding site and on properties of the solution. The latter is important when discussing applications of AFPs, but also an important factor to keep in mind when comparing AFP studies.

In **Chap. 7**, Dr. Carsten Budke and Prof. Dr. Thomas Koop address the AFPs' ability to inhibit recrystallization of ice, as well as their relevance in especially freeze-tolerant organisms. The chapter contains the theory behind recrystallization including the driving force and rates as well as a description of methods used to quantify recrystallization. Efficacy of AFPs and non-AFP molecules in regard to ice recrystallization inhibition are compared and discussed.

The ability of AFPs, or ice-binding proteins in general, to inhibit recrystallization has great interest as it makes them a valuable tool in, e.g., cryopreservation (see Chap. 11), where recrystallization is one of the greatest hazards, or in the making of cold hardy transgenic crops. Therefore, investigation in this aspect of the proteins is also very interesting. However, while we know much about the recrystallization inhibition, the issues now lies in how to utilize the proteins in the different applications.

In **Chap. 8**, Prof. Dr. Hans Ramløv and Dr. Dennis Steven Friis present the current reportings on other functions of AFPs than their primary function. The primary function of antifreeze proteins is with no doubt their ability to inhibit ice growth and recrystallization by interacting with the ice surface. However, a number of studies have revealed that AF(G)Ps may have secondary functions which at least in some cases are of adaptive value to the organism, i.e., interaction with membranes, anti-virulence in arthropods, and inhibition of plant pathogens.

The most studied secondary function is the interaction of AF(G)Ps with membranes. In these studies it has been shown that AF(G)Ps may change the phase transition temperature (T_m) of artificial model membranes as well as inhibiting leakage through the membranes during phase transition. In some interesting studies of an AFGP from the tick *Ixodes scapularis* (IAFGP) it is shown that this AFGP inhibits the formation of biofilm and thereby renders bacteria vulnerable to clearing out by the organisms' immune system. These properties were even extended to transgenic invertebrates and vertebrates. In plants, antifreeze proteins rarely show high levels of antifreeze activity and are probably evolved from pathogen-related proteins to inhibit recrystallization. At least some plant antifreeze proteins have retained the anti-pathogen property and are thus involved in combatting psychrophilic pathogens such as snow molds. The studies are few and varied; so more studies are needed both to elucidate the mechanisms behind these secondary properties and to elucidate the evolutionary questions of origin and adaptiveness.

In **Chap. 9**, Dr. Johannes Lørup Buch presents different methods used in various aspects of the field of AFPs. The chapter covers different techniques for quantification of antifreeze activity and inhibition of recrystallization. Furthermore, more qualitative techniques for detecting AFPs are presented, e.g., by observing crystal morphology during freezing of samples, mapping of hydration shell using tetrahertz spectroscopy, or by using, e.g., immunohistochemistry for visualizing the presence of AFPs in situ in different tissues.

Many different techniques are used in AFP research, and have been driven by a significant amount of creativity, which has helped developing the research. However, there are no standard assays or techniques to, e.g., quantify AFP activity, which is highly sensitive method, e.g., in regard to sample sizes, freezing rates, or solution composition. This also means that data comparison between scientific groups should be made with great care and not without reservation.

1.3 Part III Applications of Antifreeze Proteins

In **Chap. 10**, Dr. Nebahat Sule Ustun and Prof. Dr. Sadettin Turhan present the current status on antifreeze proteins in food. Many dietary sources naturally contain AFPs, as many different fish and plants express them. However, the AFPs are also used as an active ingredient in several processed foods, due to their beneficial effect during cold/freezing. The different ways that AFPs can be utilized, and what the effect in foods are, is presented, including lowering of the freezing point and recrystallization inhibition, mechanisms which are presented earlier in the book.

Besides the current studies on, and current uses of, AFPs in food, the authors also address the issue of AFPs as a reliable food additive, as this is often discussed when active ingredients are added to foods, as well as other factors that affect the usage of AFPs in foods. Here, the possibilities of producing and isolating vast amounts of AFPs, and the costs connected to it, are some main obstacles discussed as well. There is much development within the area of large-scale production of recombinant protein, and with the big potential the AFPs have, it is likely that AFPs will be an additive in more foods in the future.

In **Chap. 11**, Prof. Dr. Kelvin G. M. Brockbank et al. present their work on how AFP's (mainly from *Dendroides canadensis*) can be used in the field of cryopreservation, and discuss their findings in relation to other current research on the field. They present studies on cellular, tissue, and organ level. There are many different ways and assays to quantify the damage on or protection of the organic matter, in relation to cryopreservation, which are also presented in this chapter.

The potential of using different types of AFPs in relation to cryopreservation of different organic materials is discussed, as both positive and negative results have been generated. Also the potential in regard to hypothermic nonfrozen storage is discussed, as AFPs have also shown effects on cells and organs at temperatures above 0 °C. Though having great potential, much research is still needed to fully understand and exploit the usage of AFPs in cryopreservation.

In **Chap. 12**, Dr. Nicolas von Solms presents the studies concerning AFPs' effect on gas hydrate inhibition. Gas hydrates are a problem in the oil and gas industry where their formation disrupts the flow of the pipeline. The chapter describes the molecular basis of gas hydrates and their similarity to ice, and the current measures of preventing their formation along with the methods that can be used to measure and quantify the formation and inhibition of gas hydrates on a small scale. The efficacy of different gas hydrate inhibitors are compared to the AFPs and the pros and cons are discussed.

In an environmental perspective, the AFPs have a great potential as gas hydrate inhibitors, as they are biodegradable and more environmental friendly than the current inhibitors. However, though preliminary studies show good results, the experiments on larger scale, and eventually life size application, are hindered by the great cost of producing the required amounts of AFPs. Also, much research into this relatively new field is needed, both in regard to upscaled experiments and deeper into the AFPs, maybe finding or designing AFP variants with great gas hydrate inhibition capability.

In **Chap. 13**, Prof. Dr. EonSeon Jin et al. present the current knowledge on creating surfaces with AFP-covered surfaces. This subject has potential in several industries where icing of surfaces is a great hazard and costly. The chapter describes the different methods of creating anti-icing surfaces, with emphasis on surface coating. The different coating materials are described and results of these and comparable studies with an AFP are discussed.

This very new field (in regard to AFPs) has great potential, as many industries are in need of a good anti-icing solution. So far very little experiments have been conducted using AFPs, but the preliminary studies induce optimism. Much research is still needed to uncover the true potential of AFP-covered surfaces.

In **Chap. 14**, Dr. Dennis Steven Friis and Prof. Dr. Hans Ramløv present the general concepts of genetic manipulation and the results of the mutation studies that have been carried out on various AFPs. As no good screening method is available for antifreeze proteins, all mutants have been made by site-directed mutagenesis and investigated separately. The purposes of the mutation studies are many and include improvement of the protein purification possibilities, creation of fusion proteins, e.g., by fluorescent tagging, localization of ice-binding domains or exploring the ice-binding mechanisms by subtle changes in amino acid structures. Some studies have also focused on increasing the potency of the AFP, or on creating transgenic organisms, aiming to making them more cold hardy.

Though the field is hampered by the lack of a suitable screening method, the tool of genetic manipulation is still very valuable and can help answer many of the questions that still lie unanswered. As the knowledge of AFPs will expand, hopefully the industrial applications to utilize the proteins will expand as well. This will unlikely bring the need to tailor AFPs for specific purposes, and in this regard the genetic manipulation will be the essential tool as well.

In **Chap. 15**, we give a final summary and perspectives on the antifreeze protein field covering both volumes of this work.

Part I
The Biochemistry and Molecular Biology of Antifreeze Proteins

Chapter 2
Characteristics of Antifreeze Proteins

Erlend Kristiansen

2.1 Introduction

Antifreeze proteins (AFPs) and antifreeze glycoproteins (AFGPs) are characterized as a group only by their common ability to prevent existing ice crystals from growing in supercooled solutions. They are found in many different life forms inhabiting cold, and often ice-laden, habitats, acting as protective means against a hostile thermal environment. Some polar unicellular organisms, including diatoms, fungi and bacteria, excrete AFPs to modify their external icy environment (Hoshino et al. 2003; Janech et al. 2006; Hanada et al. 2014), and an Antarctic bacterium use a membrane-bound AFP to adhere onto floating ice, allowing it to reside in the nutrient-rich upper part of the water column (Bar Dolev et al. 2016). Many freeze-tolerant organisms, that adaptively allow their extracellular body fluids to freeze, produce proteins that are classified as AFPs, since they cause a separation of the melting and freezing temperatures of ice in vitro. Such organisms include many plants (Urrutia et al. 1992; Duman and Olsen 1993; Worrall et al. 1998) and arthropods (Tursman and Duman 1995; Duman et al. 2004; Wharton et al. 2009; Walters et al. 2009). These proteins presumably function to control the shape and distribution of the endogen extracellular ice mass.

AF(G)Ps act as antifreeze agents in freeze-avoiding organisms, i.e. animals that die if endogenous ice is formed and that consequently rely on supercooling of their body fluids to survive. They have been shown to stabilize the supercooled state by inactivating structures within the body fluids that could initiate freezing and by preventing ice from penetrating through the body wall of the animal (Olsen and Duman 1997a, b; Olsen et al. 1998; Duman 2002). They enable hypoosmotic bony fish to occupy the cold polar waters, where these fishes may spend their entire lives

E. Kristiansen (✉)
NTNU University Library, Trondheim, Norway
e-mail: erlendkr@ntnu.no

H. Ramløv, D. S. Friis (eds.), *Antifreeze Proteins Volume 2*,
https://doi.org/10.1007/978-3-030-41948-6_2

in a supercooled state, often in contact with external ice (DeVries 1982). The evolution of the AF(G)Ps of polar fish has been driven by the cooling of the Arctic and Antarctic waters, processes that resulted in subfreezing water temperatures being reached some 5–14 million years ago in the Antarctic, and 13–18 million years ago in the Arctic (Kennett 1977; Eastman 1993).

They are also found in many freeze-avoiding terrestrial arthropods, including insects and spiders (Husby and Zachariassen 1980; Duman et al. 2004) and collembolans (Graham and Davies 2005; Hawes et al. 2014). Even in these terrestrial life forms, they may provide protection against lethal freezing throughout the supercooling range of the animal, on occasion down to –30 °C or below (Zachariassen and Husby 1982). Thus, these structures have common functions in diverse organisms associated with life in a cold environment.

AF(G)Ps are categorized as hyperactive or moderately active, based on their potency to cause antifreeze activity at equimolar concentrations. In addition to the distinct differences in antifreeze potency, the shape of the ice crystals that forms in the presence of moderately active and hyperactive AF(G)Ps are also characteristic: hexagonal bipyramids (e.g. Baardsnes et al. 2001; Loewen et al. 1998; Ewart et al. 1998) and flattened hexagonal discs, respectively (e.g. Liou et al. 2000; Graether et al. 2000). The underlying structural cause of the differences between these two activity groups appears to be differences in their ice-binding sites (IBS).

The intention of this chapter is to point to some structural, physiological and evolutionary characteristics of the AF(G)Ps found in freeze-avoiding polar fish and arthropods. It is by no means exhaustive, and it is referred to Chaps. 5 and 6 of Vol. 1 for further discussion of fish and insect AF(G)Ps and Chaps. 7 and 8 of Vol. 1 for AFPs in plants and other species. Chapter 9 of Vol. 1 and Chap. 4 of this volume give more in-depth analysis of evolutionary aspects and the interaction between AF (G)Ps and ice, respectively, and Chap. 6 of this volume focuses on the antifreeze mechanism.

2.2 Structure

The independent evolution of AF(G)Ps in various taxa has resulted in structural diversity within this functionally defined group (Graether et al. 2000; Fletcher et al. 2001; Graham and Davies 2005; Graham et al. 2007; Kiko 2010; Lin et al. 2011; Hawes et al. 2014). However, structural similarities are also abundant.

2.2.1 Polar Fish

There are currently reported five distinct kinds of antifreeze proteins in polar fish: AFGP and AFP type I–IV. However, the categorization of AFP type IV as a functional AFP has recently been questioned (see below). Table 2.1 shows the

Table 2.1 Taxonomic listing of the AF(G)Ps of polar fish

Subdivision Teleostei	Family	Genus/species	Type
Infradivision Clupeomorpha	Clupeidae	Herring	II (+ Ca^{2+})
Infradivision Euteleostei			
Superorder Protocanthopterygii	Osmeridae	Smelt	II (+ Ca^{2+})
Superorder Paracanthopterygii	Gadidae	Northern cods	AFGP
Superorder Acanthopterygii			
Order Scorpaeniformes			
Suborder Cottoidei			
Superfamily Cottoidea	Cottidae	Sculpins	I/IV
	Hemitripteridae	Sea raven	II (− Ca^{2+})
	Agonidae	Longsnout poacher	II (− Ca^{2+})
Superfamily Cyclopteroidea	Cyclopteridae	Snailfish	I
Order Perciformes			
Suborder Labridae	Labridae	Cunner	I
Suborder Zoarcoidei	Zoarcidae	Eelpouts	III
	Anarhichadidae	Wolf fish	III
Suborder Notothenioidei	5 families		AFGP/IV
Order Pleuronectiformes	Pleuronectidae	Right-eyed flounders	I

taxonomic occurrence of the AF(G)Ps, and their structures are illustrated in Fig. 2.1. As can be seen from the table, similar types of AF(G)Ps are scattered among distantly related groups of teleosts. These patterns of distribution have for the different kinds been attributed to convergent evolution (Chen et al. 1997a, b; Graham et al. 2013), to lateral gene transfer (Graham et al. 2008a, 2012) and to development from a common ancestor (Graham et al. 2013). Most fish AF(G)Ps are reportedly moderately active, with the exception of some large variants that are hyperactive.

2.2.1.1 Type I

The type I AFPs are α-helical proteins (Yang et al. 1988), see Fig. 2.1a. There are three kinds of AFP type I, based on their genetics and the size of the mature proteins. The overall structure is amphipathic, with the ice-binding side somewhat hydrophobic (Baardsnes et al. 2001). They are widely distributed among bony fishes, having been identified in members of four superfamilies in three different orders, namely the Pleuronectiformes (in flounders), Perciformes (in cunners) and Scorpaeniformes (in snailfish and sculpins) (Hew et al. 1980; Evans and Fletcher 2001; Hobbs et al. 2011), see Table 2.1.

There are two subsets of type I AFP within each species examined, coded by two different gene families; the liver-type AFPs have signal peptides, and these isoforms are secreted into the blood stream (Gourlie et al. 1984). The skin-type, in contrast, lack such signal peptides and are mostly located within skin and other peripheral

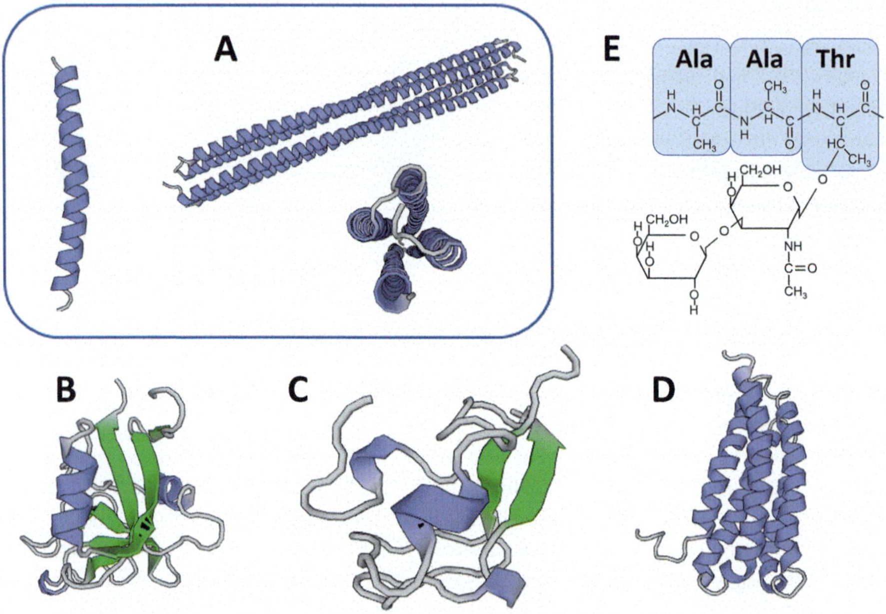

Fig. 2.1 The five different kinds of AF(G)Ps in polar fish. (**a**) Type I (PDB 1WFA) together with the hyperactive AFP maxi (PDB 4KE2). (**b**) Type II (PDB 2PY2). (**c**) Type III (PDB 1HG7). (**d**) Type IV, the illustration is of Apolipophorin III, a structural homologue of AFP type IV (PDB 1LS4). (**e**) The principal AAT repeat unit of AFGPs showing its o-link to its disaccharide. The different illustrations do not show correct proportions to each other. Colour codes: *Grey*: peptide backbone. *Blue*: α-helix. *Green*: β-strands

tissues (Gong et al. 1996; Low et al. 1998; Evans and Fletcher 2006). Both these kinds of isoforms are small peptides with masses of about 3.3–4.5 kDa. The circulating liver-type AFPs of the flounders (Gourlie et al. 1984; Graham et al. 2008a) and the cunner (Hobbs et al. 2011) are constructed from 3–4 repeats of an 11-amino acid sequence TxxD/Nxxxxxxx, where x is usually Ala (Chao et al. 1996), whereas the circulating liver-type in snailfish lacks such a basic repeat (Evans and Fletcher 2005a). The skin-type of flounders, longhorn sculpins and cunner are very similar to each other and constructed from the same 11-amino acid repeat seen in the liver-type of flounder and cunner (Low et al. 2001). In addition, shorthorn sculpin has a larger 95 amino acid skin-type isoform that lacks repeat pattern (Low et al. 1998), and the skin-types of snailfish, as is the case of its liver-type, lack the 11-amino acid repeat (Evans and Fletcher 2005a).

A third kind of AFP type I is found in several Pleuronectiformes and is characterized by being much larger than the other skin- and liver-types. In addition, this kind is hyperactive. Winter flounder (*Pseudopleuronectes americanus*), yellowtail flounder (*Limanda ferruginea*) and American plaice (*Hippoglossoides platessoides*) each contains a large hyperactive isoform of type I (Gauthier et al. 2005; Graham et al. 2008b). The best studied of these is that of the winter flounder, and this variant is denoted Maxi, see Fig. 2.1a. Such a large type I AFP is the sole AFP known from the blood of American plaice (Gauthier et al. 2005). These 17 kDa molecules are constructed from similar 11 residue repeats seen in many of the smaller forms (Graham et al. 2008b). They are dimers in solution of mass 34 kDa, and each monomer folds back onto itself, resulting in a four-helix bundle (Sun et al. 2014). Interestingly, comparable folding patterns have also been proposed for an AFP from a fungus (Badet et al. 2015) and from a Hymenopteran insect (Xu et al. 2018), hinting to an effective configuration for ice binding.

Graham et al. (2013) proposed that the wide phylogenetic distribution of type I AFP is the result of independent evolution of these proteins within each of the four superfamilies they are found. This proposal was based on studies of their genetic sequences, that revealed differences in both codon usage and non-coding regions, strongly suggesting different progenitors in the four groups. Gauthier et al. (2005) suggested that the smaller isoforms of flounders may have evolved from the larger AFP I types in this group. This was based on the observation that American plaice only contain a single large isoform. Evans and Fletcher (2005b) suggested that the AFPs of snailfish may have resulted from a shift in the reading frame of genes coding for eggshell proteins or keratin.

2.2.1.2 Type II

Type II AFPs are homologue to the carbohydrate recognition domain of Ca^{2+}-dependent (C-type) lectins (Ewart et al. 1998; Loewen et al. 1998). They are found in species from four different families from three distantly related groups of teleosts (see Table 2.1). Herring (Clupeidae) is from the infradivision Clupeomorpha, whereas smelt (Osmeridae), sea raven (Hemitripteridae) and poacher (Agonidae) are from different groups within the infradivision Euteleostei. The latter two are from the same superfamily, whereas smelt is from a different superorder.

Type II AFPs have masses varying from 14 to 24 kDa and an overall globular structure consisting of two α-helixes and nine β-strands in two β-sheets (Gronwald et al. 1998, see also Fig. 2.1b). The observed three-dimensional folding pattern is very similar to rat mannose-binding protein, a member of the family of C-type lectins from which they are likely derived. Type II AFPs are unique in having five internal SS bonds rather than 2–4 such bonds found in C-type lectins.

There are two distinct kinds of Type II AFPs; those isolated from smelt (Osmeridae) and herring (Clupeidae) require Ca^{2+} as a cofactor for activity, whereas those isolated from sea raven (Hemitripteridae) and poacher (Agonidae) are fully active in the absence of this cofactor. The IBS of these Ca^{2+}-dependent and Ca^{2+}-independent forms are located at different parts of their surfaces. Those that require Ca^{2+} for activity have IBS corresponding to the carbohydrate-binding site of C-type lectins (Ewart et al. 1998), whereas the IBS of the Ca^{2+}-independent variants are located outside this region (Loewen et al. 1998).

All AFP II have a unique SS-bond pattern not seen in related proteins and they also share great (>85%) identity in both amino acid sequence and conserved genetic sequences, including intron and exon regions. Due to this great similarity among the AFP type II, Graham et al. (2008a) and Sorhannus (2012) proposed that their scattered phylogenetic pattern of distribution is unlikely to be the result of convergent evolution, as in the case of type I AFPs. Instead, it is probably the result of a transfer of genes between the different groups of AFP type II-producing fish. Such so-called lateral gene transfer may have occurred during events of mass spawning. In the case of the Ca^{2+}-dependent AFP type II, Graham et al. (2012) found evidence to suggest that smelt was the recipient of genetic material from herring.

2.2.1.3 Type III

Type III AFPs are 7 kDa globular proteins only found in the two closely related families Zoarcidae (eelpouts) and Anarhichadidae (wolf fish) in the suborder Zoarcoidei, see Fig. 2.1c. The primary sequence has no obvious repeats and the folding pattern is complex, involving several short strands paired in two antiparallel β-sheets, in addition to several helixes.

Type III AFPs are found in two structural variants that are categorized by their isoelectric points (Chao et al. 1993). One group, the QAE forms, has pI below 7 and are consequently anionic at physiological pH, whereas the other group, the SP forms, has pI above 7 and are therefore cationic at physiological pH. Both QAE and SP forms are present in the animal. The SP forms reportedly have a lower activity than the QAE forms (Nishimiya et al. 2005). Takamichi et al. (2009) reported that the addition of minute amounts of a fully active QAE form to an inactive SP form isolated from the Japanese fish *Zoarces elongatus Kner* resulted in the SP form obtaining the same activity as the QAE form. These findings suggest that these two forms may cooperate in vivo. A natural 14 kDa intramolecular dimer has been identified, where two monomeric AFP III are linked by a short strand (Miura et al. 2001).

Since the occurrence of AFP type III is confined only to two closely related families of fishes, these forms presumably originated in a common ancestor (Graham et al. 2013). Baardsnes and Davies (2001) reported that the protein sequence of a type III AFP showed about 40% identity and 50% similarity to parts of the C-terminal domain of sialic acid synthase, an enzyme that binds carbohydrate as part of its function. Deng et al. (2010) elaborated on the evolutionary events that presumably preceded the development of today's type III AFP. Apparently, the N-terminal part of a functional sialic acid synthase molecule, that showed rudimentary antifreeze activity associated with its C-terminal, was replaced by a signal peptide. This caused the AFP-precursor to be secreted from the cells, and this molecular de-coupling of the enzymatic and antifreeze functions allowed selective pressure to act solely towards the antifreeze function.

2.2.1.4 Type IV

Type IV AFP is a 12 kDa lipoprotein-like protein with about 60% α-helix content, see Fig. 2.1d. Its proposed structure consists of four amphipathic α-helixes of similar length folded in a four-helix bundle (Deng and Laursen 1998). Type IV AFP has been found in many species, including Arctic longhorn sculpin (*Myoxocephalus octodecemspinosus*) and shorthorn sculpin (*M. scorpius*) (Deng and Laursen 1998; Gauthier et al. 2008) and two Antarctic nototheniids, *Pleuragramma antarcticum* and *Notothenia coriiceps* (Lee et al. 2011; Lee and Kim 2016). However, its role as a functional AFP has been questioned, since it is a very weak AFP, causing only 0.07 °C thermal hysteresis at a concentration of 0.5 mg/mL, and is present in blood in concentrations less than 100 μg/mL, far too low to protect these fishes against freezing in icy waters (Gauthier et al. 2008; Lee and Kim 2016). Its ability to cause thermal hysteresis could therefore be incidental. Gauthier et al. (2008) proposed that, although type IV has the potential to develop into a functional AFP, it has not been selected for this purpose due to the presence of other functional AFPs. This is supported by the presence of type IV AFP in temperate, subtropical and tropical species, including species living in fresh water (Liu et al. 2009; Xiao et al. 2014; Lee et al. 2011; Lee and Kim 2016). These species have no need for any freeze protection, and type IV AFP may instead be involved in embryogenesis, since several of its homologues are essential in this process.

2.2.1.5 AFGPs

AFGPs are found in two distantly related and geographically separate groups of teleost fish, the Arctic cods (family Gadidae of the superorder Paracanthopterygii) and the Antarctic Nototheniids, (suborder Notothenioidei of the superorder Acanthopterygii). They contain a varying number of the tripeptide AAT, where the hydroxyl group of each Thr is O-linked to a disaccharide (β-D-galactosyl-(1,3)-α-D-*N*-acetylgalactosamine), see Fig. 2.1e for an illustration of the basic unit. In this unit, the carbohydrate moiety makes up about 60% of the mass. The smallest variants contain only 4 of these repeat units and have a mass of about 2.6 kDa and the largest contain about 50 repeat units with a mass of 33 kDa. The differently sized AF(G)Ps are arranged into eight distinct size groups (DeVries 1982), and each group contains a number of isoforms (Wu et al. 2001).

The secondary structure of AFGPs has been difficult to elucidate. There is mounting evidence to suggest that they obtain a type II polyproline helix, but only at low temperatures (Franks and Morris 1978; Bush et al. 1984; Mimura et al. 1992; Tachibana et al. 2004). In this configuration, each triplet AAT makes one turn in the coil, resulting in the carbohydrate units being in a regular arrangement on one side of the molecule. Such an arrangement gives the molecule and overall amphipathic character, where the carbohydrate side is more hydrophilic, and the protein backbone with the methyl group of Ala, is more hydrophobic. The shape of the ice crystals that form in the presence of AFGPs also suggests a regular configuration; these ice crystals are hexagonal bipyramids, exposing only a single crystal plane to the surrounding solution onto which the AFGPs are adsorbed. Such crystal plane specificity likely requires that all adsorbed molecules have the same configuration.

Wöhrmann (1996) reported that an exceptionally large 150 kDa AFGP from the nototheniid *Pleuragramma antarcticum* was hyperactive. No other AFGP is known to be hyperactive.

The AF(G)Ps found in Gadoids and nototheniids, members of different superorders of teleosts, have evolved independently (Chen et al. 1997a). Those of the Antarctic nototheniids apparently evolved from a trypsinogen gene (Chen et al. 1997b) some 5–14 million years ago, whereas those of the Arctic gadoids evolved from a non-coding part of their DNA some 13–18 million years ago (Baalsrud et al. 2018). The timing of their independent emergence coincides well with the reported time the Antarctic and Arctic waters reached subfreezing temperatures (Kennett 1977; Eastman 1993).

2.2.2 *Arthropods*

Table 2.2 shows a taxonomic listing of known or tentative arthropod AFPs with some structural features indicated. The table suggests that AFPs in closely related species are homologue structures with a common progenitor. Almost all arthropod AFPs are constructed as shorter repetitive segments in series and almost all contain variations of the tripeptide pattern TxT within the repeats. The table also shows the high prevalence of the β-helical folding pattern, a feature that undoubtedly has evolved by convergent evolution in distantly related groups (Liou et al. 2000; Graether et al. 2000; Graether and Sykes 2004). Some of the variants of AFPs found in arthropods are illustrated in Fig. 2.2.

2.2.2.1 Insects

There is structural information available on AFPs or putative AFPs from five orders of insects, Coleoptera, Hymenoptera, Lepidoptera, Diptera and Hemiptera.

Coleoptera The beetles within the superfamily Tenebrionidea all have AFPs with very similar sequences that most likely are homologue structures (Table 2.2). These AFPs are constructed of 5–7 tandem repeats of the 12 or 13-mer consensus amino acid sequence TCTxSxxCxxAx. Notably, the Thr in position 1 and 3 and the Cys in position 2 and 8 in the repeat are highly conserved in isoforms within and between species.

The conserved positions of the Cys within the 12-mer repeat structure observed in the AFPs identified from species within the superfamily Tenebrionidea results in every sixth residue in the sequence being occupied by a Cys. The two Cys within

Table 2.2 Taxonomic listing and structural features of known and putative AFPs from arthropods

Phylum Arthropoda	Family	Species	Code	MW (kDa)	Primary repeat	Secondary
Class Entognatha						(D) Antiparallel L-h PPII helixes,
Order Collembola	Hypogastruridae	*Hypogastrura harveyi*[1,2]	sfAFP	6.5 and 15.7	Gxx	stacked in two sets.
		Gomphiocephalus hodgsoni[21]	GomphyAFP	9	? Rich in Gly and Cys	?
Class Insecta						
Order Coleoptera						
Intra order Cucujiformia						
Superfamily Tenebrionoidea	Tenebrionidae	*Tenebrio molitor*[6,7,19]	TmAFP	8.3–12	TCTxSxxCxxAx (x)	(D) R-h β-helix
		Dendroides canadensis[8]	DAFP	7.3–12.4	”	(A) ” (sim. to TmAFP)
		Microdera punctipennis[9]	MpAFP	12.7	”	(A) ” (”)
		Pterocoma loczyi[c,10]	PLAFP	~ 12	”	(A) ” (”)
		Anatolica polita[c,11]	ApAFP	10.9 and 11.4	”	(A) ” (”)
Superfamily Cucujoidea	Cucujidae	*Cucujus clavipes*[a]			”	(A) ” (”)
Superfamily Chrysomeloidea	Cerambycidae	*Rhagium inquisitor*[12,13]	RiAFP	13	TxTxTxT + x_{9-15}	(D) Flattened β-helix
		R. mordax[14]	RmAFP	13	”	(A) ” (sim. to RiAFP)
Infra order Scarabaeiformia						
Superfamily Scarabeaoidea	Lucanidae	*Dorcus curvidens*[b]		11.4–14.3	TCTxSxxCxxAx (x)	(A) R-h β-helix (sim. to TmAFP)
Order Hymenoptera						

(continued)

Table 2.2 (continued)

Phylum Arthropoda	Family	Species	Code	MW (kDa)	Primary repeat	Secondary
Sub order Apocrita	Apidae	*Apis cerena cerena*[22]	AcerAFP	60		(M) 3 α-helixes looped together
Order Diptera	Chironomidae	Sp. "Lake Ontario midge"[16]		5.7–10.4	xxCxGxYCxG. Glyco.	(M) L-h solenoid coil
Order Hemiptera						
Suborder Heteroptera	Scutelleridae	*Eurygaster maura*[c,15]	EmAFP	10.2	TxT + x_{10}	(M) L-h β-helix
Order Lepidoptera						
Superfamily Tortricoidea	Tortricidae	*Choristoneura fumiferana*[3,4]	CfAFP	9–12	TCT + x_{12}	(D) L-h β-helix
		and sister species				
Superfamily Geometroidea	Geometridae	*Campaea perlata*[5,6]	iwAFP	3.5 & 8.3	TxTxTxTxTxxx	(M) R-h flattened β-helix
Class Arachnida						
Order Ixodida	Ixodidae	*Ixodes scapularis*[c,17]	IAFGP	~ 23	TAA Probably Glyco.	?
Order Trombidiformes	Tetranychidae	*Tetranychus urticae*[c,18]		10–21	NCTxCxxCxNCx	(M) β-helix
Class Maxillopoda						
Order Calanoida	Stephidae	*Stephos longipes*[20]		26 kDa	No apparent repeat	β-helix with a parallel α-helix

Abbreviations: (A): assumed by this author based on sequence similarity. (D): determined. (M): modelled. L-h: Left-handed. R-h: right-handed. Sim. to: similar to. Glyco.: Glycosylated. [a]Mentioned in Duman (2015). [b]Sequence only published in NCBI. [c]Only assumed to be an AF(G)P, as no hysteresis activity is reported. (1) Graham and Davies (2005); (2) Pentelute et al. (2008); (3) Tyshenko et al. (2005); (4) Graether et al. (2000); (5) Lin et al. (2011); (6) Graham et al. (2007); (7) Liou et al. (2000); (8) Andorfer and Duman (2000); (9) Qiu et al. (2010); (10) Ma et al. (2008); (11) Ma et al. (2012); (12) Kristiansen et al. (2011); (13) Hakim et al. (2013); (14) Kristiansen et al. (2012); (15) Guz et al. (2014); (16) Basu et al. (2015); (17) Neelakanta et al. (2010); (18) Bryon et al. (2013); (19) Liou et al. (1999); (20) Kiko (2010); (21) Hawes et al. (2014); (22) Xu et al. (2018)

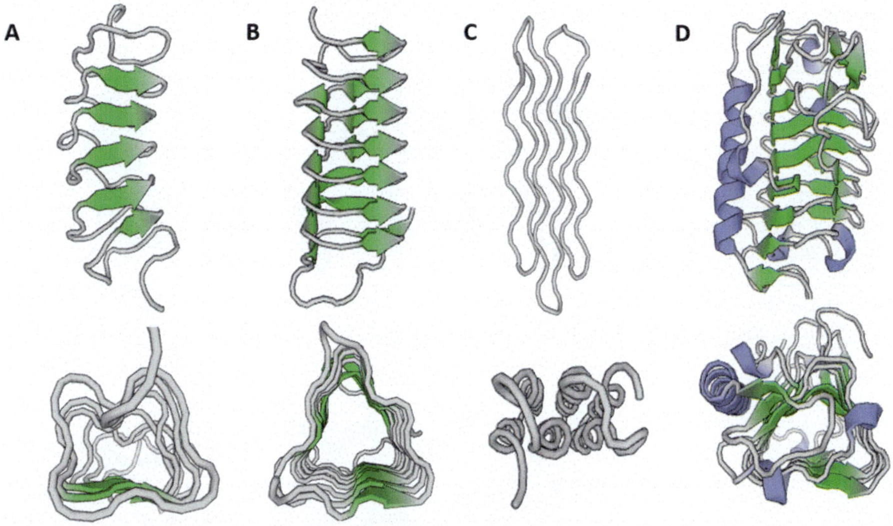

Fig. 2.2 Some different kinds of AFPs from Arthropods. (**a**) TmAFP from the coleopteran *T. molitor* (PDB 1L1I). (**b**) CfAFP from the lepidopteran *C. fumiferana* (PDB 1M8N). (**c**) An AFP from the collembolan *Hypogastrura harveyi* (PDB 2PNE). (**e**) A crustacean AFP from *Stephos longipes*. The illustration is of the AFP from *Colwellia* sp., a structural homologue (PDB 3WP9). The upper illustrations are frontal views, the lower illustrations are views from the top. The different illustrations do not show correct proportions to each other. Colour codes: *Grey*: peptide backbone. *Blue*: α-helix. *Green*: β-strands

each repeat form an SS bond (Li et al. 1998a; Liou et al. 2000). Liou et al. (2000) showed that the AFPs of *Tenebrio molitor*, TmAFP, fold as a tight regular right-handed solenoid, where each 12-mer repeat segment form one full turn in the coil. Each segment forms β-strands and the strands form β-sheets. This folding pattern results in a β-helix where the Thr residues in position 1 and 3 in each repeat are stacked on one side of the structure and form a highly regular ladder of 5–7 TCT motifs. The side chains of the Thr residues within each motif point outward from the structure, whereas the SS bonds between position 2 and 8 within each repeat cross the coil in a regular manner, contributing to the tightness and stability of the structure. Li et al. (1998a) found that the disulphide pattern in AFPs from the closely related *Dendroides canadensis*, DAFP, is similar to that of TmAFP. Li et al. (1998b) reported high content of β-sheet also in DAFP, and Jia and Davies (2002) and Wang et al. (2009) modelled DAFP according to the folding pattern of TmAFP. Other tenebrionid species that reportedly have the same consensus sequence as *T. molitor* and *D. canadensis* are *Microdera punctipennis* (Qiu et al. 2010), *Pterocoma loczyi* (Ma et al. 2008) and *Anatolica polita* (Ma et al. 2012). Given the degree of sequence similarity between AFPs of different species within Tenebrionidea (Table 2.2), there is little doubt that they fold into the same configuration as TmAFP. An illustration of the folding pattern of TmAFP is shown in Fig. 2.2a.

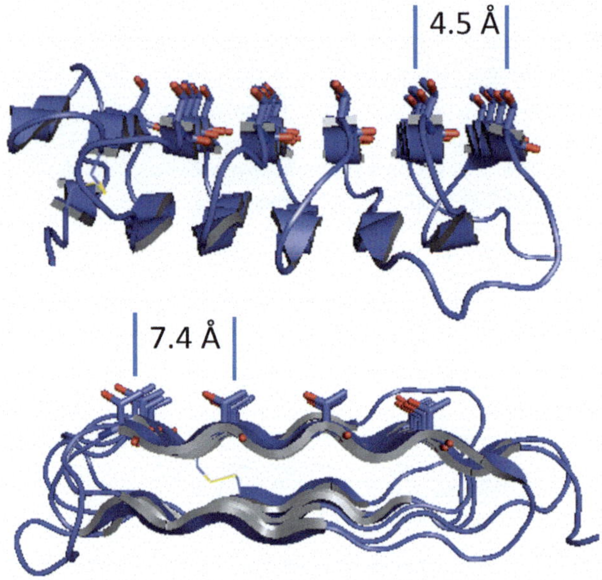

Fig. 2.3 The flatness and regularity of IBSs. RiAFP from the cerambycid beetle *Rhagium inquisitor* (PDB 4DT5) oriented to depict the flatness and regularity of the IBS and the distances between Thr residues in the TxTxTxT motifs within and between the β-stands in the IBS. The side chains of the Thr residues are protruding upwards from the β-sheet

The two closely related species of longhorn beetles, *Rhagium inquisitor* and *R. mordax*, express AFPs, RiAFP and RmAFP, respectively, which contain an expanded version of the TxT motif seen in the Tenebrionidea AFPs. The consensus sequence of RiAFP and RmAFP is the repeat TxTxTxT interrupted by stretches of 13–20 residues that do not have any obvious pattern (Kristiansen et al. 2011, 2012). Six of these segments fold into a flattened β-helical configuration with the TxTxTxT motifs stacked on one side in a regular ladder (Kristiansen et al. 2012; Hakim et al. 2013). In the case of the longhorn beetles, there are only two cysteines present (Kristiansen et al. 2011), and these form a single SS bond at the N-terminal of the molecule (Hakim et al. 2013). An illustration of RiAFP is given in Fig. 2.3.

The beetle *Dorcus curvidens* belongs to the family Lucanidae in the intraorder Scarabaeiformia. Nevertheless, its reported nucleotide sequences coding for AFPs (Nishimiya et al. 2007) is very similar to those of the tenebrionids of the intraorder Cucujiformia. A BLAST search of one of these sequences (AB264320.1) showed 86% identity to a nucleotide sequence coding an isoform of *Tenebrio molitor* (AF159114.1), and a BLASTp showed that the identity was 75% at the amino acid level, higher than that between several of the *D. curvidens* isoforms. This is quite noteworthy, given the fact that these species are more distantly related than the tenebrionid and cerambycid beetles, that share no sequence similarity between their AFPs.

Hymenoptera Xu et al. (2018) reported on an AFP from the Chinese honeybee, *Apis cerena cerena*, denoted AcerAFP. This 60 kDa AFP consists of 365 amino acids, is rich in alanine and contains 11 repeats of the four residues AAxA. The recombinant protein expressed a 0,5 °C antifreeze activity and was found to have

63–96% sequence similarity to gene sequences from 9 other species spanning several suborders of Hymenoptera, reported in the NCBI database (Xu et al. 2018), suggesting a wide hymenopteran distribution of AcerAFP. Some 96.4% of the protein consists of α-helixes and the remainder is loops, and the proposed tertiary structure consists of three α-helical regions of the protein that is folded onto each other. Interestingly, this tertiary structure is quite similar to that of the hyperactive Maxi fish type I AFP found in winter flounder (Sun et al. 2014).

Lepidoptera The repetitive occurrence of two Thr residues spaced one residue apart seen in the coleopteran AFPs is also found starting at every 15th position throughout the sequence of CfAFP, the AFPs found in the lepidopteran genus, *Choristoneura*. There is no apparent consensus repeat pattern in CfAFP beyond the TxT motif. This is analogue to the situation with RiAFP from the beetle *R. inquisitor*, were the wider TxT motif is separated by stretches devoid of any clear consensus sequence. Nevertheless, these AFPs have been shown to fold into a β-helix configuration in a manner similar to that of the coleopteran TmAFP (Graether et al. 2000). Each turn in the helix is composed of 15 residues, resulting in the repetitive TxT motifs being stacked on one side of the helix to form a ladder of TxT motifs, as seen in TmAFP. In the case of CfAFP, the helix is left handed rather than right handed, and although these AFPs are also stabilized by many internal SS bonds crossing the helix, these do not form the highly regular pattern seen in TmAFP (Gauthier et al. 1998; Graether et al. 2000). Figure 2.2b shows an illustration of the folding pattern of CfAFP. Tyshenko et al. (2005) suggested that isoforms found in *Choristoneura fumiferana* and closely related species in the same genus emerged from a common progenitor prior to species divergence, about 3.2–3.7 million years ago. This time frame corresponds to the cold period preceding the Pleistocene ice ages that started some 3 million years ago.

Lin et al. (2011) reported that AFPs from the lepidopteran inchworm *Campaea perlata*, CpAFP, are constructed of a series of the basic consensus repeat TxTxTxTxTxxx. Different isoforms were identified that formed two subsets, four small isoforms of ~ 3.5 kDa and five isoforms with masses of ~ 8.3 kDa. One of the larger isoforms was modelled as a flattened β-helix, where four motifs of the wider TxTxTxTxT repeat is stacked into a ladder on one side of the flattened helix (Lin et al. 2011), analogue to the structure determined in the coleopteran RiAFP (Hakim et al. 2013).

Diptera Basu et al. (2015) reported that a midge from the family Chironomidae produces an AFP consisting of repeats of the consensus 10 residue sequence xxCxGxYCxG. This 9.1 kDa protein has an even higher content of cysteine than TmAFP, DAFP and CfAFP. An energy-stabilized model was constructed based on the helical configuration, where each of the eight turns in the construction consists of only 10 residues. The two cysteines within each 10-residue repeat form an internal SS bond and these bonds cross the coil in a regular manner akin to the pattern seen in the coleopteran TmAFP. In this construction, one side of the molecule consists of a regular ladder of stacked YCx motifs. The position x is usually occupied by Thr or Val. The side chains of the residues flanking the Cys in the motif point outward and are the suspected ice-binding site. The coiled structure is not likely to form β-sheets, and its configuration was therefore described as a solenoid (Basu et al. 2015). Several isoforms appear to be present in the species, ranging from 5.7 to 10 kDa.

Hemiptera Guz et al. (2014) identified a putative AFP, EmAFP, in the sun pest *Eurygaster maura*. Although antifreeze activity was not explicitly reported, it was interpreted as being an AFP based on sequence features and its association with the overwintering stage. The 10 kDa protein shows 52% similarity with the Lepidopteran CfAFP and has a repetitive pattern of TxT spaced 12–13 residues throughout the sequence. It contains four Cys resides suspected of forming two internal SS bonds. It was proposed to fold as a left-handed helix, leaving the TxT motif as a regular ladder on one flat side of the protein, as reported for TmAFP and CfAFP.

2.2.2.2 Collembola

Graham and Davies (2005) discovered a glycine-rich hyperactive AFP, sfAFP, from the collembolan snow flea, *Hypogastrura harveyi*. The primary sequence is a repeat of the triplet Gxx, where the first x-position is often also a Gly. The protein exists as two isoforms, a small 6.5 kDa variant and a 15.7 kDa variant. The smaller form has two internal SS bonds whereas the larger has only one. Their sequences are not very similar, suggesting that their separation is ancient. The smaller isoform has been shown to fold into six short polyproline helixes, where each triplet makes one turn in the helix (Lin et al. 2007; Pentelute et al. 2008). Interestingly, the type II polyproline helix fold is also the likely configuration of AFGPs of polar fish. The overall arrangement of these helixes in sfAFP is a structure consisting of two flat sheets, where each sheet consists of three parallel type II polyproline helixes and the three helixes in each of the two sheets run antiparallel to each other. This folding pattern results in the overall structure having two flat sides, one more hydrophobic than the other. Mok et al. (2010) modelled the larger isoform according to the same folding pattern. In this form, there are 13 type II polyproline helixes where 12 of these form two flat sheets, each made up of six helixes. An illustration of the folding pattern of the smaller isoform of sfAFP is given in Fig. 2.2c.

Hawes et al. (2014) reported on the amino acid composition of a 9 kDa AFP from the Antarctic springtail, *Gomphiocephalus hodgsoni*, denoted GomphyAFP. Even though *G. hodgsoni* and *H. harvey* belong to the same family of springtails, the composition of these collembolan AFPs is distinctively different. GomphyAFP contains far less glycine than sfAFP (~12%, vs. ~50%) and far more cysteine than sfAFP (~14% vs. 1–5%). The content of glycine is high compared to the known non-collembolan AFPs, whereas the high content of cysteine suggests a structure stabilized by many disulphide bonds, as seen in most of the known insect AFPs.

2.2.2.3 Arachnida

Neelakanta et al. (2010) reported on a putative antifreeze protein in the tick *Ixodes scapularis,* of the order Ixodida. The protein has about 70% sequence identity to the protein scaffold of AFGPs of polar fish, consisting of long stretches of the triplet AAT, and was subsequently named IAFGP. No information was provided to show that this protein is an AF(G)P or if it is glycosylated in a manner akin to that seen in the AFGPs of polar fish. Expression of IAFGP in *I. scapularis* is upregulated by the presence of the bacterium *Anaplasma phagocytophilum*, a human pathogen to which the tick is a host and vector. This was interpreted as reflecting a symbiotic relationship, since it implies that the bacteria induce increased cold tolerance in its host.

Bryon et al. (2013) reported upregulation of genes that code for putative AFPs in diapausing individuals of the mite *Tetranychus urticae*, from the order Trombidiformes. These proteins were examined only in silico, and identity as AFPs was only inferred, based on comparison to structural features of known AFPs from insects. The predicted AFPs consist of 92–210 residues with the identifiable consensus 12-residue repeat pattern NCTxCxxCxNCx. This pattern contains two more Cys residues than those of the tenebrionid beetles and the lepidopteran *C. fumiferana*. Automatic generation of 3D configuration suggests that they fold in a manner similar to the AFPs of *T. molitor*, where a stack of the tripeptide motif NCT forms a β-sheet that comprises the tentative IBS of the protein. In this proposed configuration, two of the Cys residues of each repeat form a disulphide pattern similar to that seen in TmAFP, whereas the two additional Cys residues in the repeat is directed inwardly and may also form SS bonds.

2.2.2.4 Crustacea

Kiko (2010) reported that the copepod *Stephos longipes* expresses two isoforms of a hyperactive AFP that shows strong homology to AFPs identified in several diatoms, bacteria and a snow mold. This wide phylogenetic distribution of an apparent homologue structure in both prokaryotes and eukaryotes is by all accounts the result of lateral gene transfer, as is apparently also the case for the type II AFPs from fish. Hanada et al. (2014) described a homologue found in the Antarctic sea ice bacterium *Colwellia* sp.; the structure consists of a β-helical domain and an α-helix aligned parallel to the β-helix. The β-helical domain folds into a left-handed helix with a triangular cross section and three parallel β-sheets. The IBS of the protein is located on one of the flat sides of the β-helix. An illustration of the folding pattern of this protein is given in Fig. 2.2d.

2.3 Isoform Diversity

As mentioned in the previous section, the phylogenetic occurrence of the various fish type AF(G)Ps are proposedly the results of independent convergent evolution (type I and AFGPs), lateral gene transfer (type II) and development from a common ancestor (type III). Among arthropods, a common progenitor is implied for many, and common secondary structural features have evolved by convergent evolution among distantly related species.

At the organismal level, there are many different isoforms of AFPs present in the body fluids, and they result from a high number of genes. These genes are generally arranged in tandem, suggesting extensive gene duplication (Scott et al. 1985; Hew et al. 1988). The AFGPs of both Antarctic nototheniids and Arctic cods are coded by polyprotein genes, where the polyprotein is post-translationally cleaved to produce the mature AFGPs (Chen et al. 1997a, b; Hsiao et al. 1990; Baalsrud et al. 2018). One such gene found in *Notothenia coriiceps neglecta* codes for 46 mature proteins (Hsiao et al. 1990). In *Dissostichus mawsoni*, Chen et al. (1997b) found 41 copies of polyprotein sequences, coding isoforms belonging to four of the eight known size groups of isoforms, and Baalsrud et al. (2018) found that the number of copies of genes in Arctic cods varied with the species according to their thermal environment. Scott et al. (1985) reported that winter flounder has about 40 genes coding for AFP I, and Hew et al. (1988) found 150 genes coding for AFP type III in ocean pout. There is a similar situation in insects; in the coleopteran *T. molitor*, there are some 30–50 gene copies (Liou et al. 1999), and some 27 isoforms of TmAFP have been described to date (Graham et al. 2007). Some 30 isoforms have been described in the related *D. canadensis* (Nickell et al. 2013). The CfAFP of the lepidopteran *C. fumiferana* is coded by about 17 different genes, each found in 2–5 copies tandemly arranged within the genome (Doucet et al. 2002). Thus, AF(G)P expression is augmented by high gene dosage caused by gene duplication in both insects and fish.

Many AF(G)Ps are constructed as repeat segments in series, and some of the variation among isoforms is caused by a varying number of repeat segments. As mentioned, the unrelated AFGPs of Antarctic nototheniids and Arctic cods have from 4 to 50 segments of the basic AAT unit. Several of the AFP type I contain three or four segments of its 11-residue repeat unit (Chao et al. 1996; Gourlie et al. 1984; Low et al. 2001; Graham et al. 2008b; Hobbs et al. 2011). The isoforms of the coleopterans *T. molitor* and *D. canadensis* vary from five to eight copies of a repeat pattern (Liou et al. 1999; Andorfer and Duman 2000), whereas those of the lepidopteran *C. fumiferana* have either five or seven segments of the repeat (Doucet et al. 2000). Thus, in both fish and insects the genes themselves coding these functional proteins apparently evolved by similar mechanisms; duplication of internal repeat patterns, resulting in groups of isoforms within the organism that differ in their number of repeats, analogous to the apparent process by which the high gene dosage evolved. In the case of the large fish type I variants found in flounders, Gauthier et al. (2005) proposed that smaller isoforms may be derived from larger precursors.

Gene duplication results in certain isoforms within the organism being more closely related to a common original gene than to others, causing isoforms to form subsets based on structural similarity. For instance, the QAE and the SP forms of AFP type III share about 50% identity whereas the similarity is about 75–90% within each group (Chao et al. 1993). As mentioned, the AFP type I found in right-eyed flounders, sculpins, snailfish and cunner are coded by two gene families; one group codes for proteins with signal peptides and are produced in the liver and secreted to the blood stream, while another group, the skin-type, mostly lacks coding for signal peptides and are produced and located in other tissues (Gong et al. 1996; Low et al. 1998; Evans and Fletcher 2006). The isoforms of the coleopteran *D. canadensis* are divided into three subsets, group I, II and III, based on sequence similarity (Andorfer and Duman 2000). In the lepidopteran *C. fumiferana*, they are also classified into three subsets, based on the length of the 3′untranslated region (UTR) of their mRNAs: those with short UTRs (9 kDa), those with intermediate UTR (12 kDa) and those with long UTRs (9 kDa). Members of each group are more structurally similar to other members of that group than to members of the other two groups of isoforms (Doucet et al. 2000).

The isoforms of closely related species of insects and of fish are homologue structures, as they most likely evolved in a common ancestor prior to species divergence. Tyshenko et al. (2005) characterized isoforms homologue to those of the lepidopteran *C. fumiferana* in three other species of *Choristoneura*; phylogenetic comparison of the sequences found in these four sister-species showed that the isoforms formed two subsets. Each subset contained isoforms from all four species. The similarities within each subset were greater than between subsets, showing that sequence similarity between some of the isoforms was greater between species than within. This is in contrast to the situation when comparing homologue isoforms from the two more distantly related tenebrionid beetles *Tenebrio molitor* and *Dendroides canadensis* (Graham et al. 2007), where the isoforms are more similar within each species.

It is not clear if the evolutionary drive towards this high number of isoforms has been a selection towards some unknown specific isoform functionality or a selection towards augmenting protein production. Scott et al. (1985) pointed out that the ~40 genes coding for AFP type I in winter flounder seems very high, since protein production could be improved by other mechanisms than gene dosage, i.e. by enhanced transcription or translation rates or increased mRNA stability. The flounders produce their AFPs over periods of several weeks, and the high gene-number appears somewhat excessive. Swanson and Aquadro (2002) suggested that isoform diversity in the coleopteran *T. molitor* is the result of functional selection at the amino acid level, suggesting specific functionality. Graham et al. (2007) did not find support for this contention and suggested that selection instead has operated on the nucleotide level towards greater AT content at the third codon position. This nucleotide selection presumably facilitates transcription at low temperature and is functionally neutral at the protein level. Thus, the selection may have been towards a more effective expression rather than specific function. This is supported by the observations that populations of polar fish inhabiting warmer waters have lower gene dosage coding AF(G)Ps (Hew et al. 1988; Desjardins et al. 2012; Baalsrud et al. 2018; Yamazaki et al. 2019). On the other hand, Duman et al. (2002) found a specific pattern of expression of different isoforms in the coleopteran *D. canadensis*, Ma et al. (2012) found differential expression of two AFP isoforms from the coleopteran *A. polita* and Doucet et al. (2000, 2002) found expression of some isoforms to be life stage specific in the lepidopteran *C. fumiferana*, hinting to differentiation in isoform function.

2.4 Synthesis and Distribution

Low temperature and short day-length are environmental cues of winter, and both conditions have been shown to stimulate production of AFPs in insects (Duman 1977; Patterson and Duman 1978; Horwath and Duman 1983a; Ma et al. 2012), a collembolan (Meier and Zettel 1997), and fish (Duman and DeVries 1974; Fourney et al. 1984; Fletcher et al. 1989a). In addition, dry conditions and starvation also stimulate AFP production in several insects (Duman 1977; Patterson and Duman 1978; Graham et al. 2000).

Short day-length seem to act by affecting hormonal control of expression. In winter flounder, expression of the liver type is strongly influenced by photoperiod, acting through the central nervous system on the pituitary gland (Fourney et al. 1984; Fletcher et al. 1989a). During the summer, long day-length causes release of growth hormone from the pituitary that blocks transcription of AFP genes. As the day-length shortens during fall, the level of growth hormone decreases, and transcription of AFP genes in the liver ensues. Removal of the pituitary in individuals during summer caused strong production of liver-type AFPs (Fourney et al. 1984; Fletcher et al. 1989a). However, such removal does not affect the levels of skin-type AFPs, suggesting that these genes are not under pituitary control (Gong et al. 1995). Since the expression of skin-type AFPs are temperature sensitive, their regulation may be post-transcriptional, with the half-life of their mRNAs being increased by low temperature (Gong et al. 1995).

In the coleopterans *D. canadensis* and *T. molitor*, short day-length apparently affects AFP production by affecting the level of juvenile hormone (Horwath and Duman 1983b; Xu and Duman 1991; Xu et al. 1992), a hormone primarily released from the *corpus allatum*. Individuals treated with juvenile hormone and kept under long day-length conditions and room temperature produced high levels of AFPs, while control individuals did not. In *D. canadensis*, addition of the anti-juvenile hormone Precocene II prevented AFPs from being expressed under short photoperiod at room temperature, while the untreated controls expressed AFPs. Precocene II also prevented expression of AFPs in individuals kept under winter conditions (Xu and Duman 1991). In isolated fat body cells, juvenile hormone induces transcription in both *T. molitor* and *D. canadensis*, but only if the individuals had been previously exposed to juvenile hormone (Xu and Duman 1991; Xu et al. 1992), suggesting that some factor(s) other than juvenile hormone is needed to induce AFP production.

In contrast to the environmental sensitivity of AFP expression seen in many species, that of the lepidopteran *C. fumiferana* seems to be strictly developmentally controlled. Individuals from different life stages expressed different levels of AFPs and these levels were quite insensitive to changing light conditions and temperatures (Doucet et al. 2002), and transcription levels are negatively affected by hormones in vitro (Qin et al. 2007).

2.4.1 Sites of Synthesis and Distribution in Polar Fish

Several sites of synthesis of AF(G)Ps have been identified. In Arctic species, a major source is the liver. These liver-type variants are exported directly into the blood stream. Contrary to longstanding belief, Cheng et al. (2006) showed that Antarctic nototheniids do not synthesize any of their AFGPs in the liver but uses the pancreas and associated tissues. Following synthesis, the AFGPs are released into the intestinal fluid via the pancreatic duct. Since the pancreas is the only identified site of production of AFGPs in Antarctic nototheniids, their circulating AFGPs have apparently entered their blood by uptake from the intestine. Cheng et al. (2006) also discovered that the pancreas was a second major site of synthesis in Arctic species producing all known types of AF(G)Ps. Since the intestinal fluid of polar fish expresses antifreeze activity (O'Grady et al. 1982; Præbel and Ramløv 2005; Cheng et al. 2006), a similar circulatory pattern relying on uptake of AF(G)Ps from the intestine may well be a second source of AF(G)Ps in the blood stream of non-nototheniid fishes, in addition to those secreted directly into their blood steam from the liver.

This indirect route from the site of synthesis via the intestinal fluid to the blood stream in Antarctic nototheniids probably reflects the importance of preventing ingested ice crystals from inoculating the intestinal fluid (Cheng et al. 2006); since the polar fishes are hypoosmotic to their environment they ingest seawater as part of their obligate osmoregulation. This potentially exposes them to ice crystals in the ingested water. In addition to the danger of direct inoculation of body fluids through the intestinal wall, such ingested ice crystals may potentially grow as salts are removed during the process of water uptake and the intestinal fluid becomes progressively hypoosmotic to seawater along the length of the intestine (O'Grady et al. 1983). The need to combat this danger has apparently caused the pancreas, with its direct connection to the intestinal fluid via the pancreatic duct, to become a major site of synthesis in diverse taxa of polar fishes and the only such site in Antarctic nototheniids.

Why do Arctic fishes rely on two major sites of synthesis of their blood-borne AF (G)Ps and the Antarctic nototheniids have only one? Perhaps it is due to differences in the need to rapidly augment the circulating levels of AF(G)Ps. The water temperatures of the Antarctic are permanently below freezing. Fishes living in these waters would have no need to rapidly augment the circulating amounts of AF (G)Ps in response to environmental changes, i.e. have hepatic synthesis with a direct excretion to the blood. Arctic fishes, on the other hand, may well need to augment their antifreeze protection due to seasonal variations or because of migration into colder waters, and the direct route from the site of synthesis in the liver to the blood may be relevant.

The skin-type isoforms of type I AFP are synthesized in tissues that are exposed to the exterior icy environment. These tissues include skin, gill filaments and dorsal fins, in addition to intestine and brain (Gong et al. 1996; Low et al. 1998; Evans and Fletcher 2006). In sculpin, there is no expression of skin-type genes in the liver (Low et al. 1998), whereas co-expression of skin-type isoforms in liver does occur in winter flounder (Gong et al. 1996).

Although all AF(G)P-producing polar fish contain AF(G)Ps in their blood, less is known about their distribution in other body fluid compartments. The Antarctic nototheniids produces AFGPs of eight distinct size groups. Ahlgren et al. (1988) reported that all size groups of AFGPs are distributed passively throughout the extracellular body fluids of two species of Antarctic nototheniids but they were not present intracellularly. No AFGPs were found in the brain or urine, attributable to the blood–brain barrier and the aglomerular kidneys of these fishes (see below). Bile contains AFGPs, and O'Grady et al. (1983) argued that this is a route for transfer of blood-borne AFGPs to enter the intestine. Evans et al. (2011) also observed injected fluorescently tagged AFGPs in most extracellular fluids, except urine and brain.

For the Arctic winter flounder and shorthorn sculpin, the genes for their skin-type AFPs lack coding regions for signal peptides, indicating that they are not excreted from the cells but function intracellularly (Gong et al. 1996; Low et al. 1998). In snailfish, however, the skin-type AFP I is identical to those circulating in blood, suggesting excretion into the blood stream after synthesis (Evans and Fletcher 2005a). Also, liver-type AFP II from sea raven, *H. americanus,* is located in skin tissue (Evans and Fletcher 2006), suggesting uptake of liver-type AFP II from the blood or synthesis of similar AFPs in skin and liver. Low et al. (1998) also found expression of skin-type AFPs in the brain of shorthorn sculpin. Thus, contrasting the findings from the Antarctic nototheniids, several Arctic non-nototheniid species have been shown to have AFPs in their cells and brain tissue.

Preventing Urinary Loss of AF(G)Ps in Polar Fish Loss of AF(G)Ps represents an energetic cost to the organism. The apparent absorption of AFGPs from the intestine in nototheniids (Cheng et al. 2006) probably reduces their loss during evacuation of the gut. AF(G)Ps circulating in the blood, however, may potentially be lost via the urine. Molecules with sizes below 68 kDa are filtered out in the glomeruli (Eastman 1993), suggesting that AF(G)Ps may become filtered out of the plasma during urine formation. Such filtration could be countered by energetically costly reabsorption of AF(G)Ps from the filtrate. In Antarctic nototheniids, this potential problem is effectively avoided by evolutionary degeneration of their glomeruli (Eastman and DeVries 1986). Formation of urine in such aglomerular species is based on secretion rather than filtration, and the loss of AFGPs is effectively avoided (Dobbs and DeVries 1975; Eastman 1993).

Eastman et al. (1987) did not find aglomerular kidneys when examining diverse taxa of Arctic teleosts that produce AF(G)Ps. Instead, Arctic fishes have an anionic repulsion barrier in the basement membrane of the nephron. This repulsion barrier operates in the same manner as the mammalian anionic repulsion barrier (Kenwar et al. 1980), where carboxyl-rich glycoproteins in the basement membrane restrict filtration of anionic molecules, including anionic AF(G)Ps (Petzel and DeVries 1980; Boyd and DeVries 1983, 1986). The type I AFPs are reportedly repelled at the basement membrane by this mechanism (Petzel and DeVries 1980; Boyd and DeVries 1983). As mentioned above, the QAE and SP variants of AFP type III have opposite charges at physiological pH and both are present in the animal. Boyd and DeVries (1986) found that the AFP type III-producing northern eelpouts have glomerular kidneys and an anionic repulsion mechanism. Thus, although retention of the anionic QAE forms may be similar to that seen for the winter flounder type I AFPs, the cationic SP forms would be expected to filter out. Many of the Arctic fishes only express AF(G)Ps during parts of the year (Scott et al. 1985; Reisman et al. 1987). In these species, a means of reducing urinary loss may be to lower their glomerular filtration during winter (Hickman 1968). Interestingly, Eastman et al. (1979) found that, contrary to the northern eelpouts, the AFP III-producing Antarctic eelpout has non-functional glomeruli. In this case, there would be no problem with potential loss of the SP variants of the AFP type III, and the urine did not contain any AFP type III (Eastman et al. 1979).

Contrary to the earlier findings (Petzel and DeVries 1980; Boyd and DeVries 1986; Eastman et al. 1987), Fletcher et al. (1989b) did find AF(G)Ps in the urine of several Arctic species. These included type I AFP in the urine of winter flounder (*Pseudopleuronectes americanus*), type II AFP in the urine of sea raven (*Hemitripterus americanus*), type III AFP in the urine of ocean pout (*Macrozoarces americanus*) and AFGPs in the urine of Atlantic cod (*Gadus morhua*). There was no AFP type I in the urine of shorthorn sculpin (*Myoxocephalus scorpius*). The levels in the urine varied substantially, and the presence of relatively high concentrations of AFPs in the urine may be a consequence of concentrating small amounts of AFP from a large volume of urine by water reabsorption (DeVries and Cheng 2005). The presence of AF(G)Ps in the urine may be functional, as they presumably afford the same freeze protection to the urine as to other fluid compartments (Fletcher et al. 1989b).

2.4.2 Sites of Synthesis and Distribution in Insects

Only a few studies provide information on the site of synthesis and/or distribution of AFPs in insects. Taken together, these studies report the presence of AFPs in one or several of the different body fluid compartments hemolymph, gut fluid, pre-urine, muscular tissue and epidermal tissue (Duman et al. 2002; Nickell et al. 2013;

Ramsay 1964; Graham et al. 2000; Kristiansen et al. 1999, 2005; Buch and Ramløv 2017; Guz et al. 2014). The fat body is the major site of protein synthesis in insects (Arrese and Soulages 2010), and all species examined have shown synthesis of AFPs in this organ. Other tissues shown to transcribe AFP genes are gut tissue, Malpighian tubules and epidermis. All species examined have several isoforms of the AFPs, and evidence exists of specific distribution of isoforms in body fluids and between life stages.

Duman et al. (2002) reported on the expression and distribution of 12 isoforms in the beetle *D. Canadensis*. These are divided into three groups, I, II and III, based on their sequence similarity. Mature isoforms belonging to group I are only located in the hemolymph whereas those of group II and III are located in the gut fluid. The genes of all isoforms are transcribed in the fat body, whereas group II and III are also transcribed in the gut tissue. In addition, there is expression of several of the isoforms belonging to group I and II, but not III, in epidermal tissue. Nickell et al. (2013) reported that 24 isoforms from *D. canadensis*, of which 18 were previously unknown, were transcribed in the Malpighian tubules. Representatives of all groups (I, II, III) were transcribed in the Malpighian tissue. Hysteresis activity in this species has been reported from Malpighian tubule fluid, excreted rectal fluid (Nickell et al. 2013), gut fluid and hemolymph (Duman et al. 2002).

Ramsay (1964) observed hysteresis activity in all extracellular fluid compartments of the closely related beetle *Tenebrio molitor*, except the fluid of the Malpighian tubules. The individuals tested by Ramsay were reared at room temperature. These extracellular compartments included gut fluid, hemolymph and perirectal fluids. Graham et al. (2000) reported transcription of AFPs in *T. molitor* in fat body, midgut and hindgut but not in ovaries or the male reproductive tract.

Kristiansen et al. (1999) studied the hysteresis activity in different body fluid compartments of the beetle *Rhagium inquisitor* and found activity in both gut fluid and hemolymph. In addition, extracts of larval tissue, where hemolymph had been washed away and fat body and gut removed by dissection, showed considerable activity. These findings strongly suggested the presence of substantial amounts of intracellular AFPs in the muscular tissues. In addition, extracts from the fat body also showed high activity. Although the complete amino acid sequence of only a single 13 kDa isoform is known from *R. inquisitor*, Kristiansen et al. (2005) observed at least six additional distinct activity peaks during ion exchange chromatography of its hemolymph, suggesting that multiple isoforms are present in the hemolymph. Buch and Ramløv (2017) used fluorescently tagged monoclonal antibodies raised against a homologue single isoform of the closely related *R. mordax* and found that the protein was present in gut tissue, gut fluid and cuticle. The pattern of fluorescence in summer individuals was indicative of cellular storage of these AFPs during summer.

Guz et al. (2014) reported that the tentative AFP, EmAFP, from the hemipteran *Eurygaster maura* only showed significant transcription levels for this protein in the gut tissue. Only trace amounts of mRNA were detected in the fat body, ovary, Malpighian tubules, trachea, heart, flight muscle or the nervous system.

2.5 Characteristics of Ice-Binding Sites

The ice-binding sites (IBS) of AFPs are reportedly very planar and more hydrophobic than the rest of the structure (Yang et al. 1988; Sönnichsen et al. 1996; Haymet et al. 1998; Yang et al. 1998; Graether et al. 2000; Liou et al. 2000). The hydrophobic character of the IBS presumably causes the protein to orient away from the solution and towards the ice surface, whereas the flatness of the IBS is probably to obtain a good structural fit to the crystal plane. The planar character of the IBS of RiAFP is illustrated in Fig. 2.3.

The residues making up the ice-binding sites of AF(G)Ps are generally organized in a repetitive manner, resulting in repetitive distances between the residues. For instance, in the helical type II polyproline helix configuration proposed for the moderately active AFGPs (Franks and Morris 1978; Bush et al. 1984; Mimura et al. 1992; Tachibana et al. 2004), the repeat distance between hydroxyl groups of the disaccharide units is about 9.31 Å (Knight et al. 1993). This distance is very close to that between oxygen atoms in the ice lattice in the primary crystal plane oriented along the *a*-axis, the experimentally determined adsorption plane and orientation of these AFGPs (Knight et al. 1993). Similarly, for the moderately active AFP Type I, the 11-residue spacing between hydroxyl groups in the side chains of Thr residues in the α-helix is 16.5 Å, matching very closely the 16.7 Å spacing of oxygen atoms along a single direction on the crystal plane they are known to adsorb (Knight et al. 1991). In the β-helical AFPs, the width between hydroxyl groups of outwardly projecting Thr residues in the TxT motifs is about 7.4 Å within each β-strand. The length between strands is about 4.5 Å (Liou et al. 2000). These distances in the IBS of RiAFP are illustrated in Fig. 2.3 and occur between water molecules in multiple orientations on several crystal planes.

Exactly how AF(G)Ps adsorb onto ice crystals has been a topic of debate (Garnham et al. 2011). A number of studies have shown that AF(G)Ps have bound water molecules arranged in an ice-like lattice at their ice-binding sites (Liou et al. 2000; Leinala et al. 2002; Garnham et al. 2011; Hakim et al. 2013; Sun et al. 2014). In all likelihood these water molecules fuse with the solidifying ice surface at temperatures below the melting point and de-couple from the ice surface as the temperature is raised above the melting point. Essentially, the AF(G)Ps "freeze" onto and "melt" off the ice, depending on the temperature (Kristiansen and Zachariassen 2005). Thus, the functionality of the specific arrangement of residues in the IBS may well be to structure the hydration water at the ice-binding site rather than interacting directly with specific oxygens in ice (Sun et al. 2014; Chakraborty and Jana 2019).

2.5.1 Moderately Active AF(G)Ps

In the moderately active fish AF(G)Ps, the ice-binding sites consist of residues organized in ways that restrict the AF(G)Ps to adsorb onto a single specific ice crystal plane and in a specific orientation on that plane. The specificity in absorption orientation was documented by Laursen et al. (1994) who showed that chiral L and D variants of AFP type I adsorbs at mirror image orientations at the same crystal plane.

Due to their single plane-specific adsorption, ice crystals in the presence of moderately active fish AF(G)Ps obtain the shape of a hexagonal bipyramid (e.g. Baardsnes et al. 2001; Loewen et al. 1998; Ewart et al. 1998). This shape is the only possible shape that exposes a single protected plane towards the surrounding solution. Characteristically, such hexagonal bipyramid crystals freeze from their apex at the hysteresis freezing point. Apparently, moderately active AF(G)Ps only weakly protect the apexes of the bipyramidal crystals, which is the probable cause of their moderate activity (Jia and Davies 2002).

2.5.2 Hyperactive AF(G)Ps

The IBS of the hyperactive AFPs, such as the β-helical forms found in many insects (Table 2.2) have both a width and a length, enabling them to adsorb onto multiple planes and in multiple orientations. The high occurrence of the β-helix folding pattern among hyperactive AFPs may reflect the good 2D-fit between internal residue-to-residue distances within the β-sheet and distances between oxygen atoms in ice (Graether and Sykes 2004). This may have been the driving force that caused today's abundance of this structural scaffold in unrelated AFPs (Table 2.2). Interestingly, both the large hyperactive Maxi variant of fish AFP type I and the hyperactive AFP from the collembolan snow flea obtain width and length of their ice-binding sites by having several helixes side by side.

Crystals that form in the presence of hyperactive AFPs express several crystal planes towards the surrounding solution. It is likely that their hyperactivity is caused by their ability to adsorb onto multiple crystal planes and thereby effectively protect the entire surface. The ability to adsorb onto the basal plane has been proposed as the root cause of their hyperactivity (Liou et al. 2000; Graether et al. 2000; Pertaya et al. 2008).

2.6 Conclusions

AF(G)Ps have independently evolved in many different groups of fish and arthropods inhabiting cold regions. Their present-day taxonomic distribution reflects complex evolutionary processes, where convergent evolution and lateral gene transfer have led to both analogue and homologue structures being found in distantly related species. The simple repetitive construction of the AFGPs, type I AFPs and many AFPs found in arthropods, as a series of shorter repeat sequences, is presumably the result of internal duplication of repeats that has resulted in functional genes. The more complex structures (AFP type II and III) are apparently derived from functional proteins originally involved in binding of carbohydrates. In the case of the repetitive structures, they all fold into helical configurations with their IBS composed of regularly spaced residues located on one side of the coil.

Both fish and insects have a high gene dosage of AF(G)Ps that apparently is the result of gene duplication. All species examined have high numbers of isoforms, and it is unclear if this is due to a selective pressure towards divergence in isoform function or exclusively towards augmenting protein production. Several sites of synthesis have been identified in both fish and insects, and isoform-specific location of expression is prevalent. In many species, expression is regulated by environmental cues acting through hormonal mechanisms, but some species appear to be insensitive to such cues and expression may be linked to developmental stage.

In polar fish, both the site(s) of synthesis and mechanism(s) to prevent urinary loss of AF(G)Ps seem to be related to the permanence of their thermal environment; The Antarctic waters are permanently cold and thermally stable, whereas the temperature of Arctic waters vary with location and season. The AFGPs of Antarctic notothenioids take an indirect ("slow") route from their pancreatic site of synthesis to the blood via the intestine, whereas Arctic AF(G)P-producing species also have hepatic synthesis, affording them an additional direct ("fast") secretion from the liver to the blood. In Antarctic species, prevention of urinary loss of AF(G)Ps is primarily achieved by degeneration of the kidney-glomeruli, a permanent physiological adaptation to a constant environment. In Arctic species, on the other hand, a charge-based repulsion mechanism in the basement membrane of the nephron prevents urinary loss of AF(G)Ps, affording these species functional kidneys year-round.

The functionality of AF(G)Ps arises from the ability of their IBS to irreversibly adsorb onto the surface of ice crystals. The IBS is reportedly more hydrophobic than the rest of the protein surface, presumably orienting the IBS towards the ice. In the presence of moderately active AF(G)Ps, bipyramidal crystals are formed that exposes only a single protected crystal plane to the surrounding solution. In the presence of hyperactive AF(G)Ps, ice crystals expose several protected planes to the solution. These crystal habits must arise from features of the IBS. In moderately active fish AFGPs and AFP type I, the helical folding results in the IBS consisting of a single row of ice-binding residues, apparently affording these proteins the ability to only adsorb onto a single plane. In the hyperactive helical arthropod AFPs the IBS is made up of several parallel such rows of residues that cause the IBS to fit several planes and orientations. In some hyperactive AFPs the IBS is formed by several inter- or intramolecular helixes side by side. This organization of the helixes results in several parallel rows of ice-binding residues and consequently provide the necessary ability of the AF(G)P to adsorb onto multiple planes and orientations similar to other hyperactive AFPs.

References

Ahlgren J, Cheng C-HC, Schrag JD, DeVries AL (1988) Freezing avoidance and the distribution of antifreeze glycopeptides in body fluids and tissues of Antarctic fish. J Exp Biol 137:549–563

Andorfer CA, Duman JG (2000) Isolation and characterization of cDNA clones encoding antifreeze proteins of the pyrochroid beetle *Dendroides canadensis*. J Insect Physiol 46:365–372

Arrese EL, Soulages JL (2010) Insect fat body: energy, metabolism, and regulation. Annu Rev Entomol 55:207–225

Baalsrud HT, Tørresen OK, Solbakken MH, Salzburger W, Hanel R, Jakobsen KS, Jentoft S (2018) *De novo* gene evolution of antifreeze glycoproteins in codfishes revealed by whole genome sequence data. Mol Biol Evol 35:593–606

Baardsnes J, Davies PL (2001) Sialic acid synthase: the origin of fish type III antifreeze protein? Trends Biochem Sci 26:468–469

Baardsnes J, Jelokhani-Niaraki M, Kondejewski LH, Kuiper MJ, Kay CM, Hodges RS, Davies PL (2001) Antifreeze protein from shorthorn sculpin: identification of the icebinding surface. Protein Sci 10:2566–2576

Badet T, Peyraud R, Raffaele S (2015) Common protein sequence signatures associate with *Sclerotinia borealis* lifestyle and secretion in fungal pathogens of the Sclerotiniaceae. Front Plant Sci 6:776

Bar Dolev M, Bernheim R, Guo S, Davies PL, Braslavsky I (2016) Putting life on ice: bacteria that bind to frozen water. J R Soc Interface 13:20160210

Basu K, Graham LA, Campbell RL, Davies PL (2015) Flies expand the repertoire of protein structures that bind ice. Proc Natl Acad Sci USA 112:737–742

Boyd RB, DeVries AL (1983) The seasonal distribution of anionic binding sites in the basement membrane of the kidney glomerulus of the winter flounder *Pseudopleuronectes americanus*. Cell Tiss Res 234:271–277

Boyd RB, DeVries AL (1986) A comparison of anionic sites in the glomerular basement membranes from different classes of fishes. Cell Tiss Res 245:513–517

Bryon A, Wybouw N, Dermauw W, Tirry L, Van Leeuwen T (2013) Genome wide gene-expression analysis of facultative reproductive diapause in the two-spotted spider mite *Tetranychus urticae*. BMC Genomics 14:815

Buch JL, Ramløv H (2017) Detecting seasonal variation of antifreeze protein distribution in *Rhagium mordax* using immunofluorescence and high resolution microscopy. Cryobiology 74:132–140

Bush CA, Ralapati S, Matson GM, Yamasaki RB, Osuga DT, Yeh Y, Feeney RE (1984) Conformation of the antifreeze glycoprotein of polar fish. Arch Biochem Biophys 232:624–631

Chakraborty S, Jana B (2019) Ordered hydration layer mediated ice adsorption of a globular antifreeze protein: mechanistic insight. Phys Chem Chem Phys 21:19298–19310

Chao H, Davies PL, Sykes BD, Sönnichsen FD (1993) Use of proline mutants to help solve the NMR solution structure of type III antifreeze protein. Protein Sci 2:1411–1428

Chao H, Hodges RS, Kay CM, Gauthier SY, Davies PL (1996) A natural variant of Type I antifreeze protein with four ice-binding repeats is a particularly potent antifreeze. Protein Sci 5:1150–1155

Chen L, DeVries AL, Cheng C-HC (1997a) Convergent evolution of antifreeze glycoproteins in Antarctic notothenioid fish and Arctic cod. Proc Natl Acad Sci USA 94:3817–3822

Chen L, DeVries AL, Cheng C-HC (1997b) Evolution of antifreeze glycoprotein gene from a trypsinogen gene in Antarctic notothenioid fish. Proc Natl Acad Sci USA 94:3811–3816

Cheng C-HC, Cziko PA, Evans CW (2006) Nonhepatic origin of notothenioid antifreeze reveals pancreatic synthesis as common mechanism in polar fish freezing avoidance. Proc Natl Acad Sci USA 103:10491–10496

Deng G, Laursen RA (1998) Isolation and characterization of an antifreeze protein from the longhorn sculpin, *Myoxocephalus octodecimspinosis*. Biochim Biophys Acta 1388:305–314

Deng C, Cheng C-HC, Ye H, He X, Chen L (2010) Evolution of an antifreeze protein by neofunctionalization under escape from adaptive conflict. Proc Natl Acad Sci USA 107:21593–21598

Desjardins M, Graham LA, Davies PL, Fletcher GL (2012) Antifreeze protein gene amplification facilitated niche exploitation and speciation in wolffish. FEBS J 279:2215–2230

DeVries AL (1982) Biological antifreeze agents in Coldwater fishes. Comp Biochem Physiol A 73:627–640

DeVries AL, Cheng CHC (2005) Antifreeze proteins and organismal freezing avoidance in polar fishes. In: Farrell AP, Steffensen JF (eds) Fish Physiology, vol XXII. Academic Press, San Diego, pp 155–201

Dobbs GH, DeVries AL (1975) Renal function in Antarctic teleost fishes: serum and urine composition. Mar Biol 29:59–70

Doucet D, Tyshenko MG, Kuiper M, Graether S, Sykes B, Daugulis A, Davies P, Walker VK (2000) Structure-function relationships in spruce budworm antifreeze protein revealed by isoform diversity. Eur J Biochem 267:6082–6088

Doucet D, Tyshenko MG, Davies PL, Walker VK (2002) A family of expressed antifreeze protein genes from the moth, *Choristoneura fumiferana*. Eur J Biochem 269:38–46

Duman JG (1977) Environmental effects on antifreeze levels in larvae of the darkling beetle, *Meracantha contracta*. J Exp Zool A 201:333–337

Duman JG (2002) The inhibition of ice nucleators by insect antifreeze proteins is enhanced by glycerol and citrate. J Comp Physiol B 172:163–168

Duman JG (2015) Animal ice-binding (antifreeze) proteins and glycolipids: an overview with emphasis on physiological function. J Exp Biol 218:1846–1855

Duman JG, DeVries AL (1974) The effects of temperature and photoperiod on antifreeze production in cold water fishes. J Exp Zool 190:89–97

Duman JG, Olsen TM (1993) Thermal hysteresis protein activity in bacteria, fungi, and phylogenetically diverse plants. Cryobiology 30:322–328

Duman JG, Verleye D, Li N (2002) Site-specific forms of antifreeze protein in the beetle *Dendroides canadensis*. J Comp Physiol B 172:547–552

Duman JG, Bennett V, Sformo T, Hochstrasser R, Barnes BM (2004) Antifreeze proteins in Alaskan insects and spiders. J Insect Physiol 50:259–266

Eastman JT (1993) Antarctic fish biology. Evolution in a unique environment. Academic Press, San Diego

Eastman JT, DeVries AL (1986) Renal glomerular evolution in Antarctic notothenioid fishes. J Fish Biol 29:649–662

Eastman JT, DeVries AL, Coalson RE, Nordquist RE, Boyd RB (1979) Renal conservation of antifreeze peptide in Antarctic eelpout, *Rhigophila dearborni*. Nature 282:217–218

Eastman JT, Boyd RB, DeVries AL (1987) Renal corpuscle development in boreal fishes with and without antifreezes. Fish Physiol Biochem 4:89–100

Evans RP, Fletcher GL (2001) Isolation and characterization of type I antifreeze proteins from Atlantic snailfish (*Liparis atlanticus*) and dusky snailfish (*Liparis gibbus*). Biochim Biophys Acta 1547:235–244

Evans RP, Fletcher GL (2005a) Type I antifreeze proteins expressed in snailfish skin are identical to their plasma counterparts. FEBS J 272:5327–5336

Evans RP, Fletcher GL (2005b) Type I antifreeze proteins: possible origins from chorion and keratin genes in Atlantic snailfish. J Mol Evol 61:417–424

Evans RP, Fletcher GL (2006) Isolation and purification of antifreeze proteins from skin tissues of snailfish, cunner and sea raven. Biochim Biophys Acta 1700:209–217

Evans CW, Gubala V, Nooney R, Williams DE, Brimble MA, DeVries AL (2011) How do Antarctic notothenioid fishes cope with internal ice? A novel function for antifreeze glycoproteins. Antarct Sci 23:57–64

Ewart KV, Li Z, Yang DS, Fletcher GL, Hew CL (1998) The ice-binding site of Atlantic herring antifreeze protein corresponds to the carbohydrate-binding site of C-type lectins. Biochemistry 37:4080–4085
Fletcher GL, Idler DR, Vaisius A, Hew CL (1989a) Hormonal regulation of antifreeze protein gene expression in winter flounder. Fish Physiol Biochem 7:387–393
Fletcher GL, King MJ, Kao MH, Shears MA (1989b) Antifreeze proteins in the urine of marine fish. Fish Physiol Biochem 6:121–127
Fletcher GL, Hew CL, Davies PL (2001) Antifreeze proteins of teleost fishes. Annu Rev Physiol 63:359–390
Fourney RM, Fletcher GL, Hew CL (1984) Accumulation of winter flounder antifreeze messenger RNA after hypophysectomy. Gen Comp Endocrinol 54:392–401
Franks F, Morris ER (1978) Blood glycoprotein from Antarctic fish. Possible conformational origin of antifreeze activity. Biochim Biophys Acta 540:346–356
Garnham CP, Campbell RL, Davies PL (2011) Anchored clathrate waters bind antifreeze proteins to ice. Proc Natl Acad Sci USA 108:7363–7367
Gauthier SY, Kay CM, Sykes BD, Walker VK, Davies PL (1998) Disulfide bond mapping and structural characterization of spruce budworm antifreeze protein. Eur J Biochem 258:445–453
Gauthier SY, Marshall CB, Fletcher GL, Davies PL (2005) Hyperactive antifreeze protein in flounder species. The sole freeze protectant in American plaice. FEBS J 272:4439–4449
Gauthier SY, Scotter AJ, Lin FH, Baardsnes J, Fletcher GL, Davies PL (2008) A re-evaluation of the role of type IV antifreeze protein. Cryobiology 57:292–296
Gong Z, King MJ, Fletcher GL, Hew CL (1995) The antifreeze protein genes of the winter flounder, *Pleuronectus americanus*, are differently regulated in liver and non-liver tissues. Biochem Biophys Res Commun 206:387–392
Gong Z, Ewart KV, Hu Z, Fletcher GL, Hew CL (1996) Skin antifreeze protein genes of the Winter flounder, *Pleuronectes americanus*, encode distinct and active polypeptides without the secretory signal and prosequences. J Biol Chem 271:4106–4112
Gourlie B, Lin Y, Price J, DeVries AL, Powers D, Huang RC (1984) Winter flounder antifreeze proteins: a multigene family. J Biol Chem 259:14960–14965
Graether SP, Sykes BD (2004) Cold survival in freeze-intolerant insects: the structure and function of beta-helical antifreeze proteins. Eur J Biochem 271:3285–3296
Graether SP, Kuiper MJ, Gagne SM, Walker VK, Jia Z, Sykes BD, Davies PL (2000) β-Helix structure and ice-binding properties of a hyperactive antifreeze protein from an insect. Nature 406:325–328
Graham LA, Davies PL (2005) Glycine-rich antifreeze proteins from snow fleas. Science 310:461
Graham LA, Walker VK, Davies PL (2000) Developmental and environmental regulation of antifreeze proteins in the mealworm beetle *Tenebrio molitor*. Eur J Biochem 267:6452–6458
Graham LA, Qin W, Lougheed SC, Davies PL, Walker VK (2007) Evolution of hyperactive, repetitive antifreeze proteins in beetles. J Mol Evol 64:387–398
Graham LA, Lougheed SC, Ewart KV, Davies PL (2008a) Lateral transfer of a Lectin-like antifreeze protein gene in fishes. PLoS One 3:e2616
Graham LA, Marshall CB, Lin F-H, Campbell RL, Davies PL (2008b) Hyperactive antifreeze protein from fish contains multiple ice-binding sites. Biochemistry 47:2051–2063
Graham LA, Li J, Davidson WS, Davies PL (2012) Smelt was the likely beneficiary of an antifreeze gene laterally transferred between fishes. BMC Evol Biol 12:190
Graham LA, Hobbs RS, Fletcher GL, Davies PL (2013) Helical antifreeze proteins have independently evolved in fishes on four occasions. PLoS One 8:e81285
Gronwald W, Loewen MC, Lix B, Daugulis AJ, Sönnichsen FD, Davies PL, Sykes B (1998) The solution structure of type II antifreeze protein reveals a new member of the lectin family. Biochemistry 37:4712–4721
Guz N, Toprak U, Dageri A, Gurkan MO, Denlinger DL (2014) Identification of a putative antifreeze protein gene that is highly expressed during preparation for winter in the sunn pest, *Eurygaster maura*. J Insect Physiol 68:30–35

Hakim A, Nguyen JB, Basu K, Zhu DF, Thakral D, Davies PL, Isaacs FJ, Modis Y, Meng W (2013) Crystal structure of an insect antifreeze protein and its implications for ice binding. J Biol Chem 288:12295–12304

Hanada Y, Nishimiya Y, Miura A, Tsuda S, Kondo H (2014) Hyperactive antifreeze protein from an Antarctic sea ice bacterium *Colwellia* sp. has a compound ice-binding site without repetitive sequences. FEBS J 281:3576–3590

Hawes TC, Marchall CJ, Warthon DD (2014) A 9 kDa antifreeze protein from the Antarctic springtail, *Gomphiocephalus hodgsoni*. Cryobiology 69:181–183

Haymet ADJ, Ward LG, Harding MM, Knight CA (1998) Valine substituted winter flounder 'antifreeze': preservation of ice growth hysteresis. FEBS Lett 430:301–306

Hew CL, Fletcher GL, Ananthanarayanan VS (1980) Antifreeze proteins from the shorthorn sculpin, *Myoxocephalus scorpius*: isolation and characterization. Can J Biochem 58:377–383

Hew CL, Wang N-C, Joshi S, Fletcher GL, Scott GK, Hayes PH, Buettner B, Davies PL (1988) Multiple genes provide the basis for antifreeze protein diversity and dosage in the ocean pout, *Macroxoarces americanus*. J Biol Chem 263:12049–12055

Hickman CP (1968) Glomerular filtration and urine flow in the euryhaline southern flounder, *Paralichthys lethostigma*, in sea water. Can J Zool 46:427–437

Hobbs RS, Shears MA, Graham LA, Davies PL, Fletcher GL (2011) Isolation and characterization of type I antifreeze proteins from cunner, *Tautogolabrus adspersus*, order Perciformes. FEBS J 278:3699–3710

Horwath KL, Duman JG (1983a) Photoperiodic and thermal regulation of antifreeze protein levels in the beetle *Dendroides canadensis*. J Insect Physiol 29:907–917

Horwath KL, Duman JG (1983b) Induction of antifreeze protein production by juvenile hormone in larvae of the beetle, *Dendroides canadensis*. J Comp Physiol 151:233–240

Hoshino T, Kiriaki M, Ohgiya S, Fujiwara M, Kondo H, Nishimiya Y, Yumoto I, Tsuda S (2003) Antifreeze proteins from snow mold fungi. Can J Bot 81:1175–1181

Hsiao KC, Cheng C-HC, Fernandes IE, Detrich HW, DeVries AL (1990) An antifreeze glycopeptide gene from the antarctic cod *Notothenia coriiceps neglecta* encodes a polyprotein of high peptide copy number. Proc Natl Acad Sci USA 87:9265–9269

Husby J, Zachariassen KE (1980) Antifreeze agents in the body fluid of winter active insects and spiders. Experientia 36:963–964

Janech MG, Krell A, Mock T, Kang J-S, Raymond JA (2006) Ice-binding protein from sea ice diatoms (Bacillariophyceae). J Phycol 42:410–416

Jia Z, Davies PL (2002) Antifreeze proteins: an unusual receptor-ligand interaction. Trends Biochem Sci 27:101–106

Kennett JP (1977) Cenozoic evolution of Antarctic glaciation, the Circum-Antarctic Ocean, and their impact on global paleoceanography. J Geophys Res 82:3843–3859

Kenwar YS, Linker A, Farquhar MG (1980) Increased permeability of the glomerular basement membrane to ferritin after removal of glycosaminoglycans (heparan sulfate) by enzyme digestion. J Cell Biol 86:688–693

Kiko R (2010) Acquisition of freeze protection in a sea-ice crustacean through horizontal gene transfer? Polar Biol 33:543–556

Knight CA, Cheng CC, DeVries AL (1991) Adsorption of α-helical antifreeze peptides on specific ice crystal surface planes. Biophys J 59:409–418

Knight CA, Driggers E, DeVries AL (1993) Adsorption to ice of fish antifreeze glycopeptides 7 and 8. Biophys J 64:252–259

Kristiansen E, Zachariassen KE (2005) The mechanism by which fish antifreeze proteins cause thermal hysteresis. Cryobiology 51:262–280

Kristiansen E, Pedersen S, Ramløv H, Zachariassen KE (1999) Antifreeze activity in the cerambycid beetle *Rhagium inquisitor*. J Comp Physiol B 169:55–60

Kristiansen E, Ramløv H, Hagen L, Pedersen SA, Andersen RA, Zachariassen KE (2005) Isolation and characterization of hemolymph antifreeze proteins from larvae of the longhorn beetle *Rhagium inquisitor* (L.). Comp Biochem Physiol B 142:90–97

Kristiansen E, Ramløv H, Højrup P, Pedersen SA, Hagen L, Zachariassen KE (2011) Structural characteristics of a novel antifreeze protein from the longhorn beetle *Rhagium inquisitor*. Insect Biochem Mol Biol 41:109–117

Kristiansen E, Wilkens C, Vincents B, Friis D, Lorentzen A, Jenssen H, Løbner-Olesen A, Ramløv H (2012) Hyperactive antifreeze proteins from longhorn beetles: some structural insights. J Insect Physiol 58:1502–1510

Laursen RA, Wen D, Knight CA (1994) Enantioselective adsorption of the D-and L-forms of an α-helical antifreeze polypeptide to the {2021} planes of ice. J Am Chem Soc 116:12057–12058

Lee JK, Kim HJ (2016) Cloning, expression, and activity of type IV antifreeze protein from cultured subtropical olive flounder (*Paralichthys olivaceus*). Fish Aquatic Sci 19:33

Lee JK, Kim YJ, Park KS, Shin SC, Kim HJ, Song YH, Park H (2011) Molecular and comparative analyses of type IV antifreeze proteins (AFPIVs) from two Antarctic fishes, *Pleuragramma antarcticum* and *Notothenia coriiceps*. Comp Biochem Physiol B 159:197–205

Leinala EK, Davies PL, Jia Z (2002) Crystal structure of beta-helical antifreeze protein points to a general ice binding model. Structure 10:619–627

Li N, Chibber BAK, Castellino FJ, Duman JG (1998a) Mapping of disulfide bridges in antifreeze proteins from overwintering larvae of the beetle *Dendroides canadensis*. Biochemistry 37:6343–6350

Li N, Kendrick BS, Manning MC, Capenter JF, Duman JG (1998b) Secondary structure of antifreeze proteins from overwintering larvae of the beetle *Dendroides Canadensis*. Arch Biochem Biophys 360:25–32

Lin F-H, Graham LA, Campbell RL, Davies PL (2007) Structural modeling of snow flea antifreeze protein. Biophys J 92:1717–1723

Lin F-H, Davies PL, Graham LA (2011) The Thr- and Ala-rich hyperactive antifreeze protein from Inchworm folds as a flat silk-like β-helix. Biochemistry 50:4467–4478

Liou YC, Thibault P, Walker VK, Davies PL, Graham LA (1999) A complex family of highly heterogeneous and internally repetitive hyperactive antifreeze proteins from the beetle *Tenebrio molitor*. Biochemistry 38:11415–11424

Liou Y-C, Tocilj A, Davies PL, Jia Z (2000) Mimicry of ice structure by surface hydroxyls and water of a β-helix antifreeze protein. Nature 406:322–324

Liu JX, Zhai YH, Gui JF (2009) Molecular characterization and expression pattern of AFPIV during embryogenesis in gibel carp (*Carassiu auratus gibelio*). Mol Biol Rep 36:2011–2018

Loewen MC, Gronwald W, Sönnichsen FD, Sykes BD, Davies PL (1998) The ice-binding site of sea raven antifreeze protein is distinct from the carbohydrate-binding site of the homologous C-type lectin. Biochemistry 37:17745–17753

Low WK, Miao M, Ewart KV, Yang DSC, Fletcher GL, Hew CL (1998) Skin-type antifreeze protein from the shorthorn sculpin, *Myoxocephalus scorpius*. Expression and characterization of a Mr 9,700 recombinant protein. J Biol Chem 273:23098–23103

Low WK, Lin Q, Stathakis C, Miao M, Fletcher GL, Hew CL (2001) Isolation and characterization of skin-type, type I antifreeze polypeptides from the longhorn sculpin, *Myoxocephalus octodecemspinosus*. J Biol Chem 276:11582–11589

Ma J, Wang Y, Liu Z-Y, Zhang F-C (2008) Cloning and activity analysis of insect antifreeze protein gene from *Pterocoma loczyi* (Coleoptera: Tenebrionidae). Acta Entomol Sin 51:480–485

Ma J, Wang J, Mao X, Wang Y (2012) Differential expression of two antifreeze proteins in the desert beetle *Anatolica polita* (Coleoptera: Tenebrionidae): seasonal variation and environmental effects. CryoLetters 33:337–348

Meier P, Zettel J (1997) Cold hardiness in *Entomobrya nivalis* (Collembola, Entomobryidae): annual cycle of polyols and antifreeze proteins, and antifreeze triggering by temperature and photoperiod. J Comp Physiol B 167:297–304

Mimura Y, Yamamoto Y, Inoue Y, Chûjô R (1992) NMR study of interaction between sugar and peptide moieties in mucin-type model glycopeptides. Int J Biol Macromol 14:242–248

Miura K, Ohgiya S, Hoshino T, Nemoto N, Suetake T, Miura A, Spyracopoulos L, Kondo H, Tsuda S (2001) NMR analysis of Type III antifreeze protein intramolecular dimer. Structural basis for enhanced activity. J Biol Chem 276:1304–1310

Mok Y-F, Lin F-H, Graham LA, Celik Y, Braslavsky I, Davies PL (2010) Structural basis for the superior activity of the large isoform of snow flea antifreeze protein. Biochemistry 49:2593–2603

Neelakanta G, Sultana H, Fish D, Anderson JF, Fikrig E (2010) Anaplasma *phagocytophilum* induces *Ixodes scapularis* ticks to express an antifreeze glycoprotein gene that enhances their survival in the cold. J Clin Invest 120:3179–3190

Nickell PK, Sass S, Verleye D, Blumenthal EM, Duman JG (2013) Antifreeze proteins in the primary urine of larvae of the beetle *Dendroides Canadensis*. J Exp Biol 216:1695–1703

Nishimiya Y, Sato R, Takamichi M, Miura A, Tsuda S (2005) Co-operative effect of the isoforms of type III antifreeze protein expressed in Notched-fin eelpout, *Zoarces elongatus* Kner. FEBS J 272:482–492

Nishimiya Y, Sato R, Miura A, Tsuda S (2007) Antifreeze protein of *Dorcus curvidens binodulus*. NCBI database, accession numbers AB264317.1–AB264322.1

O'Grady SM, Ellory JC, DeVries AL (1982) Protein and glycoprotein antifreezes in the intestinal fluid of polar fishes. J Exp Biol 98:429–438

O'Grady SM, Ellory JC, DeVries AL (1983) The role of low molecular weight antifreeze glycoproteins in the bile and intestinal fluid of Antarctic fishes. J Exp Biol 104:149–162

Olsen TM, Duman JG (1997a) Maintenance of the supercooled state in overwintering pyrochroid beetle larvae, *Dendroides canadensis*: role of hemolymph ice nucleators and antifreeze proteins. J Comp Physiol B 167:105–113

Olsen TM, Duman JG (1997b) Maintenance of the supercooled state in the gut fluid of overwintering pyrochroid beetle larvae, *Dendroides canadensis*: role of ice nucleators and antifreeze proteins. J Comp Physiol B 167:114–122

Olsen TM, Sass SJ, Li N, Duman JG (1998) Factors contributing to seasonal increases in inoculative freezing resistance in overwintering fire-colored beetle larvae *Dendroides canandensis* (Pyrochroidae). J Exp Biol 201:1585–1594

Patterson JL, Duman JG (1978) The role of the thermal hysteresis factor in *Tenebrio molitor* larvae. J Exp Biol 74:37–45

Pentelute BL, Gates ZP, Tereshko V, Dashnau JL, Vanderkooi JM, Kossiakoff AA, Kent SBH (2008) X-ray structure of snow flea antifreeze protein determined by racemic crystallization of synthetic protein enantiomers. J Am Chem Soc 130:9695–9701

Pertaya N, Marshall CB, Celik Y, Davies PL, Braslavsky I (2008) Direct visualization of spruce budworm antifreeze protein interacting with ice crystals: basal plane affinity confers hyperactivity. Biophys J 95:333–341

Petzel DH, DeVries AL (1980) Renal handling of peptide antifreeze in northern fishes. Bull Mt Desert Isl Biol Lab 19:17–19

Præbel K, Ramløv H (2005) Antifreeze activity in the gastrointestinal fluids of *Arctogadus glacialis* (Peters 1874) is dependent on food type. J Exp Biol 208:2609–2613

Qin W, Doucet D, Tyshenko M, Walker V (2007) Transcription of antifreeze protein genes in *Choristoneura fumiferana*. Insect Mol Biol 16:423–434

Qiu L, Wang Y, Wang J, Zhang F, Ma J (2010) Expression of biologically active recombinant antifreeze protein His-MpAFP149 from the desert beetle (*Microdera punctipennis dzungarica*) in *Escherichia coli*. Mol Biol Rep 37:1725–1732

Ramsay JA (1964) The rectal complex of the mealworm *Tenebrio molitor* L. (Coleoptera, Tenebrionidae). Philos Trans R Soc B 348:279–314

Reisman HM, Fletcher GL, Kao MH, Shears MA (1987) Antifreeze proteins in the grubby sculpin, *Myoxocephalus aenaeus* and the tomcod, *Microgadus tomcod*: comparisons of seasonal cycles. Environ Biol Fish 18:295–301

Scott GK, Hew CL, Davies PL (1985) Antifreeze protein genes are tandemly linked and clustered in the genome of the winter flounder. Proc Natl Acad Sci USA 82:2613–2617

Sönnichsen FD, DeLuca CI, Davies PL, Sykes BD (1996) Refined solution structure of type III antifreeze protein: hydrophobic groups may be involved in the energetics of the protein-ice interaction. Structure 4:1325–1337
Sorhannus U (2012) Evolution of type II antifreeze protein genes in teleost fish: a complex scenario involving lateral gene transfers and episodic directional selection. Evol Bioinformatics Online 8:535–544
Sun T, Lin FH, Campbell RL, Allingham JS, Davies PL (2014) An antifreeze protein folds with an interior network of more than 400 semi-clathrate waters. Science 343:795–798
Swanson WJ, Aquadro CF (2002) Positive Darwinian selection promotes heterogeneity among members of the antifreeze protein multigene family. J Mol Evol 54:403–410
Tachibana Y, Fletcher GL, Fujitani N, Tsuda S, Monde K, Nishimura S-I (2004) Antifreeze glycoproteins: elucidation of the structural motifs that are essential for antifreeze activity. Angew Chemie 43:856–862
Takamichi M, Nishimiya Y, Miura A, Tsuda S (2009) Fully active QAE isoform confers thermal hysteresis activity on a defective SP isoform of type III antifreeze protein. FEBS J 276:1471–1479
Tursman D, Duman JG (1995) Cryoprotective effects of thermal hysteresis protein on survivorship of frozen gut cells from the freeze-tolerant centipede *Lithobius forficatus*. J Exp Zool A 272:249–257
Tyshenko MG, Doucet D, Walker VK (2005) Analysis of antifreeze proteins within spruce budworm sister species. Insect Mol Biol 14:319–326
Urrutia ME, Duman JG, Knight CA (1992) Plant thermal hysteresis proteins. Biochim Biophys Acta 1121:199–206
Walters KR, Sformo T, Barnes BM, Duman JG (2009) Freeze tolerance in an arctic Alaska stonefly. J Exp Biol 212:305–312
Wang S, Amornwittawat N, Juwita V, Kao Y, Duman JG, Pascal TA, Goddard WA, Wen X (2009) Arginine, a key residue for the enhancing ability of an antifreeze protein of the beetle *Dendroides canadensis*. Biochemistry 48:9696–9703
Wharton D, Pow B, Kristensen M, Ramløv H, Marshall C (2009) Ice-active proteins and cryoprotectants from the New Zealand alpine cockroach, *Celatoblatta quinquemaculata*. J Insect Physiol 55:27–31
Wöhrmann APA (1996) Antifreeze glycopeptides and peptides in Antarctic fish species from the Weddel Sea and the Lazarev Sea. Mar Ecol Prog Ser 130:47–59
Worrall D, Elias L, Ashford D, Smallwood M, Sidebottom C, Lillford P, Telford J, Holt C, Bowles D (1998) A carrot leucine-rich repeat protein that inhibits ice recrystallization. Science 282:115–117
Wu Y, Banoub J, Goddard SV, Kao MH, Fletcher GL (2001) Antifreeze glycoproteins: relationship between molecular weight, thermal hysteresis and the inhibition of leakage from liposomes during thermotropic phase transition. Comp Biochem Physiol B 128:265–273
Xiao Q, Xia JH, Zhang XJ, Li Z, Wang Y, Zhou L, Gui JF (2014) Type-IV antifreeze proteins are essential for epiboly and convergence in gastrulation of zebrafish embryos. Int J Biol Sci 10:715–732
Xu L, Duman JG (1991) Involvement of juvenile hormone in the induction of antifreeze protein production by the fat body of larvae of the beetle *Dendroides canadensis*. J Exp Zool 258:288–293
Xu L, Duman JG, Wu DW, Goodman WG (1992) A role for juvenile hormone in the induction of antifreeze protein production by the fat body in the beetle *Tenebrio molitor*. Comp Biochem Physiol B 101:105–109
Xu K, Niu Q, Zhao H, Du Y, Guo L, Jiang Y (2018) Sequencing and expression characterization of antifreeze protein maxi-like in *Apis cerana cerana*. J Insect Sci 18:1–11
Yamazaki A, Nishimiya Y, Tsuda S, Togashi K, Munehara H (2019) Freeze tolerance in sculpins (Pisces; cottoidea) inhabiting north pacific and arctic oceans: antifreeze activity and gene sequences of the antifreeze protein. Biomol Ther 9(Art no. 139):1–13

Yang DSC, Sax M, Chakrabartty A, Hew CL (1988) Crystal structure of an antifreeze polypeptide and its mechanistic implications. Nature 333:232–237

Yang DSC, Hon W-C, Bubanko S, Xue Y, Seetharaman J, Hew CL, Sicheri F (1998) Identification of the ice-binding surface on a type III antifreeze protein with a "flatness function" algorithm. Biophys J 74:2142–2151

Zachariassen KE, Husby JA (1982) Antifreeze effect of thermal hysteresis agents protects highly supercooled insects. Nature 298:865–867

Chapter 3
Physicochemical Properties of Antifreeze Proteins

Dennis Steven Friis and Hans Ramløv

3.1 Introduction

As stated in the preceding chapters, the antifreeze proteins (AFPs) are found in various species where they in several cases have evolved convergently (described in detail in Chap. 9 of Vol. 1), and have been found in different shapes and sizes. The only common denominator of the antifreeze proteins is apparently their affinity for ice. The vast diversity of the AFPs also implies a diversity of their physicochemical properties, described in this chapter. The physicochemical properties of proteins are a wide field and cover numerous parameters, and far from all of these have been investigated for the AFPs. In this chapter, the investigation on the field is presented, covering size and activity of AFPs, their solubility and hydrophobicity and lastly their stability in regard to temperature and pH.

3.2 Weight and Activity of Antifreeze Proteins

The weights of the AFPs are often one of the first determined properties when a new AFP is discovered, together with its activity. The weights of known AFPs vary a lot between the different groups of species they are found in, and within the types they are grouped in. This section focuses on AFPs from fish and arthropods, as these are the most investigated proteins. In general, the AFPs are small (3–30 kDa) monomeric proteins. In the following, AFPs of different types or groups of species are

D. S. Friis (✉)
Copenhagen, Denmark

H. Ramløv
Department of Natural Sciences, Roskilde University, Roskilde, Denmark
e-mail: hr@ruc.dk

H. Ramløv, D. S. Friis (eds.), *Antifreeze Proteins Volume 2*,
https://doi.org/10.1007/978-3-030-41948-6_3

discussed in regard to the hysteresis they evoke in relation to the weight of the AFPs. Within each group, the AFPs can differ a lot, not only in size, but also in, e.g., peptide sequences or the specific ice plane they have bind to, as they origin from different species. Furthermore, though all thermal hysteresis activities presented here have been measured on a nanoliter osmometer or similar device (unless stated otherwise), there can still be differences in the measuring procedures among laboratories. These differences can affect the activity measurements and impair direct comparisons, and making the size–activity relation harder to evaluate. In the results below, data from across laboratories and different AFPs in each subgroup is presented together, and the overall conclusion is presented in the end of the subchapter.

3.2.1 AFPs from Fish

The fish AFPs are divided in five types, based on their structure, as described in detail in Chap. 5 of Vol. 1, given as type I through IV and antifreeze glycoproteins (AFGPs); however, it can be discussed whether the type IV AFP serves as an antifreeze in vivo due to low concentration. Several isoforms of the AFPs are observed within each group, differing in both weight and activity, while maintaining the overall structure. The data presented here are divided into the fish AFP types, focusing on the different isoforms and their weight and activity within each group.

Antifreeze glycoproteins (AFGPs), the glycosylated AFPs, are found in a wide range of sizes, and encompass both some of the smallest and the biggest AF(G)Ps found. The smallest AFGP reported is found in bald rockcod, *Pagothenia borchgrevinki*, weigh 2.6 kDa and consist of four of the AAT repeats with a disaccharide bound to the threonine that makes up the AFGPs in general (Feeney 1974). In the same species the thus far largest AFGP is found, being 32 kDa, consisting of 50 AAT repeats (DeVries et al. 1970). Though found in various species, the structural basis is the same—glycosylated AAT repeats, with only a few variations in the bound sugar group as well as a few substitutions of alanine with proline in the amino acid sequence.

Figure 3.1a and b depict the activities of various sized AFGPs from *Gadus ogac* (rock cod), compared both on a weight basis (Fig. 3.1a) and a molar basis (Fig. 3.1b). The small-sized AFGPs (2.6-10 kDa) are less active than the three largest (13.7–24.1 kDa) AFGPs, both at a total protein amount and at the molecular concentration (Wu et al. 2001). In general, a tendency of higher activity for larger AFGPs can be seen, and is most clear when comparing molar concentrations. This is also evident by the study on *P. borchgrevinki* AFGPs (Fig. 3.1c) where the activity of different sizes of AFGPs at 1 mM concentration are compared (DeVries 1988). Though correlated, the activity of AFGP does not seem to be solely dependent on size, and factors such as the type of glycosylation as well as amount of proline substitutions are likely to influence the activity of AFGPs. Interestingly, while the

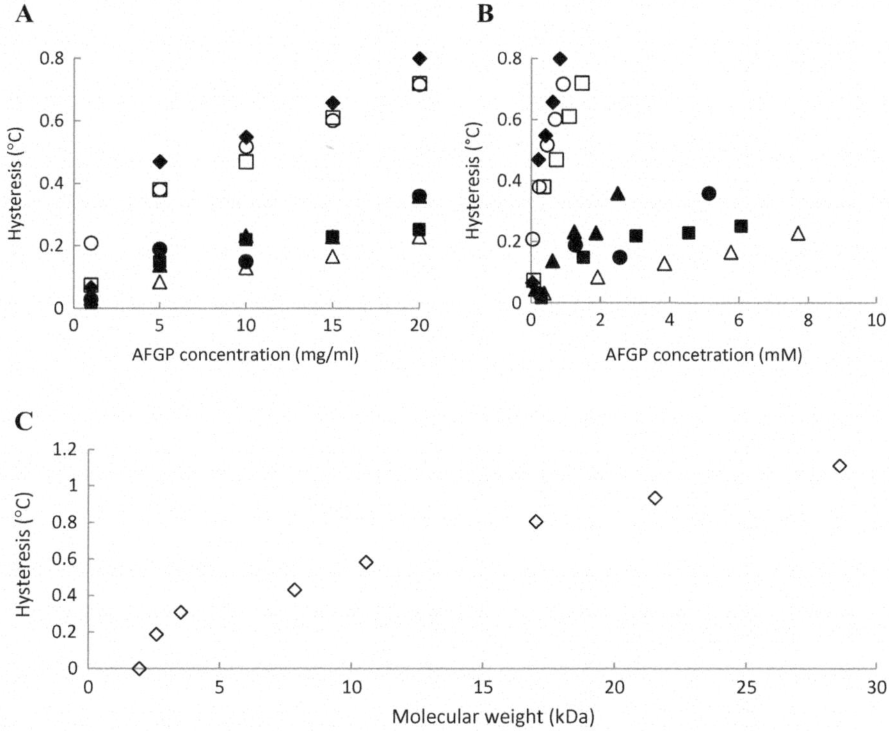

Fig. 3.1 AFGP size and activity. Hysteresis evoked by AFGP from *Gadus ogac* correlated to concentration in mg/l (**a**) or mM (**b**). Legend: open triangle: 2.6 kDa, filled square: 3.3 kDa, filled circle: 3.9 kDa, filled triangle: 6-10 kDa, open square: 13.7 kDa, open circle:21.7 kDa, filled diamond: 24.1 kDa (Wu et al. 2001). (**c**) Hysteresis evoked at 1 mM concentration of *Pagothenia borchgrevinki* AFGPs of different sizes. The data point at 2 kDa is the 2.6 kDa AFGP which has been shortened by one glycotripeptide unit by Edman degradation (DeVries 1988)

small AFGPs are the least active, these are the far most abundant AFGPs in fish blood (Burcham et al. 1984).

Type I AFPs, the simple α-helix structure, are amongst the smallest of the AFPs. The smallest protein in this group is found in shorthorn sculpin, *Myoxocephalus scorpius* (Hew et al. 1985). It consists of 33 amino acids and weighs 2.9 kDa. The largest single helix Type I AFP is found in the yellowtail flounder, *Limanda ferruginea*, (Scott et al. 1987), consisting of 48 amino acids and weighs 4.5 kDa. Intermediate sizes are found in winter flounder, *Pseudopleuronectes americanus*, weighing around 3.3 kDa, called HPLC-6 and HPLC-8, where the former is one of the most studied AFPs. Recently, a new Type I AFP showing the same helical structure has also been found in the *P. americanus*, but it consists of a homodimer of two 196 amino acids long subunits each weighing 16.7 kDa arranged in four parallel helices in its dimeric form (PDB ID: 4KE2) (Sun et al. 2014). This type I AFP shows high antifreeze activity, and is called hyperactive (Marshall et al. 2004b), a term

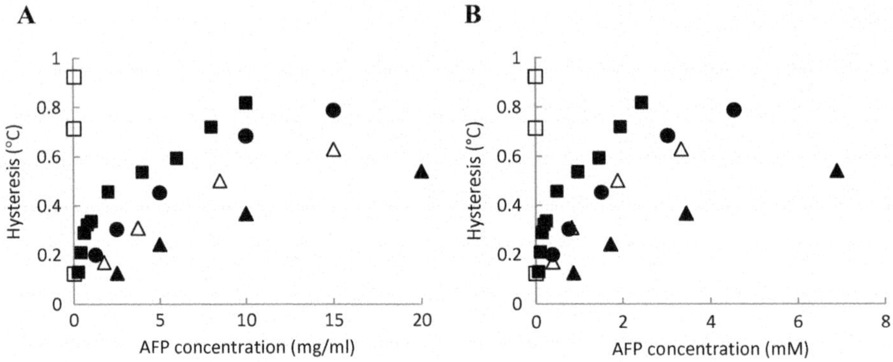

Fig. 3.2 Type I AFP size and activity. Hysteresis evoked by AFP Type I of different sizes correlated to concentration in mg/ml (**a**) or mM (**b**). Legend: open triangle: *Limanda ferruginea*, 4.5 kDa (Scott et al. 1987), filled square: *Tautogolabrus adspersus*, 4.1 kDa (Hobbs et al. 2011), filled circle: *Pseudopleuronectes americanus*, 3.3 kDa (Scott et al. 1987), filled triangle: *Myoxocephalus scorpius*, 2.9 kDa (Kao et al. 1986), open square: *P. americanus* dimer, 33.4 kDa (Marshall et al. 2005)

recently reserved to insect antifreeze proteins. The activities of Type I AFPs correlated to their weight and molar concentrations are presented in Fig. 3.2.

The by far largest Type I AFP (*P. americanus* dimer 33.4 kDa) is also the most active. This AFP shows hysteresis of more than 2 °C at mere 0.4 mg/ml (Marshall et al. 2004b). Also, the smallest Type I AFP (*M. scorpius*, 2.9 kDa) shows the lowest activity, indicating some correlation between their sizes and their activity. However, the *L. ferruginea* AFP (4.5 kDa), which has an extra helical turn compared to the *P. americanus* AFP (3.3 kDa), is not more active than the latter. The reason for this could very well be connected to the sequence of the proteins, as the *L. ferruginea* does not have as many of the threonines that are believed to drive the binding to ice, despite of the increased size (Scott et al. 1987). Considering this, there seems to be a good correlation between the size of the type I AFP and the activity it provides, supposing the ice-binding motifs are intact.

Type II AFPs are large globular antifreeze proteins, with known sizes spanning from 13.8 kDa in the Longsnout Poacher, *Brachyopsis rostratus*, (PDB ID: 2ZIB) (Nishimiya et al. 2008) to 24.0 kDa in the Rainbow Smelt, *Osmerus mordax* (Ewart and Fletcher 1990). Intermediate sizes are found in Sea Raven, *Hemitripterus americanus*, (14.0 kDa, PDB ID: 2AFP) (Gronwald et al. 1998), Atlantic Herring, *Clupea harengus* (14.6 kDa, PDB ID, 2PY2) (Liu et al. 2007), and the Japanese Smelt, *Hypomesus nipponensis* (16.8 kDa) (Yamashita et al. 2003). While *O. mordax* and *C. harengus* are dependent on Ca^{2+} ions to be active, the other three Type II AFPs function independently of Ca^{2+} ions (Liu et al. 2007; Venketesh and Dayananda 2008).

At first glance, the correlation between size and activity for the Type II AFPs seems to be negative (Fig. 3.3). While the largest Type II AFP (*C. harengus*, 24 kDa) seems to be one of the least active, especially when comparing activity with total

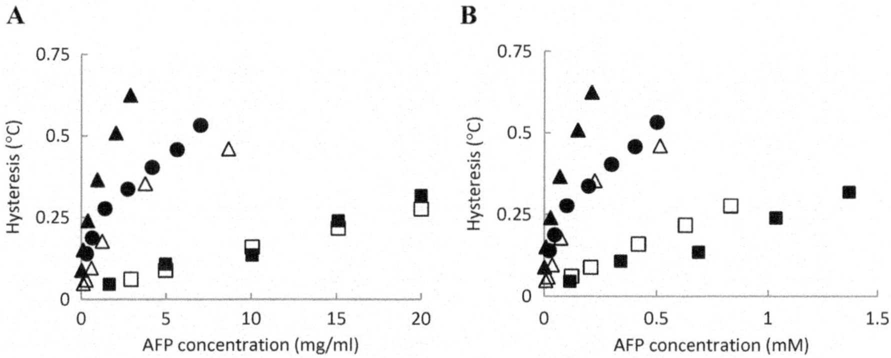

Fig. 3.3 Type II AFP size and activity. Hysteresis evoked by AFP Type II of different sizes correlated to concentration in mg/ml (**a**) or mM (**b**). Legend: open square: *Clupea harengus*, 24 kDa (Ca^{2+}) (Ewart and Fletcher 1990), open triangle: *Hypomesus nipponensis*, 16.8 kDa (Yamashita et al. 2003), filled square: *Osmerus mordax*, 14.6 kDa (Ca^{2+}) (Ewart and Fletcher 1990), filled circle: *Hemitripterus americanus*, 14 kDa (Slaughter et al. 1981), filled triangle: *Brachyopsis rostratus*, 13.8 kDa (Nishimiya et al. 2008)

protein concentration (Fig. 3.3a), the smallest Type II AFP (*B. rostratus*, 13.9 kDa) is the most active. However, the activity of the Type II AFPs seems more to be grouped after their dependence on Ca^{2+} ions. The activity of both *C. harengus* and *O. mordax* AFPs (squares in Fig. 3.3) are much lower than the calcium-independent Type II AFPs. The mechanism for this correlation is not clear. When looking at the calcium-independent Type II AFPs only, the correlation between activity and size still seems to be negative, with the largest AFP evoking the lowest degree of hysteresis, contrary to the tendency seen for the other AFPs presented in the other activity curves in this chapter; however, the sizes of these AFPs are roughly the same.

Type III AFPs are small globular proteins with roughly the same size around 6.9 kDa. They are divided into two groups, QAE and SP, depending on which of these chromatographic resins (Quarternary Amines or Sulfopropyl, respectively) they bind to (Desjardins et al. 2012; Wilkens et al. 2014). The type III AFPs are found in Atlantic wolffish, *Anarhichas lupus*, (64 amino acids, 6.9 kDa) (Scott et al. 1988), ocean pout, *Macrozoarces americanus*, (66 amino acids, 7.0 kDa, (Hew et al. 1988), PDB ID: 1KDF (Sönnichsen et al. 1996)), European eelpout, *Zoarces viviparus* (66 amino acids, 6.9 kDa (Sørensen et al. 2006), PDB ID: 4UR4 (Wilkens et al. 2014)), and Antarctic eelpout, *Lycodichthys dearborni*, (64 amino acids, 6.9 kDa, PDB ID: 1UCS (Ko et al. 2003)). An AFP consisting of virtually two AFP type III subunits joint with a linker has been found in *L. dearborni* as well. This protein consists of 134 amino acids and weighs 14.5 kDa (PDB ID: 1C8A) (Miura et al. 2001). Not much is published about the activity of the native Type III AFPs. In Fig. 3.4 are the activity curves of the 7.0 kDa *M. americanus* and 14.5 kDa *L. dearborni* AFPs presented, along with a recombinant "monomer" of the latter (74 amino acids, 7.4 kDa).

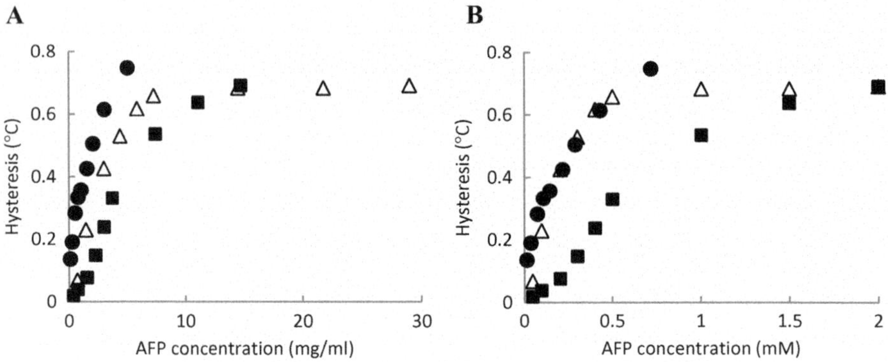

Fig. 3.4 Type III AFP size and activity. Hysteresis evoked by AFP Type III of different sizes correlated to concentration in mg/ml (**a**) or mM (**b**). Legend: filled circle: *Macrozoarces americanus*, 7 kDa (Sönnichsen et al. 1993), filled square: *Lycodichthys dearborni* 7.4 kDa (Miura et al. 2001), open triangle: *Lycodichthys dearborni* 'dimer', 14.5 kDa (Miura et al. 2001)

For these Type III AFPs, the size of the proteins also seems to have an impact on the hysteresis when comparing the *L. dearborni* AFP (14.5 kDa) with the recombinant monomeric version of the protein (7.4 kDa), especially when comparing the activities plotted against the molar concentrations Fig. 3.4b). Here the large *L. dearborni* AFP is much more efficient in evoking antifreeze activity than its monomeric counterpart is. However, even though a big difference in activity is seen between these proteins, compared to the small 7.0 kDa *M. americanus* AFP, the 14.5 kDa AFP evokes almost identical hysteresis, based on the molar concentration (Fig. 3.4b). The 7.0 kDa *M. americanus* AFP is ca. 82% identical to the N-terminal half of the 14.5 kDa *L. dearborni* AFP, but despite of the strong similarity, the *M. americanus* AFP seems more active than the other Type III AFPs.

In a mutational study on the *L. dearborni*, focusing on both protein size and the area of the ice-binding site, the tandemer, monomer, and tandemer with one ice-binding site knocked out was investigated (Bar Dolev et al. 2016). On a molar basis the tandemer showed twice the activity as the monomer, while the tandemer with one ice-binding site knocked out was only slightly more active that the monomer. These results show that though size itself can increase the activity of the proteins, the size of ice-binding area is probably a more important factor.

Type IV AFPs is a relatively newly described type of AFP (Deng et al. 1997) and is described as helix bundles (Deng and Laursen 1998). As stated earlier, the Type IV AFPs are not thought to serve as an antifreeze agent in its host fish, and its relevance can therefore be discussed. However, the antifreeze observations made on isolations of these proteins are presented here.

Not many Type IV AFPs have been described; however, many fish species have gene sequences resembling the gene sequences of known type IV AFPs (Lee et al. 2011). Four AFP type IV have been described, originating from *M. scorpius* (Gauthier et al. 2008), longhorn sculpin, *Myoxocephalus octodecimspinosis*, (Deng et al. 1997), Antarctic silverfish, *Pleuragramma antarcticum*, and Antarctic black

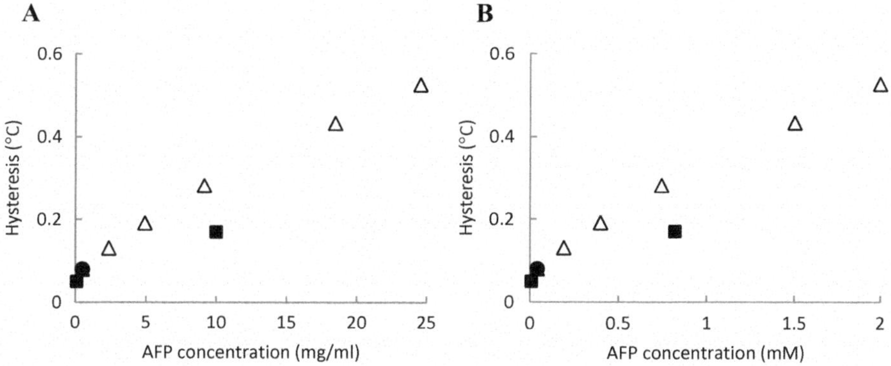

Fig. 3.5 Type IV AFP size and activity. Hysteresis evoked by AFP Type IV of different sizes correlated to concentration in mg/ml (**a**) or mM (**b**). Legend: open triangle: *Myoxocephalus octodecimspinosis*, 12.3 kDa (Deng et al. 1997), filled square: *Myoxocephalus scorpius*, 12.2 kDa (Gauthier et al. 2008), filled circle: *Pleuragramma antarcticum*, 12.1 kDa (Lee et al. 2011), filled triangle: *Notothenia coriiceps*, 12.2 kDa (Lee et al. 2011)

rock cod, *Notothenia coriiceps* (Lee et al. 2011). All of these consists of 108 amino acids in their mature form, and weigh 12.1–12.3 kDa, and are thus a very homogeneous group of proteins. The obtained activities are presented in Fig. 3.5. However, it is difficult to conclude on the activity versus weight of the Type IV AFPs, both because of the lack of actual activity curves, but also because the AFPs have almost identical weights.

Further studies comparing the different AFP types from fish in relation to their weights have been performed by Kao et al. 1986. They conclude a general positive correlation between antifreeze activity and protein molar concentration. However, increases in molecular weight above 4 kDa resulted in a decline in antifreeze activity per mg of protein (Kao et al. 1986).

3.2.2 AFPs from Arthropods

Antifreeze proteins are found in body fluids from various groups of the arthropods, including springtails, spiders, centipedes, midges, mites, moths, and beetles. However, most investigation has been made within the latter. The AFPs are probably very abundant in arthropods living in cold regions. A study on 75 species collected in Alaska found that the body fluids of 26% of these could evoke thermal hysteresis (Duman et al. 2004). The sizes of known native arthropod AFPs span from around 8 to 16 kDa. With a few exceptions the arthropod's AFPs are much alike in their overall structure—a β-helix or β-solenoid form. The helixes can be both left- or right-handed and vary in the coil length and the number of coils that compose the protein. In insects, which are the most investigated group of arthropods, the AFPs generally consist of a flat side consisting of repeating TxT motifs (the ice-binding

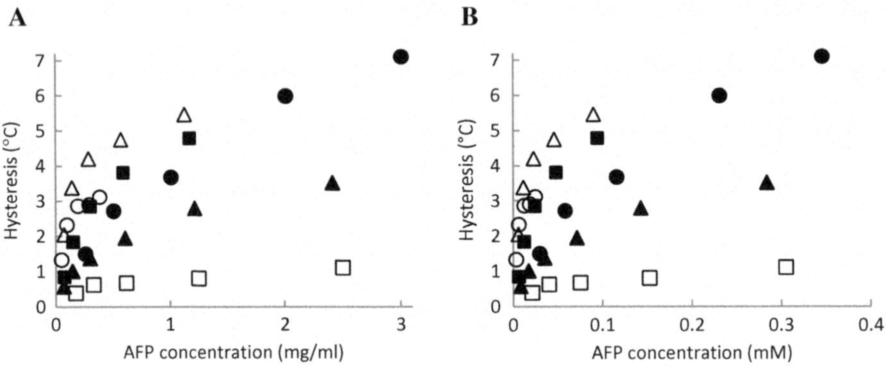

Fig. 3.6 Arthropod AFP size and activity. Hysteresis evoked by AFP from arthropods of different sizes correlated to concentration in mg/ml (**a**) or mM (**b**). Legend: open triangle: *Choristoneura fumiferana* (Leinala et al. 2002), 12.5 kDa, filled square: *Rhagium mordax*, 12.4 kDa (Kristiansen et al. 2012), open circle: *Hypogastrura harveyi*, 15.7 kDa (Mok et al. 2010), filled circle: *Dendroides canadensis*, 8.7 kDa (Wang and Duman 2005), filled triangle: *Tenebrio molitor*, 8.5 kDa (Liou et al. 2000), open square: Lake Ontario Midge, 8.2 kDa (Basu et al. 2016)

site). The variations among the AFPs are mainly among the amino acids that are not part of the TxT motifs, the number of coils in the β-helix, the extent of sulfide bridges, and the width of the TxT motif. Several species have different AFP isoforms, and in some cases the isoforms of a species differ in the number of coils in the β-helix that comprise the protein.

Though AFPs, or at least thermal hysteresis, have been detected in a wide range of species, activity measurements of the purified proteins have in many cases still not yet been produced. Thus, it is not possible to discuss the antifreeze activity of the tick (*Ixodes scapularis*, 23,2 kDa AFGP) (Neelakanta et al. 2010), the centipede (*Lithobius forficatus*) (Tursman and Duman 1995), the spider (*Bolyphantes index*) (Husby and Zachariassen 1980), and the mite (*Alaskozetes antarcticus*) (Block and Duman 1989) in relation to their sizes. However, the activities of several beetle AFPs are well investigated, and activity curves of one isoform from each investigated species are presented in Fig. 3.6, where activities of a moth AFP, a midge AFP, and a collembolan AFP are depicted as well.

The difference among the activities evoked by the arthropod AFPs is big, with spruce budworm, *Choristoneura fumiferana*, AFP showing more than fivefold higher activity than a Lake Ontario midge (Chironomidae family) AFP. When looking at the sizes of the AFPs, there seems to be a positive correlation with the activity. The larger AFPs (12.5–15.7 kDa) evoke higher activities than the smaller (8.2–8.7 kDa) AFPs. The lower activity of the midge AFP, however, is due to the difference in the nature of the ice-binding site, as this protein does not have TxT motifs, and the ice-binding pattern, i.e., the ice crystal planes it binds to, and its activity is therefore not comparable (Basu et al. 2016). The snow flea, *Hypogastrura harveyi*, AFP differs from the rest as it is not a beta-solenoid (PDB ID: 2PNE), but the activity seems comparable to the insect AFPs (Mok et al. 2010). The pyrochroid

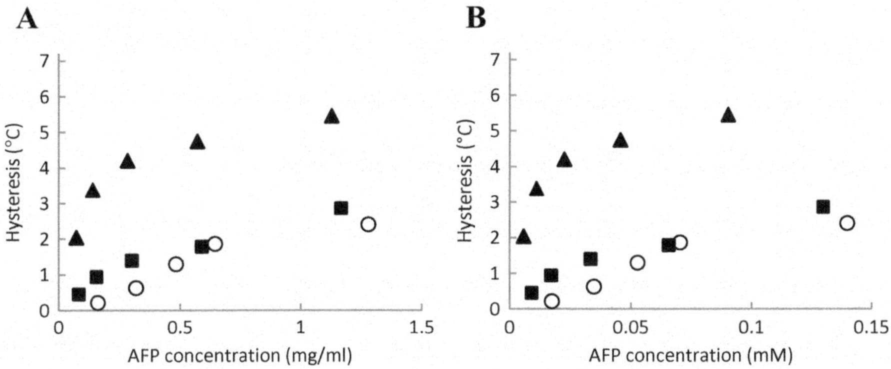

Fig. 3.7 Activity of *C. fumiferana*. The activity of a large natural AFP isoform (Cf501), a small natural AFP isoform (Cf339), and a mutant AFP where two coils have been exerted from the large isoform, to make it the size of the small natural isoform. The activities are correlated to AFP concentration in mg/ml (**a**) or mM (**b**). Legend: filled square: Cf501 isoform, 12.5 kDa, filled triangle: Cf339 isoform, 9.0 kDa, open circle: Cf501-mutant, 9.2 kDa (Leinala et al. 2002). All measurements are performed by the capillary technique

beetle, *Dendroides canadensis*, AFP activity is measured using a capillary technique and not a nanoliter osmometer (Wang and Duman 2005), which might bias the comparison.

The repetitive structure (each coil) has given rise to isoforms with different number of coils in many of the arthropod AFPs (Liou et al. 1999; Nickell et al. 2013; Lin et al. 2011; Ma et al. 2012; Qiu et al. 2010b; Doucet et al. 2002). However, only in a few cases has the activity of the isoforms been compared. The repetitive structure also makes mutational studies easier, as mutants of different sized AFP can be constructed. An advantage of these studies is that they are made by the same investigators, using the same method, equipment, and buffers to measure the activity, which makes the observations less biased and more suitable for comparisons. Leinala et al. have studied two isoforms of *C. fumiferana* AFPs—the CfAFP-501 (121 amino acids, 12.5 kDa) and CfAFP-337 (89 amino acids, 9.0 kDa). The larger isoform consist of two coils more than the smaller (Leinala et al. 2002). The activity, shown in Fig. 3.7, of the larger isoform is clearly higher both on a molar level, but also in regard to total protein amount, than the smaller. The study also included a mutant form of CfAFP501 where two coils have been removed, making it 90 amino acids and 9.2 kDa, to make it comparable to the small CfAFP-337 isoform. The activity of the mutant is roughly the same as the small CfAFP-337 isoform, even though the amino acid identity is only 72%, indicating that the difference in activity between the two native forms is mainly due to the different sizes and not the different sequence (Leinala et al. 2002).

The same tendency is observed in the fish type I AFP from *P. americanus*, where the activity of the 4.3 kDa isoform exceeds that of the 3.3 kDa isoform, and evoking hysteresis approximately twice as high at the same protein amount (Chao et al. 1996) (data is not shown here).

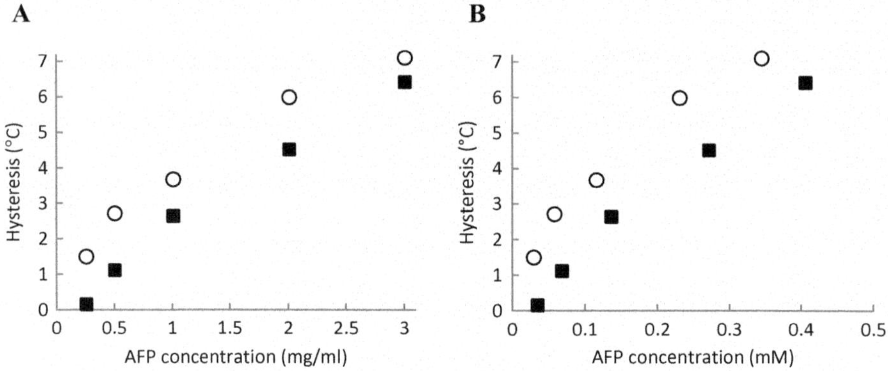

Fig. 3.8 Activity of *D. canadensis* isoforms. The activity of a large AFP isoform (DAFP-1), a small AFP isoform (DAFP-4) correlated to AFP concentration in mg/ml (**a**) or mM (**b**). Legend: open circle: DAFP-1, 8.7 kDa, filled square: DAFP-4, 7.4 kDa (Wang and Duman 2005)

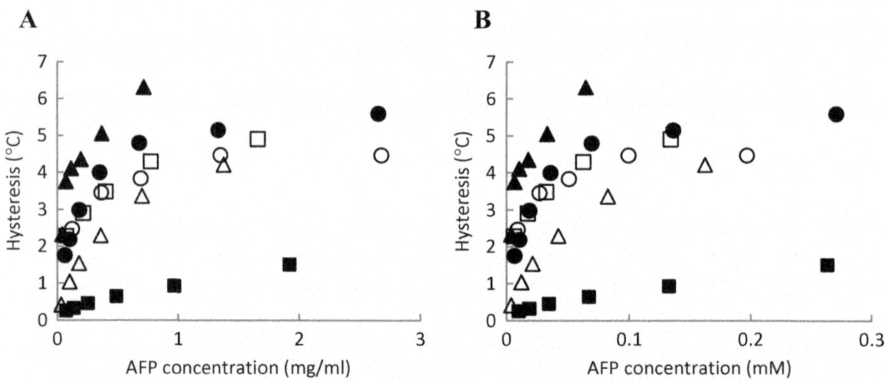

Fig. 3.9 Activity of *T. molitor* mutants. The activity of the *T. molitor* isoform TmAFP4-9 and various mutants with removal or addition of coils correlated to concentration in mg/ml (**a**) or mM (**b**). Legend: open triangle: TmAFP4-9 wild type, 8.5 kDa, filled circle: +1Coil mutant, 9.8 kDa, filled triangle: +2Coil mutant, 11.1 kDa, open square: +3Coil mutant, 12.4 kDa, open circle: +4Coil mutant, 13.6 kDa, filled square: −1Coil mutant, 7.3 kDa (Marshall et al. 2004a)

Wang and Duman have studied the isoforms of *D. canadensis* AFPs. They investigated four different isoforms with either six (DAFP-4, DAFP-6) or seven (DAFP-1, DAFP-2) coils. When comparing DAFP-1 (83 AA, 8.7 kDa) and DAFP-4 (71 AA, 7.4 kDa), as seen in Fig. 3.8, the difference in the activity is small; however, the larger DAFP-1 isoform had slightly higher activity. The extra coils aside, the two isoforms only differ in three amino acids (Wang and Duman 2005).

Marshall et al. have performed a thorough mutational study on the mealworm, *Tenebrio molitor*, AFP "4-9" (7 coils, 85 AA, 8,5 kDa), where they investigated the effect of removing one coil of the protein or added 1–6 extra coils. The activities of the TmAFP4-9 and its mutants are shown in Fig. 3.9. Here they observed that

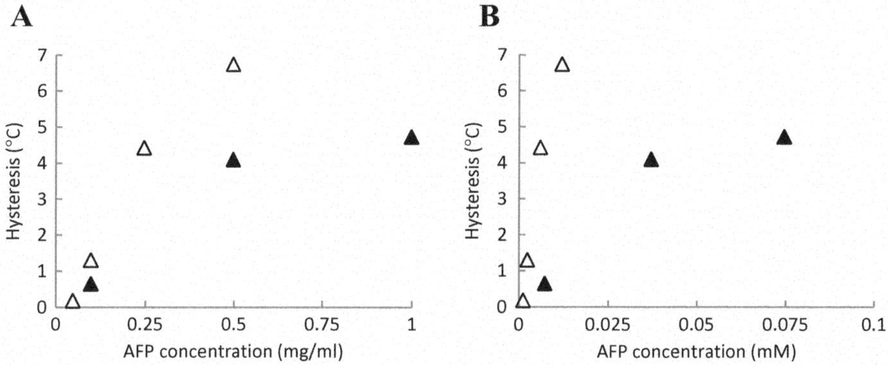

Fig. 3.10 Activity of RiAFP. The activity of the native *R. inquisitor* AFP and an AFP-GFP fusion protein correlated to concentration in mg/ml (**a**) or mM (**b**). Legend: filled triangle: RiAFP wild type, 13.4 kDa, open triangle: RiAFP-GFP fusion, 41.4 kDa (Hakim et al. 2013)

removing a coil diminished the activity, while adding one and especially two coils, the activity was increased. Addition of three and four coils caused a decreased activity (compared to the +2 coil mutant). Addition of six coils resulted in a low protein yield and was not investigated further. No mutant with five extra coils was made. This study shows a clear correlation between size and activity of TmAFP, albeit only in the low size span. This could indicate that the ice-binding possibilities is not increased further, which is speculated to be due to an accumulation of a small mismatch between the binding locations of the ice/protein and/or an increased flexibility of the long AFP (Marshall et al. 2004a).

A similar study is made on the blackspottet pliers support beetle, *Rhagium mordax*, AFP1, which has a wider TxTxTxT ice-binding motif (Friis et al. 2014b). Here a mutant with an extra coil was constructed, which showed increased activity compared to the wild type.

Interestingly, the increased antifreeze activity does not seem to only depend on a larger ice-binding surface. Hakim et al. observed that the pine borer, *Rhagium inquisitor*, AFP fused to a green fluorescent protein (RiAFP-GFP, 41.4 kDa) had significantly higher activity than the native protein (13.4 kDa), as seen in Fig. 3.10 (Hakim et al. 2013). The same pattern is observed by DeLuca et al., studying a Type III fish AFP fused to thioredoxin (12 kDa) or maltose-binding protein (42 kDa). The mechanisms causing the enhanced activity of a fusion protein could be that each bound AFP fusion protein occupies a larger area on the ice surface than the native protein, thus increasing the curvature of the ice surface between the proteins (DeLuca et al. 1998). Another theory implies that the solubility of the fusion protein is lower than the native, driving more proteins to the ice/water interface, and thus a binding of more proteins to the ice, effectuating an increased antifreeze activity. This mechanism is described in depth in Chap. 6 of this volume.

3.2.3 Concluding the Weight and Activity Relation

The AFPs differ in structure and bind to different planes of the ice surface, which in itself evokes different degrees of hysteresis, and makes antifreeze activity comparisons biased. Here annealing times, heating rates, or differences in initial crystal size, which have great impact on the measures of thermal hysteresis, can vary. Thus the strongest results are obtained when comparing measurements from one AFP isoform made in one lab (preferably measured by the same person), and changing the size by mutational studies, or by comparing AFP isoforms with high identity but different sizes. Doing this, it is evident that the size of the AFPs plays a role on the activity. The larger the protein, the higher thermal hysteresis. The effect seems to level off when the protein reaches a certain size, and may even decrease again. Interestingly, the increase in size itself has an effect, as fusions with non-AFPs have provided enhanced activities; however, fusing with other AFPs, thus also increasing the ice-binding area, contributes even more to the effect.

It can seem redundant for an organism to have other AFPs than the most effective isoform, but an explanation could be that interactions of the different isoforms lead to a synergistic effect (Lin et al. 2010; Wang and Duman 2005). This enhancement could be a direct interaction between the proteins, leading to a more efficient binding to the ice, or maybe that the different isoforms bind to different planes on the ice crystal, thus making a better overall coverage. The small and less active AFPs seem to have an important role as it is often the small isoforms that are found in higher quantities and are the dominant AFPs (Marshall et al. 2004b; Graham and Davies 2005). It is possible that this is connected to the higher solubility or a more rapid diffusion of the smaller AFPs.

3.3 Solubility and Hydrophobicity

Though being a key feature of proteins, not much research has focused on the solubility of AFPs. Generally, the AFPs are quite soluble due to their relatively small size and low content of very hydrophobic amino acids. Only in a few cases has the concentration of AFP in the different organisms been reported. For example, the concentration of AFP type III in the *M. americanus* is estimated to be up to 25 mg/ml in the blood (Fletcher et al. 1985), whereas for the type IV AFP in the *M. octodecimspinosis* it is estimated to be less than 0.1 mg/ml (Gauthier et al. 2008). For type III AFP, three isoforms are estimated to be present in the blood at 20 mg/ml, though at this point the large isoform starts to precipitate (Wang et al. 1995). The AFGPs (of various sizes) are estimated to make up approximately 3.5 (W/V) of the blood of notothenioid fishes of Antarctica (DeVries 1983).

Many of the AFP-producing fish live in sea water with temperatures down to −1.9 °C. As the colligative freezing point of the blood is around −0.9 °C, the AFPs have to evoke hysteresis of around 1 °C to prevent lethal ice growth. To accommodate this, especially the least effective AFPs have to be present in high concentrations in the blood. However, as many species have several isoforms of AFPs, which also might have a synergistic effect on each other, the concentration of a single AFP isoform could be brought down. With regard to insects and other terrestrial species producing AFPs, these organisms are subjected to much lower temperatures than fish. This is accommodated by the means of more efficient AFPs (i.e., higher antifreeze activity per protein), rather than immense concentrations.

Many AFPs are now produced by heterologous gene expression usually in yeast or bacteria, instead of recovering the proteins directly from the original source. These expression organisms are designed to produce single proteins at a very high concentration.

So far, no experiments have been focusing on the precipitation points of AFPs, challenging their solubility. However, the addition of salts to AFP solutions has shown to increase the activity of AFPs (Kristiansen et al. 2008). A theory explaining this observation is that the proteins are being "salted out," i.e., lowering the solubility, driving the AFPs to the ice/water interface. This is explained in depth in Chap. 6 of this volume.

The hydrophobicity of a protein surface is also related to the solubility, and this subject has had more focus, as it has been a great topic regarding the mechanism of binding between ice and AFP. The prevailing theory of the ice-binding mechanism is called anchored clathrate mechanism which is described in detail in Chap. 4 of this volume. In brief, it states that water molecules are arranged in an ice-like structure around the hydrophobic ice-binding site of the AFP, which freezes to ice surface (Cheng and Merz 1997; Garnham et al. 2011). The AFPs are generally amphiphilic, having a relatively hydrophobic surface at the flat ice-binding region of the protein, while the rest of the proteins' solvent exposed surface is regarded as relatively hydrophilic (Davies et al. 2002). The amino acids facilitating the ice binding are generally alanine (fish) or threonine (arthropods, bacteria). While these amino acids are not considered particularly hydrophobic, relative to the rest of the protein they are, and their ordered structure in an, often, flat surface brings the water molecules into an ordered ice-like arrangement.

Though the ice-binding site is regarded as hydrophobic, the AFPs are likely overall regarded as hydrophilic (Davies et al. 2002). Hydrophobicity is difficult to quantify and many different scales for this exist (Cornette et al. 1987). A method to compare proteins' hydrophobicity is by thermal gravity, where a protein's weight is measured in a large temperature span. As temperature rises, loosely bound water molecules evaporate first (bound to hydrophobic parts of the protein's surface), while tightly bound water molecules evaporate at higher temperatures (bound to hydrophilic parts). Thus, during heating (e.g., from 20 to 150 °C), the rate of the weight loss gives a relative measure of the hydrophobicity of the protein. Though this method has not been widely used on AFPs, a study on *Anatolica polita* AFP showed that the proteins had a higher hydrophilicity than the control protein, and generally hydrophilic protein, bovine serum albumin (Mao et al. 2011). However, the high hydrophilicity might be limited to insect AFPs (Graham et al. 1997), as fish AFPs, especially type II, have hydrophobic surfaces (Sönnichsen et al. 1995). As the

AFPs are a very diverse group of proteins, and evoking different levels of thermal hysteresis, their hydrophilic features, and how they affect the ice binding, might not be the same. The different amino acids found in the active sites also indicate this. Threonine, an important amino acid in many ice-binding sites has both a hydrophobic part, a methyl group, and a hydrophilic part, a hydroxyl group, both of which is believed to play a role in a successful ice interaction (Bar Dolev et al. 2016). A detailed description of the interaction between the antifreeze protein and the surrounding water and with ice is found in the next Chaps. 4 and 5 of this volume.

3.4 Stability

The investigations on the stability of the antifreeze proteins have mainly concerned their heat stability, and to some extent their pH stability. This has often been relevant to investigate in regard to choosing optimal purification methods when isolating the protein, or when discussing their relevance in industrial applications. Besides, when studying the secondary structure of proteins using circular dichroism (CD) spectroscopy, the experiment can easily be expanded to include a determination of the protein melting point, which could be the case for some of the AFP studies as well. The method is described in further detail later in this chapter.

The stability of proteins can be understood as thermodynamic stability or kinetic stability (Sanchez-Ruiz 2010). Thermodynamic stability is defined in respect to equilibrium between the active native form of the protein (usually termed *N*) and various non-native conformations in an often unfolded form (termed *U*). Transitions between the two forms are described by first-order rate constants, k_+ for the transition from *N* to *U*, and k_- from *U* to *N*. The thermodynamic stability is defined as the equilibrium constant K given as k_+/k_-. The reaction course can be described by the formula:

$$N \underset{k_-}{\overset{k_+}{\rightleftarrows}} U \tag{3.1}$$

Regarding kinetic stability, an irreversible inactivated form (termed *I*) is also accounted for. Usually, this entails that the native form must go through the unfolded state before it is prone to inactivation. The transition to the inactivated form, *U* to *I*, is described by another first-order constant, k_{ag}, but as the reaction is irreversible, no transitions run in the other direction, *I* to *U*. In its simplest form, the reaction course can be described as (Lumry and Eyring 1954):

$$N \underset{k_-}{\overset{k_+}{\rightleftarrows}} U \underset{k_{ag}}{\rightarrow} I \tag{3.2}$$

It is common for proteins that the route to the inactive form is aggregation. However, some proteins can avoid this, and hence transit back into the native form. This ability is frequently referred to as colloidal stability (Chi et al. 2003). Hence, the stability of proteins is more than just the transition to an unfolded state (thermodynamic stability) but also entails the ability to fold back into a native protein (kinetic or colloidal stability).

3.4.1 Thermostability

Though only a few studies have focused on stability of AFPs proteins (Friis et al. 2014a; García-Arribas et al. 2007; Chi et al. 2003), several studies have included investigations of the protein melting point, or noted observations relating to the stability, in connection with their study. Many AFP studies have included investigations of the secondary protein structure using CD spectroscopy. This technique can provide much information about a protein, e.g., van't Hoff enthalpy, entropy of unfolding, or the free energy of folding (Greenfield 2006). However, it has usually just been used for determining the midpoint of the unfolding transition, corresponding to the "melting point" of the protein. In this chapter the transition midpoint, shortened Tm (not to be confused with the melting point of solution, which is a term often used when studying AFPs), refers to the temperature where half of the proteins will be in the native (N) state and the other half in the unfolded (U) state.

In short, CD spectroscopy analyzes the difference between the absorption of left-handed and right-handed circularly polarized light, and is often measured in molar ellipticity [θ] (degrees cm^2 dmol^{-1}). Folded proteins often have structural elements which are highly asymmetric, such as α-helices or β-helices. These structures will thus have characteristic CD spectra (Greenfield 2006), which usually span from 190 to 250 nm. Proteins are composed of various secondary structural elements, giving each protein a distinct CD spectrum. The CD spectra of a broad variety of AFPs are shown in Fig. 3.11. From the CD spectra, the structural elements constituting the protein can be estimated (Greenfield 2006), and for example, the spectra for fish AFP type I and type IV are in great concordance with the expected spectra for proteins with predominant α-helical structure (Fig. 3.11a). Likewise, insect AFPs generally have a high content of β-sheets which show a dip around 218 nm, which is also the general picture in Fig. 3.11b.

When proteins unfold, e.g., by heating, they lose their ordered structure and thus their CD spectrum changes. By monitoring the change of state by the CD spectroscopy during heating, the thermodynamics of the process, hereunder the midpoint transition, can be determined. This is illustrated in Fig. 3.12 for the *R. mordax* AFP1 (Friis et al. 2014a). CD spectra are obtained at various temperatures (here 1–49 °C), spanning an interval where all proteins are in the folded state (N, 1 °C, dark blue line) to where the entire protein population is in the unfolded state (U, 49 °C, light blue line). A suitable single wavelength containing a dip or a hill, which changes during

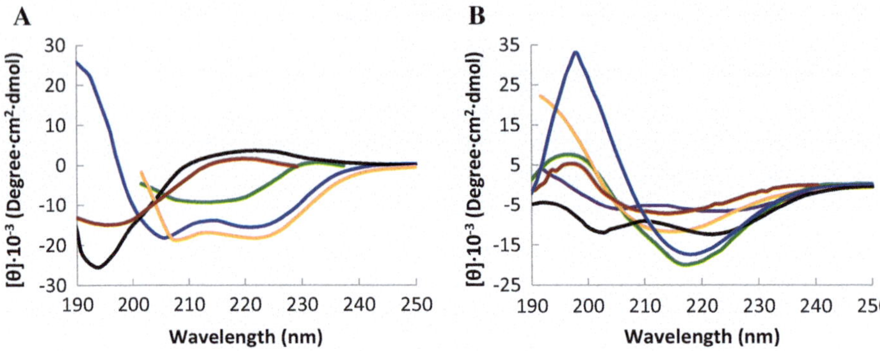

Fig. 3.11 CD spectra of AFPs. Molar ellipticity in the far UV range CD spectra of various AFPs. (**a**) CD spectra of fish AFPs; Blue: Type I (Fairley et al. 2002), Green: Type II (Slaughter et al. 1981), Black: Type III (Li et al. 1991), Orange: Type IV (Deng and Laursen 1998), and Red: AFGP (DeVries et al. 1970). (**b**) CD spectra of insect AFPs; Blue: *R. mordax* (Friis et al. 2014a), Green: *C. perlata* (Lin et al. 2011), Black: *D. canadensis* (Li et al. 1998b), Orange: *C. fumiferana* (Gauthier et al. 1998), Red: A. polita (Mao et al. 2011), and Purple: *T. molitor* (Liou et al. 2000)

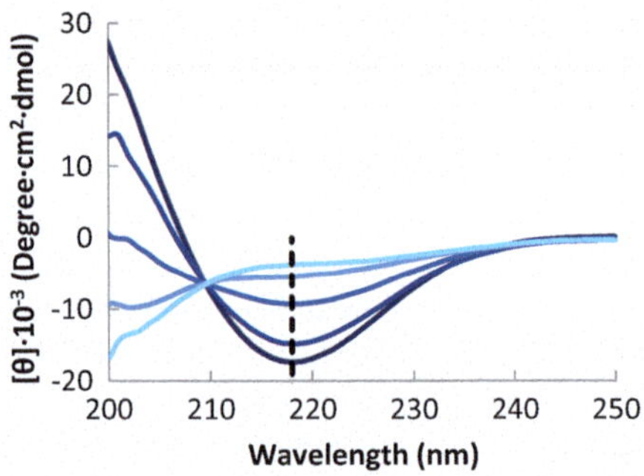

Fig. 3.12 CD spectra and temperature. The figure shows an example of CD spectra obtained from the same protein sample (*R. mordax*) at different temperatures. As the temperature rises from 1 °C (dark blue) to 25 °C, 30 °C, 33 °C, and 49 °C more proteins will be in the unfolded state. This transition can be quantified by monitoring the [θ] at a protein-specific hot spot, here 218 nm, marked with a dashed line. These values are transformed and plotted in Fig. 3.13

temperature increase, is chosen for the calculations of N and U state estimates. In the case in Fig. 3.12, 218 nm is chosen.

The fraction of native folded protein (F_N) can be calculated as:

$$F_N = \frac{\theta_t - \theta_U}{\theta_N - \theta_U} \tag{3.3}$$

where θ_t, θ_U, and θ_N are the ellipticity at the measured temperature, the temperature where all the proteins are fully unfolded, and the temperature where all the proteins are fully folded, respectively. A graphic illustration of this (using the study

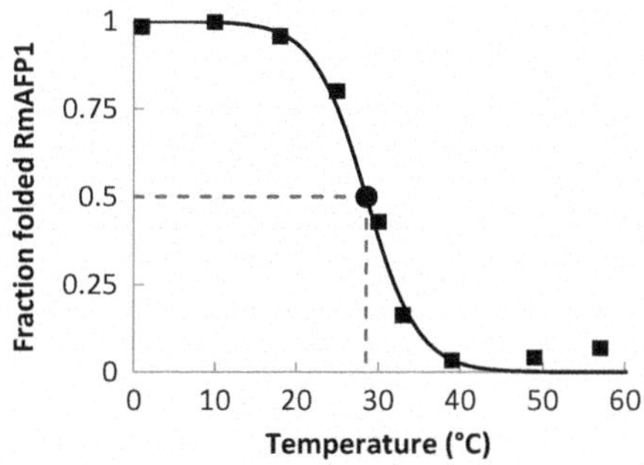

Fig. 3.13 Determining transition midpoint temperature. Extreme [θ] values, at a specific wavelength, obtained at temperatures where all proteins are either in its folded or unfolded state are set to 1 and 0, respectively. Intermediate values obtained at intermediate temperatures are standardized according to the extreme values ($0 < x < 1$). The figure shows data from the example in Fig. 3.12 (218 nm). The Tm can then be read as the temperature where N = U, corresponding to 0.5. In this case Tm = 28.5 °C (Friis et al. 2014a)

from Fig. 3.12), with θ_N set to 0 and θ_U set to 1, is illustrated in Fig. 3.13. From this the Tm can be derived as the temperature where $F_N = F_U = ½$ (the amount of *N* equals the amount of *U* stated proteins), which in the example in the figures correspond to 28.5 °C. This method, or derivations thereof, has been used to determine the Tm of various AFPs, which are collected in Table 3.1, together with other observations related to AFP Tm.

It has been observed that the desert beetle, *Microdera punctipennis,* MpAFP698, 18 cysteines, maintained almost all its activity after heating to 100 °C for 5 min, indicating either a very high Tm, or a high degree of refolding (Qiu et al. 2010a). Similarly, the *R. inquisitor* RiAFP, 2 cysteines, maintains activity after heating to 70 °C for 5 min (Kristiansen et al. 2005).

Though cysteine bridges provide stability to proteins, no clear correlation is seen between the number of cysteine bridges and the Tm of the AFPs (see Table 3.1). However, in general the fish AFPs are without cysteines (except type II) and have intermediate Tms around 40–50 °C. The insect AFPs with high cysteine content have very high Tms of >60 °C and the low cysteine content insect AFPs have low Tms of <30 °C.

With respect to colloidal stability, the ability to refold back into a native protein, the investigations are quite sparse. While several researchers have reported a high degree of refolding after heat denaturation (García-Arribas et al. 2007; Salvay et al. 2007; Basu et al. 2016; Bar et al. 2008; Marshall et al. 2005), only in a few cases has the refolding into a native active form been confirmed by performing follow-up activity measurements; The DAFP-4 regained all activity after heating to 100 °C for 5 min (Li et al. 1998a) and the RmAFP1 regained almost all activity after several heat cycles to 70 °C (Friis et al. 2014a); however, after heat cycles to 80 °C the regained activity gradually diminished.

Table 3.1 Tm of various AFPs

Antifreeze protein	Species	Tm (°C)	Cysteines
Fish			
AFP9 (Type I) (Chao et al. 1996)	*Pseudopleuronectes americanus*	55	0
17kDA (Type I) (Marshall et al. 2005)	*Pseudopleuronectes americanus*	18	0
HPLC12 (Type III) (Salvay et al. 2007)	*Macrozoarces americanus*	50	0
HPLC6 (Type III) (Li et al. 1991)	*Macrozoarces americanus*	40	0
AB1 (Type III) (Cheng and DeVries 1989)	*Austrolycicthys brachycephalus*	43	0
LS-12 (Type IV) (Deng and Laursen 1998)	*Myoxocephalus octodecimspinosis*	45	0
Insect			
DAFP-4 (Li et al. 1998a)	*Dendroides canadensis*	84	14
YL-1 (Bar et al. 2008)	*Tenebrio molitor*	66	16
RmAFP1 (Friis et al. 2014a)	*Rhagium mordax*	29	2
Mixture (Lin et al. 2011)	*Campaea perlata*	<22	0–1
Other			
Midge AFP (Basu et al. 2016)	*Chironomidae*	≈ 40[a]	16
sfAFP (snow flea) (Graham and Davies 2005)	*Hypogastrura harveyi*	<22	4
Bacterium AFP (Wang et al. 2016)	*Flavobacteriaceae*	53.5	2

[a]Estimated by formula (3)

Though there are also examples of AFPs that do not refold and lose activity after heat exposure (Deng and Laursen 1998; Lin et al. 2011; Graham and Davies 2005; Kawahara et al. 2007), they generally seem to have a high Tm or high colloidal stability, especially when considering that their native function is restricted to 0 °C and below. If this is just a relic from the proteins the AFPs have evolved from, side effect from matching with ice or because some AFPs have different functions, is not known.

3.4.2 *pH Stability*

As for the thermostability, not much research has focused on pH stability of the AFPs. However, a series of studies has included either antifreeze activity measurements or CD spectra of the AFPs at different pH values. These observations are presented in Table 3.2. From here it can be seen that the AFPs are generally active (or folded in their native state for CD observations) in a wide pH range. In some measurements the activity was not impaired in the entire measured pH spectrum.

Table 3.2 pH stability of various AFPs

Antifreeze protein	Species	pH range	Method
Fish			
HPLC-6 (Type I) (Chakrabartty and Hew 1991)	*Pseudopleuronectes americanus*	3–11	Activity + CD
rQAE m1.1 (Type III) (Chao et al. 1994)	*Macrozoarces americanus*	2–11	Activity
LS-12 (Type IV) (Deng and Laursen 1998)	*Myoxocephalus octodecimspinosis*	6–10	Activity + CD
Insect			
DAFP-H1 + H2 (Wu et al. 1991)	*Dendroides canadensis*	4–9.6	Activity
DAFP-4 (Li et al. 1998a)	*Dendroides canadensis*	2–11	Activity
YL-1 (Bar et al. 2008)	*Tenebrio molitor*	6–12.5[a]	CD
RiAFPH4 (Kristiansen et al. 2005)	*Rhagium inquisitor*	3–12[a]	Activity
sbwTHP (Gauthier et al. 1998)	*Choristoneura fumiferana*	6–12	Activity

[a]Active/folded in the entire measured pH span

The high pH stability of the AFPs could be caused by the relatively few charged amino acids generally seen in the proteins. Also, no charged amino acids are involved in the interaction with ice, and the binding mechanism is therefore not affected by pH changes. The pH changes can affect the structural stability of the AFPs that contain salt bridges, and the differences in pH stability that after all is seen among the AFPs are probably related to this. The large drop in activity seen for most AFPs at high pH could likely be due to a titration of the hydroxyl group (pKs around 13) of threonine (often associated with ice interaction) (Wu et al. 1991), obstructing the binding to ice and thus impairing the activity.

The biological reason to why the AFPs have such high pH stability is not clear. However, in the case of *D. canadensis*, the AFP is found both in the neutral hemolymph and in the acidic midgut, thus there might have been a selection pressure for this AFP to be functional in a large pH range (Wu et al. 1991). This could also apply for other AFPs.

3.5 Conclusions

Overall, the physicochemical properties of antifreeze proteins are yet a field with many unexplored paths. The basic properties such as size, activity, and sequence are explored for most of the known AFPs. From here, a positive correlation between size and activity can be seen. Due to the differences of the AFPs, even within each species, the correlation is most clear between isoforms with high sequence identity (but different sizes). Subsequent mutational studies have confirmed the correlation, and though an increased ice-binding site is shown to increase the activity, fusing AFPs with non-AFPs also increases the activity.

The temperature stability of several AFPs has also been investigated. Here, however, the results are remarkably divergent, with some AFPs being very thermo labile and start losing activity around room temperature and some withstanding short periods at 100 °C. The thermostability is to some degree correlated with the amount of cysteines in the AFPs which vary a lot.

The AFPs are generally stable and active at a broad pH range, likely because no charged amino acids are directly involved in the ice-binding mechanism.

The details about the energy of binding and reaction kinetics of AFPs are yet to be fully uncovered.

The antifreeze proteins are a quite diverse group of proteins, differing a lot in size, activity, shapes, and amino acid composition. It is therefore also difficult to generalize results and observations. The only common property of the AFPs is their ability to bind to ice, which can be correlated to the area of the ice-binding site. The prevailing theories on mechanism of ice binding are addressed in the following chapter.

References

Bar Dolev M, Braslavsky I, Davies PL (2016) Ice-binding proteins and their function. Annu Rev Biochem 85:515–542. https://doi.org/10.1146/annurev-biochem-060815-014546

Bar M, Scherf T, Fass D (2008) Two-dimensional surface display of functional groups on a beta-helical antifreeze protein scaffold. Protein Eng Des Sel 21(2):107–114. https://doi.org/10.1093/protein/gzm070

Basu K, Wasserman SS, Jeronimo PS, Graham LA, Davies PL (2016) Intermediate activity of midge antifreeze protein is due to a tyrosine-rich ice-binding site and atypical ice plane affinity. FEBS J 283(8):1504–1515. https://doi.org/10.1111/febs.13687

Block W, Duman JG (1989) Presence of thermal hysteresis producing antifreeze proteins in the antarctic mite, *Alaskozetes antarcticus*. J Exp Zool 250(2):229–231. https://doi.org/10.1002/jez.1402500215

Burcham TS, Osuga DT, Chino H, Feeney RE (1984) Analysis of antifreeze glycoproteins in fish serum. Anal Biochem 139(1):197–204

Chakrabartty A, Hew CL (1991) The effect of enhanced alpha-helicity on the activity of a winter flounder antifreeze polypeptide. Eur J Biochem 202(3):1057–1063

Chao H, Sönnichsen FD, DeLuca CI, Sykes BD, Davies PL (1994) Structure-function relationship in the globular type III antifreeze protein: identification of a cluster of surface residues required for binding to ice. Protein Sci 3(10):1760–1769. https://doi.org/10.1002/pro.5560031016

Chao H, Hodges RS, Kay CM, Gauthier SY, Davies PL (1996) A natural variant of type I antifreeze protein with four ice-binding repeats is a particularly potent antifreeze. Protein Sci 5 (6):1150–1156. https://doi.org/10.1002/pro.5560050617

Cheng CH, DeVries AL (1989) Structures of antifreeze peptides from the Antarctic eel pout, *Austrolycicthys brachycephalus*. Biochim Biophys Acta 997(1–2):55–64

Cheng A, Merz KM (1997) Ice-binding mechanism of winter flounder antifreeze proteins. Biophys J 73(6):2851–2873. https://doi.org/10.1016/S0006-3495(97)78315-2

Chi EY, Krishnan S, Randolph TW, Carpenter JF (2003) Physical stability of proteins in aqueous solution: mechanism and driving forces in nonnative protein aggregation. Pharm Res 20 (9):1325–1336

Cornette JL, Cease KB, Margalit H, Spouge JL, Berzofsky JA, DeLisi C (1987) Hydrophobicity scales and computational techniques for detecting amphipathic structures in proteins. J Mol Biol 195(3):659–685

Davies PL, Baardsnes J, Kuiper MJ, Walker VK (2002) Structure and function of antifreeze proteins. Philos Trans R Soc Lond Ser B Biol Sci 357(1423):927–935. https://doi.org/10.1098/rstb.2002.1081

DeLuca CI, Comley R, Davies PL (1998) Antifreeze proteins bind independently to ice. Biophys J 74(3):1502–1508. https://doi.org/10.1016/S0006-3495(98)77862-2

Deng G, Laursen RA (1998) Isolation and characterization of an antifreeze protein from the longhorn sculpin, *Myoxocephalus octodecimspinosis*. Biochim Biophys Acta 1388(2):305–314

Deng G, Andrews DW, Laursen RA (1997) Amino acid sequence of a new type of antifreeze protein, from the longhorn sculpin *Myoxocephalus octodecimspinosis*. FEBS Lett 402(1):17–20

Desjardins M, Graham LA, Davies PL, Fletcher GL (2012) Antifreeze protein gene amplification facilitated niche exploitation and speciation in wolffish. FEBS J 279(12):2215–2230. https://doi.org/10.1111/j.1742-4658.2012.08605.x

DeVries AL (1983) Antifreeze peptides and glycopeptides in cold-water fishes. Annu Rev Physiol 45:245–260. https://doi.org/10.1146/annurev.ph.45.030183.001333

DeVries AL (1988) The role of antifreeze glycoproteins and peptides in the freezing avoidance of antarctic fishes. Comp Biochem Physiol B Biochem Mol Biol 90(3):611–621. https://doi.org/10.1016/0305-0491(88)90302-1

DeVries AL, Komatsu SK, Feeney RE (1970) Chemical and physical properties of freezing point-depressing glycoproteins from Antarctic fishes. J Biol Chem 245(11):2901–2908

Doucet D, Tyshenko MG, Davies PL, Walker VK (2002) A family of expressed antifreeze protein genes from the moth, *Choristoneura fumiferana*. Eur J Biochem 269(1):38–46

Duman JG, Bennett V, Sformo T, Hochstrasser R, Barnes BM (2004) Antifreeze proteins in Alaskan insects and spiders. J Insect Physiol 50(4):259–266. https://doi.org/10.1016/j.jinsphys.2003.12.003

Ewart KV, Fletcher GL (1990) Isolation and characterization of antifreeze proteins from smelt (*Osmerus-mordax*) and atlantic herring (*Clupea-harengus-harengus*). Can J Zool Revue Can Zool 68(8):1652–1658. https://doi.org/10.1139/z90-245

Fairley K, Westman BJ, Pham LH, Haymet AD, Harding MM, Mackay JP (2002) Type I shorthorn sculpin antifreeze protein: recombinant synthesis, solution conformation, and ice growth inhibition studies. J Biol Chem 277(27):24073–24080. https://doi.org/10.1074/jbc.M200307200

Feeney RE (1974) A biological antifreeze. Am Sci 62(6):712–719

Fletcher GL, Hew CL, Li X, Haya K, Kao MH (1985) Year-round presence of high levels of plasma antifreeze peptides in a temperate fish, ocean pout (*Macrozoarces americanus*). Can J Zool 63 (3):488–493. https://doi.org/10.1139/z85-070

Friis DS, Johnsen JL, Kristiansen E, Westh P, Ramløv H (2014a) Low thermodynamic but high kinetic stability of an antifreeze protein from *Rhagium mordax*. Protein Sci 23(6):760–768. https://doi.org/10.1002/pro.2459

Friis DS, Kristiansen E, von Solms N, Ramløv H (2014b) Antifreeze activity enhancement by site directed mutagenesis on an antifreeze protein from the beetle *Rhagium mordax*. FEBS Lett 588 (9):1767–1772. https://doi.org/10.1016/j.febslet.2014.03.032

García-Arribas O, Mateo R, Tomczak MM, Davies PL, Mateu MG (2007) Thermodynamic stability of a cold-adapted protein, type III antifreeze protein, and energetic contribution of salt bridges. Protein Sci 16(2):227–238. https://doi.org/10.1110/ps.062448907

Garnham CP, Campbell RL, Davies PL (2011) Anchored clathrate waters bind antifreeze proteins to ice. Proc Natl Acad Sci USA 108(18):7363–7367. https://doi.org/10.1073/pnas.1100429108

Gauthier SY, Kay CM, Sykes BD, Walker VK, Davies PL (1998) Disulfide bond mapping and structural characterization of spruce budworm antifreeze protein. Eur J Biochem 258 (2):445–453

Gauthier SY, Scotter AJ, Lin FH, Baardsnes J, Fletcher GL, Davies PL (2008) A re-evaluation of the role of type IV antifreeze protein. Cryobiology 57(3):292–296. https://doi.org/10.1016/j.cryobiol.2008.10.122

Graham LA, Davies PL (2005) Glycine-rich antifreeze proteins from snow fleas. Science 310 (5747):461. https://doi.org/10.1126/science.1115145

Graham LA, Liou YC, Walker VK, Davies PL (1997) Hyperactive antifreeze protein from beetles. Nature 388(6644):727–728. https://doi.org/10.1038/41908

Greenfield NJ (2006) Using circular dichroism collected as a function of temperature to determine the thermodynamics of protein unfolding and binding interactions. Nat Protoc 1(6):2527–2535. https://doi.org/10.1038/nprot.2006.204

Gronwald W, Loewen MC, Lix B, Daugulis AJ, Sönnichsen FD, Davies PL, Sykes BD (1998) The solution structure of type II antifreeze protein reveals a new member of the lectin family. Biochemistry 37(14):4712–4721. https://doi.org/10.1021/bi972788c

Hakim A, Nguyen JB, Basu K, Zhu DF, Thakral D, Davies PL, Isaacs FJ, Modis Y, Meng W (2013) Crystal structure of an insect antifreeze protein and its implications for ice binding. J Biol Chem 288(17):12295–12304. https://doi.org/10.1074/jbc.M113.450973

Hew CL, Joshi S, Wang NC, Kao MH, Ananthanarayanan VS (1985) Structures of shorthorn sculpin antifreeze polypeptides. Eur J Biochem 151(1):167–172

Hew CL, Wang NC, Joshi S, Fletcher GL, Scott GK, Hayes PH, Buettner B, Davies PL (1988) Multiple genes provide the basis for antifreeze protein diversity and dosage in the ocean pout, *Macrozoarces americanus*. J Biol Chem 263(24):12049–12055

Hobbs RS, Shears MA, Graham LA, Davies PL, Fletcher GL (2011) Isolation and characterization of type I antifreeze proteins from cunner, *Tautogolabrus adspersus*, order Perciformes. FEBS J 278(19):3699–3710. https://doi.org/10.1111/j.1742-4658.2011.08288.x

Husby JA, Zachariassen KE (1980) Antifreeze agents in the body-fluid of winter active insects and spiders. Experientia 36(8):963–964. https://doi.org/10.1007/bf01953821

Kao MH, Fletcher GL, Wang NC, Hew CL (1986) The relationship between molecular weight and antifreeze polypeptide activity in marine fish. Can J Zool Revue Can Zool 64(3):578–582. https://doi.org/10.1139/z86-085

Kawahara H, Iwanaka Y, Higa S, Muryoi N, Sato M, Honda M, Omura H, Obata H (2007) A novel, intracellular antifreeze protein in an antarctic bacterium, *Flavobacterium xanthum*. Cryo Letters 28(1):39–49

Ko TP, Robinson H, Gao YG, Cheng CH, DeVries AL, Wang AH (2003) The refined crystal structure of an eel pout type III antifreeze protein RD1 at 0.62-A resolution reveals structural microheterogeneity of protein and solvation. Biophys J 84(2 Pt 1):1228–1237. https://doi.org/10.1016/S0006-3495(03)74938-8

Kristiansen E, Ramløv H, Hagen L, Pedersen SA, Andersen RA, Zachariassen KE (2005) Isolation and characterization of hemolymph antifreeze proteins from larvae of the longhorn beetle *Rhagium inquisitor* (L.). Comp Biochem Physiol B Biochem Mol Biol 142(1):90–97. https://doi.org/10.1016/j.cbpc.2005.06.004

Kristiansen E, Pedersen SA, Zachariassen KE (2008) Salt-induced enhancement of antifreeze protein activity: a salting-out effect. Cryobiology 57(2):122–129. https://doi.org/10.1016/j.cryobiol.2008.07.001

Kristiansen E, Wilkens C, Vincents B, Friis D, Lorentzen AB, Jenssen H, Løbner-Olesen A, Ramløv H (2012) Hyperactive antifreeze proteins from longhorn beetles: some structural insights. J Insect Physiol 58(11):1502–1510. https://doi.org/10.1016/j.jinsphys.2012.09.004

Lee JK, Kim YJ, Park KS, Shin SC, Kim HJ, Song YH, Park H (2011) Molecular and comparative analyses of type IV antifreeze proteins (AFPIVs) from two Antarctic fishes, *Pleuragramma antarcticum* and *Notothenia coriiceps*. Comp Biochem Physiol B Biochem Mol Biol 159 (4):197–205. https://doi.org/10.1016/j.cbpb.2011.04.006

Leinala EK, Davies PL, Doucet D, Tyshenko MG, Walker VK, Jia Z (2002) A beta-helical antifreeze protein isoform with increased activity. Structural and functional insights. J Biol Chem 277(36):33349–33352. https://doi.org/10.1074/jbc.M205575200

Li XM, Trinh KY, Hew CL (1991) Expression and characterization of an active and thermally more stable recombinant antifreeze polypeptide from ocean pout, *Macrozoarces americanus*, in *Escherichia coli*: improved expression by the modification of the secondary structure of the mRNA. Protein Eng 4(8):995–1002

Li N, Andorfer CA, Duman JG (1998a) Enhancement of insect antifreeze protein activity by solutes of low molecular mass. J Exp Biol 201(Pt 15):2243–2251

Li N, Kendrick BS, Manning MC, Carpenter JF, Duman JG (1998b) Secondary structure of antifreeze proteins from overwintering larvae of the beetle *Dendroides canadensis*. Arch Biochem Biophys 360(1):25–32. https://doi.org/10.1006/abbi.1998.0930

Lin X, O'Tousa JE, Duman JG (2010) Expression of two self-enhancing antifreeze proteins from the beetle *Dendroides canadensis* in *Drosophila melanogaster*. J Insect Physiol 56(4):341–349. https://doi.org/10.1016/j.jinsphys.2009.11.005

Lin FH, Davies PL, Graham LA (2011) The Thr- and Ala-rich hyperactive antifreeze protein from inchworm folds as a flat silk-like β-helix. Biochemistry 50(21):4467–4478. https://doi.org/10.1021/bi2003108

Liou YC, Thibault P, Walker VK, Davies PL, Graham LA (1999) A complex family of highly heterogeneous and internally repetitive hyperactive antifreeze proteins from the beetle *Tenebrio molitor*. Biochemistry 38(35):11415–11424. https://doi.org/10.1021/bi990613s

Liou YC, Daley ME, Graham LA, Kay CM, Walker VK, Sykes BD, Davies PL (2000) Folding and structural characterization of highly disulfide-bonded beetle antifreeze protein produced in bacteria. Protein Expr Purif 19(1):148–157. https://doi.org/10.1006/prep.2000.1219

Liu Y, Li Z, Lin Q, Kosinski J, Seetharaman J, Bujnicki JM, Sivaraman J, Hew CL (2007) Structure and evolutionary origin of Ca(2+)-dependent herring type II antifreeze protein. PLoS One 2(6): e548. https://doi.org/10.1371/journal.pone.0000548

Lumry R, Eyring H (1954) Conformation changes of proteins. J Phys Chem 58(2):110–120. https://doi.org/10.1021/j150512a005

Ma J, Wang J, Mao XF, Wang Y (2012) Differential expression of two antifreeze proteins in the desert beetle *Anatolica polita* (Coleoptera: Tenebriondae): seasonal variation and environmental effects. Cryo Letters 33(5):337–348

Mao X, Liu Z, Li H, Ma J, Zhang F (2011) Calorimetric studies on an insect antifreeze protein ApAFP752 from *Anatolica polita*. J Therm Anal Calorim 104(1):343–349. https://doi.org/10.1007/s10973-010-1067-3

Marshall CB, Daley ME, Sykes BD, Davies PL (2004a) Enhancing the activity of a beta-helical antifreeze protein by the engineered addition of coils. Biochemistry 43(37):11637–11646. https://doi.org/10.1021/bi0488909

Marshall CB, Fletcher GL, Davies PL (2004b) Hyperactive antifreeze protein in a fish. Nature 429 (6988):153. https://doi.org/10.1038/429153a

Marshall CB, Chakrabartty A, Davies PL (2005) Hyperactive antifreeze protein from winter flounder is a very long rod-like dimer of alpha-helices. J Biol Chem 280(18):17920–17929. https://doi.org/10.1074/jbc.M500622200

Miura K, Ohgiya S, Hoshino T, Nemoto N, Suetake T, Miura A, Spyracopoulos L, Kondo H, Tsuda S (2001) NMR analysis of type III antifreeze protein intramolecular dimer. Structural basis for enhanced activity. J Biol Chem 276(2):1304–1310. https://doi.org/10.1074/jbc.M007902200

Mok YF, Lin FH, Graham LA, Celik Y, Braslavsky I, Davies PL (2010) Structural basis for the superior activity of the large isoform of snow flea antifreeze protein. Biochemistry 49 (11):2593–2603. https://doi.org/10.1021/bi901929n

Neelakanta G, Sultana H, Fish D, Anderson JF, Fikrig E (2010) *Anaplasma phagocytophilum* induces *Ixodes scapularis* ticks to express an antifreeze glycoprotein gene that enhances their survival in the cold. J Clin Invest 120(9):3179–3190. https://doi.org/10.1172/JCI42868

Nickell PK, Sass S, Verleye D, Blumenthal EM, Duman JG (2013) Antifreeze proteins in the primary urine of larvae of the beetle *Dendroides canadensis*. J Exp Biol 216(Pt 9):1695–1703. https://doi.org/10.1242/jeb.082461

Nishimiya Y, Kondo H, Takamichi M, Sugimoto H, Suzuki M, Miura A, Tsuda S (2008) Crystal structure and mutational analysis of Ca2+−independent type II antifreeze protein from longsnout poacher, *Brachyopsis rostratus*. J Mol Biol 382(3):734–746. https://doi.org/10.1016/j.jmb.2008.07.042

Qiu LM, Ma J, Wang J, Zhang FC, Wang Y (2010a) Thermal stability properties of an antifreeze protein from the desert beetle *Microdera punctipennis*. Cryobiology 60(2):192–197. https://doi.org/10.1016/j.cryobiol.2009.10.014

Qiu L, Wang Y, Wang J, Zhang F, Ma J (2010b) Expression of biologically active recombinant antifreeze protein His-MpAFP149 from the desert beetle (*Microdera punctipennis dzungarica*) in *Escherichia coli*. Mol Biol Rep 37(4):1725–1732. https://doi.org/10.1007/s11033-009-9594-3

Salvay AG, Santos J, Howard EI (2007) Electro-optical properties characterization of fish type III antifreeze protein. J Biol Phys 33(5-6):389–397. https://doi.org/10.1007/s10867-008-9080-5

Sanchez-Ruiz JM (2010) Protein kinetic stability. Biophys Chem 148(1–3):1–15. https://doi.org/10.1016/j.bpc.2010.02.004

Scott GK, Davies PL, Shears MA, Fletcher GL (1987) Structural variations in the alanine-rich antifreeze proteins of the pleuronectinae. Eur J Biochem 168(3):629–633

Scott GK, Hayes PH, Fletcher GL, Davies PL (1988) Wolffish antifreeze protein genes are primarily organized as tandem repeats that each contain two genes in inverted orientation. Mol Cell Biol 8 (9):3670–3675

Slaughter D, Fletcher GL, Ananthanarayanan VS, Hew CL (1981) Antifreeze proteins from the sea raven, *Hemitripterus americanus*. Further evidence for diversity among fish polypeptide antifreezes. J Biol Chem 256(4):2022–2026

Sönnichsen FD, Sykes BD, Chao H, Davies PL (1993) The nonhelical structure of antifreeze protein type III. Science 259(5098):1154–1157

Sönnichsen FD, Sykes BD, Davies PL (1995) Comparative modeling of the three-dimensional structure of type II antifreeze protein. Protein Sci 4(3):460–471. https://doi.org/10.1002/pro.5560040313

Sönnichsen FD, DeLuca CI, Davies PL, Sykes BD (1996) Refined solution structure of type III antifreeze protein: hydrophobic groups may be involved in the energetics of the protein-ice interaction. Structure 4(11):1325–1337

Sørensen TF, Cheng CH, Ramløv H (2006) Isolation and some characterisation of antifreeze protein from the European eelpout *Zoarces viviparus*. Cryo Letters 27(6):387–399

Sun T, Lin FH, Campbell RL, Allingham JS, Davies PL (2014) An antifreeze protein folds with an interior network of more than 400 semi-clathrate waters. Science 343(6172):795–798. https://doi.org/10.1126/science.1247407

Tursman D, Duman JG (1995) Cryoprotective effects of thermal hysteresisprotein on survivorship of frozen gut cells from the freeze-tolerant centipede Lithobius forficatus. J Exp Zool 272 (4):249–257. https://doi.org/10.1002/jez.1402720402

Venketesh S, Dayananda C (2008) Properties, potentials, and prospects of antifreeze proteins. Crit Rev Biotechnol 28(1):57–82. https://doi.org/10.1080/07388550801891152

Wang L, Duman JG (2005) Antifreeze proteins of the beetle *Dendroides canadensis* enhance one another's activities. Biochemistry 44(30):10305–10312. https://doi.org/10.1021/bi050728y

Wang X, DeVries AL, Cheng CH (1995) Antifreeze peptide heterogeneity in an antarctic eel pout includes an unusually large major variant comprised of two 7 kDa type III AFPs linked in tandem. Biochim Biophys Acta 1247(2):163–172

Wang C, Oliver EE, Christner BC, Luo BH (2016) Functional analysis of a bacterial antifreeze protein indicates a cooperative effect between its two ice-binding domains. Biochemistry 55 (28):3975–3983. https://doi.org/10.1021/acs.biochem.6b00323

Wilkens C, Poulsen JC, Ramløv H, Lo Leggio L (2014) Purification, crystal structure determination and functional characterization of type III antifreeze proteins from the European eelpout *Zoarces viviparus*. Cryobiology 69(1):163–168. https://doi.org/10.1016/j.cryobiol.2014.07.003

Wu DW, Duman JG, Cheng C-HC, Castellino FJ (1991) Purification and characterization of antifreeze proteins from larvae of the beetle *Dendroides canadensis*. J Comp Physiol B 161 (3):271–278. https://doi.org/10.1007/BF00262308

Wu Y, Banoub J, Goddard SV, Kao MH, Fletcher GL (2001) Antifreeze glycoproteins: relationship between molecular weight, thermal hysteresis and the inhibition of leakage from liposomes during thermotropic phase transition. Comp Biochem Physiol B Biochem Mol Biol 128 (2):265–273

Yamashita Y, Miura R, Takemoto Y, Tsuda S, Kawahara H, Obata H (2003) Type II antifreeze protein from a mid-latitude freshwater fish, Japanese smelt (*Hypomesus nipponensis*). Biosci Biotechnol Biochem 67(3):461–466

Chapter 4
Structure–Function of IBPs and Their Interactions with Ice

Maya Bar-Dolev, Koli Basu, Ido Braslavsky, and Peter L. Davies

4.1 Introduction

One of the key factors in understanding how a protein functions is to know its three-dimensional (3-D) structure at atomic resolution. In many cases, the protein structure provides enough clues to deduce its mechanism of action. Yet in the case of ice-binding proteins (IBPs), although high resolution structures of some IBPs were experimentally determined over two decades ago, their interactions with ice, their natural ligand, is still a matter of debate and extensive research. After the three-dimensional structure of type I AFP from flounder was published in 1988 (Yang et al. 1988), many experimental and theoretical studies were conducted to elucidate its ice-binding site (IBS) and find out how it adheres to ice at the molecular level. The publication of other AFP structures, and the discovery of AFPs in other biological kingdoms, have increased interest in understanding the driving force for ice recognition and the crucial elements involved in freezing point depression. The mystery of protein–ice interactions results from several issues. One is the ambiguity of the interface between ice and water, which makes the ligand difficult to define at the molecular level (Guo et al. 2012; Vance et al. 2014). Another is that the difference between bulk water and ice, the ligand, is only in the spatial organization of the water molecules. These aspects make ice recognition a challenging issue to unravel. Another complication lies in the diversity of IBPs structures and functions, making it difficult to formulate generalized conclusions. This chapter describes the structural diversity of IBPs with emphasis on the structures of the IBSs and their specificities

M. Bar-Dolev · I. Braslavsky
The Hebrew University of Jerusalem, Rehovot, Israel
e-mail: maya.bar1@mail.huji.ac.il; ido.braslavsky@mail.huji.ac.il

K. Basu · P. L. Davies (✉)
Queen's University, Kingston, ON, Canada
e-mail: 0kb15@queensu.ca; peter.davies@queensu.ca

H. Ramløv, D. S. Friis (eds.), *Antifreeze Proteins Volume 2*,
https://doi.org/10.1007/978-3-030-41948-6_4

for different planes of ice. The topics of ice shaping under different temperature regimes, which is a consequence of IBP specificities for different ice planes, the effects of protein size on activity, and protein engineering studies on IBPs are presented. We discuss the opposing functions of ice-nucleating proteins (INPs) and AFPs in the context of the structural differences between these proteins. Furthermore, we explain the interactions of IBPs with water and ice at the molecular level with emphasis on recent theories about the importance of the hydration shell, the anchored clathrate water mechanism, the issue of reversible/irreversible binding, and dynamic aspects of ice binding by different types of IBPs.

The term "ice-binding proteins" is used in this chapter to embrace all proteins that bind ice. These include those that function as freezing point depressors (AFPs) and others such as ice-recrystallization inhibitors, INPs, and ice adhesins.

4.2 Structural Diversity of IBPs

IBP structures are incredibly diverse considering they all share the same function of binding to ice (Davies 2014). That their primary and tertiary structures are radically different attests to the fact that they have evolved on many separate occasions from different progenitors in various branches of the tree of life (Bar Dolev et al. 2016b) (discussed in detail in Chap. 9 of Vol. 1). The structures of fish type I, II, and III AFPs as well as some of the insect and plant IBPs have been described in detail and extensively reviewed (Bar Dolev et al. 2016b; Davies 2014; Duman 2001; Feeney et al. 1986; Venketesh and Dayananda 2008). Some branches (like arthropods, plants, and bacteria) have only been sparsely surveyed for ice-binding activity, and the likelihood of new IBP structures being found here is high. IBPs are typically small, single-domain proteins, the majority of which are repetitive in structure. Details of solved and convincingly modeled IBP 3-D structures are presented in Table 4.1. These include the three types of AFPs found in fish: short (3–4 kDa) α-helical type I AFPs that have independently arisen four times in fishes (Graham et al. 2013), as well as a long-type I AFP dimer that folds into a four-helix bundle (Maxi), and the globular type II and type III AFPs. Several types of β-solenoid IBPs are found in insects, plants, and microorganisms, and a polyproline type II helix bundle has been described in snow fleas (Collembola), a primitive arthropod. The IBP DUF3494 fold, which is widely dispersed in microorganisms due to lateral gene transfer (Raymond and Kim 2012), has a discontinuous β-solenoid with a supporting α-helix (see Table 4.1). Additional IBPs have been modeled with high confidence (Basu et al. 2015; Lin et al. 2011). The regularity of repeating structures might facilitate interaction with a crystal lattice and help explain the frequent evolution of the beta-solenoid fold in IBPs. IBPs function at 0 °C or lower. A few of them rely almost entirely on hydrogen bonding for the stability of their folds (see snow flea and type I AFPs in Table 4.1) and are easily denatured by warming to room temperature. Some IBPs are stabilized by disulfide bridges, like type II AFP and the AFP from the insect *Tenebio molitor* (*Tm*AFP), and some by coordinating Ca^{2+} ions, like type II

Table 4.1 Structures and main properties of IBPs

Name and structure	Details
Type I AFP 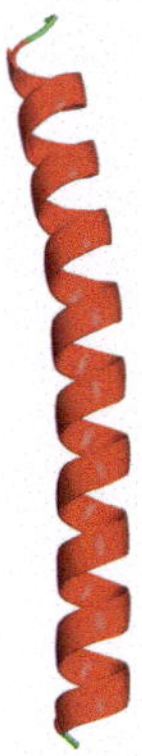 PDB: 1wfa	*Organisms*: Several branches of fishes: righteye flounders (Duman and DeVries 1976; Knight et al. 1991), sculpins (Hew et al. 1985; Low et al. 2001), snailfish (Evans and Fletcher 2001), cunner (Hobbs et al. 2011). *Structure*: Amphipathic α-helix with 50–65% Ala. Despite apparent homology based on Ala-richness, type I AFPs have independently evolved on at least four occasions (Graham et al. 2013). The helix has an 11-residue periodicity, with 3.7 aa/turn. Size: ~55 × 6 × 6 Å. *Isoforms*: Many, including tissue-specific. Most are 33–42 aa peptides. *IBS*: Involves the indicated Thr and Ala in TxxxAxxxAxx repeats that project on one side of the helix. *Ice planes*: Flounder and plaice AFPs bind to the $[20\bar{2}1]$ pyramidal plane. Sculpin AFP binds to the $[11\bar{2}0]$ secondary prism plane. *Activity*: Moderate TH activity. Produce bipyramidal ice shapes with a constant *a*:*c* axis ratio of 3.3 (Wen and Laursen 1992b).
Long type I AFP, Maxi 	*Organisms*: Winter flounder (Sun et al. 2014). *Structure*: A four-helix bundle homodimer. Each α-helix monomer has a hairpin fold and the two hairpins align in an antiparallel fashion. The two helices are homologous to type I AFP but 5 times longer. Size: ~145 × 20 × 20 Å. *Isoforms*: (Short) type I AFP. *IBS*: Ice-like waters organized by the aqueous core extend to the protein surface between the helices. *Ice planes*: Multiple including the basal plane. *Activity*: Hyperactive.
Type II AFP 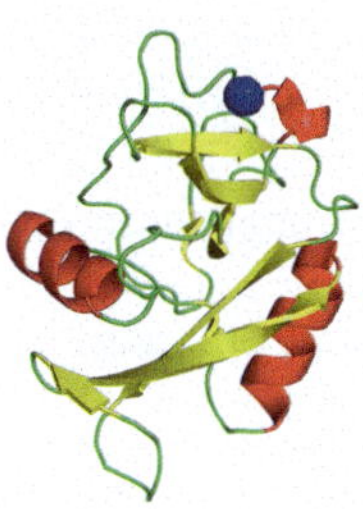 PDB: 2py2	*Organisms*: Several species of fish; sea raven (Patel and Graether 2010), rainbow and Japanese smelt (Yamashita et al. 2003), herring (Ewart and Fletcher 1990) (Liu et al. 2007), longsnout poacher (Nishimiya et al. 2008). *Structure*: 14–24 kDa globular protein with a C-type lectin fold. Contains five disulfide bonds. Occurs as Ca^{2+}-dependent and -independent forms. The protein forms a dimer in rainbow smelt (Achenbach and Ewart 2002). Size: ~30 × 30 × 40 Å. *Isoforms*: Many, including organ-specific (Liu et al. 2007). *IBS*: Ill-defined. In the Ca^{2+}-dependent type II AFPs the metal ion is part of the ice-binding site (Liu et al. 2007), but neighboring regions of the protein are also involved in ice binding (Loewen et al. 1998). *Ice planes*: Non-basal. *Activity*: Produces bipyramidal ice shapes and has moderate TH activity.

(continued)

Table 4.1 (continued)

Name and structure	Details
Type III AFP  PDB: 5MSI	*Organisms*: Several families of fishes from one suborder; e.g., ocean pout (Li et al. 1985) and other eelpouts (Cheng and DeVries 1989). *Structure*: 7 kDa, Globular and rigid with short imperfect β-strands and a single-helix turn (Jia et al. 1996). Size: ~30 × 20 × 20 Å. *Isoforms*: Numerous with 50% identity (Hew et al. 1988; Nishimiya et al. 2005). A dimer (Wilkens et al. 2014) and a natural tandem repeat have been seen (Wang et al. 1995). *IBS*: Compound. Two adjacent ice-binding surfaces lie inclined to each other and bind different ice planes (Garnham et al. 2010). *Ice planes*: Several—including primary prism plane [$10\bar{1}0$], pyramidal [$20\bar{2}1$], and perhaps some additional planes inclined by a small rotation from these planes (Antson et al. 2001). *Activity*: QAE isoforms have moderate TH activity. SP isoforms shape ice but only stop it growing in the presence of the QAE isoforms (Nishimiya et al. 2005). Produce bypiramidal ice shapes with varying *a*:*c* axis ratios (DeLuca et al. 1996).
AFGPs  (Jia and Davies 2002)	*Organisms*: Two distinct branches of fishes: Antarctic toothfish (DeVries and Wohlschlag 1969) and cods (Chen et al. 1997). *Structure*: From 4 to 50 repeats of the TA/PA tripeptide with all hydroxyl groups of the threonines glycosylated by the disaccharide 3-*O*-(β-D-galactosyl)-(α1 → 3)-*N*-acetylgalactosamine. Produced from much larger polyproteins. Most likely fold as an amphipathic helix of the PPII type (Tachibana et al. 2004) *Isoforms*: In the rock cod (*Gadus ogac*) the AFGP isoforms range in size from 2.6 to 24 kDa as measured by electrospray mass spectrometry (Wu et al. 2001). *IBS*: unknown. *Planes bound*: Primary prism planes (Knight et al. 1993). *Activity*: Low TH activity for the smallest isoforms; moderate activity for the longer isoforms.
*Tm*AFP 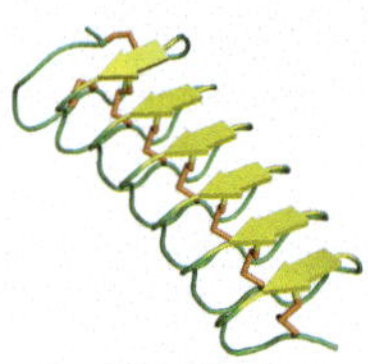 PDB:1EZG	*Organisms*: Insect. Yellow mealworm (*Tenebrio molitor*) (Graham et al. 1997; Liou et al. 2000), pyrochid beetle *(Dendroides Canadensis)* (Duman et al. 1998). *Structure*: 8–9 kDa, tight β-solenoid with a rectangular cross section. The coils are composed of the tandem 12-aa repeat sequence TCTXSXXCXXAX. Eight disulfide bonds cross-link the coils and rigidify the structure that lacks a hydrophobic core (Liou et al. 2000). Most isoforms contain a consensus *N*-glycosylation site near the C-terminus (nonessential for activity and folding) (Liou et al. 1999). Size: ~32 × 14 × 12 Å. *Isoforms*: Many. Most have 7 coils; a few have 8–11 coils (84–120 aa) (Liou et al. 1999). *IBS*: A flat β-sheet with regularly spaced Thr-Cys-Thr motifs (Liou et al. 2000). *Planes bound*: Multiple, including basal (Bar Dolev et al. 2016b; Scotter et al. 2006). *Activity*: High TH (hyperactive). Produce lemon-shaped ice crystals.

(continued)

Table 4.1 (continued)

Name and structure	Details
sbwAFP 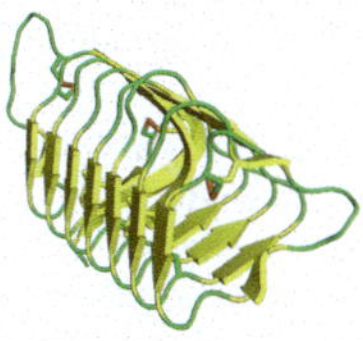 PDB: 1M8N	*Organisms*: Insect. Spruce budworm (*Choristoneura fumiferana*) (Graether et al. 2000). *Structure*: 9–12.5 kDa, tight β-solenoid of ~15-residue coils with a triangular cross section and a TXT motif. 4–5 inter-stand disulfide bonds. Size: ~36 × 20 × 18 Å. *Isoforms*: Several with 5 or 7 repeats, in some there are 16 or 17 aa/coil (Doucet et al. 2000). *IBS*: Flat β-sheet of TXT motifs (Graether et al. 2000). *Planes bound*: Primary prism plane and basal plane (Graether et al. 2000; Pertaya et al. 2008). *Activity*: High TH (hyperactive). Produces flat-tipped, hexagonal-shaped ice crystals (Graether et al. 2000; Pertaya et al. 2008).
*Ri*AFP 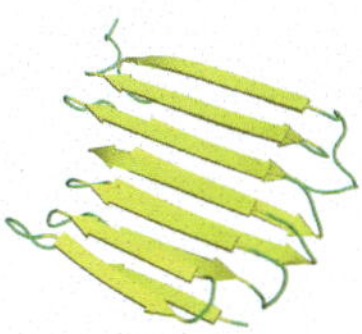 PDB: 4DT5	*Organisms*: Insect. *Rhagium inquisitor* (Hakim et al. 2013) and longhorn beetle *Rhagium mordax* (Kristiansen et al. 2011, 2012). *Structure*: 13 kDa, Flat β-sandwich with a rectangular cross section. The structure consists of 7 loops of 13–20 aa making two staggered β-sheets 6 Å apart, with Ser and Ala side-chains interdigitating inside, making the tight packing similar to silk fibers, without a hydrophobic core. One S–S bond (Hakim et al. 2013). Size: ~35 × 30 × 6 Å. *Isoforms*: Five isoforms with the same number of loops in the beetle *Rhagium mordax* (Kristiansen et al. 2011, 2012). *IBS*: Predicted to be one of the two β-sheets, which is composed of an array of five expanded strands of the motif T*X*T*X*T*X*T (Hakim et al. 2013). *Planes bound*: Multiple, including primary prism and basal planes (Hakim et al. 2013) *Activity*: High TH (hyperactive).
sfAFP PDB: 2PNE 	*Organisms*: Primitive arthropod, Collembola *Hypogastrura harveyi* (snow flea) (Graham and Davies 2005). *Structure*: Six antiparallel left-handed polyproline type II coils, tightly packed in two sets of three, making flat surfaces on both sides, one hydrophobic and one hydrophilic (Pentelute et al. 2008). Composed of GXY repeats. In the small isoform there are two S–S bonds, and one in the large isoform. Extremely thermolabile (Graham and Davies 2005). Size: ~40 × 10 × 5 Å. *Isoforms*: One small (6.5 kDa) and one large (15.7 kDa). *IBS*: putatively the hydrophobic side of the flat disk. *Planes bound*: Multiple, including the basal (Mok et al. 2010). *Activity*: High TH (hyperactive). Produces rice grain-shaped crystals (Graham and Davies 2005; Mok et al. 2010).
*Lp*IBP 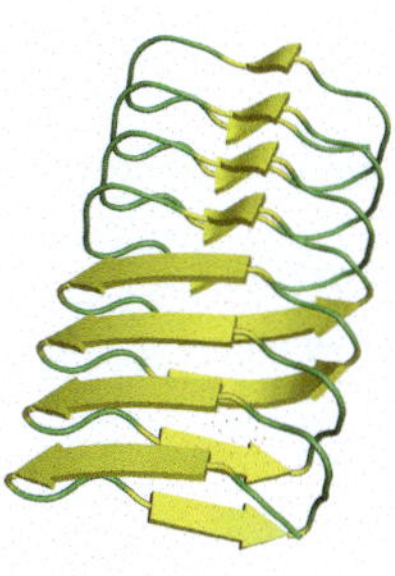 PDB: 3ult	*Organisms*: Plant: rye grass (*Lolium perenne).* *Structure*: 12-kDa β-solenoid with 8 coils composed of tandem repeats of the 7-aa consensus sequence XXNXVXG, repeating twice in each coil. The first three coils contain one more residue per loop, which make a bulge on one side of the solenoid (Middleton et al. 2012). Size: ~33 × 20 × 10 Å. *Isoforms*: Several, most are linked to a N-terminal Leucine-rich repeat domain of unknown function. One contains an extra loop (Kumble et al. 2008). *IBS*: A flat β-sheet with imperfect rank of TXT motifs (~30% Thr, most other residues are Ser, Ala, and Val) (Kumble et al. 2008). *Planes bound*: Primary prism and basal planes (Middleton et al. 2012). *Activity*: Low TH, high ice recrystallization inhibition. Produces flat-tipped, hexagonal-shaped ice crystals that grow in all directions once the freezing point is exceeded.

(continued)

Table 4.1 (continued)

Name and structure	Details
*Tis*AFP and *Le*AFP 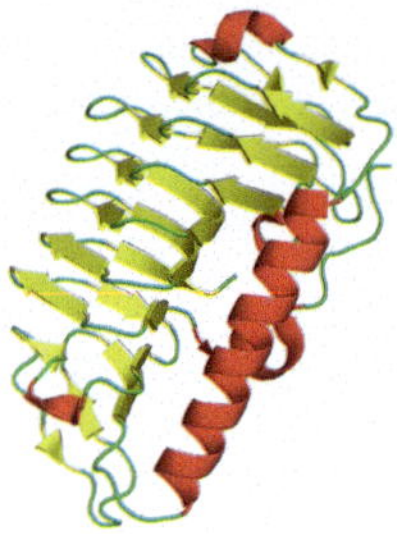 PDB:3UYU	*Organisms*: Microorganisms. Examples: snow mold fungus *Typhula ishikariensis* (*Tis*AFP) (Kondo et al. 2012); Arctic yeast *Leucosporidium* sp. (*Le*AFP) (Lee et al. 2012). Many homologs in bacteria, fungi, algae, and diatoms. *Structure*: 23 kDa, right-hand β-solenoid with a triangular cross section composed of 6 coils with 18 (or more) aa per coil alongside an α-helix. The sequence is non-repetitive with no consensus motif (Kondo et al. 2012). *Le*AFP has a glycosylated residue and a C-terminal loop that makes it a dimer (Lee et al. 2012). Size: ~47 × 30 × 27 Å. *Isoforms*: Several isoforms with poorly conserved IBSs and range of activities from moderate to hyperactive. *IBS*: A flat but irregular face on the solenoid with no repeating motif and no conserved residues in homologs. Probed by mutagenesis in *Tis*AFP6 (Kondo et al. 2012) *Planes bound*: Various depending on isoform studied; e.g., basal plane and additional sixfold plane that is not primary or secondary prism for *Tis*AFP6 (Kondo et al. 2012). *Activity*: Produces crystal shapes that can be pitted bipyramids. Most variants have moderate TH and burst along the *c*-axis. At least one homolog (*Tis*AFP8 isoform) is hyperactive and bursts normal to the *c*-axis (Kondo et al. 2012).
*Mp*IBP-RIV 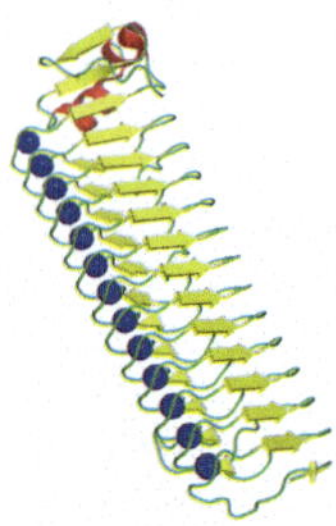 PDB:3P4G	*Organisms*: Antarctic bacterium: *Marinomonas primoryensis* (Garnham et al. 2011a). *Structure*: The ice-binding domain *Mp*IBP-RIV of the bacterial adhesin is a 34 kDa, β-solenoid with 13 tandem repeats of 19-residue coils. Ca^{2+} ions inside the coils rigidify the IBS. Size: ~70 × 27 × 17 Å. *Isoforms*: None detected. *IBS*: Two parallel rows of outward projecting residues, one Thr, the other Asx. *Planes bound*: Multiple, including basal. *Activity*: The recombinant *Mp*IBP-RIV is hyperactive but functions as a bacterial ice adhesin rather than an antifreeze (Bar Dolev et al. 2016a; Guo et al. 2012).
Midge AFP 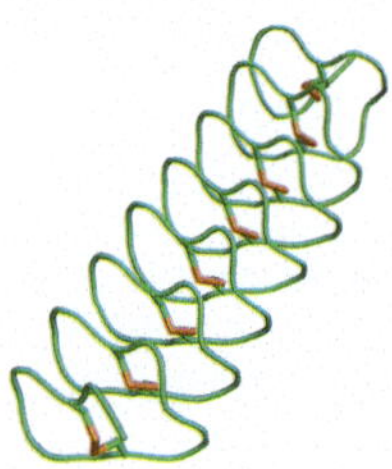 (model)	*Organisms*: Midge (*Chironomidae*) (Basu et al. 2016). *Structure*: Tight left-handed solenoid composed of 8 coils of 10-residue tandem repeats of CXGXYCXGXX. Eight disulfide bonds crosslink the coils. The solenoid is so tight that there is no β-strand character. Size: ~40 × 20 × 15 Å. *Isoforms*: Four isoforms with minor sequence differences and variations in the number of coils. *IBS*: Predicted to be a row of Tyr stacked 4.5 Å apart on one face of the solenoid. *Planes bound*: Pyramidal plane intermediate between basal and prism planes. Produces bipyramidal ice crystals. *Activity*: Intermediate TH activity. Burst normal to the *c*-axis.

Color code: alpha-helix, red; beta-strands, yellow; coil, green; disulfide bonds, orange; Ca^{2+}, blue

AFPs and the bacterial ice adhesin from *Marinomonas primoryensis*. Many of the IBPs have multiple isoforms that vary in size and activity. In general, the larger the isoform, the higher is its TH activity. Although there is no crystal structure for the antifreeze glycoproteins (AFGPs), it is likely they fold as a polyproline type II helix because this will fit the three-residue periodicity (Table 4.1) and produce an amphipathic helix that seems so important to the functioning of type I AFP (Baardsnes et al. 1999; Graham et al. 2013). Fish type IV AFPs are not listed in the table because recent evidence suggests they do not function as IBPs in nature (Gauthier et al. 2008).

Some of the recently published IBP structures possess interesting characteristics that have led to new questions and a broadening of the scope of the structure–function relationship in IBPs. One example is the hyperactive IBP from winter flounder: Maxi. This protein forms a homodimer in solution, with no stable monomer form. The monomer of Maxi is like a fivefold longer version of the small type I AFP from the same fish. One wonders how this giant version of type I AFP evolved and whether it preceded or followed on from the small single helix (Sun et al. 2014). What is especially puzzling in this context is that the role of the ice-binding residues in type I AFP changes in Maxi into forming and maintaining the interior network of 400 clathrate waters that hold the four-helix bundle structure together. Another interesting example of structure–function relationships is the IBP from the Antarctic bacterium *Marinomonas primoryensis* (*Mp*IBP). The molecular weight of *Mp*IBP is >1.5 MDa, consisting of ~130 domains. The structure of this protein can be broken down into five regions, only one of which (*Mp*IBP-RIV) binds ice. *Mp*IBP-RIV is a single domain that folds into a β-solenoid with a row of Ca^{2+} ions stabilizing the helical structure (Garnham et al. 2011a). When this region was recombinantly expressed, it had the high TH values and ice-shaping characteristics of a hyperactive IBP (Garnham et al. 2008). However, *Mp*IBP is an ice adhesin in its natural context (Bar Dolev et al. 2016a), and its other regions have other specific roles (Guo et al. 2012, 2017; Vance et al. 2014). This is an unusual function for an IBP (Bar Dolev et al. 2016a) but there is now evidence that the DUF3494 fold has been co-opted into this adhesion function in another marine bacterium (Vance et al. 2018).

4.3 Identification and Mapping of Ice-Binding Sites

Each IBP has a surface—the IBS—that has evolved to dock the protein to ice (Fig. 4.1). The key to experimentally defining this surface has been to produce the IBP as a recombinant protein (Chao et al. 1993), solve its 3-D structure (Sonnichsen et al. 1993), and then probe the extent of the IBS using site-directed mutagenesis (Chao et al. 1994). This process is nicely illustrated by studies on type III AFP, the first IBP for which the IBS was experimentally defined. Type III AFP is a good case in point because the IBS of this AFP could not be initially predicted by flatness and regularity as it had been for type I AFP (DeVries and Lin 1977) and subsequently was for AFPs like *Tm*AFP (Liou et al. 2000) and *Rhagium inquisitor* AFP (*Ri*AFP)

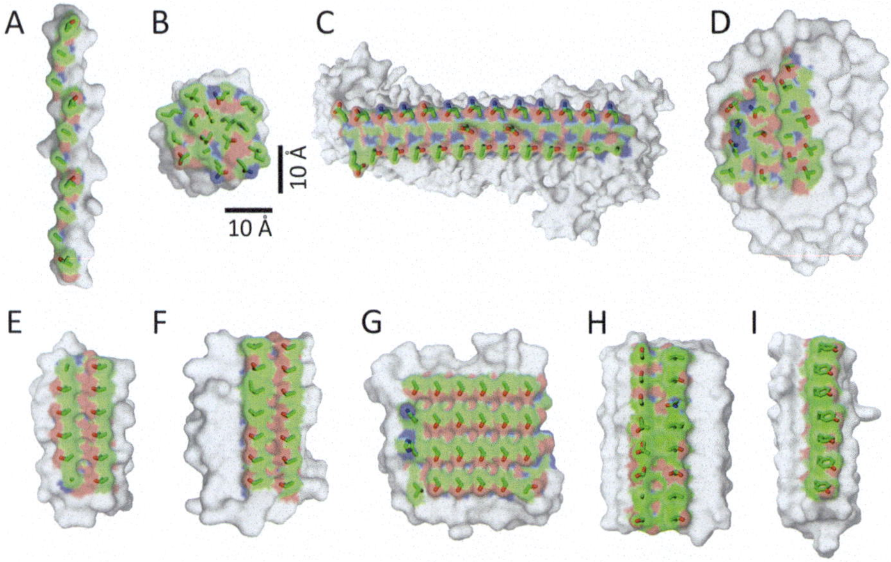

Fig. 4.1 Ice-binding sites of IBPs. IBP structures are shown in surface presentation and are to scale. Residues of the experimentally determined or putative IBSs are colored (carbon atoms in green, oxygen in red, and nitrogen in blue). Non-IBS residues are in gray. Stick representation of the IBS residue side chains are overlaid on each structure. (**a**) Type I AFP from winter flounder. PDB 1WFA (Sicheri and Yang 1995). (**b**) Type III AFP from ocean pout, isoform HPLC12. PDB 1MSI (Jia et al. 1996). (**c**) Region IV of the ice adhesin from *Marinomonas primoryensis* (*Mp*IBP-RIV). PDB 3P4G (Garnham et al. 2011a). (**d**) Snow mold fungus IBP from *Typhula ishikariensis*, moderately active isoform (*Tis*AFP6). PDB 3VN3 (Kondo et al. 2012). (**e**) *Tenebrio molitor* AFP (*Tm*AFP), isoform 4–9. PDB 1EZG (Liou et al. 2000). (**f**) Spruce budworm AFP (sbwAFP), isoform 501. PDB 1M8N (Leinala et al. 2002). (**g**) *Rhagium inquisitor* AFP (*Ri*AFP). PDB 4DT5 (Hakim et al. 2013). (**h**) *Lolium perenne* AFP (*Lp*AFP). PDB 3ULT (Middleton et al. 2012). (**i**) Model of midge AFP (Basu et al. 2015)

(Hakim et al. 2013). The only other indicator for IBS function in this initial study was the conservation of the residues that make up the IBS when aligned to a dozen different isoforms. Conserved internal residues are probably essential for the protein fold. But conserved surface residues, especially those that form a patch, are likely to be involved in the function of the protein, which in this case is ice binding. The conserved surface residues of type III AFP first targeted for mutagenesis included T18, N14, and Q44. The initial rationale for the mutagenesis choices, when it was still thought that IBPs bound to ice through a hydrogen-bonding network (DeVries and Lin 1977; Wen and Laursen 1992a), was to disturb this hydrogen-bonding pattern, as for example with the effective mutants T18N, N14S, and Q44T.

When it was realized that the IBS was the most hydrophobic surface of the IBP (Sonnichsen et al. 1996) and the hydrophobic effect was speculated to be the binding force for holding the IBP on ice, the success of these mutagenesis experiments was attributed to steric hindrance to binding when replacing a small side chain with a larger one (DeLuca et al. 1996), or to spoiling the "snug" fit when the replacement

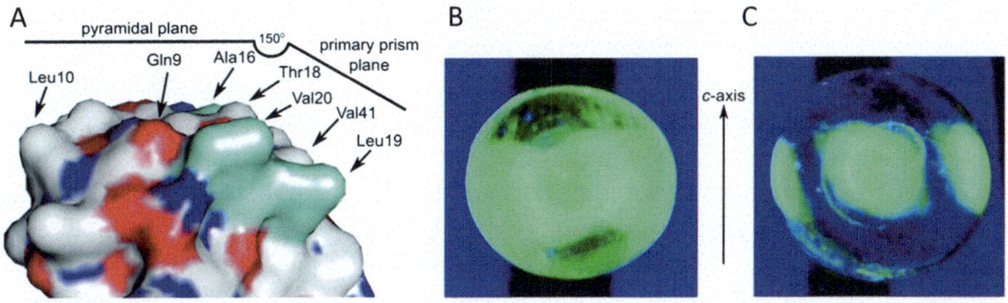

Fig. 4.2 Compound ice-binding site of type III AFP from ocean pout (QAE1 isoform, also called HPLC12). (**a**) The ice-binding site of type III AFP showing its two adjacent subsites, each binding a different plane or set of planes of ice. (**b–c**) FIPA analysis of (**b**) a GFP-labeled wild-type protein and (**c**) a GFP-labeled A16H mutant. The single-crystal ice hemispheres were mounted with the secondary prism plane oriented normal to the long axis of the cold finger. The large fluorescence signal in **b** indicates binding to more than the primary prism plane of ice. The reduced fluorescence in **c** demonstrates that the mutant has reduced ice-plane activity. There is no binding on the top and bottom of the hemispheres because the proteins do not bind the basal plane of ice. Reprinted (adapted) with permission from (Garnham et al. 2010) Copyright (2010) American Chemical Society

was smaller than the original residue (Baardsnes and Davies 2002). In the latter study, a series of replacements of hydrophobic residues on the IBS of type III AFP with generally smaller side chains were detrimental to antifreeze activity as measured by thermal hysteresis (TH). Currently, the binding mechanism of IBPs to ice is thought to be due to the anchored clathrate water hypothesis (Garnham et al. 2011a) where the hydrophobic groups on the IBS are caged by water molecules that are linked and stabilized by hydrogen bonding to nearby hydrophilic groups (discussed in Sect. 4.10). In retrospect, changing the shape and hydrogen bonding capabilities of residues on the IBS would also interfere with ice binding by this anchored clathrate water mechanism.

Another complication added to the difficulty of defining the IBS of type III AFP is that the most active isoforms have two adjacent IBSs on an angle to each other. Together they form a compound IBS (Fig. 4.2) (Garnham et al. 2010). These QAE1 isoforms bind to both the primary prism and a pyramidal plane. However, the SP and QAE2 isoforms bind only to the pyramidal plane, and as such can slow the growth of ice but not stop it. These adjacent IBSs have been defined by site-directed mutagenesis, and have been validated by fluorescence-based ice plane affinity (FIPA) analysis. The proof of principle has come from an engineering study where an inactive QAE2 isoform was converted into a fully active form by as few as four surface mutations (Garnham et al. 2012).

What should have been a simpler study system for defining the IBS has been the type I AFP from righteye flounders (Sicheri and Yang 1995). This single α-helix was initially predicted to bind ice by hydrogen bonding from the regularly spaced Thr and Asx residues (Chou 1992; Wen and Laursen 1992a). The periodicity of these Thr and Asx places them on the same side of the helix at 11 residues apart each. The 37-residue HPLC-6 isoform of type I AFP is small enough for production in a good yield and at reasonable cost by solid-phase peptide synthesis. Thus, numerous variants were made to test the binding hypothesis. Replacement of the putative ice-binding Thr by Ser or Val proved to be informative (Chao et al. 1997; Haymet et al. 1998, 1999; Zhang and Laursen 1998). Change of the central two Thr to Ser caused a major loss of activity whereas the switch to Val had minimal effect (Chao et al. 1997). This emphasized the importance of the Thr methyl groups relative to the hydroxyls. Again, alignment of isoforms and orthologues was highly informative in defining the IBS to be conserved Thr and Ala residues on the same side of the helix. The role of the Ala residues was confirmed by the synthesis of steric substitutions where Leu replaced Ala. The role of Thr and adjacent Ala in ice binding fits well with the anchored clathrate water hypothesis. Waters around the methyl groups of these ice-binding residues can be anchored to the Thr OH group or to the peptide backbone that is accessible to solvent waters due to the high Ala content (65%) of the helix. Although this clathrate water pattern was not seen in the original X-ray crystal structure of winter flounder type I AFP because the protein was crystallized in acetone (Sicheri and Yang 1995), this clathrate arrangement has been recently seen in the crystal structure of Maxi, an extremely divergent isoform of type I AFP. Maxi can serve as a surrogate to show approximately what the anchored clathrate waters might look like on the ice-binding residues of type I AFP (Sun et al. 2014).

The identification and mapping of an IBS by site-directed mutagenesis has even worked well with models of an IBP before its structure was determined by crystallography or nuclear magnetic resonance (NMR). A convincing example of this method was the systematic mutagenesis of the ice-binding domain of the giant ice adhesin from the Antarctic bacterium *Marinomonas primoryensis*. Having modeled this domain as an extended beta-solenoid, a series of outward pointing steric mutations (and one inward pointing one) were made at intervals around the circumference of the solenoid. Mutations on the outer surface that curve around the inner row of Ca^{2+} ions severely attenuated TH activity, whereas those elsewhere on the surface had no significant effect. The internal mutation that tried to place an arginine side chain into the protein interior (V93R) was also highly detrimental. Thus the mutation study first validated the modeled protein fold by confirming the outward and inward pointing residues, and then revealed that the IBS was composed of the two parallel rows of Thr and Asx that run the length of the solenoid on one side (Garnham et al. 2008).

4.4 Planes Bound by AFPs

One outcome of the structural diversity of IBPs is they can have different ice plane binding preferences. That their IBSs have different functional residues and different spacing between exposed chemical groups grants them specificity for particular planes of ice in specific orientations. When an IBP molecule sticks to its energetically favored site on ice, ice growth in the vicinity slows down. Because of the periodicity of ice crystals, multiple IBP molecules of the same type bind to adjacent, equivalent sites on the ice plane. Consequently, a facet is developed. This binding specificity is remarkable because the IBP targets a specific pattern of crystalline water molecules among many similar possibilities in a situation where the ice is surrounded by a huge excess of liquid water in which IBPs are freely soluble. Also, the ice surface is not directly exposed to IBPs but is coated by a thin layer of quasi-liquid water (Hayward and Haymet 2001; Limmer 2016). The answer to how each IBP binds to a specific plane at the molecular level is still somewhat speculative. Ice-etching studies (see Chap. 9 of this volume for details on the method) conducted in the early 1990s have shown that type I AFPs, which are short repetitive peptides, from the winter flounder (*Pseudopleuronectes americanus*) and the closely related Alaskan plaice (*Pleuronectes quadritaberulatus*) bind to the $[\bar{2}\ 0\ 2\ 1]$ pyramidal planes of ice. However, type I AFP (SS-8 isoform) from shorthorn sculpin (*Myoxocephalus scorpius*) adsorbs onto $[2\ \bar{1}\ \bar{1}0]$, the secondary prism planes (Knight et al. 1991). Although these two versions of type I AFP share high sequence identity, they are not homologues. They arose independently to form similar IBPs by a remarkable case of convergent evolution (Graham et al. 2013). Indeed, the fact that the flounder and sculpin type I AFPs bind to different planes of ice is an additional argument for their independent origins. Ice etching of two type III isoforms (AB1 isoform of *A. brachycephalus* and the HPLC12 isoform of *M. americanus)* revealed a more complex pattern, suggesting that both proteins bind to several ice planes including the primary prism plane $[1\ 0\ \bar{1}\ 0]$, the pyramidal plane $[2\ 0\ \bar{2}\ 1]$, and some additional planes inclined by a small rotation from these planes (Antson et al. 2001). Thus, an AFP that binds to several ice planes can still have moderate TH activity. The compound IBS of type III AFP which consists of two adjacent parts juxtaposed at an angle of 150° to each other (Fig. 4.2) clearly demonstrates how the protein binds to more than one crystallographic plane (Garnham et al. 2010).

A growing number of IBPs have been shown to bind the basal plane of ice. Most of these proteins have particularly high TH activities, for which they are termed "hyperactive." In fact, the hyperactivity of IBPs has been linked to their ability to adhere to the basal plane of ice in addition to other planes, and stop ice growth along both the *a* and the *c* directions (Scotter et al. 2006; Pertaya et al. 2008). Ice etching studies of the *Choristoneura fumiferana* (spruce budworm) AFP (sbwAFP) showed that it binds to the primary-prism plane and the basal plane (Graether et al. 2000). The fluorescence-based version of the ice etching method, FIPA analysis (Basu et al. 2014), yielded an ice crystal covered from all directions when grown in *Tm*AFP (Basu et al. 2014) and *Ri*AFP (Hakim et al. 2013) solutions (Fig. 4.3e and g, respectively), as did a more potent isoform of sbwAFP, isoform 501 (Fig. 4.3f). This suggests that these proteins can bind to multiple planes of ice. The abovementioned three IBPs have repetitive structures with well-defined IBSs, determined from 3-D crystal structures, modeling, and, in the cases of *Tm*AFP, by extensive surface mutagenesis studies (Marshall et al. 2002). The IBSs of these proteins are flat and composed of multiple arrays of outward pointing threonine residues that presumably bind to several ice planes. This is in contrast to the IBS of

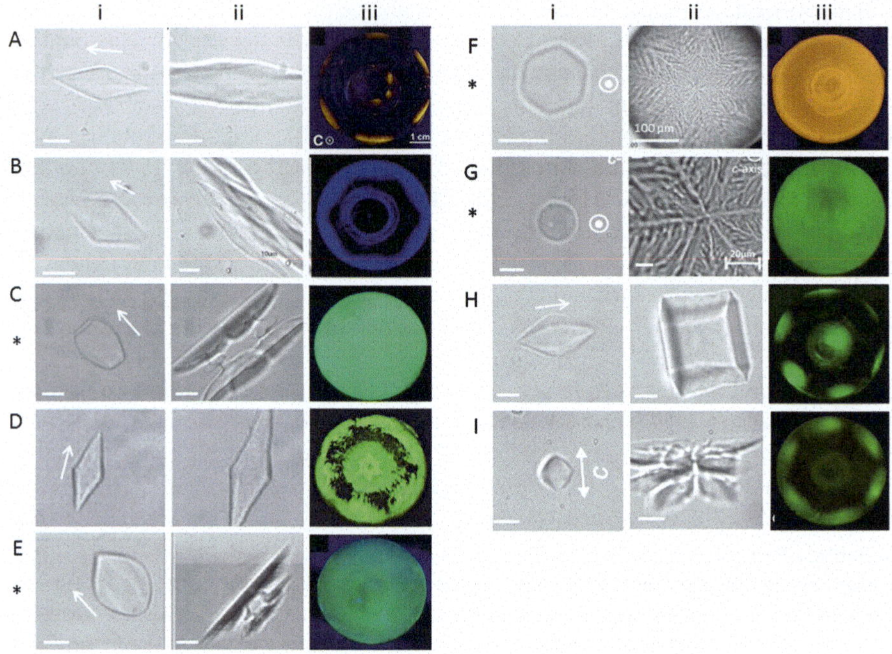

Fig. 4.3 Comparison of ice-binding properties of IBPs. Typical morphology of IBP-bound ice (i) within the TH gap and (ii) just after the burst. The white arrow indicates the direction of the *c*-axis. (iii) FIPA analysis with the *c*-axis normal to the field of view. (**a**) Type I AFP from winter flounder (Bar Dolev et al. 2012; Basu et al. 2014). (**b**) Type III AFP from ocean pout, isoform HPLC12 (Bar Dolev et al. 2012; Basu et al. 2014). (**c**) *Mp*IBP-RIV (Bar Dolev et al. 2012; Basu et al. 2014). (**d**) Snow mold fungus IBP from *Typhula ishikariensis*, moderate isoform (*Tis*AFP6) (Kondo et al. 2012; Xiao et al. 2010). (**e**) *Tm*AFP, isoform 4–9 (Bar et al. 2008a; Basu et al. 2014). (**f**) sbwAFP, isoform 501 (Bar Dolev et al. 2012; Basu et al. 2014). (**g**) *Ri*AFP (Hakim et al. 2013). (**h**) *Lp*AFP (Middleton et al. 2012). (**i**) Midge AFP (Basu et al. 2016). Images reprinted from the above references with permission. An asterisk on the left indicates that the IBP is hyperactive. Scale bars are 10 µm unless otherwise specified

type III AFP, which binds different planes from distinct positions on its compound IBS (Garnham et al. 2010). Another interesting example is the IBP from ryegrass (*Lp*IBP), which binds to both the basal plane and the primary prism plane (Middleton et al. 2012), the same planes bound by sbwAFP. However, sbwAFP is hyperactive with a TH activity that exceeds 5 °C at 1 mg/ml concentration (for the 501 isoform). On the contrary, *Lp*AFP has low TH activity, in the range of 0.3 °C at a concentration of 2 mg/ml. The two proteins are approximately of same size (~12 kDa), but the ice-binding site of *Lp*IBP is less repetitive than that of sbwAFP, and the ice hemisphere grown in *Lp*IBP solution (Fig. 4.3h) is less covered by protein compared to the hemispheres grown with the insect proteins (Fig. 4.3e–g). The recently characterized IBP from midge binds to a pyramidal-ice plane positioned intermediate between the basal and primary prism planes. This plane is different from those observed with other moderate IBPs, and it seems to be at least partially responsible for the intermediate TH activity of this protein, higher than the moderately active IBPs, but lower than that of the hyperactive proteins (Basu et al. 2016). Another example of unusual ice binding is EfcIBP (a DUF3494-type protein) that shows basal plane affinity without prism plane affinity and has moderate TH activity (Kaleda et al. 2019; Mangiagalli et al. 2018). This points to the possibility of a broader spectrum of yet undefined IBPs that have activities between moderate and hyperactive.

4.5 Ice Shaping

The fundamental ability of IBPs to bind to ice can cause outward growth of the bound surface to cease as dictated by the Gibbs–Thomson effect (Wilson 1993). This growth inhibition leads to the formation of ice shapes distinct from the disk shape of ice in pure water (Kawahara 2013). Examples of shapes induced by IBPs are shown in Fig. 4.3(i). The same effect causes inhibition of melting, which also leads to ice shaping (Bar Dolev et al. 2012; Liu et al. 2012; Pertaya et al. 2007a). The specificity of IBPs for different ice planes and the ice-binding rates of each protein to each ice plane dictate the particular ice shapes, characteristic of an IBP type. We describe ice shaping in the presence of IBPs in three separate temperature regimes: below the hysteresis freezing point (the "burst" pattern), above the hysteresis freezing point (within the TH gap), and above the melting point (melting pattern).

4.5.1 Ice Shaping Below the Hysteresis Freezing Point

When a supercooled ice crystal is brought to its freezing point, sudden fast growth is observed. This "burst" temperature defines the freezing point of the crystal and the lower limit of the TH gap. Proteins with high TH values, such as the hyperactive insect IBPs, can supercool for several degrees, so the burst is abrupt with a dendritic growth pattern (Fig. 4.3(ii), c, e–g). Proteins with low TH can produce milder burst forms, including just a steady increase in the size of the crystal in all dimensions, an example of which is provided by the plant IBP, *Lp*AFP. The direction to which the crystal grows during the burst is dependent of the IBP-covered planes. In general, IBPs that can bind the basal planes will direct the ice burst normal to the *c*-axis, while IBPs that do not bind the basal plane induce the ice burst along the *c*-axis (Scotter et al. 2006). Still, there are differences in the burst pattern among the non-basal binding IBPs. For example, in type I and type II fish AFPs solutions the ice bursts as a single sharp needle emerging from the tip of the crystals, while in type III AFP solutions many small crystals emerge from the original one at the burst (Fig. 4.3(ii), a–b). An interesting case is *Lp*AFP, which binds to both basal and prism planes and has low TH activity (Middleton et al. 2012). This leads to the growth of ice shaped like a hexagonal box (Bar Dolev et al. 2012; Middleton et al. 2012) (Fig. 4.3(ii), h).

The recently characterized midge IBP is an example of a protein that does not bind the ice basal plane and yet causes the ice crystal to burst perpendicular to the *c*-axis (Fig. 4.3(ii), i) (Basu et al. 2016). Another interesting example is type I AFP from a righteye flounder, the barfin plaice (bpAFP), that has exceptionally high solubility (Mahatabuddin et al. 2017). At low concentrations, bpAFP behaves as a typical type I AFP with pyramidal plane binding, low TH activity, and directs crystal burst along the *c*-axis. At high concentrations, TH values can reach 3 °C and the ice crystal bursts perpendicular to the *c*-axis. In this regard, bpAFP behaves more like Maxi, and there is indeed evidence for oligomerization. Recently, a Saturn-like shape has been observed as the burst growth pattern of ice in efcIBP solution. This growth pattern is consistent with the affinity of efcIBP for the basal plane of ice without affinity for the prism plane (Kaleda et al. 2019).

4.5.2 Ice Shaping Within the TH Gap

The bipyramidal ice shapes in solutions of moderate fish AFPs are widely documented (Bar Dolev et al. 2012). However, a careful examination showed that below and close to the melting point, the crystals have the form of truncated bipyramid, with some exposed basal plane. At lower temperatures, the crystals continue to grow in the *c* direction, and the basal planes shrink until they reach a critical size on which new layers of water cannot be incorporated (Knight and DeVries 2009). At this stage, the crystal is populated by IBP molecules on all 12 equivalent surfaces of the hexagonal bipyramid, which is the basis for their ability to arrest ice growth. At lower temperatures, the critical size for ice nucleation is smaller, so the tips continue to grow and become sharper. These tips are protected less effectively relative to the crystal planes and the burst usually starts from them. In the case of the hyperactive IBPs, which bind to the basal planes, the ice crystals are covered by proteins from all directions (Pertaya et al. 2008), so they are protected from growth (and melting) throughout the TH gap. Therefore, the growth of pyramidal tips is not apparent, and their shapes remain constant (Bar Dolev et al. 2012).

In the early 1990s, researchers noted that at varying low concentrations of type I AFP from winter flounder ice crystals always grew with a constant *c*:*a* axis ratio of 3.3:1. This ratio remained constant also in a series of mutants with lower TH activities relative to the wild type. However, the axis ratio of ice crystals in dilute solutions of type III AFP from ocean pout (QAE1 isoform) changed as the crystal continued to grow and changed with AFP concentration. In addition, partially inactivated type III AFP mutants produced different axis ratios (DeLuca et al. 1996). A theory developed to explain this phenomenon suggests that the flounder AFP, which binds only to one pyramidal plane of ice, is aligned on the pyramidal planes without forming steps. Type III AFP binds to prismatic planes, so it was reasoned that steps must form in order to get pyramidal shapes. The length of each step is concentration dependent, which leads to variation in the axis ratio at different protein concentrations (DeLuca et al. 1996). However, another type I AFP from sculpin adsorbs on the secondary prism planes (Knight et al. 1991), which seem to require a step-growth mechanism to achieve a bipyramidal ice crystal shape. Later studies showed that type III binds to more than the prismatic plane (Antson et al. 2001; Garnham et al. 2010) adding complexity to this phenomenon. A plausible scenario to explain the changes in axis ratios due to concentration or mutagenesis is that ice growth velocities depend on the adsorption rates of each protein to a particular ice plane (Drori et al. 2014a; Knight and DeVries 2009).

4.5.3 Ice Shaping at Melting

Bipyramidal ice crystals in moderate IBP solutions always start to melt from their tips, where ice coverage is less effective. The melting advances along the *c*-axis, exposing more basal surface until the crystal obtains an eye shape with tips normal to the *c*-axis (Bar Dolev et al. 2012). Hyperactive IBPs, which effectively bind the basal planes, result in characteristic ice shaping during melting (see Fig. 4.3(i)—c, e, f, and g). This is counterintuitive because the crystal is retreating. However, it was shown that IBPs can slow down ice melting, and even completely inhibit melting, resulting in superheated ice (Celik et al. 2010; Cziko et al. 2014; Knight and DeVries 1989). Furthermore, the methods used to observe crystal shaping include flash freezing of the whole solution, so the melting shapes are possibly formed due to protein trapped in the crystal. The variations in the shapes obtained with each IBP during melting are related to differences in the melting velocities on the *a* and the *c* axial directions. A model of ice melting that considers low melting velocity in the *c* direction relative to the *a* direction successfully predicted the formation of the lemon shape structure, characteristic of *Tm*AFP (Liu et al. 2012).

4.6 Size and Cooperative Effects of IBP Activity

4.6.1 Size of AFP Molecule and Cooperativity

One puzzle in understanding IBP activity was the observation that TH appears only above a certain protein concentration threshold. Below this concentration, AFPs were sufficient to shape ice but not arrest its growth. An early model proposed that at high protein concentrations there is a cooperative effect: protein molecules form side-by-side interactions that create an AFP patch with several IBS aligned on the ice surface (Wen and Laursen 1992a). To test this hypothesis, the 7-kDa type III AFP was fused to 12-kDa or 42-kDa proteins, increasing its overall size to ~20 and ~50 kDa, respectively. It was expected that the TH activity of the fusion protein would be lower than the TH of the wild type because the fused proteins impose a steric intervention and disrupt any possible side-by-side AFP interactions. However, the opposite was observed. Larger (bulkier) conjugates resulted in greater TH activity (DeLuca et al. 1998). These findings rule out the hypothesis of the cooperative binding effect of AFP patches and they are supported by the adsorption–inhibition model (Raymond and DeVries 1977) and the Kelvin effect (Wilson 1993). Increasing the size of AFPs reduces the distance between adjacent molecules since each molecule covers more of the ice surface. This lowers the probability of water addition on the ice surface and lowers the freezing point. We have tested green-fluorescent protein (GFP) fusions of AFP type III and the hyperactive *Tm*AFP (9 kDa), where the GFP molecule (26 kDa) increases the overall size of the proteins ~fourfold. We observed that GFP fusions are >twofold more active than unconjugated *Tm*AFP on a molar basis. Another study demonstrated that addition of low concentrations of polyclonal antibodies raised against *Tm*AFP or its homolog from the beetle *Dendroides canadensis* (*Dc*AFP) to the antigen solution enhanced the TH activity severalfold. The addition of secondary antibodies (goat anti-rabbit) that bound to the primary antibodies raised the TH even more (Wu et al. 1991). Clearly, larger IBPs lead to higher TH levels. These observations raise the question, why did IBPs evolve to be small, highly expressed proteins? Low-level expression of bigger proteins might have been metabolically favored. It may be that smaller molecules have increased accessibility to other areas in the body through extravasation from the circulatory system (Bar Dolev et al. 2016b). In addition, small molecules diffuse faster than larger ones. Notably, although side-by-side interactions between IBP molecules can be ruled out in the abovementioned examples, cooperative effects are still possible in certain cases. In a study of AFP type III from Notched-fin eelpout (nfeAFP), a significant increase in the TH activity of a barely active isoform was observed upon addition of low concentrations of a more active isoform. This may be due to a form of cooperativity between the two isoforms (Nishimiya et al. 2005; Takamichi et al. 2009) that likely involves the stabilization of binding planes for the less active isoform (Berger et al. 2019). In many fish and insects, low-active AFP isoforms have been identified in addition to moderate and hyperactive ones (Hew et al. 1988; Liou et al. 1999). Multiple isoforms may cooperate to provide better overall protection from ice growth.

4.6.2 Size of IBS

The intuitive concept that IBPs with bigger IBSs will have higher antifreeze activities is consistent with both the reversible and irreversible ice-binding models (discussed in Sect. 4.11). IBPs with a larger IBS have a better chance to bind ice than those with a small IBS. It is also consistent with the hydration shell/anchored clathrate hypothesis (see Sect. 4.10), since a large IBS has more positions for binding water, which increases the likelihood of formation of a quorum of ice-like water molecules. This eventually increases the possibility of the clathrate waters on the IBP merging into the quasi-liquid layer at the ice–water interface and turning into ice. The advantage of a large IBS has been observed in organisms producing IBP isoforms with different IBS sizes from either repetitive or non-repetitive structures, moderate and hyperactive.

Correlations between IBP size and TH activity are described in Chap. 3 of this volume. Here we specifically look at this relationship in isoforms of the same type where the structure of the IBSs is the same, but their sizes are different. One straightforward example is AFGPs. The small isoforms of AFGPs are much less potent than the larger ones, both natural (Wu et al. 2001) and synthetic (Tachibana et al. 2004), as shown in Fig. 3.1. Another example is the abundant 3.3 kDa α-helical type I AFPs of the winter flounder, which consist of 3 repeats of the 11 amino acid (aa) consensus sequence. One isoform of this protein (AFP9) has 4 repeats of the consensus sequence and, therefore, a larger IBS surface area. This isoform had almost double the TH activity of the 3-repeat protein, although its size is only ~30% larger (Chao et al. 1996). Consistent with this size-to-activity relationship, a synthetic peptide consisting of only one repeating unit from the same AFP did not show any TH activity. The peptide did, however, produced ice shaping, indicating that it was still able to bind ice.

Two well-characterized isoforms of the hyperactive, left-handed β-helical AFP from spruce budworm (sbwAFP) are the 9 kDa (isoform-337) (Graether et al. 2000) and the 12 kDa (isoform 501), consisting of 5 and 7 helical loops, respectively. Parallel TXT motifs comprise the ice-binding face of both isoforms, but in isoform 501 there are two positions where Thr are replaced: in one case by Ile, and Val in the other. Although the overall IBS is only ~30% larger in isoform 501, its TH is three- to fourfold higher than that of the smaller isoform (Leinala et al. 2002), as shown in Fig. 3.7. In snow fleas, the TH of a larger isoform (15.7 kDa) is double the TH activity of a smaller isoform (6.5 kDa) at low concentrations (<0.2 mg/ml) (Graham and Davies 2005).

The Antarctic eelpout, *Lycodichthys dearborni,* produces a large type III AFP isoform called RD3 that consists of two consecutive units of the 7-kDa protein. The monomers are similar to each other in structure (Miura et al. 2001) and activity (Wang et al. 1995), and they are connected by a flexible 9-amino-acid linker. NMR studies indicated that this linker allows the simultaneous binding ice of both IBSs (Holland et al. 2008). A comparison of the TH activity of this tandemer relative to the monomer is presented in Chap. 3 (Fig. 3.4). At low concentrations (<0.5 mM), the TH of the dimer reaches sixfold the activity of the monomers on a molar basis, although the overall size increased only twofold. Cooperative effects of the two units explained this significant enhancement (Miura et al. 2001; Wang et al. 1995). Notably, in this case both the size of the IBP and the size/number of the IBS were doubled. In a study of a recombinant model based on the RD3 dimer, where mutagenesis was used to knock out one of the IBSs, the IBS area contributed 80% of the increased TH activity, and the larger size accounted for the remaining 20% (Baardsnes et al. 2003). Previous studies with type III AFP fusion proteins had shown that TH activity increases with the overall size of the complex. A 20% increase in activity on doubling protein size was consistent with the range of increases seen with other naturally occurring isoforms (Chao et al. 1996; Leinala et al. 2002) and fusion proteins.

4.7 Protein Engineering of Better IBPs

IBPs have evolved on many occasions in different organisms to serve specific functions (Davies 2014). Those that serve as AFPs to prevent fish from freezing in icy seawater have a defined lower limit of ~1.2 °C freezing point depression to achieve. This they accomplish with little leeway (Scotter et al. 2006), and only by producing high concentrations (mM) of AFPs in their blood. In several fishes these 10–30 mg/ml concentrations have required massive amplification of the AFPs genes to meet the concentration demand (Chen et al. 1997; Hew et al. 1988; Hsiao et al. 1990; Scott et al. 1988). Some terrestrial insects must overwinter at temperatures of −30 °C without freezing. Based on in vitro assays of their AFPs it is likely that other factors besides TH contribute to this freezing point depression. Nevertheless, insect AFPs are considerably more potent at TH than fish AFPs. We attribute this hyperactivity to the ability of insect AFPs to bind the basal plane of ice in addition to other planes, as discussed in Sect. 4.4. Although there are hyperactive isoforms known for type I AFP, in general fish AFPs appear to be underachievers by not evolving basal plane binding. Why has evolution not made fish AFPs of greater potency, in lesser amounts, at a reduced metabolic cost? Although nature has not worked this way, hyperactive AFPs transgenically introduced into fish could potentially provide a biotechnological solution to the aquaculture of salmon in seawater areas where superchill mortality is a problem (Hew et al. 1992).

Aside from the lesson of basal plane binding, nature has provided other hints about how to improve on antifreeze activity. When there are multiple isoforms of a repetitive AFP, the bigger isoforms are invariably better at TH when compared on a molar basis (see discussion above, Sect. 4.6.2). This principle was confirmed in a protein engineering study where coils were added and subtracted from the β-solenoid structure of *Tm*AFP. Removal of just one coil from the seven found in the most abundant *Tm*AFP isoform caused a huge loss in activity; whereas the insertion of one or two coils had the opposite effect (Marshall et al. 2004a). How can this be rationalized when binding of an IBP to ice is necessarily irreversible? We can cite similar arguments used to explain why TH is a function of AFP concentration. Stopping a seed ice crystal from growing at supercooling temperatures requires diffusion and surface binding of sufficient AFPs to harness the Gibbs–Thompson effect. But for a productive contact that leads to binding between the IBP and ice, there must be a good match between the "anchored" clathrate waters on the IBS and the quasi-liquid water layer coating the ice surface. Having a larger IBS makes it statistically more likely that a match will be found.

Artificial constructs that multimerize IBPs have produced significant increases in antifreeze activity (Can and Holland 2011). These increases are optimally presented by the lower concentrations needed to achieve the same TH as free IBPs. The attachment of type I and type III AFPs to scaffolds like dendrimers demonstrates the potential of multimerization as a protein engineering approach to antifreeze enhancement, but it suffers from incomplete reaction (Stevens et al. 2015). A more controlled approach has been achieved through the use of self-assembling protein cages (King et al. 2014; Padilla et al. 2001) to attach a fixed number of IBPs in a defined orientation with their IBSs projecting outward (Phippen et al. 2016). When TH activity is compared on a molar basis, a multimer displaying 12 IBPs is an order of magnitude more active than the monomer. This increase in freezing point depression is mirrored by a similar increase in the ability of the multimers to inhibit ice recrystallization. Following this approach there are many possibilities to form 1- 2- and 3-D arrays of different IBPs types and mixtures thereof to design ways to control and shape ice growth. Further details on this subject are given in Chap. 14 of this volume.

4.8 INPs: Ice-Nucleating Proteins

Ice-nucleating agents are widely dispersed in nature and serve to raise the temperature at which ice freezes by organizing an ice nucleus of sufficient size to promote its rapid growth (Pummer et al. 2015). Of relevance to this chapter are the biological ice-nucleating agents found on the surface of some bacteria. These INPs cluster together as aggregates on the bacterial surface that can promote freezing at temperatures as high as −2 °C (Guriansherman and Lindow 1993; Kawahara 2002; Kieft 1988; Wolber and Warren 1989). The primary sequence of these >120 kDa bacterial INPs is composed of three regions. The central region is the largest and is a series of tandem 16-amino-acid repeats of consensus sequence GYGSTxTAxxxSxLxA, which promote nucleation (Green et al. 1988) (Warren and Corotto 1989). This repetitive region is flanked by shorter non-repetitive regions, the N-terminal one of which is thought to anchor the INP to the outer membrane of the bacterium (Kawahara 2002; Wolber and Warren 1989). Although there are no experimentally solved structures of INPs, some structural models for the repetitive region have been predicted (Guriansherman and Lindow 1993). Inspired by the β-helical folds of some insect AFPs (Graether et al. 2000; Liou et al. 2000), the repetitive regions of INPs from *P. syringae* (*Ps*INP) (Graether and Jia 2001) and a close relative *P. borealis,* (*Pb*INP) (Garnham et al. 2011b) were modeled as β-solenoids. The logic of this choice is that both IBPs and INPs have repeats of a similar length and that a solenoid fold places repeating motifs in line on the same face of the helix. In IBPs, the solenoid fold aligns the TXT ice-binding motifs into a two-dimensional array that functions as the IBS. As described in Sect. 4.10 below, the anchored clathrate water hypothesis for ice binding suggests that the IBS functions by organizing waters into an ice-like pattern sufficient to merge with the quasi-liquid waters on ice and in turn become ice. The fact that a solenoid fold for INP can potentially form an even longer array of TXT motifs strongly suggests a similar mechanism of surface water ordering. This hypothesis is strengthened by simulations of water molecules around the TXT arrays of the *Pb*INP model, showing that the INP could order water on this site (Garnham et al. 2011b). In INPs, the exaggerated length of the water-organizing region goes far beyond the six to eight coils needed for ice binding to a length where the excess ordered water can promote ice nucleation. Models have predicted a total water-organizing area of 4200 Å^2 in *Ps*INP (for a monomer) (Graether and Jia 2001), and 25,600 Å^2 in *Pb*INP (for a dimer) (Garnham et al. 2011b). Taking into account that one INP molecule is necessary to obtain an ice nucleus at −12 °C (Govindarajan and Lindow 1988), the overall size of the water-organizing area of the INP models is roughly in agreement with the necessary size of ice nuclei at this temperature, which is 20,100 Å^2. This calculation, together with several studies that showed INPs aggregate to facilitate nucleation at elevated temperatures (~ −2 °C) (Guriansherman and Lindow 1993), support the idea that INPs form ice by the same anchored clathrate water mechanism used by AFPs to bind ice (Garnham et al. 2011a). Thus, the basic difference between AFPs and INPs is their size. The large IBS of INPs can bind enough water molecules in an ice-like organization, sufficient to serve as a heterogeneous ice nucleus rather than suppress ice growth. Accordingly, some IBPs have slight ice nucleation activity, such as type I AFP at high concentrations (Wilson et al. 2010). Recent studies show low ice nucleation activity for type III AFP, *Tm*AFP, and sfAFP that match the proteins' small size (Bissoyi et al. 2019; Eickhoff et al. 2019). Computational studies supported these findings (Qiu et al. 2019). Another study showed truncated INP has antifreeze activity (Kobashigawa et al. 2005). These results provide further support that INPs and AFPs bind ice by the same mechanism. On an issue of semantics, we include INPs in the group of IBPs not just because of their predicted common mechanism but because at the instant that INPs form ice they are effectively bound to its surface.

4.9 The Molecular Basis to Protein–Ice Interactions

The mechanism by which IBPs bind to ice at the molecular level has been an intensely debated topic in IBP research. The interface between protein and ice in an aqueous medium is difficult to probe experimentally and theoretically. Ice in water does not have clear boundaries between the crystalline and liquid states that can define a binding site at the molecular level. Instead, there is a thin zone (quasi-liquid layer) of water that is intermediate in state between solid and liquid water and effectively blurs the boundary between the two states (Hayward and Haymet 2001; Limmer 2016). Moreover, it is fascinating that IBPs can actually select particular ice surfaces, made up of only ordered water molecules, when they are surrounded by <55 M of bulk water. Adding to these challenges is determining the IBS of many IBPs where there are no obvious motifs to help identify it and suggest a match to the ice. Even in the cases where the IBS are flat repetitive surfaces like *Tm*AFP (Liou et al. 2000) and *Ri*AFP (Hakim et al. 2013), there are numerous different ice planes that can be developed during growth when inhibitors such as IBPs are present. The differences between many of these planes (energetically or sterically) may be small enough to be practically indistinguishable.

Since the early studies of IBP mechanisms, many theories to explain the molecular basis for ice recognition by IBPs have been suggested (reviewed in (Davies et al. 2002; Vrielink et al. 2016; Yeh and Feeney 1996)). An early explanation was that the abundant hydrophilic moieties on AFGPs immobilize water molecules in their vicinity and reduce the amount of water available for ice formation. This idea was disputed by NMR studies showing the amount of bound water molecules on AFGP surfaces is small (DeVries and Price 1984). It was then suggested that AFPs bind to ice through hydrogen bonds between protein side chains (or disaccharide moieties in the case of AFGPs) and available water molecules on the ice surface (DeVries and Price 1984; Knight et al. 1991). The low number of hydrogen bonds and their weakness in comparison to covalent bonds seemed insufficient to account for irreversible binding of IBPs to ice. An expansion of this idea based on an excellent match between the IBS and the bound ice plane was that hydroxyl groups on the protein surface could be incorporated into the ice lattice (Knight et al. 1993). However, mutagenesis studies of the IBS of type I AFP (Chao et al. 1997; Haymet et al. 1998; Zhang and Laursen 1998) and *Tm*AFP (Bar et al. 2008b) showed that the hydrophobic moieties on the IBS are more important than the surface hydroxyls for maintaining TH activity. Also, theoretical works showed that there is no gain of hydrogen bonds upon binding of type I AFP to ice (Madura et al. 2000; Wierzbicki et al. 2007).

4.10 The Hydration Shell Theory

Molecular dynamic simulations proposed that the first hydration layer on the IBS of type I (Yang and Sharp 2005) and type III (Smolin and Daggett 2008; Yang and Sharp 2004) AFPs makes a considerable contribution to the binding of the proteins to ice. These studies showed that apolar grooves in the IBSs of the proteins are filled with water molecules that are hydrogen-bonded to the polar groups on the IBS. These water molecules are arranged in a tetrahedral organization, similar to the organization of water molecules in ice. The quasi-ice layer bound to the protein facilitates merging of the protein to the quasi-liquid layer on ice in aqueous media (Gallagher and Sharp 2003; Wierzbicki et al. 2007). One experimental verification of this concept came from the crystal structure of ocean pout type III AFP, which was determined by combined X-ray and neutron diffraction data. Four water molecules arranged in an ice-like pattern were noted in the combined structure, suggesting that they were anchored in this organization to the protein in solution. However, only three of these water molecules were sufficiently well ordered to locate the hydrogen atoms in the crystal structure. The fourth water molecule in this water quorum was solved with partial occupancy (Fig. 4.4) (Howard et al. 2011).

Clearer evidence for ice-like water organization came from the crystal structure of the ice-binding domain of *Mp*IBP (*Mp*IBP-RIV). An array of ~50 water molecules arranged in an ice-like structure was present on the IBS of this protein (Fig. 4.5) (Garnham et al. 2011a). These waters formed cages around the surface methyl groups and this clathrate array was "anchored" to nearby hydroxyl and peptide backbone amide groups. Although ice-like ranks of water molecules were noted before in crystal structures of other IBPs (Liou et al. 2000), their IBSs were always

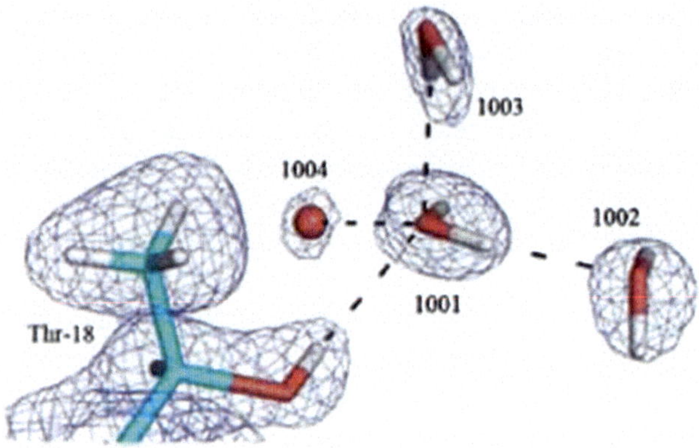

Fig. 4.4 Ordered surface waters on the IBS of type III AFP. The tetrahedral water cluster model is superimposed on the σ_A-weighted $2F_o - F_c$ nuclear scattering density map (contour level = 1 r.m.s.) for the tetrahedral water cluster. Note that while there are large nuclear scattering peaks for the three water molecules 1001, 1002, and 1003, the nuclear scattering peak for water 1004 is much weaker, indicating disorder. As such, it can only be modeled as an oxygen atom. Reprinted from (Howard et al. 2011). Copyright (2011) WILLEY

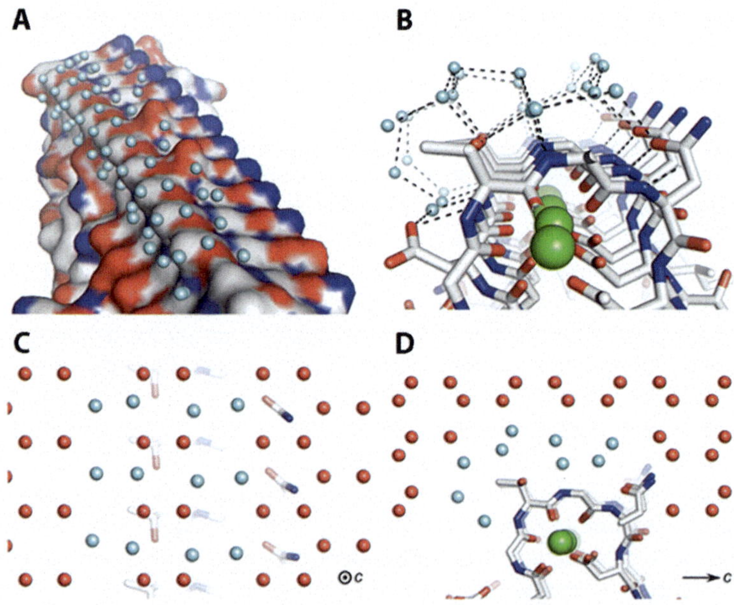

Fig. 4.5 Ordered surface waters on the IBS of *Mp*AFP-RIV. (**a**) Surface representation of the IBS of one of the protein molecules freely exposed to solvent in the unit cell. Carbon atoms are white, nitrogen are blue, and oxygen are red. Ordered surface waters are represented as aqua spheres. (**b**) Ordered surface waters showing some hydrogen bonded directly to the IBS. Structural Ca^{2+} ions are in green. (**c–d**) The organized surface waters make an excellent 3-D match to both the basal (**c**) and primary prism (**d**) planes of ice. The direction of the *c*-axis is indicated in both figures. Modified from (Garnham et al. 2011a)

directly opposed to the IBS of another IBP molecule, so the organization of any remaining water could be interpreted as a crystal-induced artifact. In the structure of *Mp*IBP-RIV, two out of the four protein molecules in the unit cell had their IBS exposed or partially exposed to the solvent, suggesting that the organization of surface waters was that seen in solution (Garnham et al. 2011a). Further support for the anchored clathrate water model has come from molecular dynamics (MD) simulations that showed ordered water structures on the IBS of *Tm*AFP (Yang et al. 2003) (Liu et al. 2016), sbwAFP (Nutt and Smith 2008), and *Pb*INP (Garnham et al. 2011b) in solution. While other simulations find only slight prestructuring of water molecules on the *Tm*AFP ice-binding surface in solution, they also support the formation of ordered water prior to the binding of AFP to ice, suggesting the need for ice to stabilize the clathrate water on the IBP surface (Hudait et al. 2018). In another experimental study, additional IBS-bound waters on type III AFP were revealed by crystallizing the protein as a fusion to maltose-binding protein that changes the orientation of the IBS toward the solvent (Sun et al. 2015). On a less active variant of the same AFP, mutations that improved activity were associated with a network of polypentagonal waters on the ice-binding face (Mahatabuddin et al. 2018)

An extension of the hydration shell theory suggests a contribution of the non-ice-binding faces of IBPs to their activity. MD simulations of the hydration shell around the three faces of the β-helical sbwAFP showed that while the ice-binding face facilitates the ordering of tetrahedral water structures, the other two faces disrupt water clusters such that ice-like water are excluded from those planes. It was concluded that the non-ice-binding faces of IBPs cooperate with the IBS by preventing overgrowth of the protein by the ice (Knight and Wierzbicki 2001; Nutt and Smith 2008).

Although the role of water molecules in the molecular recognition of ice by AFPs is thought to be generic to all IBPs, some controversies need to be resolved. Vibrational sum-frequency generation spectroscopy, which is similar to Raman scattering but sensitive only to molecules in nonsymmetric environments such as surfaces was used to observe the surface water molecules on the IBS of some IBPs. A clear peak attributed to the ice-like waters bound to the protein IBS in solution was present in the spectra of eelpout type III AFP but was absent in the spectra of an inactive mutant of the same protein. This peak was noted also in the spectrum of the wild-type protein taken at room temperature, although with lower intensity relative to that seen at low temperatures, indicating the stability of the ice-like water structure is dependent on temperatures (Meister et al. 2014). However, similar experiments with the hyperactive *Dc*AFP from beetles showed no such water organization. The authors concluded that the highly ordered β-helical *Dc*AFP does not need a water clathrate for ice recognition (Meister et al. 2015). Further contradictory data on *Tm*AFP, a homolog of *Dc*AFP, suggested the presence (Yang et al. 2003) or lack of (Modig et al. 2010) ice-like water molecules on its IBS. Another unresolved issue concerns the range of water molecules involved in ice recognition. An MD simulation on ocean pout type III AFP predicted that the water clathrate around the IBS of the protein consists of water molecules of the primary hydration layer but not beyond

(Smolin and Daggett 2008). However, terahertz absorption spectroscopy conducted to probe the long-range interactions of AFGPs (Ebbinghaus et al. 2010), type I AFPs (Ebbinghaus et al. 2012), and the hyperactive insect *Dc*AFP (Meister et al. 2013) with water suggest that IBPs can retard hydrogen bond dynamics up to 20 Å from the protein surface. Although in the case of type I AFP a weakly active mutant had the same effect on the solution as the highly active wild-type protein (Ebbinghaus et al. 2012), the authors concluded that the long-range perturbation of solution dynamics is essential for ice recognition by IBPs. In a theoretical study of the first and second hydration shells of type III AFP it was found that the wild-type protein and two mutants with 10% and 54% of the wild-type activity had the same hydration properties. In this case it was suggested that there is no correlation between the effects of the proteins on the solvation water and the antifreeze activity (Grabowska et al. 2016). Clearly, the discrepancies between the abovementioned studies need to be resolved.

4.11 The Reversible–Irreversible Binding Conflict

The adsorption–inhibition theory (Raymond and DeVries 1977) suggests that IBPs bind to the surface of an ice crystal and pin it such that the ice can grow only between the bound protein molecules. This local pinning results in surface curvature that increases as the ice grows, leading to reduction of the freezing point due to the Gibbs–Thomson effect, and subsequently to inhibition of ice growth. The observation that ice growth can be completely stopped under supercooled conditions (in the TH interval) indicates that the IBPs bind irreversibly to the ice surface. If binding was reversible, ice would have grown at any position where an IBP molecule desorbed from the surface. However, this basic description does not explain the observation that the measured TH of IBPs is a function of their concentration in solution, typically proportional to the square root of the concentration (Wen and Laursen 1992b). If the irreversibility assumption is dismissed, such proportionality can be supported by assuming that the surface concentration of the IBP on ice is a result of an equilibrium between adsorption and desorption of the IBPs. Several equilibrium models have been suggested, where the density of IBPs on the ice surface is a function of the concentration of the IBPs in solution. The distance between molecules can be calculated from the surface density and is related to the TH by the Gibbs–Thomson equation (Yeh and Feeney 1996); but in some instances, surface density was related to TH without justification (Jorov et al. 2004; Liu and Li 2006).

The irreversibly of IBP binding to ice was tested by a series of experiments using fluorescently labeled IBPs (Celik et al. 2013; Drori et al. 2014b, 2015; Haleva et al. 2016; Meister et al. 2018; Pertaya et al. 2007b; Zepeda et al. 2008). Fluorescence recovery after photobleaching showed that type III AFP molecules are bound to an ice crystal surface and do not exchange with the surrounding protein in solution during 20 h of observation. This finding led to the estimation of the fastest IBP–ice

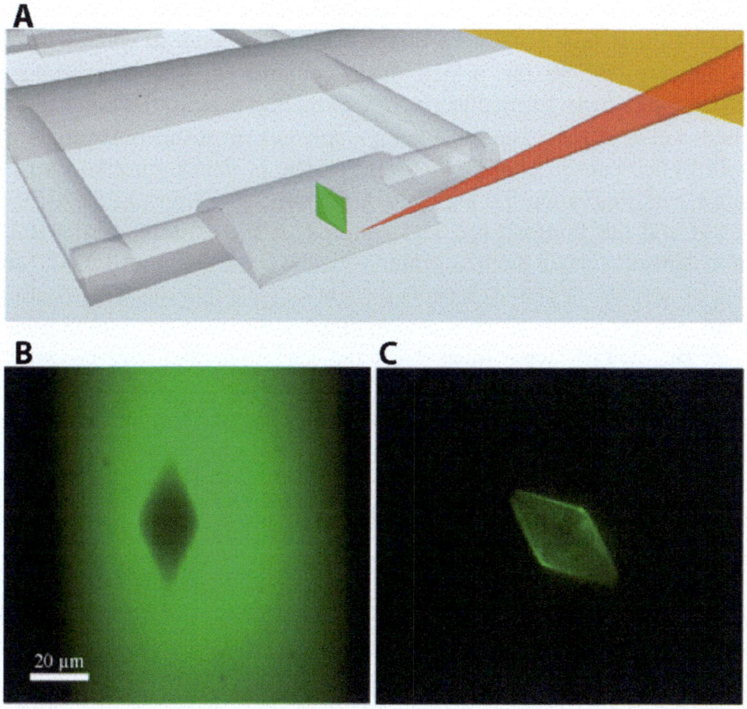

Fig. 4.6 Ice crystal in a microfluidic device. (**a**) An illustration of an ice crystal coated by GFP-tagged IBPs in a microfluidic channel. (**b–c**) Imaging of an ice crystal coated with GFP-labeled type III AFP in the microfluidic channel. The green color is the fluorescent signal of the GFP-labeled protein. (**b**) The ice crystal is held in a solution containing protein. The crystal is darker than the solution because it does not contain protein inside it. (**c**) The same crystal in **b** after washing the protein solution from the microfluidic channel. The labeled protein remains bound to the ice surface. Reprinted (adapted) with permission from (Drori et al. 2015) Copyright (2015) American Chemical Society

off-rate being a week^{-1} and the slowest being infinity—essentially quasi-permanent binding (Pertaya et al. 2007b). Experiments in microfluidic devices that allow the exchange of the solution around ice crystals provided further support for irreversible binding. Ice crystals coated with IBPs were washed such that the solution around the crystals was replaced with solution containing almost no protein. It was shown that the fluorescence signal from the ice surfaces was not reduced after the washing (Fig. 4.6) (Celik et al. 2013; Drori et al. 2014b, 2015). Further still, freezing hysteresis by hyperactive AFP (*Tm*AFP) was maintained even when the only proteins in the system were those on the ice surface and not in the solution.

In two situations where fluorescently labeled IBPs were allowed to accumulate on an ice crystal surface, after which ice growth was forced on the system, the fluorescence signal was lost or reduced. With a mixture of AFGPs 4 to 6 (70% being AFGP 6) that were labeled at the N terminus with fluorescein isothiocyanate, the fluorescence on the ice surface vanished when a layer of ice overgrew the area that was exposed to the AFPs (Zepeda et al. 2008). However, with hyperactive IBPs, although the fluorescence of their GFP tags was greatly reduced when a layer of ice overgrew the area where the protein had accumulated, the fluorescence signal returned when the newly formed ice layer was melted back to the point of accumulation (Haleva et al. 2016). The most likely explanation for this phenomenon is that overgrowth by ice caused enough distorting of the protein fluorophore to spoil its fluorescence, and that when the ice melted, the structural stress in the GFP moiety was relieved and the fluorescence was restored. It seems unlikely that this same explanation can account for the loss of fluorescence when the tagged AFGP area was overgrown by ice. While this experiment might suggest that the AFGPs are reversible binders, it might also indicate that under the condition of ice growth used in that experiment, the AFGPs were pushed off the ice rather than overgrown. On the contrary, other experiments with AFGPs support their irreversible binding to ice (Meister et al. 2018). Overall, these results offer additional support for the irreversible binding of hyperactive AFPs and hint at irreversible binding of other IBPs as well.

Irreversible binding of IBPs to ice is the basis for ice affinity purification (Adar et al. 2018; Garnham et al. 2010; Kuiper et al. 2003; Marshall et al. 2004b, 2016), ice etching (Knight et al. 1991, 2001), and the FIPA modification of ice etching (Basu et al. 2014; Garnham et al. 2010). In these methods ice is grown slowly in solutions of IBPs, around chilled cold fingers, or on a shell of ice formed around a round-bottomed flask (Marshall et al. 2016), or on vertical cold plate on which the solution is flowing (Adar et al. 2018). The slowly growing ice front rejects all solutes except IBPs that are incorporated into it. Thus, the binding of IBPs to ice allows them to resist rejection by the growing ice layers. Overall, the experimental results indicate that most IBPs, at least the moderate and hyperactive ones, irreversibly bind to ice surfaces.

To reconcile the concept of irreversible ice binding with the dependence of TH on protein solution concentration, Kristiansen and Zachariassen suggested a two-step binding model. In the first step, the surface densities of the proteins equilibrate near the melting point at the ice–water interface. Upon cooling, the IBPs are "locked" to their position on the ice surface (Kristiansen and Zachariassen 2005). However, this explanation does not have experimental support. For example, it was shown that in addition to ice growth arrest, IBPs also inhibit ice melting (Celik et al. 2010; Cziko et al. 2014; Knight and DeVries 1989). Therefore, supercooling is not necessary for "locking" IBP molecules to ice surfaces, and the irreversibility model is relevant at temperatures close to the melting point. Moreover, fluorescence measurements showed that IBPs accumulate on the ice surface at temperatures lower than the melting point (Drori et al. 2014a, 2015; Haleva et al. 2016; Takamichi et al. 2007; Zepeda et al. 2008). Recent concepts suggest the coexistence of irreversible binding and the measured dependence of TH on protein solution concentration are related to the kinetics of IBP binding to ice surfaces, as discussed in the next session on the binding kinetics of IBPs to ice.

4.12 The Dynamics of Binding

If TH stems from a surface phenomenon, one should expect that the dynamics of the adsorption process would have a crucial role (Burcham et al. 1986; Kubota 2011). Experiments to detect accumulation of AFGPs on ice surfaces were first attempted by ellipsometry and yielded evidence that there is a time scale of minutes for accumulation of AFGPs to ice surfaces (Wilson 1993). Chapsky and Rubinsky used a unidirectional capillary-based TH measurement of type I AFP to evaluate the dynamic nature of the TH activity. In these experiments, ice propagation in capillaries was monitored in a controlled-temperature gradient. The authors found that in the presence of the AFPs, ice stops growing at a temperature just below the melting point, and resumes growing at lower temperatures. This additional supercooling needed for growth was time dependent. The TH increased over time up to fivefold when the ice was held at sub-melting temperatures for an hour without growth. The authors concluded that the observed time dependence is too slow to be limited by the binding kinetics, and the increase in TH was a result of rearrangement of the surface of the ice and the proteins that bind to it (Chapsky and Rubinsky 1997). The long incubation time of ice crystals in solutions of type III AFP in a nanoliter osmometer was found to increase the thermal hysteresis up to 2.5-fold over a period of 2 h if the crystal was held at high supercooling (close to the freezing point) (Takamichi et al. 2007). Fig. 4.7a represents the course of such an experiment using a nanoliter osmometer. A much stronger time dependence was found for hyperactive AFPs, as shown in Fig. 4.7b (Braslavsky and Drori 2013; Drori et al. 2014a; Xiao et al. 2014). When the crystal was allowed to incubate just below the melting point for ~ 16 h in a solution of a hyperactive AFP, TH increased 40-fold over the value obtained after just a few seconds incubation (Drori et al. 2014a) (Fig. 4.7b). On the contrary, moderate AFPs such as type III AFP achieved most of their full activity even at short exposure times of a few seconds (Drori et al. 2014a; Takamichi et al. 2007). The difference in the dynamic response between hyperactive and moderate AFPs was shown also in sonocrystallization experiments, in which a 1-ml IBP solution was cooled to several degrees below the melting point before it was nucleated by an acoustic pulse. After nucleation, the solution stabilized at a temperature lower than the melting point, and the difference from the melting point was determined as the TH. It was shown that the TH values of type III AFP were similar when measured by a nanoliter osmometer and sonocrystallization, but hyperactive AFP had a very small TH values in the sonocrystallization assay compared to the nanoliter osmometer (Olijve et al. 2016).

Measurements of the fluorescence signal from labeled IBPs on ice crystals allow determination of their accumulation rates as well as identification of the planes on which IBPs gather (Celik et al. 2010; Drori et al. 2014a, b, 2015; Haleva et al. 2016; Kaleda et al. 2019; Pertaya et al. 2008). It was shown that hyperactive AFPs such as *Tm*AFP, sbwAFP, *Mp*IBP-RIV, and *Ri*AFP, accumulate on the basal plane of ice crystals, in addition to other planes. For some of the hyperactive AFPs, the accumulation continues for hours, with no clear end point (Drori et al. 2014a, b; Haleva

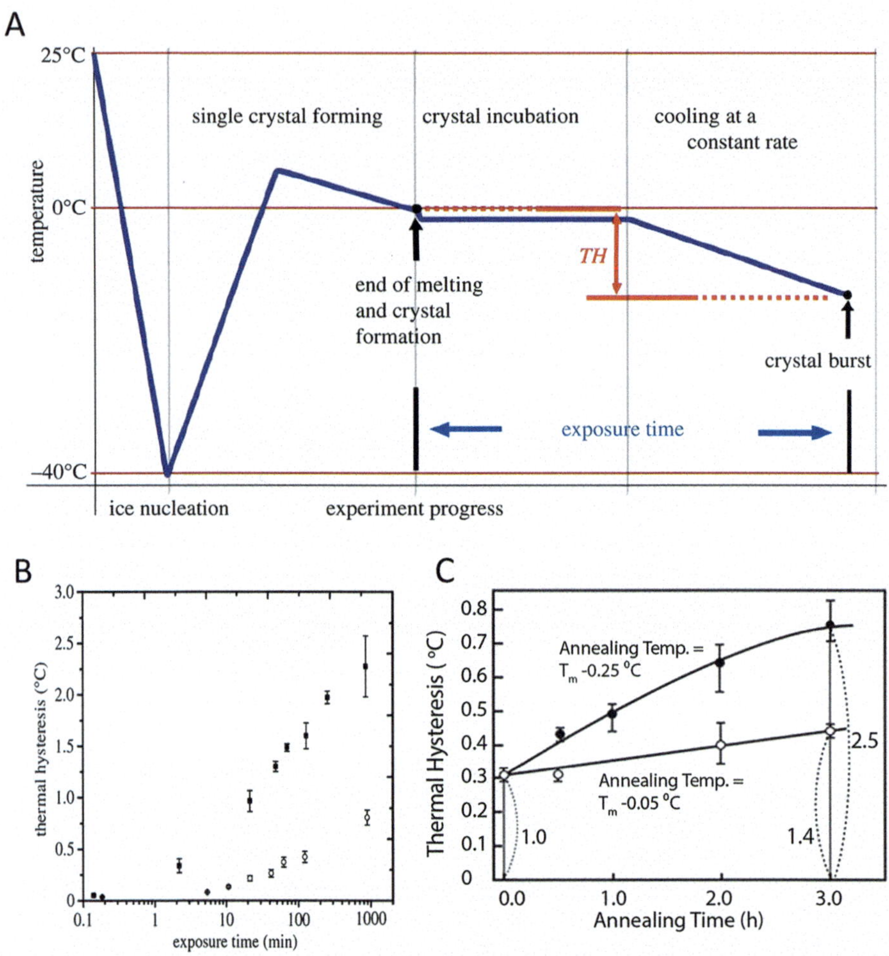

Fig. 4.7 Time dependence of TH activity. (**a**) Representation of the experimental procedure conducted using a nanoliter osmometer. (**b**) TH as a function of the exposure time for a solution of 4 μM (open circles) and 8 μM (filled squares) sbwAFP annealed at −0.05 °C below the melting point (Modified from (Drori et al. 2014a)). (**c**) TH as a function of the exposure time for a solution of 100 μM of type III AFP (nfeAFP) annealed at −0.25 °C (filled circles) or −0.05 °C (open circles) below the melting point (reprinted from (Takamichi et al. 2007). Copyright (2007) WILLEY)

et al. 2016; Pertaya et al. 2008). In order to fit the fluorescence intensity as a function of a time scale that spans from few seconds to an hour, three exponents were necessary (Drori et al. 2014a). In contrast, the accumulation of moderate AFPs on non-basal planes fits a single exponent, with an on-rate constant of $K_{\text{on}} = 0.008\ \mu\text{M}^{-1}\ \text{S}^{-1}$. Using this rate constant, we can calculate for a solution with an AFP concentration of $C = 50\ \mu\text{M}$, for example, a typical time for accumulation of the proteins as $\tau = \frac{1}{CK_{\text{on}}} = 2.5\ s$. Thus the observed weak dependence of TH of type III AFP on much longer times of exposure is probably not related to the rate of accumulation of the proteins on the ice surface (Drori et al. 2014a).

A theory developed by Knight and DeVries based on experiments on ice growth in the presence of AFGPs gives a good explanation for the dependence of TH on the concentration of type III AFPs (Knight and DeVries 2009). According to this theory, the TH limit is determined by the ability of AFP to kinetically block new ice that grows on the unprotected face of the ice crystal—typically the basal plane for moderately active AFPs. The AFPs bind irreversibly to ice prism and pyramidal planes, but additional AFPs are needed in solution in order to inhibit further growth that may emerge from the basal plane. In accordance with this theory, experiments by Drori et al. using type III AFP in microfluidic devices showed that when the solution around protein-bound ice crystals was exchanged with a solution containing only traces of protein, the originally bound proteins stayed on the ice (Fig. 4.6), but the TH was somewhat reduced (Drori et al. 2015).

If the observed solution concentration dependence stems from IBP adsorption rate and not from the density of accumulated protein on the ice surface, the observed square root dependence of the solution concentration on TH remains an open question. Sander and Tkachenko developed a kinetic pinning theory for the inhibition of ice growth by IBPs. They assumed that the binding is irreversible, with an option for the proteins to be engulfed by the growing ice if the angle of contact between the ice and the protein becomes too large. Under these assumptions, they showed that for a certain protein concentration and supercooling, the velocity of the ice growth comes to a halt. They obtained a square root relation between the TH and the protein concentration (Sander and Tkachenko 2004).

The combination of the theories by Sander and Knight gives a plausible explanation for the experimental findings for the moderate IBPs that do not bind to ice basal planes. However, there is clear experimental evidence that hyperactive AFPs bind to basal planes and inhibit their growth, as discussed in Sect. 4.4. It was also shown that the TH of hyperactive AFPs was not diminished when there was no protein in solution (Celik et al. 2013), as opposed to type III AFP (Drori et al. 2015). The observation that the TH increases with longer exposure times of the ice to the proteins (Drori et al. 2014a), the insensitivity of the TH to the protein concentration in solution (Celik et al. 2013), as well as the long accumulation times of hyperactive AFPs (Drori et al. 2014a; Haleva et al. 2016), led to the conclusion that there is a direct connection between the surface protein concentration and the measured TH in the case of AFPs that bind the basal planes in addition to prism planes. It is interesting to note that while TH as a function of concentration of a variety of moderate AFPs can be well matched with the kinetic pinning model, there is not a good match for hyperactive AFPs (Chasnitsky and Braslavsky 2019; Kozuch et al. 2018). Estimation of the surface density of the proteins and the TH as a function of time is in agreement with the basic inverse relation between TH and molecules distance as in the Gibbs–Thomson equation. Still, TH measurements deviate from the TH calculated using this equation. Drori et al. measured a distance of a few nanometers between bound protein molecules on an ice surface (Drori et al. 2014b). The calculated TH for such surface concentration is much higher. This disagreement was noted before and was speculated to be related to the angle of contact between the proteins and the ice (Acker et al. 2001; Drori et al. 2014b; Higgins and Karlsson 2013; Karlsson et al. 2019; Mazur 1965). Another aspect in evaluating the relationship between molecule distance separation and TH is the arbitrary assumption that the proteins are evenly distributed on the ice surface. This assumption can be modified to a random distribution (Hansen-Goos et al. 2014).

While the advance in the understanding of the activity of IBPs is significant, many aspects remain to be explained. For example, what is the microscopic arrangement of IBPs on a particular ice plane? Why do hyperactive AFPs continue to accumulate on the ice surface for such a long time? What determines the recrystallization inhibition limits? What is the correlation between ice recrystallization inhibition and TH? To answer these and further questions, additional experimental approaches with higher resolution should be implemented, along with elaborate simulations of ice, water, and IBPs.

4.13 Conclusions

IBPs from many biological kingdoms have been characterized. New IBPs with novel structures await discovery, and the range of their specific activities and natural functions may yet increase. Clearly, we have not yet plumbed the complexity of this protein class. Many theories regarding the ice-binding mechanism of various types of IBPs have been advanced and ruled out by experiments. In this chapter, we discussed the advances in the understanding of IBP structures and their mechanism of ice recognition, both for specific IBP types and more generally for all IBPs. The most plausible mechanism that explains the way IBPs interact with ice is the "anchored clathrate water" model. Despite the dramatic differences between IBP types, this theory can apply to all IBPs. Nevertheless, the way each IBP holds water molecules on its ice-binding face, and the number of water molecules in the clathrate can be type specific.

Acknowledgments This work was supported by grants from the Canadian Institutes of Health Research (P.L.D.) and the Israel Science Foundation (I.B.). P.L.D. holds the Canada Research Chair in Protein Engineering.

References

Achenbach JC, Ewart KV (2002) Structural and functional characterization of a C-type lectin-like antifreeze protein from rainbow smelt (*Osmerus mordax*). Eur J Biochem 269:1219–1226

Acker JP, Elliott JAW, McGann LE (2001) Intercellular ice propagation: experimental evidence for ice growth through membrane pores. Biophys J 81:1389–1397

Adar C, Sirotinskaya V, Bar Dolev M, Friehmann T, Braslavsky I (2018) Falling water ice affinity purification of ice-binding proteins. Sci Rep 8:11046

Antson AA, Smith DJ, Roper DI, Lewis S, Caves LSD, Verma CS, Buckley SL, Lillford PJ, Hubbard RE (2001) Understanding the mechanism of ice binding by type III antifreeze proteins. J Mol Biol 305:875–889
Baardsnes J, Davies PL (2002) Contribution of hydrophobic residues to ice binding by fish type III antifreeze protein. Biochim Biophys Acta 1601:49–54
Baardsnes J, Kondejewski LH, Hodges RS, Chao H, Kay C, Davies PL (1999) New ice-binding face for type I antifreeze protein. FEBS Lett 463:87–91
Baardsnes J, Kuiper MJ, Davies PL (2003) Antifreeze protein dimer: when two ice-binding faces are better than one. J Biol Chem 278:38942–38947
Bar Dolev M, Celik Y, Wettlaufer JS, Davies PL, Braslavsky I (2012) New insights into ice growth and melting modifications by antifreeze proteins. J R Soc Interface 9:3249–3259
Bar Dolev M, Bernheim R, Guo SQ, Davies PL, Braslavsky I (2016a) Putting life on ice: bacteria that bind to frozen water. J R Soc Interface 13:20160210
Bar Dolev M, Braslavsky I, Davies PL (2016b) Ice-binding proteins and their function. Annu Rev Biochem 85:515–542
Bar M, Celik Y, Fass D, Braslavsky I (2008a) Interactions of beta-helical antifreeze protein mutants with ice. Cryst Growth Des 8:2954–2963
Bar M, Scherf T, Fass D (2008b) Two-dimensional surface display of functional groups on a beta-helical antifreeze protein scaffold. Protein Eng Des Sel 21:107–114
Basu K, Garnham CP, Nishimiya Y, Tsuda S, Braslavsky I, Davies P (2014) Determining the ice-binding planes of antifreeze proteins by fluorescence-based Ice plane affinity. J Vis Exp 83: e51185
Basu K, Graham LA, Campbell RL, Davies PL (2015) Flies expand the repertoire of protein structures that bind ice. Proc Natl Acad Sci USA 112:737–742
Basu K, Wasserman SS, Jeronimo PS, Graham LA, Davies PL (2016) Intermediate activity of midge antifreeze protein is due to a tyrosine-rich ice-binding site and atypical ice plane affinity. FEBS J 283:1504–1515
Berger T, Meister K, DeVries AL, Eves R, Davies PL, Drori R (2019) Synergy between antifreeze proteins is driven by complementary ice-binding. J Am Chem Soc 141:19144–19150
Bissoyi A, Reicher N, Chasnitsky M, Arad S, Koop T, Rudich Y, Braslavsky I (2019) Ice nucleation properties of ice-binding proteins from snow fleas. Biomol Ther 9:532
Braslavsky I, Drori R (2013) LabVIEW-operated novel nanoliter osmometer for ice binding protein investigations. J Vis Exp 72:e4189
Burcham TS, Osuga DT, Yeh Y, Feeney RE (1986) A kinetic description of antifreeze glycoprotein activity. J Biol Chem 261:6390–6397
Can O, Holland NB (2011) Conjugation of type I antifreeze protein to polyallylamine increases thermal hysteresis activity. Bioconjug Chem 22:2166–2171
Celik Y, Graham LA, Mok YF, Bar M, Davies PL, Braslavsky I (2010) Superheating of ice crystals in antifreeze protein solutions. Proc Natl Acad Sci USA 107:5423–5428
Celik Y, Drori R, Pertaya-Braun N, Altan A, Barton T, Bar-Dolev M, Groisman A, Davies PL, Braslavsky I (2013) Microfluidic experiments reveal that antifreeze proteins bound to ice crystals suffice to prevent their growth. Proc Natl Acad Sci USA 110:1309–1314
Chao H, Davies P, Sonnichsen F, Sykes B (1993) Solution structure of a novel, non-helical type-iii antifreeze protein via proline replacement. Protein Eng 6:31–31
Chao H, Sonnichsen FD, DeLuca CI, Sykes BD, Davies PL (1994) Structure-function relationship in the globular type III antifreeze protein: identification of a cluster of surface residues required for binding to ice. Protein Sci 3:1760–1769
Chao H, Hodges RS, Kay CM, Gauthier SY, Davies PL (1996) A natural variant of type I antifreeze protein with four ice-binding repeats is a particularly potent antifreeze. Protein Sci 5:1150–1156
Chao H, Houston ME, Hodges RS, Kay CM, Sykes BD, Loewen MC, Davies PL, Sönnichsen FD (1997) A diminished role for hydrogen bonds in antifreeze protein binding to Ice. Biochemistry 36:14652–14660

Chapsky L, Rubinsky B (1997) Kinetics of antifreeze protein-induced ice growth inhibition. FEBS Lett 412:241–244

Chasnitsky M, Braslavsky I (2019) Ice-binding proteins and the applicability and limitations of the kinetic pinning model. Philos Trans R Soc A Math Phys Eng Sci 377:20180391

Chen L, DeVries AL, Cheng CH (1997) Convergent evolution of antifreeze glycoproteins in Antarctic notothenioid fish and Arctic cod. Proc Natl Acad Sci USA 94:3817–3822

Cheng CHC, DeVries AL (1989) Structures of antifreeze peptides from the antarctic eel pout, Austrolycicthys-Brachycephalus. Biochim Biophys Acta 997:55–64

Chou K-C (1992) Energy-optimized structure of antifreeze protein and its binding mechanism. J Mol Biol 223:509–517

Cziko PA, DeVries AL, Evans CW, Cheng C-HC (2014) Antifreeze protein-induced superheating of ice inside Antarctic notothenioid fishes inhibits melting during summer warming. Proc Natl Acad Sci USA 111:14583–14588

Davies PL (2014) Ice-binding proteins: a remarkable diversity of structures for stopping and starting ice growth. Trends Biochem Sci 39:548–555

Davies PL, Baardsnes J, Kuiper MJ, Walker VK (2002) Structure and function of antifreeze proteins. Philos Trans R Soc Lond Ser B Biol Sci 357:927–935

DeVries AL, Price TJ (1984) Role of glycopeptides and peptides in inhibition of crystallization of water in polar fishes [and discussion]. Philos Trans R Soc Lond B Biol Sci 304:575–588

DeLuca CI, Chao H, Sonnichsen FD, Sykes BD, Davies PL (1996) Effect of type III antifreeze protein dilution and mutation on the growth inhibition of ice. Biophys J 71:2346–2355

DeLuca CI, Comley R, Davies PL (1998) Antifreeze proteins bind independently to ice. Biophys J 74:1502–1508

DeVries AL, Lin Y (1977) Structure of a peptide antifreeze and mechanism of adsorption to ice. Biochim Biophys Acta 495:388–392

DeVries AL, Wohlschlag DE (1969) Freezing resistance in some Antarctic fishes. Science 163:1073–1075

Doucet D, Tyshenko MG, Kuiper MJ, Graether SP, Sykes BD, Daugulis AJ, Davies PL, Walker VK (2000) Structure–function relationships in spruce budworm antifreeze protein revealed by isoform diversity. Eur J Biochem 267:6082–6088

Drori R, Celik Y, Davies PL, Braslavsky I (2014a) Ice-binding proteins that accumulate on different ice crystal planes produce distinct thermal hysteresis dynamics. J R Soc Interface 11:20140526

Drori R, Davies PL, Braslavsky I (2014b) Experimental correlation between thermal hysteresis activity and the distance between antifreeze proteins on an ice surface. RSC Adv 5:7848

Drori R, Davies PL, Braslavsky I (2015) When are antifreeze proteins in solution essential for ice growth inhibition? Langmuir 31:5805–5811

Duman JG (2001) Antifreeze and ice nucleator proteins in terrestrial arthropods. Annu Rev Physiol 63:327–357

Duman JG, DeVries AL (1976) Isolation, characterization, and physical properties of protein antifreezes from the winter flounder, *Pseudopleuronectes americanus*. Comp Biochem Physiol B 54:375–380

Duman JG, Li N, Verleye D, Goetz FW, Wu DW, Andorfer CA, Benjamin T, Parmelee DC (1998) Molecular characterization and sequencing of antifreeze proteins from larvae of the beetle *Dendroides canadensis*. J Comp Physiol B Biochem Syst Environ Physiol 168:225–232

Ebbinghaus S, Meister K, Born B, DeVries AL, Gruebele M, Havenith M (2010) Antifreeze glycoprotein activity correlates with long-range protein—water dynamics. J Am Chem Soc 132:12210–12211

Ebbinghaus S, Meister K, Prigozhin MB, DeVries AL, Havenith M, Dzubiella J, Gruebele M (2012) Functional importance of short-range binding and long-range solvent interactions in helical antifreeze peptides. Biophys J 103:L20–L22

Eickhoff L, Dreischmeier K, Zipori A, Sirotinskaya V, Adar C, Reicher N, Braslavsky I, Rudich Y, Koop T (2019) Contrasting behavior of antifreeze proteins: ice growth inhibitors and ice nucleation promoters. J Phys Chem Lett 10:966–972

Evans RP, Fletcher GL (2001) Isolation and characterization of type I antifreeze proteins from Atlantic snailfish (*Liparis atlanticus*) and dusky snailfish (*Liparis gibbus*). Biochim Biophys Acta 1547:235–244

Ewart KV, Fletcher GL (1990) Isolation and characterization of antifreeze proteins from smelt (Osmerus-Mordax) and atlantic herring (Clupea-Harengus-Harengus). Can J Zool 68:1652–1658

Feeney RE, Burcham TS, Yeh Y (1986) Antifreeze glycoproteins from polar fish blood. Annu Rev Biophys Biophys Chem 15:59–78

Gallagher KR, Sharp KA (2003) Analysis of thermal hysteresis protein hydration using the random network model. Biophys Chem 105:195–209

Garnham CP, Gilbert JA, Hartman CP, Campbell RL, Laybourn-Parry J, Davies PL (2008) A Ca2 +−dependent bacterial antifreeze protein domain has a novel beta-helical ice-binding fold. Biochem J 411:171–180

Garnham CP, Natarajan A, Middleton AJ, Kuiper MJ, Braslavsky I, Davies PL (2010) Compound ice-binding site of an antifreeze protein revealed by mutagenesis and fluorescent tagging. Biochemistry 49:9063–9071

Garnham CP, Campbell RL, Davies PL (2011a) Anchored clathrate waters bind antifreeze proteins to ice. Proc Natl Acad Sci USA 108:7363–7367

Garnham CP, Campbell RL, Walker VK, Davies PL (2011b) Novel dimeric beta-helical model of an ice nucleation protein with bridged active sites. BMC Struct Biol 11:36

Garnham CP, Nishimiya Y, Tsuda S, Davies PL (2012) Engineering a naturally inactive isoform of type III antifreeze protein into one that can stop the growth of ice. FEBS Lett 586:3876–3881

Gauthier SY, Scotter AJ, Lin FH, Baardsnes J, Fletcher GL, Davies PL (2008) A re-evaluation of the role of type IV antifreeze protein. Cryobiology 57:292–296

Govindarajan AG, Lindow SE (1988) Size of bacterial ice-nucleation sites measured in situ by radiation inactivation analysis. Proc Natl Acad Sci USA 85:1334–1338

Grabowska J, Kuffel A, Zielkiewicz J (2016) Structure of solvation water around the active and inactive regions of a type III antifreeze protein and its mutants of lowered activity. J Chem Phys 145:075101

Graether SP, Jia Z (2001) Modeling *Pseudomonas syringae* ice-nucleation protein as a beta-helical protein. Biophys J 80:1169–1173

Graether SP, Kuiper MJ, Gagne SM, Walker VK, Jia Z, Sykes BD, Davies PL (2000) Beta-helix structure and ice-binding properties of a hyperactive antifreeze protein from an insect. Nature 406:325–328

Graham LA, Davies PL (2005) Glycine-rich antifreeze proteins from snow fleas. Science 310:461

Graham LA, Liou YC, Walker VK, Davies PL (1997) Hyperactive antifreeze protein from beetles. Nature 388:727–728

Graham LA, Hobbs RS, Fletcher GL, Davies PL (2013) Helical antifreeze proteins have independently evolved in fishes on four occasions. PLoS One 8:e81285

Green RL, Corotto LV, Warren GJ (1988) Deletion mutagenesis of the ice nucleation gene from *Pseudomonas syringae* S203. Mol Gen Genet 215:165–172

Guo S, Garnham CP, Whitney JC, Graham LA, Davies PL (2012) Re-evaluation of a bacterial antifreeze protein as an adhesin with ice-binding activity. PLoS One 7:e48805

Guo S, Stevens CA, Vance TDR, Olijve LLC, Graham LA, Campbell RL, Yazdi SR, Escobedo C, Bar-Dolev M, Yashunsky V et al (2017) Structure of a 1.5-MDa adhesin that binds its Antarctic bacterium to diatoms and ice. Sci Adv 3:e1701440

Guriansherman D, Lindow SE (1993) Bacterial ice nucleation – significance and molecular-basis. FASEB J 7:1338–1343

Hakim A, Nguyen JB, Basu K, Zhu DF, Thakral D, Davies PL, Isaacs FJ, Modis Y, Meng W (2013) Crystal structure of an insect antifreeze protein and its implications for ice binding. J Biol Chem 288(17):12295–12304

Haleva L, Celik Y, Bar-Dolev M, Pertaya-Braun N, Kaner A, Davies PL, Braslaysky I (2016) Microfluidic cold-finger device for the investigation of ice-binding proteins. Biophys J 111:1143–1150

Hansen-Goos H, Thomson ES, Wettlaufer JS (2014) On the edge of habitability and the extremes of liquidity. Planet Space Sci 98:169–181

Haymet ADJ, Ward LG, Harding MM, Knight CA (1998) Valine substituted winter flounder 'antifreeze': preservation of ice growth hysteresis. FEBS Lett 430:301–306

Haymet ADJ, Ward LG, Harding MM (1999) Winter flounder "antifreeze" proteins: synthesis and ice growth inhibition of analogues that probe the relative importance of hydrophobic and hydrogen-bonding interactions. J Am Chem Soc 121:941–948

Hayward JA, Haymet ADJ (2001) The ice/water interface: molecular dynamics simulations of the basal, prism, {20(2)over-bar1}, and {2(11)over-bar0} interfaces of ice Ih. J Chem Phys 114:3713–3726

Hew CL, Joshi S, Wang N-C, Kao M-H, Ananthanarayanan VS (1985) Structures of shorthorn sculpin antifreeze polypeptides. Eur J Biochem 151:167–172

Hew CL, Wang NC, Joshi S, Fletcher GL, Scott GK, Hayes PH, Buettner B, Davies PL (1988) Multiple genes provide the basis for antifreeze protein diversity and dosage in the ocean pout, *Macrozoarces americanus*. J Biol Chem 263:12049–12055

Hew CL, Davies PL, Fletcher G (1992) Antifreeze protein gene transfer in Atlantic salmon. Mol Mar Biol Biotechnol 1:309–317

Higgins AZ, Karlsson JOM (2013) Effects of intercellular junction protein expression on intracellular ice formation in mouse insulinoma cells. Biophys J 105:2006–2015

Hobbs RS, Shears MA, Graham LA, Davies PL, Fletcher GL (2011) Isolation and characterization of type I antifreeze proteins from cunner, *Tautogolabrus adspersus*, order Perciformes. FEBS J 278:3699–3710

Holland NB, Nishimiya Y, Tsuda S, Sönnichsen FD (2008) Two domains of RD3 antifreeze protein diffuse independently. Biochemistry 47:5935–5941

Howard EI, Blakeley MP, Haertlein M, Haertlein IP, Mitschler A, Fisher SJ, Siah AC, Salvay AG, Popov A, Dieckmann CM et al (2011) Neutron structure of type-III antifreeze protein allows the reconstruction of AFP–ice interface. J Mol Recognit 24:724–732

Hsiao KC, Cheng CH, Fernandes IE, Detrich HW, DeVries AL (1990) An antifreeze glycopeptide gene from the Antarctic cod *Notothenia coriiceps* neglecta encodes a polyprotein of high peptide copy number. Proc Natl Acad Sci USA 87:9265–9269

Hudait A, Moberg DR, Qiu Y, Odendahl N, Paesani F, Molinero V (2018) Preordering of water is not needed for ice recognition by hyperactive antifreeze proteins. Proc Natl Acad Sci USA 115:8266–8271

Jia Z, Davies PL (2002) Antifreeze proteins: an unusual receptor-ligand interaction. Trends Biochem Sci 27:101–106

Jia Z, DeLuca CI, Chao H, Davies PL (1996) Structural basis for the binding of a globular antifreeze protein to ice. Nature 384:285–288

Jorov A, Zhorov BS, Yang DSC (2004) Theoretical study of interaction of winter flounder antifreeze protein with ice. Protein Sci 13:1524–1537

Kaleda A, Haleva L, Sarusi G, Pinsky T, Mangiagalli M, Bar Dolev M, Lotti M, Nardini M, Braslavsky I (2019) Saturn-shaped ice burst pattern and fast basal binding of an ice-binding protein from an Antarctic bacterial consortium. Langmuir 35:7337–7346

Karlsson JOM, Braslavsky I, Elliott JAW (2019) Protein-water-ice contact angle. Langmuir 35:7383–7387

Kawahara H (2002) The structures and functions of ice crystal-controlling proteins from bacteria. J Biosci Bioeng 94:492–496

Kawahara H (2013) Characterizations of functions of biological materials having controlling-ability against ice crystal growth. In: Sukarno F (ed) Advanced topics on crystal growth. InTech, Rijeka, pp 119–143

Kieft TL (1988) Ice nucleation activity in lichens. Appl Environ Microbiol 54:1678–1681

King NP, Bale JB, Sheffler W, McNamara DE, Gonen S, Gonen T, Yeates TO, Baker D (2014) Accurate design of co-assembling multi-component protein nanomaterials. Nature 510:103–108
Knight CA, DeVries AL (1989) Melting inhibition and superheating of ice by an antifreeze glycopeptide. Science 245:505–507
Knight CA, DeVries AL (2009) Ice growth in supercooled solutions of a biological "antifreeze", AFGP 1-5: an explanation in terms of adsorption rate for the concentration dependence of the freezing point. Phys Chem Chem Phys 11:5749–5761
Knight CA, Wierzbicki A (2001) Adsorption of biomolecules to ice and their effects upon ice growth. 2. A discussion of the basic mechanism of "antifreeze" phenomena. Cryst Growth Des 1:439–446
Knight CA, Cheng CC, DeVries AL (1991) Adsorption of alpha-helical antifreeze peptides on specific ice crystal-surface planes. Biophys J 59:409–418
Knight CA, Driggers E, DeVries AL (1993) Adsorption to ice of fish antifreeze glycopeptide-7 and glycopeptide-8. Biophys J 64:252–259
Knight CA, Wierzbicki A, Laursen RA, Zhang W (2001) Adsorption of biomolecules to ice and their effects upon ice growth. 1. Measuring adsorption orientations and initial results. Cryst Growth Des 1:429–438
Kobashigawa Y, Nishimiya Y, Miura K, Ohgiya S, Miura A, Tsuda S (2005) A part of ice nucleation protein exhibits the ice-binding ability. FEBS Lett 579:1493–1497
Kondo H, Hanada Y, Sugimoto H, Hoshino T, Garnham CP, Davies PL, Tsuda S (2012) Ice-binding site of snow mold fungus antifreeze protein deviates from structural regularity and high conservation. Proc Natl Acad Sci USA 109:9360–9365
Kozuch DJ, Stillinger FH, Debenedetti PG (2018) Combined molecular dynamics and neural network method for predicting protein antifreeze activity. Proc Natl Acad Sci USA 115:13252–13257
Kristiansen E, Zachariassen KE (2005) The mechanism by which fish antifreeze proteins cause thermal hysteresis. Cryobiology 51:262–280
Kristiansen E, Ramløv H, Højrup P, Pedersen SA, Hagen L, Zachariassen KE (2011) Structural characteristics of a novel antifreeze protein from the longhorn beetle *Rhagium inquisitor*. Insect Biochem Mol Biol 41:109–117
Kristiansen E, Wilkens C, Vincents B, Friis D, Lorentzen AB, Jenssen H, Lobner-Olesen A, Ramlov H (2012) Hyperactive antifreeze proteins from longhorn beetles: some structural insights. J Insect Physiol 58:1502–1510
Kubota N (2011) Effects of cooling rate, annealing time and biological antifreeze concentration on thermal hysteresis reading. Cryobiology 63:198–209
Kuiper MJ, Lankin C, Gauthier SY, Walker VK, Davies PL (2003) Purification of antifreeze proteins by adsorption to ice. Biochem Biophys Res Commun 300:645–648
Kumble KD, Demmer J, Fish S, Hall C, Corrales S, DeAth A, Elton C, Prestidge R, Luxmanan S, Marshall CJ et al (2008) Characterization of a family of ice-active proteins from the Ryegrass, *Lolium perenne*. Cryobiology 57:263–268
Lee JH, Park AK, Do H, Park KS, Moh SH, Chi YM, Kim HJ (2012) Structural basis for antifreeze activity of ice-binding protein from arctic yeast. J Biol Chem 287:11460–11468
Leinala EK, Davies PL, Doucet D, Tyshenko MG, Walker VK, Jia Z (2002) A beta-helical antifreeze protein isoform with increased activity – structural and functional insights. J Biol Chem 277:33349–33352
Li XM, Trinh KY, Hew CL, Buettner B, Baenziger J, Davies PL (1985) Structure of an antifreeze polypeptide and its precursor from the ocean pout, Macrozoarces-Americanus. J Biol Chem 260:2904–2909
Limmer DT (2016) Closer look at the surface of ice. Proc Natl Acad Sci USA 113:12347–12349
Lin FH, Davies PL, Graham LA (2011) The Thr- and Ala-rich hyperactive antifreeze protein from inchworm folds as a flat silk-like beta-helix. Biochemistry 50:4467–4478

Liou Y-C, Thibault P, Walker VK, Davies PL, Graham LA (1999) A complex family of highly heterogeneous and internally repetitive hyperactive antifreeze proteins from the beetle *Tenebrio molitor*. Biochemistry 38:11415–11424

Liou YC, Tocilj A, Davies PL, Jia Z (2000) Mimicry of ice structure by surface hydroxyls and water of a beta-helix antifreeze protein. Nature 406:322–324

Liu JJ, Li QZ (2006) Theoretical model of antifreeze protein-ice adsorption: binding of large ligands to a two-dimensional homogeneous lattice. Chem Phys Lett 422:67–71

Liu Y, Li Z, Lin Q, Kosinski J, Seetharaman J, Bujnicki JM, Sivaraman J, Hew CL (2007) Structure and evolutionary origin of Ca2+−dependent herring type II antifreeze protein. PLoS One 2: e548

Liu JJ, Qin YZ, Bar Dolev M, Celik Y, Wettlaufer JS, Braslavsky I (2012) Modelling the influence of antifreeze proteins on three-dimensional ice crystal melt shapes using a geometric approach. Proc R Soc A Math Phys Eng Sci 468:3311–3322

Liu K, Wang C, Ma J, Shi G, Yao X, Fang H, Song Y, Wang J (2016) Janus effect of antifreeze proteins on ice nucleation. Proc Natl Acad Sci USA 113:14739–14744

Loewen MC, Gronwald W, Sonnichsen FD, Sykes BD, Davies PL (1998) The ice-binding site of sea raven antifreeze protein is distinct from the carbohydrate-binding site of the homologous C-type lectin. Biochemistry 37:17745–17753

Low WK, Lin Q, Stathakis C, Miao M, Fletcher GL, Hew CL (2001) Isolation and characterization of skin-type, type I antifreeze polypeptides from the longhorn sculpin, *Myoxocephalus octodecemspinosus*. J Biol Chem 276:11582–11589

Madura JD, Baran K, Wierzbicki A (2000) Molecular recognition and binding of thermal hysteresis proteins to ice. J Mol Recognit 13:101–113

Mahatabuddin S, Hanada Y, Nishimiya Y, Miura A, Kondo H, Davies PL, Tsuda S (2017) Concentration-dependent oligomerization of an alpha-helical antifreeze polypeptide makes it hyperactive. Sci Rep 7:42501

Mahatabuddin S, Fukami D, Arai T, Nishimiya Y, Shimizu R, Shibazaki C, Kondo H, Adachi M, Tsuda S (2018) Polypentagonal ice-like water networks emerge solely in an activity-improved variant of ice-binding protein. Proc Natl Acad Sci USA 115:5456–5461

Mangiagalli M, Sarusi G, Kaleda A, Bar Dolev M, Nardone V, Vena VF, Braslavsky I, Lotti M, Nardini M (2018) Structure of a bacterial ice binding protein with two faces of interaction with ice. FEBS J 285:1653–1666

Marshall CB, Daley ME, Graham LA, Sykes BD, Davies PL (2002) Identification of the ice-binding face of antifreeze protein from *Tenebrio molitor*. FEBS Lett 529:261–267

Marshall CB, Daley ME, Sykes BD, Davies PL (2004a) Enhancing the activity of a beta-helical antifreeze protein by the engineered addition of coils. Biochemistry 43:11637–11646

Marshall CB, Tomczak MM, Gauthier SY, Kuiper MJ, Lankin C, Walker VK, Davies PL (2004b) Partitioning of fish and insect antifreeze proteins into ice suggests they bind with comparable affinity. Biochemistry 43:148–154

Marshall CJ, Basu K, Davies PL (2016) Ice-shell purification of ice-binding proteins. Cryobiology 72:258–263

Mazur P (1965) Role of cell membranes in freezing of yeast and other single cells. Ann N Y Acad Sci 125:658–676

Meister K, Ebbinghaus S, Xu Y, Duman JG, DeVries A, Gruebele M, Leitner DM, Havenith M (2013) Long-range protein–water dynamics in hyperactive insect antifreeze proteins. Proc Natl Acad Sci USA 110:1617–1622

Meister K, Strazdaite S, DeVries AL, Lotze S, Olijve LLC, Voets IK, Bakker HJ (2014) Observation of ice-like water layers at an aqueous protein surface. Proc Natl Acad Sci USA 111:17732–17736

Meister K, Lotze S, Olijve LLC, DeVries AL, Duman JG, Voets IK, Bakker HJ (2015) Investigation of the ice-binding site of an insect antifreeze protein using sum-frequency generation spectroscopy. J Phys Chem Lett 6:1162–1167

Meister K, DeVries AL, Bakker HJ, Drori R (2018) Antifreeze glycoproteins bind irreversibly to ice. J Am Chem Soc 140:9365–9368
Middleton AJ, Marshall CB, Faucher F, Bar-Dolev M, Braslavsky I, Campbell RL, Walker VK, Davies PL (2012) Antifreeze protein from freeze-tolerant grass has a beta-roll fold with an irregularly structured ice-binding site. J Mol Biol 416:713–724
Miura K, Ohgiya S, Hoshino T, Nemoto N, Suetake T, Miura A, Spyracopoulos L, Kondo H, Tsuda S (2001) NMR analysis of type III antifreeze protein intramolecular dimer. Structural basis for enhanced activity. J Biol Chem 276:1304–1310
Modig K, Qvist J, Marshall CB, Davies PL, Halle B (2010) High water mobility on the ice-binding surface of a hyperactive antifreeze protein. Phys Chem Chem Phys 12:10189–10197
Mok YF, Lin FH, Graham LA, Celik Y, Braslavsky I, Davies PL (2010) Structural basis for the superior activity of the large isoform of snow flea antifreeze protein. Biochemistry 49:2593–2603
Nishimiya Y, Sato R, Takamichi M, Miura A, Tsuda S (2005) Co-operative effect of the isoforms of type III antifreeze protein expressed in Notched-fin eelpout, *Zoarces elongatus* Kner. FEBS J 272:482–492
Nishimiya Y, Kondo H, Takamichi M, Sugimoto H, Suzuki M, Miura A, Tsuda S (2008) Crystal structure and mutational analysis of Ca2+-independent type II antifreeze protein from longsnout poacher, *Brachyopsis rostratus*. J Mol Biol 382:734–746
Nutt DR, Smith JC (2008) Function of the hydration layer around an antifreeze protein revealed by atomistic molecular dynamics simulations. J Am Chem Soc 130:13066–13073
Olijve LLC, Meister K, DeVries AL, Duman JG, Guo S, Bakker HJ, Voets IK (2016) Blocking rapid ice crystal growth through nonbasal plane adsorption of antifreeze proteins. Proc Natl Acad Sci USA 113:3740–3745
Padilla JE, Colovos C, Yeates TO (2001) Nanohedra: using symmetry to design self assembling protein cages, layers, crystals, and filaments. Proc Natl Acad Sci USA 98:2217–2221
Patel SN, Graether SP (2010) Structures and ice-binding faces of the alanine-rich type I antifreeze proteins. Biochem Cell Biol 88:223–229
Pentelute BL, Gates ZP, Tereshko V, Dashnau JL, Vanderkooi JM, Kossiakoff AA, Kent SB (2008) X-ray structure of snow flea antifreeze protein determined by racemic crystallization of synthetic protein enantiomers. J Am Chem Soc 130:9695–9701
Pertaya N, Celik Y, DiPrinzio CL, Wettlaufer JS, Davies PL, Braslavsky I (2007a) Growth-melt asymmetry in ice crystals under the influence of spruce budworm antifreeze protein. J Phys Condens Matter 19:412101
Pertaya N, Marshall CB, DiPrinzio CL, Wilen L, Thomson ES, Wettlaufer JS, Davies PL, Braslavsky I (2007b) Fluorescence microscopy evidence for quasi-permanent attachment of antifreeze proteins to ice surfaces. Biophys J 92:3663–3673
Pertaya N, Marshall CB, Celik Y, Davies PL, Braslavsky I (2008) Direct visualization of spruce budworm antifreeze protein interacting with ice crystals: basal plane affinity confers hyperactivity. Biophys J 95:333–341
Phippen SW, Stevens CA, Vance TDR, King NP, Baker D, Davies PL (2016) Multivalent display of antifreeze proteins by fusion to self-assembling protein cages enhances ice-binding activities. Biochemistry 55:6811–6820
Pummer BG, Budke C, Augustin-Bauditz S, Niedermeier D, Felgitsch L, Kampf CJ, Huber RG, Liedl KR, Loerting T, Moschen T et al (2015) Ice nucleation by water-soluble macromolecules. Atmos Chem Phys 15:4077–4091
Qiu Y, Hudait A, Molinero V (2019) How size and aggregation of ice-binding proteins control their ice nucleation efficiency. J Am Chem Soc 141:7439–7452
Raymond JA, DeVries AL (1977) Adsorption inhibition as a mechanism of freezing resistance in polar fishes. Proc Natl Acad Sci USA 74:2589–2593
Raymond JA, Kim HJ (2012) Possible role of horizontal gene transfer in the colonization of sea ice by algae. PLoS One 7:e35968

Sander LM, Tkachenko AV (2004) Kinetic pinning and biological antifreezes. Phys Rev Lett 93:128102
Scott GK, Hayes PH, Fletcher GL, Davies PL (1988) Wolffish antifreeze protein genes are primarily organized as tandem repeats that each contain two genes in inverted orientation. Mol Cell Biol 8:3670–3675
Scotter AJ, Marshall CB, Graham LA, Gilbert JA, Garnham CP, Davies PL (2006) The basis for hyperactivity of antifreeze proteins. Cryobiology 53:229–239
Sicheri F, Yang DSC (1995) Ice-binding structure and mechanism of an antifreeze protein from Winter Flounder. Nature 375:427–431
Smolin N, Daggett V (2008) Formation of ice-like water structure on the surface of an antifreeze protein. J Phys Chem B 112:6193–6202
Sonnichsen FD, Sykes BD, Chao H, Davies PL (1993) The nonhelical structure of antifreeze protein type-iii. Science 259:1154–1157
Sonnichsen FD, DeLuca CI, Davies PL, Sykes BD (1996) Refined solution structure of type III antifreeze protein: hydrophobic groups may be involved in the energetics of the protein-ice interaction. Structure 4:1325–1337
Stevens CA, Drori R, Zalis S, Braslavsky I, Davies PL (2015) Dendrimer-linked antifreeze proteins have superior activity and thermal recovery. Bioconjug Chem 26:1908–1915
Sun T, Lin FH, Campbell RL, Allingham JS, Davies PL (2014) An antifreeze protein folds with an interior network of more than 400 semi-clathrate waters. Science 343:795–798
Sun TJ, Gauthier SY, Campbell RL, Davies PL (2015) Revealing surface waters on an antifreeze protein by fusion protein crystallography combined with molecular dynamic simulations. J Phys Chem B 119:12808–12815
Tachibana Y, Fletcher GL, Fujitani N, Tsuda S, Monde K, Nishimura SI (2004) Antifreeze glycoproteins: elucidation of the structural motifs that are essential for antifreeze activity. Angew Chem Int Ed 43:856–862
Takamichi M, Nishimiya Y, Miura A, Tsuda S (2007) Effect of annealing time of an ice crystal on the activity of type III antifreeze protein. FEBS J 274:6469–6476
Takamichi M, Nishimiya Y, Miura A, Tsuda S (2009) Fully active QAE isoform confers thermal hysteresis activity on a defective SP isoform of type III antifreeze protein. FEBS J 276:1471–1479
Vance TDR, Olijve LLC, Campbell RL, Voets IK, Davies PL, Guo S (2014) Ca2+−stabilized adhesin helps an Antarctic bacterium reach out and bind ice. Biosci Rep 34:e00121
Vance TDR, Graham LA, Davies PL (2018) An ice-binding and tandem beta-sandwich domain-containing protein in *Shewanella frigidimarina* is a potential new type of ice adhesin. FEBS J 285:1511–1527
Venketesh S, Dayananda C (2008) Properties, potentials, and prospects of antifreeze proteins. Crit Rev Biotechnol 28:57–82
Vrielink ASO, Aloi A, Olijve LLC, Voets IK (2016) Interaction of ice binding proteins with ice, water and ions. Biointerphases 11:018906
Wang X, DeVries AL, Cheng CH (1995) Antifreeze peptide heterogeneity in an Antarctic eel pout includes an unusually large major variant comprised of two 7 kDa type III AFPs linked in tandem. Biochim Biophys Acta 1247:163–172
Warren G, Corotto L (1989) The consensus sequence of ice nucleation proteins from Erwinia-Herbicola, Pseudomonas-Fluorescens and Pseudomonas-Syringae. Gene 85:239–242
Wen DY, Laursen RA (1992a) A model for binding of an antifreeze polypeptide to ice. Biophys J 63:1659–1662
Wen DY, Laursen RA (1992b) Structure-function-relationships in an antifreeze polypeptide – the role of neutral, polar amino-acids. J Biol Chem 267:14102–14108
Wierzbicki A, Dalal P, Cheatham Iii TE, Knickelbein JE, Haymet ADJ, Madura JD (2007) Antifreeze proteins at the ice/water interface: three calculated discriminating properties for orientation of type i proteins. Biophys J 93:1442–1451

Wilkens C, Poulsen JCN, Ramløv H, Lo Leggio L (2014) Purification, crystal structure determination and functional characterization of type III antifreeze proteins from the European eelpout *Zoarces viviparus*. Cryobiology 69:163–168

Wilson PW (1993) Explaining thermal hysteresis by the kelvin effect. Cryo Letters 14:31–36

Wilson PW, Osterday KE, Heneghan AF, Haymet ADJ (2010) Type i antifreeze proteins enhance ice nucleation above certain concentrations. J Biol Chem 285:34741–34745

Wolber P, Warren G (1989) Bacterialice-nucleation proteins. Trends Biochem Sci 14:179–182

Wu DW, Duman JG, Xu L (1991) Enhancement of insect antifreeze protein-activity by antibodies. Biochim Biophys Acta 1076:416–420

Wu Y, Banoub J, Goddard SV, Kao MH, Fletcher GL (2001) Antifreeze glycoproteins: relationship between molecular weight, thermal hysteresis and the inhibition of leakage from liposomes during thermotropic phase transition. Comp Biochem Physiol B Biochem Mol Biol 128:265–273

Xiao N, Suzuki K, Nishimiya Y, Kondo H, Miura A, Tsuda S, Hoshino T (2010) Comparison of functional properties of two fungal antifreeze proteins from *Antarctomyces psychrotrophicus* and *Typhula ishikariensis*. FEBS J 277:394–403

Xiao N, Hanada Y, Seki H, Kondo H, Tsuda S, Hoshino T (2014) Annealing condition influences thermal hysteresis of fungal type ice-binding proteins. Cryobiology 68:159–161

Yamashita Y, Miura R, Takemoto Y, Tsuda S, Kawahara H, Obata H (2003) Type II antifreeze protein from a mid-latitude freshwater fish, Japanese smelt (*Hypomesus nipponensis*). Biosci Biotechnol Biochem 67:461–466

Yang C, Sharp KA (2004) The mechanism of the type III antifreeze protein action: a computational study. Biophys Chem 109:137–148

Yang C, Sharp KA (2005) Hydrophobic tendency of polar group hydration as a major force in type I antifreeze protein recognition. Proteins Struct Funct Bioinf 59:266–274

Yang DSC, Sax M, Chakrabartty A, Hew CL (1988) Crystal structure of an antifreeze polypeptide and its mechanistic implications. Nature 333:232–237

Yang Z, Zhou Y, Liu K, Cheng Y, Liu R, Chen G, Jia Z (2003) Computational study on the function of water within a beta-helix antifreeze protein dimer and in the process of ice-protein binding. Biophys J 85:2599–2605

Yeh Y, Feeney RE (1996) Antifreeze proteins: structures and mechanisms of function. Chem Rev 96:601–618

Zepeda S, Yokoyama E, Uda Y, Katagiri C, Furukawa Y (2008) In situ observation of antifreeze glycoprotein kinetics at the ice interface reveals a two-step reversible adsorption mechanism. Cryst Growth Des 8:3666–3672

Zhang W, Laursen RA (1998) Structure-function relationships in a type I antifreeze polypeptide – the role of threonine methyl and hydroxyl groups in antifreeze activity. J Biol Chem 273:34806–34812

Chapter 5
Interaction of Antifreeze Proteins with Water

Ilja Karina Voets and Konrad Meister

5.1 Introduction

Controlling ice crystal growth is a grand scientific challenge with major technological ramifications for settings as diverse as oil fields, wind turbines, road surfaces, and frozen food products (Cochet and Widehem 2000; Hassas-Roudsari and Goff 2012; Koop and Zobrist 2009; Perfeldt et al. 2014; Tonelli et al. 2015; Zeng et al. 2006). The crystallization of water into ice is further lethal to most organisms and the process of ice recrystallization upon thawing is one of the major contributors to cell death (Briard et al. 2016; Chaytor et al. 2012; Huebinger et al. 2016). Cold-adapted organisms living at subzero temperatures have evolved an elegant macromolecular solution to cope with this problem. They produce antifreeze proteins that depress the freezing point of water in a non-colligative manner, enabling the survival of a variety of organisms in freezing and subfreezing habitats (DeVries 1971; DeVries and Wohlschlag 1969). Antifreeze proteins are believed to act by an adsorption–inhibition mechanism in which the proteins recognize and irreversibly bind to an ice surface and thereby prevent further growth of the ice crystal (Raymond and DeVries 1977). As a result, ice growth is limited to the regions between adsorbed AF(G)Ps, and thus the ice is forced to grow in energetically unfavorable curved fronts. This

I. K. Voets
Laboratory of Self-Organizing Soft Matter, Laboratory of Macromolecular and Organic Chemistry, Department of Chemical Engineering and Chemistry, and Institute for Complex Molecular Systems, Eindhoven University of Technology, Eindhoven, The Netherlands
e-mail: I.Voets@tue.nl

K. Meister (✉)
Natural Sciences Department, University of Alaska Southeast, Juneau, AK, USA

Molecular Spectroscopy Department, Max Planck Institute for Polymer Research, Mainz, Germany
e-mail: kkmeister@alaska.edu; Meisterk@mpip-mainz.mpg.de

H. Ramløv, D. S. Friis (eds.), *Antifreeze Proteins Volume 2*,
https://doi.org/10.1007/978-3-030-41948-6_5

effect will eventually lead to the termination of crystal growth, and a lowering of the freezing point, a phenomenon which is known as the Kelvin effect (Kristiansen and Zachariassen 2005).

Unlike man-made machines, nature's proteins work in liquid water and rely on water's active role in determining their structure and functionality. The idea of water being not just a passive spectator but an active player in dynamical processes in biosystems has recently gained ground in both experiment and theory. It was shown that proteins need a minimal amount of hydration water in order to retain catalytic activity and that internal water molecules act as bridges between different protein residues (Careri et al. 1980; Levy and Onuchic 2006). Internal water molecules also contribute to the folding speed and stability of a protein while the unfavorable interaction between water and hydrophobic groups is one of the main driving forces of protein folding (Dill 1990; Papoian et al. 2004). Understanding the subtle interplay between water and biomolecules is hence an essential prerequisite for understanding their function and to design synthetic analogues for applications.

Antifreeze proteins play an extraordinary role in the field of protein–solvent interactions. In order to fulfill their task, AFPs must specifically recognize and bind nascent ice crystals within the vast excess of 55 M liquid water (Sharp 2011). Unlike other proteins, AF(G)Ps cannot use chemical differences to drive ligand binding. They are solely sensitive to structural and/or dynamical differences between the two water phases (ice, water) which arguably is one of the most remarkable recognition problems in biology.

5.2 Structures of Antifreeze Proteins

The specific task of a protein is determined by its three-dimensional structure which can be solved with high resolution using X-ray crystallography or nuclear magnetic resonance spectroscopy (NMR). Traditionally AFPs are categorized into nonfish and fish AFPs, which are further subdivided into a class of antifreeze glycoproteins (AFGP) and four classes of antifreeze proteins (types I–IV). The original subdivision of AFPs is based on differences in amino acid sequences and structural characteristics.

AFGPs consist predominately of a repetitive three amino acid unit (Ala-Ala-Thr)$_n$ with the disaccharide β-D-galactosyl-(1–3)-α-D-*N*-acetylgalactosamine attached to the hydroxyl oxygen of each threonine residue. Their molar mass varies between 2.6 and 33.7 kDa, which correspond to $n = 4$–50 repetitions of the glycosylated tripeptide unit (Harding et al. 2003; Haridas and Naik 2013). AFGPs are grouped into eight size classes based on their electrophoretic properties, with AFGP$_{(1)}$ representing the largest and AFGP$_{(8)}$ the smallest glycoprotein (DeVries et al. 1970). A simplified version of the classification divides the AFGPs in small AFGP$_{(6–8)}$ and large AFGP$_{(1–5)}$ versions. To date, AFGPs have not been successfully crystallized and cannot be expressed in cell lines, an aspect which prevents systematic mutation and structure–function studies. No solution structure is available for AFGPs, since they are polydisperse, flexible, and relatively disordered. The other four classes of fish AFPs are small peptides and proteins without attached sugars. Type I AFPs have well-defined α-helical structures and contain highly repetitive amino acid sequences. The widely investigated AFP type I from winter flounder for instance contains an 11-residue periodicity that is rich in alanine and threonine (Sicheri and Yang 1995). Type II and type III AFPs are nonrepetitive and have an overall globular fold. Type II AFPs are 11–24 kDa, cysteine-rich proteins found in highly divergent fish (*Osmerus mordax*, *Clupea harengus*, and *Hemitripterus americanus*). NMR structural characterization revealed type II AFPs to be β-structured proteins with a few disulfide bridges (Nishimiya et al. 2008). Type III AFPs are 7 kDa and 14 kDa globular proteins found in the closely related *Macrozoarces americanus*, *Lycodichthys dearborni*, and *Anarhichas lupus*. Their crystal structures have been determined with high resolution using X-ray crystallography and NMR (Jia et al. 1996). Type III AFPs are often further divided in quaternary aminoethyl (QAE) and sulfopropyl (SP) isoforms (Li et al. 1985), based on their sequence similarity and isoelectric point. QAE isoforms adhere to QAE sephadex ion exchange resins, while SP isoforms adhere to SP sephadex ion exchange resins. The most commonly investigated type III AFP from ocean pout belongs to the QAE isoforms (Lotze et al. 2014; Meister et al. 2014b). Type IV AFPs are 12.3 kDa and are predicted to have a helix bundle structure. They have only been discovered in *Myoxocephalus octodecimspinosis* and its concentration in the blood is relatively low (Gauthier et al. 2008).

Nowadays antifreeze proteins are not limited to polar fish and have been identified in insects, plants, fungi, and various microorganisms (Duman 2001; Yeh and Feeney 1996). The most common structural motif in AFPs from arthropods, fungi, and bacteria is the β-solenoid fold. In *Mp*AFP from *Marinomonas primoryensis*, *CfAFP* from *Choristoneura fumiferana*, *Tm*AFP from *Tenebrio molitor*, and *Lp*AFP from *Lolium perenne* are the β-turns formed by regular repeats (Garnham et al. 2011; Leinala et al. 2002; Liou et al. 2000). On the contrary, in *Tis*AFP from the snow mold fungus *Typhula ishikariensis*, *Col*AFP from the bacterium *Colwellia* strain SLW05, and *Le*AFP from the yeast *Leucosporidium*, loops are of different lengths and arranged in an irregular order (Hanada et al. 2014; Kondo et al. 2012; Lee et al. 2012). An α-helix lies alongside the β-helix axis. Yet other ice-binding protein (IBP) structures are found in the insects *Rhagium inquisitor Ri*AFP and *Hypogastrura harveyi* (snow flea) *Sf*AFP. The structure of *Ri*AFP and *Rm*AFP1 of the beetle *Rhagium inquisitor* and *Rhagium mordax* is formed by two closely packed β-sheets (Hakim et al. 2013; Kristiansen et al. 2011, 2012). Six antiparallel polyproline type II (PPII) left-handed helices are observed in the structure of *Sf*AFP from snow flee (Pentelute et al. 2008). When comparing the astonishing variety of structures associated with antifreeze proteins (α-helices, β-solenoids, four helix bundles, polyproline type II helix bundles, and small globular proteins) it becomes clear that nature came up with a variety of solutions to fulfill the common function of adhering to ice. The structural diversity also points toward independent and recent evolution origins in the different animal kingdoms. In some cases, like the AFGPs in Arctic cod and Antarctic notothenioids, structural solutions have evolved,

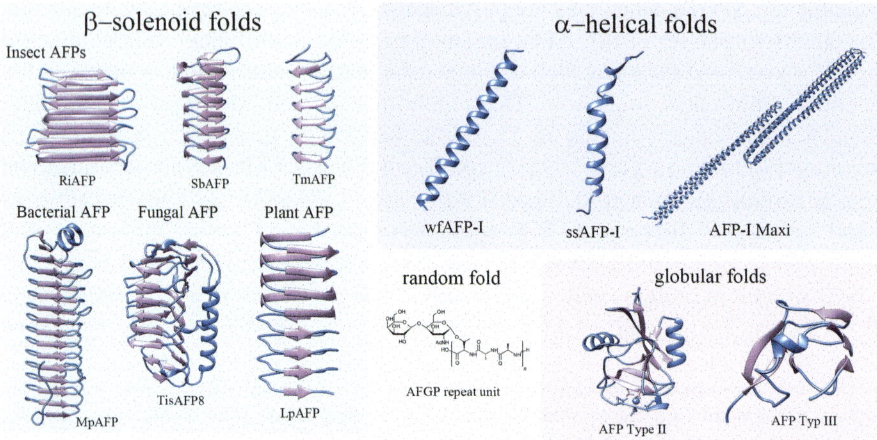

Fig. 5.1 Different classes of antifreeze proteins based on their core unit structure and the corresponding Protein Data Base (PDB) accession codes. α-helical folds include type I AFP form *Pseudopleuronectes americanus* (PDB: 1WFA); type I AFP from *Myoxocephalus scorpius* (PDB: 1Y03), and AFP-1 maxi from *Pseudopleuronectes americanus* (PDB: 4KE2). Globular folds include type II AFP from *Clupea harengus* (PDB: 2PY2) and type III AFP from *Zoarces americanus* (PDB: 1MSI). β-solenoid AFPs from different organisms: *Ri*AFP from *Rhagium inquisitor* (PDB: 4DT5), *Cf*AFP from *Choristoneura fumiferana* (PDB: 1M8N), and *Tm*AFP from *Tenebrio molitor* (PDB: 1EZG) are found in insects. *Mp*AFP is found in the bacterium *Marinomonas primoryensis* (PDB: 3P4G), *Tis*AFP8 in the fungus *Typhula ishikariensis* (PDB: 5B5H), and *Lp*AFP stems from the ryegrass *Lolium perenne* (PDB: 3ULT)

that are virtually the same (Chen et al. 1997). Cold-adapted organisms further often produce AFPs from different structural classes or multiple isoforms of the same type (Fletcher et al. 2001; Griffith and Yaish 2004). Figure 5.1 gives an overview of the major structural classes that can be found among antifreeze proteins. The classification here is based on their core unit structure. In comparison to non-antifreeze proteins it seems that the stabilization of AFP tertiary structures relies less on a traditional hydrophobic core but is rather performed by arrays of hydrogen and disulfide bonds.

5.3 Structure of the Ice-Binding Site

Despite their disparate structures, the AFPs all perform similar functions and share some characteristic features. Most antifreeze proteins are relatively small and have an active part that is able to interact with ice. The functional part that binds the ice surface is commonly referred to as the ice-binding site (IBS) of the protein (Strom et al. 2005).

The extreme diversity of structural motifs among AFPs has made it difficult to discern which structural features are important to bind to ice. The most common

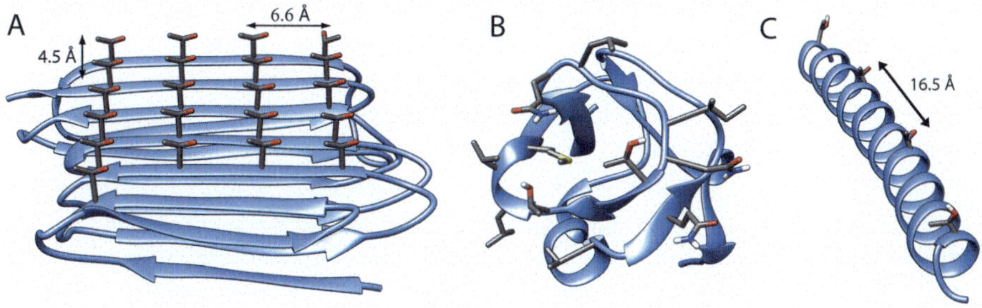

Fig. 5.2 Ice-binding sites of structurally different antifreeze protein classes. (**a**) Highly regular ordered IBS of *Ri*AFP where four rows of threonine residues are arranged with a ~6.66 × 4.7 Å spacing. (**b**) Top view of the ice-binding site of AFP type III. The IBS seems random and contains hydrophilic and hydrophobic amino acids that are mandatory for activity. (**c**) Ice-binding site of AFP type I from *Pseudopleuronectes americanus*. A single row of threonine residues is formed with 16.5 Å interresidue spacing

experimental approach to identify important amino acid residues of the ice-binding site is through a combination of mutagenesis and thermal hysteresis measurements as described in detail in Chap. 14 (Garnham et al. 2010). Typically mutations of amino acids which are essential for ice binding result in a large reduction of thermal hysteresis activity (Baardsnes et al. 1999; Hanada et al. 2014). A second useful tool for identifying ice-binding sites has been to do an alignment of isoform and orthologue sequences for a particular AFP type (Kondo et al. 2012). Conserved residues are normally either critical for stabilizing the protein fold or are involved in their activity of binding to ice.

Common characteristics of the known ice-binding sites are that they are extensive, relatively flat, and somewhat hydrophobic. They also rarely contain charged residues and often have repeating motifs (Garnham et al. 2011; Liou et al. 2000; Sicheri and Yang 1995; Sun et al. 2014). In Fig. 5.2, we show an overview of the ice-binding sites of three of the most commonly investigated AFPs. The ice-binding site of AFP type I is located on the flat and hydrophobic side of the helix and contains a single row of ordered Thr-residues with a spacing of 16.5 Å. This spacing matches well with the 16.7 Å repeat in the {2 0 2 1} pyramidal plane (Baardsnes et al. 1999). Replacing threonine to serine (loss of γ-methyl group) caused a reduction in antifreeze activity, while a mutation to valine (loss of OH-group) had little effect (Harding et al. 1999). This finding initially suggests that the γ-methyl group rather than hydrogen bonding of the OH group of threonine is important for binding to ice. However, amino acid substitutions can also lead to conformational changes and in the case of type I AFP the straight helix conformation disappears and changes in the more stable conformation of a kinked helix, devoid of ice-binding ability (Chakraborty and Jana 2017; Ebbinghaus et al. 2012).

Threonine residues are also widely found in the ice-binding sites of most of the β-solenoid folds where they are typically arranged in a remarkably high order. For example, *Tm*AFP and *Cf*AFP have two arrays of Thr-residues (Liou et al. 2000) that are typically referred to as (Thr-X-Thr) ice-binding motifs. Similarly, a row of Thr residues and a row of Asx (mostly Asn) residues appear in the IBS of *Mp*AFP (Garnham et al. 2011). The Thr-side chains have the same rotameric position and the spacing between side chains (7.4 × 4.6 Å) matches the ice lattice on the basal plane (7.83 × 4.52 Å) as well as on the prism plane (7.35 × 4.52 Å). Furthermore, the crystal structure of *Ri*AFP from the beetle *Rhagium inquisitor*, and the putative structure of *Rhagium mordax*, showed four instead of two arrays of threonine residues with a 6.66 × 4.73 Å spacing, creating a broad (Thr-X-Thr-X-Thr-X-Thr) motif (Friis et al. 2014; Hakim et al. 2013). On the contrary, examples of ice-binding sites without apparent structure are also known, even in IBPs with β-solenoid structure, such as *Tis*AFP, *Lp*IBP, and *Col*AFP (Hanada et al. 2014; Kondo et al. 2012; Lee et al. 2012). Despite this lack of regularity in the IBS, these proteins are able to bind both basal and prism planes.

In contrast to the highly regular (Thr-X-Thr) ice-binding sites of the β-solenoid fold AFPs is the ice-binding site AFP type III. The IBS of type III AFPs has an unbiased amino acid composition with both hydrophobic and hydrophilic residues (Garnham et al. 2010). It is further composed of two regions positioned at an angle of roughly 150° with respect to each other. One binds the primary prism plane of ice; the other a pyramidal plane (Garnham et al. 2010). Strikingly many crystal structures of AFPs also revealed ordered water layers associated with the ice-binding site which are believed to play a mandatory role in ice binding (Garnham et al. 2011; Hakim et al. 2013; Sun et al. 2015).

5.4 Solution Behavior of Antifreeze Proteins

Atomic resolution structures of AFPs provide important clues about their structures and determinants of ice binding. However, studying only the X-ray diffraction structure of flash-frozen crystals strongly suppresses any dynamical information or solvent interaction of the protein of interest. Studying the structure and behavior of AFPs in conditions close to their natural environment is hence crucial for deciphering the function of these extraordinary systems. Circular Dichroism (CD) and Infrared (IR) spectroscopic studies revealed that the structure of *D*AFP-1 from *Dendroides canadensis* in aqueous solution is similar to the modeled three-dimensional structure (Li et al. 1998). Lowering the solution temperature close to the biological relevant working temperature further led to a slight increase in the structural order and rigidity of the β-solenoid helix. Similar solution and crystal structures were also found for *Tm*AFP and *Cf*AFP using NMR techniques (Daley et al. 2002). Furthermore, conserved side chains, like the threonine residues of the ice-binding plane, also revealed an increased rigidity upon lowering the temperature (Daley and Sykes 2004). Small angle X-ray scattering (SAXS) revealed that the hypAFP1 adopts a long cylindrical shape in solution excluding the possibility of a fully extended helical dimer in solution and revising the original structural model of a flat threonine-rich IBS extending down the length of the protein (Olijve et al. 2013).

Two-dimensional infrared spectroscopy (2D-IR) revealed the prevalence of β-sheet character in the amide I response in agreement with the crystal structure of AFP type III (Lotze et al. 2014). Fast temperature-independent conformational fluctuations of the β-sheet of the protein further suggested that the protein remains relatively flexible even at low temperature. This notion is corroborated by the absence of detectable secondary structure elements by CD spectroscopy, while the X-ray structure with strongly suppressed dynamics reveals a compact fold with two β-sheets, a β-bridge, and two small helices (Lotze et al. 2014).

A variety of techniques have been used to study the solution conformation of AFGPs including CD, Raman, Fourier Transform Infrared (FTIR), and NMR spectroscopy as well as light scattering techniques (Feeney et al. 1986; Tsvetkova et al. 2002). Early CD spectra on AFGP suggested an extended random coil structure in solution. Follow-up studies using a combination of NMR and CD experiments proposed a left-handed helical conformation that is similar to a polyproline II helix (Franks and Morris 1978). Further CD studies, quasielastic light scattering, and Raman spectroscopy all suggested the presence of some folded structural elements in $AFGP_{1-5}$ (Bush et al. 1981; Feeney et al. 1986). For the low molecular weight AFGPs NMR experiments revealed that the overall structure is temperature dependent. At higher temperatures, AFGPs showed a random coil formation which changed into a three-folded left-handed PPII helix at lower temperatures (Bush and Feeney 1986). Up to now the definite solution structure of the AFGPs remains a subject of intense debate and more recent studies include effects like supercooling, protein–water interactions, or the presence of ice crystals when deciphering the absolute solution conformation of AFGP (Groot et al. 2016; Tsvetkova et al. 2002; Uda et al. 2007).

5.4.1 *Interaction of Antifreeze Proteins with Water*

The interaction with surrounding waters impacts the structure, stability, and activity of many if not all proteins (Dill and MacCallum 2012; Thirumalai et al. 2012). Naturally, the interaction with waters and ice has long been a focal point of research on antifreeze proteins. In the following sections, we will address recent experimental and theoretical studies on antifreeze protein hydration and its relation to their activity.

5.4.1.1 Adsorption–Inhibition Model

Common antifreezes, like ethylene glycol, salts, and ethanol, depress the freezing point T_f in a colligative fashion. This colligative freezing point depression $\Delta T_f = K_F \times b \times I$ is proportional to the solution osmolality b, and does not depend on the type of antifreeze. By contrast, in the presence of ice, antifreeze proteins lower the hysteresis freezing point (hFP) in a non-colligative manner, which is much more effective. A 1 mM solution of ethanol in water gives ΔT_f ~0.002 °C, while a 1 mM type III AFP solution gives hFP = −0.6 °C: a 300-fold larger effect.

To explain this large non-colligative impact of AF(G)Ps on ice crystal growth, Raymond and DeVries introduced an adsorption–inhibition model (Raymond and DeVries 1977). Simply put, this step-pinning model proposes that the antifreeze proteins bind onto nascent ice crystals in the path of a growth step in an irreversible manner. Thus, the permanently bound AF(G)Ps force ice to grow between the binding sites. Above a threshold surface coverage; i.e., below a critical separation between bound proteins, the curvature of the growing ice front is increased, which lowers the local freezing point via the Gibbs–Thomson or Kelvin effect (Kristiansen and Zachariassen 2005). This generates a subzero temperature interval between the hysteresis freezing point and the freezing point of the solution without AF(G)Ps (Tf), wherein steps are immobile and further ice growth is halted. This is known as the freezing hysteresis gap (Cziko et al. 2014). Similarly, the melting point is elevated in the presence of AF(G)Ps, which generates a melting hysteresis between the hysteresis melting point (hMP, in the presence of antifreeze proteins) and the melting point (Tm) in the absence of AF(G)Ps (Celik et al. 2010; Cziko et al. 2014; Knight and DeVries 1989). The thermal hysteresis (TH) gap is the sum of the melting and freezing hysteresis gaps (Cziko et al. 2014; Knight and DeVries 1989), i.e., the difference between the melting and freezing point. The magnitude of the TH gap is dependent on the type and concentration of the antifreeze protein, ice crystal size, and other parameters and can be as large as 6 °C for some AFPs as determined by nanoliter cryoscopy (Drori et al. 2014; Takamichi et al. 2007; Zachariassen et al. 2002).

5.4.1.2 Computational Studies Predict Unusual Ice-Like Hydration at the Ice-Binding Site

It is well-established that binding ice is a prerequisite for the non-colligative freezing point depression by antifreeze (glyco)proteins, but the molecular details of how and why AF(G)Ps bind ice remains an unresolved question to date. Early studies suggested that AFPs may perturb the structure and/or dynamics of a large fraction of waters, reducing thereby significantly the amount of waters that could join the ice lattice and hence freeze (DeVries 1971; DeVries et al. 1970). This hypothesis was proven invalid by NMR experiments, which showed that only small amounts of waters were surface bound (Haschemeyer et al. 1977). Major advances in computational power and molecular dynamics algorithms enabled Molecular Dynamics (MD) simulations on AFPs in explicit solvent models, which led to an alternative explanation for the activity of antifreeze proteins: unusual, ice-like hydration at the ice-binding site.

Sharp and coworkers monitored the hydration of wild-type and mutant fish type I and type III AFPs by studying the water-water hydrogen bond angle distribution functions obtained from MD simulations as shown in Fig. 5.3 (Gallagher and Sharp 2003a; Yang and Sharp 2004, 2005). The distribution is bimodal in the first

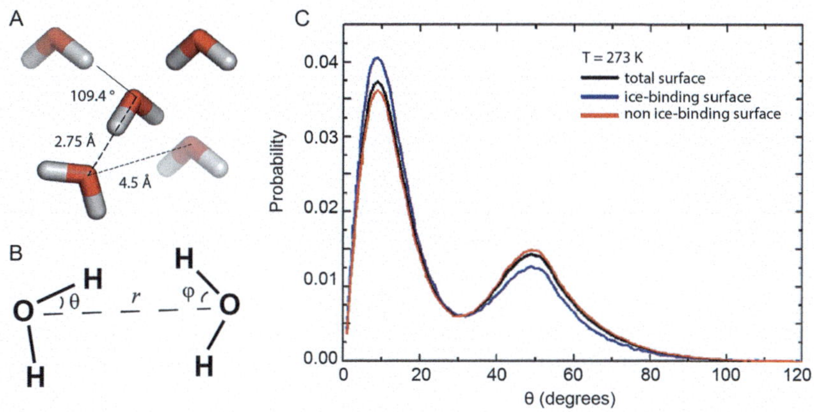

Fig. 5.3 Description of the water–water hydrogen bond angle distribution function in MD simulations. (**a**) Tetrahedral arrangement in the ice Ih lattice. (**b**) Geometric definitions for a water molecule pair: r is the distance between the two oxygen atoms, θ is the smallest of the four H–O–O angles, and φ is the corresponding smallest H–O–O angle of the other water molecule. (**c**) The water–water hydrogen bond angle distribution function for the first hydration shell of AFP III. The low-angle peak corresponds to the ice-like water arrangement. Figures adapted from Sharp et al. (Sharp and Vanderkooi 2010) and Smolin et al. (Smolin and Daggert 2008)

hydration layer, with a low- and high-angle population centered around ~10° (ice-like water arrangement) and ~50° in pure water, respectively. The tetrahedral order parameter corresponds to the ratio of the peak probabilities of the two populations, which is ~27% in pure water. Solutes that induce water structuring largely increase the tetrahedral order parameter relative to this value (Gallagher and Sharp 2003b; Sharp and Vanderkooi 2009). Small differences were observed in the tetrahedral order parameter of the waters close to and far away from the IBS of wild-type AFPs.

IBS-associated waters were more tetrahedral, and ice-like. By contrast, non-IBS waters were less tetrahedral and more heterogeneous. Furthermore, the tetrahedral signature was absent for less active (or inactive) mutants, suggesting ice-like hydration is important for activity and may facilitate ice binding of AFPs. A series of recent simulations confirm the earlier observation of ice-like waters in the first hydration layer in the vicinity of the IBS of wild-type fish type I and type III AFPs (Chakraborty and Jana 2017; Sharp 2014; Smolin and Daggett 2008; Wierzbicki et al. 2007). Herein tetrahedricity is quantified differently, on the basis of a tetrahedricity parameter θ. The tetrahedricity parameter measures the deviation of a tetrahedron from an ideal one, making it simpler to distinguish "bulk" and "ice-like" water (Smolin and Daggett 2008).

Experimental (Liou et al. 2000) and simulation studies (Yang et al. 2003) have also revealed regularly arranged waters at the IBS of β-helical AFPs from insects, like TmAFP. Yang et al. proposed that these IBS-associated water act as a "ligand" prior to ice adhesion to facilitate ice recognition and binding, but are finally excluded for entropic and steric reasons (Yang et al. 2003).

5.4.1.3 Molecular Dynamics Simulations Reveal a Dual Function of the Hydration Shell and Local Melting of Ice

Nutt and Smith realized that the hydration layer around β-helical proteins has a dual function (Nutt and Smith 2008). To adhere to ice and force further ice growth to take place in between bound proteins, AF(G)Ps must bind ice at the IBS only. Elsewhere, at the non-IBS, freezing must be unfavorable to prevent engulfment of the AFPs (Fig. 5.4). The authors proposed that waters at the IBS are preconfigured into ice-like architectures to promote ice growth in the 10–20 Å thick interfacial region between ice and liquid water (Nutt and Smith 2008). They further suggested that the entropic gain associated with this "zipper mechanism" of pre-organized IBS waters merging with ice leads to a sufficiently large reduction in the free energy to induce quasi-irreversible binding.

The adsorption–inhibition model proposed that irreversible binding of AF(G)Ps onto ice results in a convex curvature of the ice front growing in between bound proteins. In recent years, various groups have studied experimentally if and when AF (G)Ps bind ice in an irreversible manner (Celik et al. 2013; Pertaya et al. 2007). But, to date, it has been impossible to examine the topology of the ice/water interface with sufficient spatial resolution to resolve whether or not the growing ice front between surface-bound AF(G)Ps is indeed concave. However, MD simulations with explicit solvent can address this open question. Todde et al. focused specifically on the impact of *Cf*AFP (Todde et al. 2014) and AFP type I on the ice–water interface (Todde et al. 2015), and observed the development of a concave curvature on the stable adsorption planes. Interestingly, *Cf*AFP triggered local melting on the bipyramidal and prism planes onto which the AFPs do not adsorb. Future studies on other

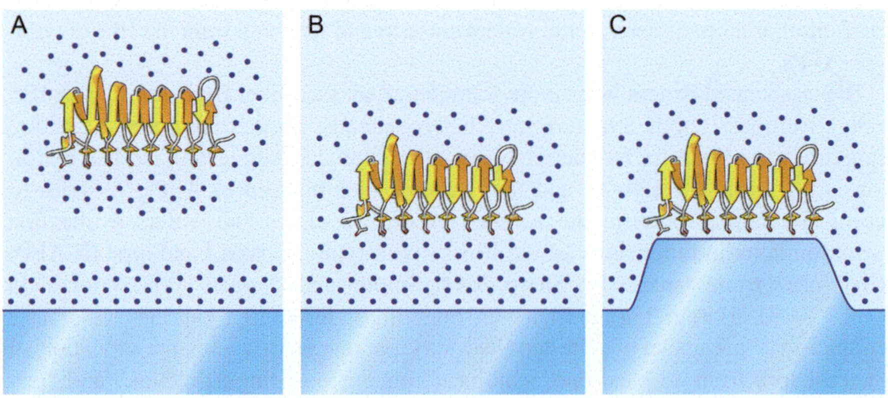

Fig. 5.4 Schematic representation of the proposed ice-binding mechanism by Nutt et al. (**a**) AFPs have ordered, ice-like water at the ice-binding site, while hydration water at the nonbinding regions is more distorted. (**b**) Overlap of the ordered hydration water of AFP with the hydration layer of the ice surface. (**c**) Binding of the AFP to the ice surface. Disruption of the water structure around the rest of the protein prevents the AFP from being overgrown by the ice. Figure adapted from Nutt et al. (Nutt and Smith 2008)

AFPs will reveal whether this is a general phenomenon, or instead, dependent on, e.g., AF(G)P type.

5.4.1.4 X-Ray and Neutron Diffraction Reveals Ordered Hydration Waters in Crystals of Antifreeze Proteins

X-ray diffraction (XRD) and neutron diffraction (ND) have offered a wealth of information on the structure and hydration of antifreeze proteins from various classes with atomistic detail (Garnham et al. 2011; Howard et al. 2011; Sun et al. 2015, 2014). An ND study on type III AFP from *Zoarces americanus* revealed a hydrogen bond interaction between a threonine residue within the IBS (Thr18) and a tetrahedral water cluster with a structural match to the primary prism plane of ice (Howard et al. 2011). The XRD structure of TmAFP showed a lattice match between an ordered array of water molecules, spaced at a distance of 4.64 ± 0.20 Å, in the vicinity of surface-exposed threonines (Liou et al. 2000). A similar lattice match was found between tens of waters bound to the β-helical MpAFP and the basal and primary prism planes of ice (Garnham et al. 2011).

The burning question that these studies could not answer is whether the bound surface waters act as "glue" joining the IBS and ice as an intermediary "ligand" or whether, instead, they are released before or during ice binding. But, this distinction between mechanistic scenarios could be made for the unusually large hyperactive type I AFP from *Pseudopleuronectes americanus* (Fig. 5.5), which is referred to as "Maxi." Surprisingly, it assembles into a dimeric, four helix bundle, with its ice-binding residues pointing inward to coordinate with two intersecting polypentagonal networks of interior waters (Sun et al. 2014). Hence, the IBS of the monomers in the dimer cannot contact the growing ice lattice directly. Nevertheless, permanent Maxi dimers obtained by chemical cross-linking are active. Hence, Maxi does not need to disassemble nor release its preorganized waters to enable ice binding, suggesting that instead binding is mediated by the surface-associated ordered waters (Sun et al. 2014). Whether Maxi serves as an antifreeze protein in its host fish is currently under debate.

5.4.1.5 Experimental Studies on the Hydration Shell Reveal Differences Between AFP Classes

High-quality crystals for XRD cannot be grown from solutions of antifreeze glycopeptides, since these are flexible macromolecules. Instead, structural information has been obtained from detailed NMR studies. These provided quantitative information on the fraction of surface-bound waters, which turned out to be modest; i.e., too small to rationalize the AFGP-induced freezing point depression in terms of a significant reduction in the amount of unperturbed waters available for freezing (Feeney et al. 1986).

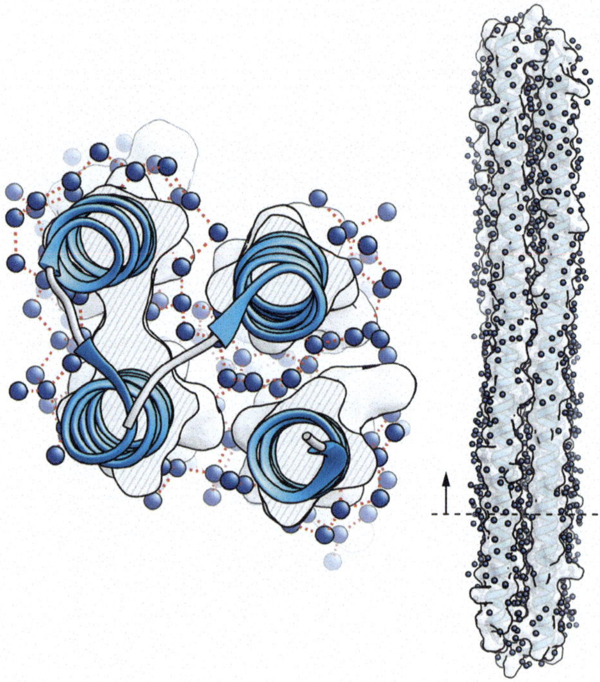

Fig. 5.5 Unusual hydration waters in the crystal structure of the large hyperactive type I AFP. The protein comprises a dimeric four helix. The tentative IBS points inward and is coordinated with over 400 semiclathrate water molecules arranged in two insecting polypentagonal networks extending from the interior to the exterior surface of the protein. Figure adapted from Sun et al. (2014)

To probe the collective motion of the water network and evaluate whether, as suggested by XRD and ND studies, tetrahedral waters that mediate ice binding are present in the hydration shell of AFPs in solution, a series of advanced spectroscopic experiments were performed. Terahertz (THz) absorption spectroscopy on AFGP (Ebbinghaus et al. 2010), AFP-I (Ebbinghaus et al. 2012), AFP-III (Xu et al. 2016), and the β-helical *D*AFP-1, a structural homologue of *Tm*AFP (Meister et al. 2013, 2014a), revealed a significant impact on the hydration shell dynamics, presumably correlated to AFP activity. A significant ability to influence the hydrogen-bonding structure of a large number of water molecules was also observed for AFP type III using a combination of FTIR-spectroscopy in the attenuated total reflection configuration and self-modeling curve resolution (Sun and Petersen 2017). On the contrary, no evidence for unusual hydration was found in ^{17}O magnetic relaxation dispersion measurements on *Tm*AFP (Modig et al. 2010). These probe selectively the rotational motion and exchange kinetics of internal water molecules and the external hydration layer. No evidence for unusual hydration was also found in polarization-resolved femtosecond infrared experiments of small and large AFGP isoforms (Groot et al. 2016). Here it was shown that water dynamics near AFGP surfaces are quite similar to the dynamics of water near non-antifreeze proteins and sugars, and that the majority of the slowly reorienting water molecules is slowed down by the AFGP peptide backbone (Groot et al. 2016). These findings are in line with the viscosity, translational diffusion, and NMR experiments (Feeney et al. 1986), which indicate that AFGPs affect a similar amount of water as

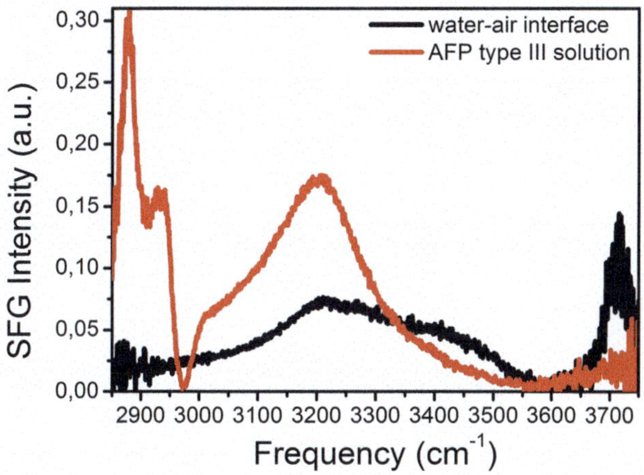

Fig. 5.6 Direct observation of ice-like water on the protein surface of type III AFP by VSFG experiments. The VSFG spectrum of the air–water interface (black) consists of a broad band between ~3200 and ~3450 cm^{-1} assigned to hydrogen-bonded liquid water and a sharp peak at ~3700 cm^{-1} that is assigned to dangling OH groups sticking out of the surface. The VSFG spectrum of a AFP-III solution (red) shows spectral features associated with the CH vibrations of the protein at frequencies <3000 cm^{-1} and a single strong relatively narrow peak at 3200 cm^{-1} assigned to the response of ice-like water layers associated with the ice-binding region. The ice-like water peak at 3200 cm^{-1} was absent in the spectrum of an inactive AFP-III mutant. Figure adapted from Meister et al. (2014)

non-antifreeze glycoproteins as well as with vibrational sum-frequency generation spectra of aqueous solutions of AFGP that show no indication of ice-like or unusually structured water around AFGP (Groot et al. 2016; Grossman et al. 2011).

No vibrational frequency shift was observed in the amide I region of the IR spectrum of AFP-I, which was attributed to the formation of ice-like waters at its IBS (Zelent et al. 2009). The first direct observation of such ice-like waters in the hydration shell of AFPs in solution at room temperature was reported in a vibrational sum-frequency generation (VSFG) spectroscopy study on AFP type III (Fig. 5.6). Bands corresponding to ice-like waters in the VSFG spectrum of wild-type AFP-III became increasingly prominent as temperatures were decreased from room temperature down to −2 °C (Meister et al. 2014b). Furthermore, these were absent in the VSFG spectrum of the inactive T18N mutant. Hence, ordered ice-like waters appear essential for the proper functioning of AFP type III, presumably mediating ice recognition and binding. However, no such ordered ice-like waters were observed in the VSFG spectrum of the potent wild-type *D*AFP-1 or AFGP, suggesting that these AF(G)Ps do not require mediation by bound waters for ice binding (Meister et al. 2015). Future work will shed light on the reason for these differences. One explanation is that *D*AFP-1's IBS has a sufficiently close structural match with the ice lattice without bound waters whereas AFP-III does not. The sugar residues of AFGP also provide an adequate structural match as the active ice-binding site, making additional ice-like hydration layers unnecessary. Alternatively, differences may be related to differences in the structure of the ice crystal adsorption planes. Interestingly, recent VSFG studies suggest that enhanced structural ordering of the adjacent water network is also relevant for ice-nucleation proteins which are present on the surface of bacteria *Pseudomonas syringae* (Pandey et al. 2016). The freezing transition is further facilitated by the effective removal of latent heat from the nucleation site, as shown by time-resolved VSFG spectroscopy experiments.

5.5 Conclusions

The sheer exuberance of structural motifs found in AFPs raises the question whether all AFPs work by the same mechanism. The fact that different AFPs target different surfaces of ice crystals adds to this question as binding basal or prism planes of ice is likely different at the molecular level. Differential mechanistic roles of AF(G)Ps might also be needed due to their diverse biological roles. AFPs produced by insects need to block ice growth when temperatures drop far below 0 °C while fish living in polar sea water face modest temperature fluctuations but all year around freezing conditions. The finding of ordered waters associated with the protein ice-binding sites at different AFP crystal structures provide the first hints for a plausible mechanism for how the structural different AFPs all can perform the same task of adsorbing to ice. However, the presence of ordered water molecules by itself only provides a short static picture to a molecular mechanism which is highly dynamic. Nowadays the properties of AF(G)Ps are being investigated with more advanced physicochemical and spectroscopic tools that enable the label-free study of the structural dynamics of the AF(G)Ps, their hydration shell, and potentially their mechanism of binding to ice. Typically only one class of AF(G)Ps is investigated or simulated, making direct comparisons or the discovery of common characteristics difficult. Direct comparison between results of hydration shells experiments of AF(G)Ps are further hampered by the fact that different techniques probe hydration dynamics on different time scales. For example, solvent density changes are mostly restricted to the first or second shell. "Hydration water" and "bulk water" are constantly exchanging on a time scale of 100 picoseconds; therefore, any method that probes static solvent properties, or motions which are slower than this diffusional exchange time, will observe different (smaller) hydration shell sizes, just as in case of too long exposure times (Conti Nibali and Havenith 2014). A series of THz experiments revealed differences in the dynamical hydration shells of different AFP classes which were interpreted to different strategies of the AF(G)Ps to optimize either their short-range interactions (such as hydroxyl bond formation by threonine) or long-range effects (protein-induced water dynamics) (Ebbinghaus et al. 2010, 2012; Meister et al. 2013). Surface-specific sum-frequency generation spectroscopy experiments revealed ice-like water layers for AFP type III while for an insect AFP (DAFP-1) highly ordered threonines with bound water molecules were observed, but the associated water molecules maintained no signature of ice (Meister et al. 2014b, 2015). Potentially differences in hydration structure might be needed to be able to distinguish and bind to the different planes of ice. Clear differences between AFP classes were also observed in their binding rate and the solution concentration required to maintain activity (Drori et al. 2014; Olijve et al. 2016).

To conclude, further advances in our understanding of the working mechanism of antifreeze proteins crucially depend on multidisciplinary studies encompassing a broad spectrum of AF(G)Ps including inactive mutants (Olijve et al. 2016). Moreover, experimental results, simulations, and analytical theories should be quantitatively compared to critically test the limitations and opportunities of the available models to describe and predict non-colligative freezing point depression by AF(G)Ps. We are confident that this fundamental insight into the (in)activity of biological cryoprotectants will support the development of bioinspired analogues. To date, artificial non-colligative antifreezes with moderate ice recrystallization inhibition activity have been produced, but none with strong thermal hysteresis activity. A knowledge-based approach may accelerate the engineering of potent mimics aiming ultimately to outcompete their own source of inspiration: native AF(G)Ps.

References

Baardsnes J, Kondejewski LH, Hodges RS, Chao H, Kay C, Davies PL (1999) New ice-binding face for type I antifreeze protein. FEBS Lett 463:87–91

Briard JG, Poisson JS, Turner TR, Capicciotti CJ, Acker JP, Ben RN (2016) Small molecule ice recrystallization inhibitors mitigate red blood cell lysis during freezing, transient warming and thawing. Sci Rep 6:23619

Bush CA, Feeney RE (1986) Conformation of the glycotripeptide repeating unit of antifreeze glycoprotein of polar fish as determined from the fully assigned proton n.m.r. spectrum. Int J Pept Protein Res 28:386–397

Bush CA, Feeney RE, Osuga DT, Ralapati S, Yeh YIN (1981) Antifreeze glycoprotein. Conformational model based on vacuum ultraviolet circular dichroism data. Int J Pept Protein Res 17:125–129

Careri G, Gratton E, Yang PH, Rupley JA (1980) Correlation of IR spectroscopic, heat capacity, diamagnetic susceptibility and enzymatic measurements on lysozyme powder. Nature 284:572–573

Celik Y, Graham LA, Mok Y-F, Bar M, Davies PL, Braslavsky I (2010) Superheating of ice crystals in antifreeze protein solutions. Proc Natl Acad Sci 107:5423–5428

Celik Y, Drori R, Pertaya-Braun N, Altan A, Barton T, Bar-Dolev M, Groisman A, Davies PL, Braslavsky I (2013) Microfluidic experiments reveal that antifreeze proteins bound to ice crystals suffice to prevent their growth. Proc Natl Acad Sci 110:1309–1314

Chakraborty S, Jana B (2017) Conformational and hydration properties modulate ice recognition by type I antifreeze protein and its mutants. Phys Chem Chem Phys 19:11678–11689

Chaytor JL, Tokarew JM, Wu LK, Leclère M, Tam RY, Capicciotti CJ, Guolla L, von Moos E, Findlay CS, Allan DS et al (2012) Inhibiting ice recrystallization and optimization of cell viability after cryopreservation. Glycobiology 22:123–133

Chen L, DeVries AL, Cheng C-HC (1997) Convergent evolution of antifreeze glycoproteins in Antarctic notothenioid fish and Arctic cod. Proc Natl Acad Sci 94:3817–3822

Cochet N, Widehem P (2000) Ice crystallization by *Pseudomonas syringae*. Appl Microbiol Biotechnol 54:153–161

Conti Nibali V, Havenith M (2014) New insights into the role of water in biological function: studying solvated biomolecules using terahertz absorption spectroscopy in conjunction with molecular dynamics simulations. J Am Chem Soc 136:12800–12807
Cziko PA, DeVries AL, Evans CW, Cheng C-HC (2014) Antifreeze protein-induced superheating of ice inside Antarctic notothenioid fishes inhibits melting during summer warming. Proc Natl Acad Sci 111:14583–14588
Daley ME, Sykes BD (2004) Characterization of threonine side chain dynamics in an antifreeze protein using natural abundance 13C NMR spectroscopy. J Biomol NMR 29:139–150
Daley ME, Spyracopoulos L, Jia Z, Davies PL, Sykes BD (2002) Structure and dynamics of a beta-helical antifreeze protein. Biochemistry 41:5515–5525
DeVries AL (1971) Glycoproteins as biological antifreeze agents in Antarctic fishes. Science 172:1152–1155
DeVries AL, Wohlschlag DE (1969) Freezing resistance in some Antarctic fishes. Science 163:1073–1075
DeVries AL, Komatsu SK, Feeney RE (1970) Chemical and physical properties of freezing point-depressing glycoproteins from Antarctic fishes. J Biol Chem 245:2901–2908
Dill KA (1990) Dominant forces in protein folding. Biochemistry 29:7133–7155
Dill KA, MacCallum JL (2012) The protein-folding problem, 50 years on. Science 338:1042–1046
Drori R, Celik Y, Davies PL, Braslavsky I (2014) Ice-binding proteins that accumulate on different ice crystal planes produce distinct thermal hysteresis dynamics. J R Soc Interface 11:20140526
Duman JG (2001) Antifreeze and ice nucleator proteins in terrestrial arthropods. Annu Rev Physiol 63:327–357
Ebbinghaus S, Meister K, Born B, DeVries AL, Gruebele M, Havenith M (2010) Antifreeze glycoprotein activity correlates with long-range protein−water dynamics. J Am Chem Soc 132:12210–12211
Ebbinghaus S, Meister K, Prigozhin MB, Devries AL, Havenith M, Dzubiella J, Gruebele M (2012) Functional importance of short-range binding and long-range solvent interactions in helical antifreeze peptides. Biophys J 103:L20–L22
Feeney RE, Burcham TS, Yeh Y (1986) Antifreeze glycoproteins from polar fish blood. Annu Rev Biophys Biophys Chem 15:59–78
Fletcher GL, Hew CL, Davies PL (2001) Antifreeze proteins of teleost fishes. Annu Rev Physiol 63:359–390
Franks F, Morris ER (1978) Blood glycoprotein from antarctic fish possible conformational origin of antifreeze activity. Biochim Biophys Acta Gen Subj 540:346–356
Friis DS, Johnsen JL, Kristiansen E, Westh P, Ramløv H (2014) Low thermodynamic but high kinetic stability of an antifreeze protein from *Rhagium mordax*. Protein Sci 23:760–768
Gallagher KR, Sharp KA (2003a) Analysis of thermal hysteresis protein hydration using the random network model. Biophys Chem 105:195–209
Gallagher KR, Sharp KA (2003b) A new angle on heat capacity changes in hydrophobic solvation. J Am Chem Soc 125:9853–9860
Garnham CP, Natarajan A, Middleton AJ, Kuiper MJ, Braslavsky I, Davies PL (2010) Compound ice-binding site of an antifreeze protein revealed by mutagenesis and fluorescent tagging. Biochemistry 49:9063–9071
Garnham CP, Campbell RL, Davies PL (2011) Anchored clathrate waters bind antifreeze proteins to ice. Proc Natl Acad Sci USA 108:7363–7367
Gauthier SY, Scotter AJ, Lin F-H, Baardsnes J, Fletcher GL, Davies PL (2008) A re-evaluation of the role of type IV antifreeze protein. Cryobiology 57:292–296
Griffith M, Yaish MWF (2004) Antifreeze proteins in overwintering plants: a tale of two activities. Trends Plant Sci 9:399–405
Groot CCM, Meister K, DeVries AL, Bakker HJ (2016) Dynamics of the hydration water of antifreeze glycoproteins. J Phys Chem Lett 7:4836–4840

Grossman M, Born B, Heyden M, Tworowski D, Fields GB, Sagi I, Havenith M (2011) Correlated structural kinetics and retarded solvent dynamics at the metalloprotease active site. Nat Struct Mol Biol 18:1102–1108

Hakim A, Nguyen JB, Basu K, Zhu DF, Thakral D, Davies PL, Isaacs FJ, Modis Y, Meng W (2013) Crystal structure of an insect antifreeze protein and its implications for ice binding. J Biol Chem 288:12295–12304

Hanada Y, Nishimiya Y, Miura A, Tsuda S, Kondo H (2014) Hyperactive antifreeze protein from an Antarctic sea ice bacterium *Colwellia* sp. has a compound ice-binding site without repetitive sequences. FEBS J 281:3576–3590

Harding MM, Ward LG, Haymet ADJ (1999) Type I 'antifreeze' proteins. Eur J Biochem 264:653–665

Harding MM, Anderberg PI, Haymet ADJ (2003) 'Antifreeze' glycoproteins from polar fish. Eur J Biochem 270:1381–1392

Haridas V, Naik S (2013) Natural macromolecular antifreeze agents to synthetic antifreeze agents. RSC Adv 3:14199–14218

Haschemeyer A, Guschlbauer W, DeVries AL (1977) Water binding by antifreeze glycoproteins from Antarctic fish. Nature 269(5623):87–88

Hassas-Roudsari M, Goff HD (2012) Ice structuring proteins from plants: Mechanism of action and food application. Food Res Int 46:425–436

Howard EI, Blakeley MP, Haertlein M, Petit-Haertlein I, Mitschler A, Fisher SJ, Cousido-Siah A, Salvay AG, Popov A, Muller-Dieckmann C et al (2011) Neutron structure of type-III antifreeze protein allows the reconstruction of AFP-ice interface. J Mol Recognit 24:724–732

Huebinger J, Han H-M, Hofnagel O, Vetter IR, Bastiaens PIH, Grabenbauer M (2016) Direct measurement of water states in cryopreserved cells reveals tolerance toward ice crystallization. Biophys J 110:840–849

Jia Z, DeLuca CI, Chao H, Davies PL (1996) Structural basis for the binding of a globular antifreeze protein to ice. Nature 384:285–288

Knight CA, DeVries AL (1989) Melting inhibition and superheating of ice by an antifreeze glycopeptide. Science 245:505–507

Kondo H, Hanada Y, Sugimoto H, Hoshino T, Garnham CP, Davies PL, Tsuda S (2012) Ice-binding site of snow mold fungus antifreeze protein deviates from structural regularity and high conservation. Proc Natl Acad Sci 109:9360–9365

Koop T, Zobrist B (2009) Parameterizations for ice nucleation in biological and atmospheric systems. Phys Chem Chem Phys 11:10839–10850

Kristiansen E, Zachariassen KE (2005) The mechanism by which fish antifreeze proteins cause thermal hysteresis. Cryobiology 51:262–280

Kristiansen E, Ramløv H, Højrup P, Pedersen SA, Hagen L, Zachariassen KE (2011) Structural characteristics of a novel antifreeze protein from the longhorn beetle *Rhagium inquisitor*. Insect Biochem Mol Biol 41:109–117

Kristiansen E, Wilkens C, Vincents B, Friis D, Lorentzen AB, Jenssen H, Løbner-Olesen A, Ramløv H (2012) Hyperactive antifreeze proteins from longhorn beetles: some structural insights. J Insect Physiol 58:1502–1510

Lee JH, Park AK, Do H, Park KS, Moh SH, Chi YM, Kim HJ (2012) Structural basis for antifreeze activity of ice-binding protein from Arctic yeast. J Biol Chem 287:11460–11468

Leinala EK, Davies PL, Doucet D, Tyshenko MG, Walker VK, Jia Z (2002) A β-helical antifreeze protein isoform with increased activity: structural and functional insights. J Biol Chem 277:33349–33352

Levy Y, Onuchic JN (2006) Water mediation in protein folding and molecular recognition. Annu Rev Biophys Biomol Struct 35:389–415

Li XM, Trinh KY, Hew CL, Buettner B, Baenziger J, Davies PL (1985) Structure of an antifreeze polypeptide and its precursor from the ocean pout, *Macrozoarces americanus*. J Biol Chem 260:12904–12909

Li N, Kendrick BS, Manning MC, Carpenter JF, Duman JG (1998) Secondary structure of antifreeze proteins from overwintering larvae of the beetle *Dendroides canadensis*. Arch Biochem Biophys 360:25–32

Liou Y-C, Tocilj A, Davies PL, Jia Z (2000) Mimicry of ice structure by surface hydroxyls and water of a β-helix antifreeze protein. Nature 406:322–324

Lotze S, Olijve LLC, Voets IK, Bakker HJ (2014) Observation of vibrational energy exchange in a type-III antifreeze protein. J Phys Chem B 118:8962–8971

Meister K, Ebbinghaus S, Xu Y, Duman JG, DeVries A, Gruebele M, Leitner DM, Havenith M (2013) Long-range protein–water dynamics in hyperactive insect antifreeze proteins. Proc Natl Acad Sci USA 110:1617–1622

Meister K, Duman JG, Xu Y, DeVries AL, Leitner DM, Havenith M (2014a) The role of sulfates on antifreeze protein activity. J Phys Chem B 118:7920–7924

Meister K, Strazdaite S, DeVries AL, Lotze S, Olijve LLC, Voets IK, Bakker HJ (2014b) Observation of ice-like water layers at an aqueous protein surface. Proc Natl Acad Sci 111:17732–17736

Meister K, Lotze S, Olijve LL, DeVries AL, Duman JG, Voets IK, Bakker HJ (2015) Investigation of the ice-binding site of an insect antifreeze protein using sum-frequency generation spectroscopy. J Phys Chem Lett 6:1162–1167

Modig K, Qvist J, Marshall CB, Davies PL, Halle B (2010) High water mobility on the ice-binding surface of a hyperactive antifreeze protein. Phys Chem Chem Phys 12:10189–10197

Nishimiya Y, Kondo H, Takamichi M, Sugimoto H, Suzuki M, Miura A, Tsuda S (2008) Crystal structure and mutational analysis of Ca2+-independent type II antifreeze protein from longsnout poacher, *Brachyopsis rostratus*. J Mol Biol 382:734–746

Nutt DR, Smith JC (2008) Dual function of the hydration layer around an antifreeze protein revealed by atomistic molecular dynamics simulations. J Am Chem Soc 130:13066–13073

Olijve LLC, Sun T, Narayanan T, Jud C, Davies PL, Voets IK (2013) Solution structure of hyperactive type I antifreeze protein. RSC Adv 3:5903–5908

Olijve LLC, Meister K, DeVries AL, Duman JG, Guo S, Bakker HJ, Voets IK (2016) Blocking rapid ice crystal growth through nonbasal plane adsorption of antifreeze proteins. Proc Natl Acad Sci 113:3740–3745

Pandey R, Usui K, Livingstone RA, Fischer SA, Pfaendtner J, Backus EHG, Nagata Y, Fröhlich-Nowoisky J, Schmüser L, Mauri S et al (2016) Ice-nucleating bacteria control the order and dynamics of interfacial water. Sci Adv 2:e1501630

Papoian GA, Ulander J, Eastwood MP, Luthey-Schulten Z, Wolynes PG (2004) Water in protein structure prediction. Proc Natl Acad Sci USA 101:3352–3357

Pentelute BL, Gates ZP, Tereshko V, Dashnau JL, Vanderkooi JM, Kossiakoff AA, Kent SBH (2008) X-ray structure of snow flea antifreeze protein determined by racemic crystallization of synthetic protein enantiomers. J Am Chem Soc 130:9695–9701

Perfeldt CM, Chua PC, Daraboina N, Friis D, Kristiansen E, Ramløv H, Woodley JM, Kelland MA, von Solms N (2014) Inhibition of gas hydrate nucleation and growth: efficacy of an antifreeze protein from the longhorn beetle *Rhagium mordax*. Energy Fuel 28:3666–3672

Pertaya N, Marshall CB, DiPrinzio CL, Wilen L, Thomson ES, Wettlaufer JS, Davies PL, Braslavsky I (2007) Fluorescence microscopy evidence for quasi-permanent attachment of antifreeze proteins to ice surfaces. Biophys J 92:3663–3673

Raymond JA, DeVries AL (1977) Adsorption inhibition as a mechanism of freezing resistance in polar fishes. Proc Natl Acad Sci USA 74:2589–2593

Sharp KA (2011) A peek at ice binding by antifreeze proteins. Proc Natl Acad Sci 108:7281–7282

Sharp KA (2014) The remarkable hydration of the antifreeze protein maxi: a computational study. J Chem Phys 141:22D510

Sharp KA, Vanderkooi JM (2009) Water in the half shell: structure of water, focusing on angular structure and solvation. Acc Chem Res 43:231–239

Sicheri F, Yang DSC (1995) Ice-binding structure and mechanism of an antifreeze protein from winter flounder. Nature 375:427–431

Smolin N, Daggett V (2008) Formation of ice-like water structure on the surface of an antifreeze protein. J Phys Chem B 112:6193–6202

Strom CS, Liu XY, Jia Z (2005) Ice surface reconstruction as antifreeze protein-induced morphological modification mechanism. J Am Chem Soc 127:428–440

Sun Y, Petersen PB (2017) Solvation shell structure of small molecules and proteins by IR-MCR spectroscopy. J Phys Chem Lett 8:611–614

Sun T, Lin F-H, Campbell RL, Allingham JS, Davies PL (2014) An antifreeze protein folds with an interior network of more than 400 semi-clathrate waters. Science 343:795–798

Sun T, Gauthier SY, Campbell RL, Davies PL (2015) Revealing surface waters on an antifreeze protein by fusion protein crystallography combined with molecular dynamic simulations. J Phys Chem B 119:12808–12815

Takamichi M, Nishimiya Y, Miura A, Tsuda S (2007) Effect of annealing time of an ice crystal on the activity of type III antifreeze protein. FEBS J 274:6469–6476

Thirumalai D, Reddy G, Straub JE (2012) Role of water in protein aggregation and amyloid polymorphism. Acc Chem Res 45:83–92

Todde G, Whitman C, Hovmöller S, Laaksonen A (2014) Induced ice melting by the snow flea antifreeze protein from molecular dynamics simulations. J Phys Chem B 118:13527–13534

Todde G, Hovmöller S, Laaksonen A (2015) Influence of antifreeze proteins on the ice/water interface. J Phys Chem B 119:3407–3413

Tonelli D, Capicciotti CJ, Doshi M, Ben RN (2015) Inhibiting gas hydrate formation using small molecule ice recrystallization inhibitors. RSC Adv 5:21728–21732

Tsvetkova NM, Phillips BL, Krishnan VV, Feeney RE, Fink WH, Crowe JH, Risbud SH, Tablin F, Yeh Y (2002) Dynamics of antifreeze glycoproteins in the presence of ice. Biophys J 82:464–473

Uda Y, Zepeda S, Kaneko F, Matsuura Y, Furukawa Y (2007) Adsorption-induced conformational changes of antifreeze glycoproteins at the ice/water interface. J Phys Chem B 111:14355–14361

Wierzbicki A, Dalal P, Cheatham TE, Knickelbein JE, Haymet A, Madura JD (2007) Antifreeze proteins at the ice/water interface: three calculated discriminating properties for orientation of type I proteins. Biophys J 93:1442–1451

Xu Y, Bäumer A, Meister K, Bischak CG, DeVries AL, Leitner DM, Havenith M (2016) Protein–water dynamics in antifreeze protein III activity. Chem Phys Lett 647:1–6

Yang C, Sharp KA (2004) The mechanism of the type III antifreeze protein action: a computational study. Biophys Chem 109:137–148

Yang C, Sharp KA (2005) Hydrophobic tendency of polar group hydration as a major force in type I antifreeze protein recognition. Proteins 59:266–274

Yang Z, Zhou Y, Liu K, Cheng Y, Liu R, Chen G, Jia Z (2003) Computational study on the function of water within a β-helix antifreeze protein dimer and in the process of ice-protein binding. Biophys J 85:2599–2605

Yeh Y, Feeney RE (1996) Antifreeze proteins: structures and mechanisms of function. Chem Rev 96:601–618

Zachariassen KE, DeVries AL, Hunt B, Kristiansen E (2002) Effect of ice fraction and dilution factor on the antifreeze activity in the hemolymph of the cerambycid beetle *Rhagium inquisitor*. Cryobiology 44:132–141

Zelent B, Bryan MA, Sharp KA, Vanderkooi JM (2009) Influence of surface groups of proteins on water studied by freezing/thawing hysteresis and infrared spectroscopy. Biophys Chem 141:222–230

Zeng H, Wilson LD, Walker VK, Ripmeester JA (2006) Effect of antifreeze proteins on the nucleation, growth, and the memory effect during tetrahydrofuran clathrate hydrate formation. J Am Chem Soc 128:2844–2850

Part II
Molecular Mechanisms Affected by Antifreeze Proteins

Chapter 6
Thermal Hysteresis

Erlend Kristiansen

6.1 Introduction

Thermal hysteresis refers to the phenomenon where antifreeze proteins (AFPs) or antifreeze glycoproteins (AFGPs) cause a separation of the freezing and melting temperature of existing ice crystals in solution. This ability to separate the melting and freezing temperature of ice is limited in that on sufficient cooling the ice crystal undergoes a sudden and rapid ice growth. Ramsay (1964) when studying a mechanism of water reabsorption in the beetle *Tenebrio molitor* first reported the phenomenon. In a footnote, he states:

> "When small ice crystals are observed under the microscope, as in the freezing-point method of Ramsay and Brown, one notices that large crystals grow at the expense of small ones and that the edges of the crystals are rounded—the natural consequences of surface tension at the water-ice interface. The change of state between solid and liquid is perfectly temperature-reversible"......"By contrast, the crystals which appear in fluid from the anterior perinephric space tend to have jagged outline and large crystals do not grow at the expense of smaller ones. Furthermore, the system is not temperature-reversible. As the temperature is raised the crystals decrease in size, but as the temperature is lowered they do not increase in size. After the temperature has been lowered by a few degrees the crystal suddenly begins to grow rapidly. On occasion undercooling of the order of 10 °C was observed (in the continued presence of small crystals) and then suddenly the whole sample appeared to solidify instantaneously."

The temperature interval between the melting and freezing temperatures is referred to as the hysteresis gap, and the lower temperature where rapid ice growth is initiated is termed the hysteresis freezing point. The quantitative difference between the melting temperature and the hysteresis freezing point is termed the hysteresis activity, or antifreeze activity.

E. Kristiansen (✉)
NTNU University Library, Trondheim, Norway
e-mail: erlendkr@ntnu.no

H. Ramløv, D. S. Friis (eds.), *Antifreeze Proteins Volume 2*,
https://doi.org/10.1007/978-3-030-41948-6_6

Thermal hysteresis reflects the role of AF(G)Ps as protectors against ice nucleation in the supercooled body fluids of freeze avoiding animals. Their presence enables hypoosmotic fish to occupy ice-laden polar waters (DeVries 1971, 1982; Raymond and DeVries 1977), and allow terrestrial arthropods, such as insects, spiders, and collembolans, to remain year-round in the cold temperate and polar areas. The body temperatures of such terrestrial animals may in some cases drop well below –30 °C in winter (Zachariassen and Husby 1982; Duman 2001; Duman et al. 2004; Graham and Davies 2005). Within the animal, AF(G)Ps are known to act by inactivating structures in the body fluids that could initiate freezing, so-called ice-nucleating agents (INAs), and by preventing ice from penetrating through the body wall (Olsen and Duman 1997a, b; Olsen et al. 1998; Duman 2002).

AF(G)Ps are categorized as being moderately active or hyperactive, based on the hysteresis activity they cause at equimolar concentrations. This distinct difference in antifreeze potency is accompanied by distinct shapes of the ice crystals that form in their presence; moderately active AF(G)Ps cause bipyramidal crystals to develop, a shape that only exposes a single crystal plane to the surrounding solution. In the presence of hyperactive AF(G)Ps, ice crystals express several crystal planes, usually in the form of hexagonal discs. A number of different factors affect the hysteresis activity, including their size and the addition of large organic macromolecules and inorganic ions. This chapter outlines current understanding of the *modus operandi* of AF(G)Ps. An attempt is made to provide some simple explanations to the antifreeze potency of AF(G)Ps, including their characteristics as moderately active or hyperactive, and how their antifreeze potency is affected by their size and by different additives. Some characteristics of INAs and their relevance in cold tolerance are also examined briefly.

6.2 A Hysteresis Mechanism: The Kelvin Effect

The vapor pressure of bulk ice is lower than that of water. Thus, below the melting point a net transfer of water molecules from the bulk water to ice occurs and the ice mass grows. However, it follows from the observable fact that ice crystals in the presence of AF(G)Ps remain unchanged within a temperature interval, that the AF (G)Ps somehow causes vapor pressure equilibrium between ice and water at all temperatures within the hysteresis gap. This must be so, since the rate by which water molecules adds onto the crystal surface must equal the rate by which they leave. Otherwise, net transfer of water molecules would result, from solution to ice or vice versa and the crystal would visibly change volume. AF(G)Ps do not lower the vapor pressure of water any more than other solutes do (Westh et al. 1997). Thus, they must act by elevating the vapor pressure of the ice to correspond to the higher vapor pressure of the surrounding solution. The difference between the vapor pressure of water and ice increases with temperature departure below the equilibrium melting temperature. Thus, the effect of the AF(G)Ps on the vapor pressure of ice

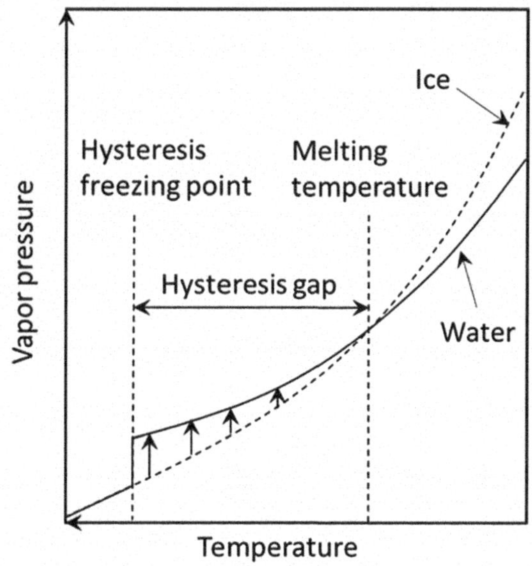

Fig. 6.1 Vapor pressure equilibrium within a temperature interval near the melting temperature. For the ice crystal to be stable within the hysteresis gap, the AF(G)Ps must elevate the vapor pressure of the ice surface to correspond to that of the surrounding supercooled solution. This elevation of the vapor pressure must increase with decreasing temperature. Adapted from Kristiansen and Zachariassen (2005)

must be temperature dependent and increase with decreasing temperature, see Fig. 6.1.

Raymond and DeVries (1977) proposed that the AF(G)Ps act by changing the microscopic growth pattern of the ice surface. Since this is achieved by the AF(G)Ps becoming irreversibly adsorbed onto the ice surface, they coined the mechanism the adsorption–inhibition mechanism. Since then, several investigators have had similar approaches to explaining the phenomenon by irreversible adsorption, including Wilson (1993) and Kristiansen and Zachariassen (2005).

Using fluorescently tagged AFPs, Celik et al. (2013) exchanged the slightly supercooled solution surrounding an ice crystal. The ice surface of the supercooled crystals remained fluorescent following the exchange of the surrounding solution, showing that AFPs were adsorbed onto the crystal surface. Further, the removal of AFPs in the surrounding solution by the exchange process did not weaken the hysteresis effect. These observations provide the most unequivocal evidence to date to show that AF(G)Ps become irreversibly adsorbed onto the ice surface and that the phenomenon is caused only by the surface-bound AF(G)Ps. Also, Chao et al. (1995) and DeLuca et al. (1998) found that AF(G)Ps principally operate as monomeric units.

Elevation of the vapor pressure of the ice by the changed microscopic surface growth pattern could occur by the so-called Kelvin effect. In the following, a brief historical outline of the Kelvin effect is provided. This is followed by a description of how the Kelvin effect is thought to operate at the ice surface.

6.2.1 The Kelvin Effect: Vapor Pressure at a Curved Interface

In 1871, Prof. William Thomson, later to become first Baron Kelvin, pointed out that the vapor pressure of water at a concave and a convex surface must be lower and higher, respectively, than at a plane surface of the water (Thomson 1871). This was deduced by considering the rise and fall of liquids in a capillary tube as a function of the curvature of the meniscus; in an atmosphere saturated with vapor, the vapor pressure decreases with height above the surface of a liquid. Consequently, since a concave interface in a capillary causes the liquid to come to rest at some fixed height above the liquid body, Thomson deduced that the vapor pressure at the elevated concave meniscus is reduced relative to the vapor pressure at the lower plane surface and must correspond to the lowered saturated atmospheric vapor pressure at that height. Otherwise, a perpetual net directional motion of water molecules would develop, as there would be continuous net evaporation at the elevated meniscus and consequently net condensation at the lower plane surface. Such directional perpetual motion of water molecules would violate the fundamental first law of thermodynamics, since energy could principally be extracted for eternity from this water movement without being supplied to the system. Convex interfaces must have the opposite effect on the vapor pressure, as such an interface comes to rest below the plane liquid body where the saturated vapor pressure is higher. The effect of a surface curvature on the vapor pressure has since become known as the Kelvin effect.

6.2.1.1 The Critical Radius of Curvature

A decade later, Prof. John Henry Poynting (1881) recognized that the effect of a surface curvature on the resultant vapor pressure in Thomson's capillary is caused by a change in the bulk pressure in the water in the capillary; a concave interface evokes a lower pressure inside the liquid water, as evident from the rise in the capillary, and hence to a lower vapor pressure, and vice versa for a convex interface. Thus, the underlying cause of the changing vapor pressure with changing curvature of an interface is an accompanying curvature-induced change in bulk pressure within the curved volume.

Poynting applied his reasoning to the melting temperature of ice. He inferred that if the bulk pressure of ice alone was elevated, then the resultant elevated vapor pressure of the ice would depress the temperature at which the vapor pressures of ice and water coincides, i.e., a pressure-induced depression of the melting temperature of the ice surface. By extension, since the pressure-elevating effect of a convexity increases with decreasing radius, there must be a convexity with a radius small enough to cause a pressure great enough for ice/water vapor pressure equilibrium to develop at any temperature below the normal melting point. The radius of this convexity at a specific temperature is referred to as the critical radius of curvature at that temperature.

6.2.2 The Kelvin Effect at the Ice Surface

It follows from the above paragraphs that AF(G)Ps that are irreversibly adsorbed onto the ice surface could evoke the Kelvin effect by causing the ice surface to grow out as many tiny convex interfaces between them. These convex interfaces would elevate the vapor pressure of the ice surface and, hence, eliminate the difference between the vapor pressures at different temperatures, as illustrated in Fig. 6.1.

The Kelvin effect implies that, at any temperature below the normal melting temperature, the growth of the convex surface zones between the adsorbed AF(G)Ps will halt when they obtain a curvature with a radius corresponding to the critical radius at that temperature. Thus, at any temperature where the phenomenon is expressed, the surface of the entire ice crystal is covered by spherical growth regions with identical convexities, i.e., identical local vapor pressures. This causes the entire ice crystal surface to be in vapor pressure equilibrium with the surrounding supercooled solution, and hence the crystal surface is at its melting temperature, see Fig. 6.2. A. Such a crystal could in principle remain unchanged indefinitely. Crystals in supercooled solutions of AF(G)Ps have been observed for many days without expressing any visible growth (DeVries 1971; Raymond and DeVries 1977; Graether et al. 2000; Fletcher et al. 2001).

As the temperature is lowered further, the many tiny surface zones expand until their convex interfaces again cause vapor pressure equilibrium with the surrounding

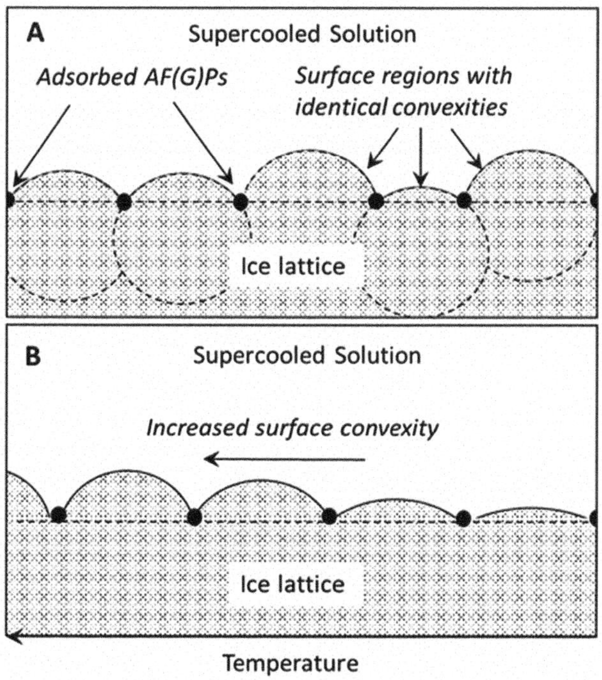

Fig. 6.2 The convexities of the growth zones within the hysteresis gap. (**a**) All growth zones must have the same convexity at a specific temperature within the hysteresis gap. (**b**) The convexities increases with decreasing temperature and elevates the vapor pressure of the ice surface in a temperature-dependent manner, as seen in Fig. 6.1. Adapted from Kristiansen and Zachariassen (2005)

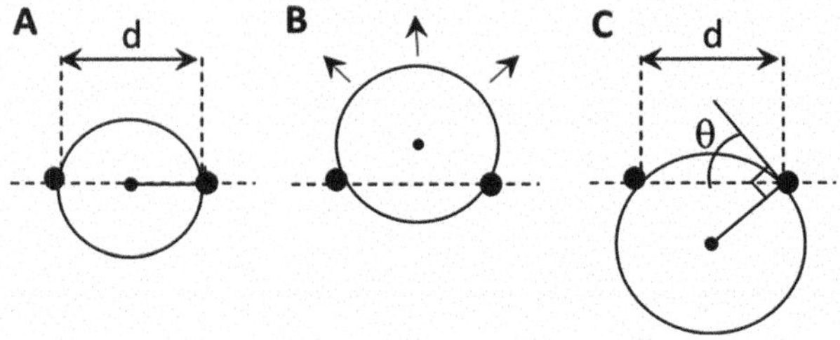

Fig. 6.3 At the hysteresis freezing point. (**a**) When one of the convexities has reached the shape of a half-sphere it has reached its maximum convexity. (**b**) Any further growth of this structure will cause the convexity to decrease and cause spontaneous growth. (**c**) The relation between adsorbent spacing, d, and the angle, θ. Adapted from Kristiansen and Zachariassen (2005)

solution. In this manner ice/water vapor pressure equilibrium is maintained across a temperature interval, the hysteresis gap, see Figs. 6.1and 6.2b.

There is a limit to how much such a crystal can be cooled, i.e., how convex the tiny curved interfaces may become; no surface zone can become more convex than that of a half-sphere. Once such a shape is reached, then any further cooling will result in the convexity of the structure to decrease on growth. The resultant drop in vapor pressure due to the reduced convexity will result in spontaneous growth. This is illustrated in Fig. 6.3a and b. This temperature is the hysteresis freezing point.

6.3 Hysteresis Activity

In the following paragraphs, an attempt is made to explain what fundamentally determines the hysteresis freezing point, based on the theory outlined above. This explanation is then extended to incorporate the characteristic difference in activity between moderately active and hyperactive kinds of AF(G)Ps.

6.3.1 The Largest Intermolecular Adsorbent Gap Determines Hysteresis Activity

If only a single one of all the tiny growth zones that protrude out at the crystal surface should fail, then the hysteresis phenomenon is terminated. Hence, the hysteresis freezing point is determined by the single growth zone that reaches the shape of a half-sphere at the highest temperature. Any further growth of this single growth zone, i.e., any further cooling, will only result in a reduction in its convexity and, consequently, the phenomenon is terminated.

Since all the surface growth zones have the same convexity, it will be the single one growth zone with the widest diameter that will reach the shape of a half-sphere at the highest temperature. Thus, the hysteresis freezing point, and therefore the hysteresis activity, is determined by the single largest intermolecular adsorbent spacing between AF(G)Ps that comprise a single growth zone at the crystal surface. Mathematically, the hysteresis activity (ΔT) as a function of the largest such adsorbent spacing, d, may be expressed as (Kristiansen and Zachariassen 2005):

$$\Delta T = \frac{4\gamma T_{\mathrm{E}} \sin\theta}{\Delta H \mathrm{d}}; \tag{6.1}$$

where d is the spacing in units of cm, γ is the ice/water interfacial tension (taken to be 32 ergs/cm^2), T_E is the normal melting temperature for a plane interface (units of K), and ΔH is the heat of fusion of water ($3.3 \cdot 10^9$ ergs/cm^3). θ is an angle describing the situation if a curvature fails before reaching the shape of a half-sphere. For a half-sphere, θ is 90° and, hence, the term (sin θ) is 1. See Fig. 6.3. C for an illustration of the angle θ.

6.3.2 Moderately Active and Hyperactive AF(G)Ps

There is a great difference in the hysteresis activities caused by different AF(G)Ps. Based on their activities at equimolar concentrations and the shape of the crystals they form in solution, they fall into two categories: hyperactive and moderately active.

Marshall et al. (2004a) found that moderately and hyperactive AFPs accumulate in ice to a similar extent. Also, experimentally determined estimates of average adsorbent spacings between AF(G)Ps on the surface of ice crystals are quite similar in the case of moderately and hyperactive AF(G)Ps; Drori et al. (2015) estimated the average adsorbent distance between hyperactive TmAFP to 7.6–35.2 nm at concentrations ranging from 31.4 to 0.4 μM. Comparable results were obtained by Celik et al. (2013) for the same protein. For the moderately active type III AFP, Drori et al. (2015) estimated the average adsorbent distance to be 8.7 to 24.7 nm at concentrations ranging from 19.8 to 1.2 μM. Others have estimated similar values for moderately active AF(G)Ps (Wilson et al. 1993; Grandum et al. 1999; Zepeda et al. 2008). Thus, the principal cause of the great difference in the activities of moderately and hyperactive AF(G)Ps do not seem to be due to differences in their preference for ice. Rather, it is likely that the distinct difference between them is the result of the single largest adsorbent gap at the ice surface for some reason is much larger in the case of moderately active AF(G)Ps.

6.3.2.1 Moderate or Hyperactive: Caused by Plane Specificity and Adsorption Pattern?

Moderate Activity A characteristic feature of moderately active AF(G)Ps is that they only adsorb onto a single crystal plane in the ice structure. Notably, none of the moderately active AF(G)Ps adsorb onto the basal plane of crystals, only onto a single prism or pyramidal plane (Knight and DeVries 1988; Knight et al. 1991). This plane-specific adsorption is apparently a consequence of structural features of their ice-binding sites (IBSs), that restricts these AF(G)Ps to only become irreversibly adsorbed onto a single plane and orientation. Laursen et al. (1994) showed this by observing that the moderately active chiral L-AFP I and D-AFP I variants resulted in adsorption on mirror image directions on the ice surface. The result of such a specific preference for a single crystal plane is a crystal that only expresses this single protected crystal plane toward the surrounding supercooled solution. Consequently, in the presence of moderately active AF(G)Ps crystals obtain a bipyramidal shape, as this is the only possible crystal shape whose entire surface consists of a single plane. At the hysteresis freezing point, these bipyramidal crystals freeze out from their apexes (Raymond and DeVries 1977; Jia and Davies 2002). The fact that they characteristically grow out of their apexes at the hysteresis freezing point strongly suggests that the antifreeze potency of moderately active AF(G)Ps are limited by a large intermolecular spacing at the apex of the bipyramidal crystal (Jia and Davies 2002). This must arise from the fact that these proteins only adsorb onto a single crystal plane.

The surface area involved in determining the hysteresis activity for moderately active AF(G)Ps is only that miniscule fraction of the total surface area of the crystal that comprises the two apexes of the bipyramid. Consequently, the hysteresis activity in the presence of moderately active AF(G)Ps should not be much affected by changing the total surface area of the ice. Consistent with this, the hysteresis activity of moderately active AF(G)Ps are reportedly rather insensitive to the amount of ice present in the sample; large variations in the ice content, i.e., large variations in total ice crystal surface area, does not appreciably affect the hysteresis activity, see Fig. 6.4 (Hansen et al. 1991; Wöhrmann 1996; Sørensen and Ramløv 2001).

Hyperactivity In contrast to the moderately active AF(G)Ps, the hyperactive AF (G)Ps have been shown to adsorb to several crystal planes that differ greatly in their orientation, such as both prism and basal planes (Graether et al. 2000; Liou et al. 2000). Structural studies have shown that hyperactive AFPs have IBS that afford the protein freedom to adsorb in different orientations and on different planes. Their ability to adsorb onto multiple crystal planes, and most notably the basal plane, is a feature that separates them from the moderately active AF(G)Ps. Basal plane adsorption has been implicated as a key feature that causes them to be hyperactive (Graether et al. 2000; Liou et al. 2000; Pertaya et al. 2008). Because of their ability to adsorb onto multiple planes, crystals formed in the presence of hyperactive AF(G)Ps expresses multiple planes to the surrounding supercooled solution and usually take the form of hexagonal discs (Graether et al. 2000; Liou et al. 2000).

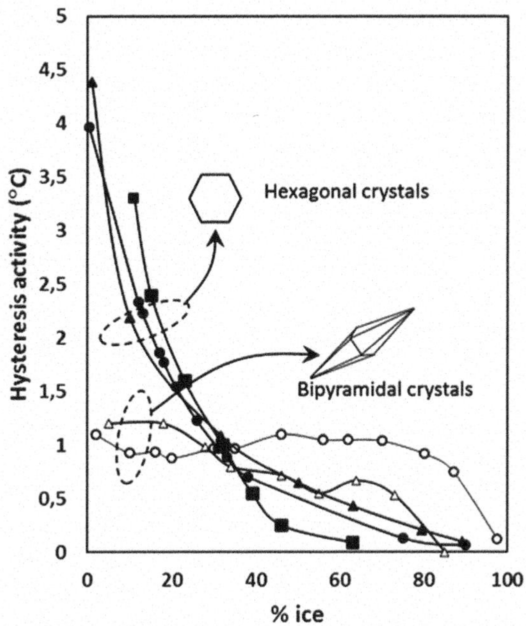

Fig. 6.4 The dependency of the hysteresis activity on the % ice in the sample. Filled symbols: hyperactive AF(G)Ps. Open symbols: moderately active AFGP. (Filled square) PAGP, a hyperactive AFGP from the nototheniid *Pleuragramma antarcticum* (Wöhrmann 1996). (Filled circle) Hemolymph from *Tenebrio molitor* (Hansen and Baust 1988). (Filled triangle) Hemolymph from *Rhagium inquisitor* (Zachariassen et al. 2002). (Open circle) Serum from *P. antarcticum*. (Open triangle) AFGP from *P. antarcticum* (Wöhrmann 1996). For explanation, see text

Because of their ability to adsorb onto different planes and at different orientations, hyperactive AFPs likely become spread out across the crystal surface in a rather random adsorption pattern. Such a random pattern should, by chance alone, result in the largest adsorption gap increasing with increasing surface area. Consequently, the hysteresis activity of hyperactive AF(G)Ps should decrease with increasing crystal surface area. Consistent with this, several investigators have reported strong dependence of hyperactive AF(G)Ps on the amount of ice present in the sample, see Fig. 6.4 (Zachariassen and Husby 1982; Hansen and Baust 1988; Wöhrmann 1996). As can be seen from the figure, "hyperactivity" is apparently a consequence of using small ice crystals in the experiment, since hyperactive AF(G)Ps have a lower hysteresis activity than their moderately active counterparts at higher contents of ice in the samples.

The Shape of the Bipyramidal Apexes When bipyramidal crystals form in the presence of moderately active AF(G)Ps, the ice crystal grows out from the basal planes. Once this bipyramidal shape is formed the crystal stops growing and it remains stable within the hysteresis gap. What is the physical shape of the apex interfaces? Since the moderately active AF(G)Ps do not adsorb onto the basal plane, is the apex a tiny unprotected flat basal plane? If so, then one could envision two-dimensional curved interfaces protruding out only in the direction of the prism planes that form the surrounding edge of the exposed apex basal plane (Raymond and DeVries 1977). The effect of these 2D curvatures that are in the prism plane direction must then also elevate the vapor pressure beyond the base of the curvature toward the center of the flat basal plane in order for vapor pressure equilibrium to persist between the flat apex interface and the surrounding solution. Another, and perhaps simpler, approach is to assume that the apexes are three-dimensional spheres protruding out in the basal plane direction. In any event, it is these areas of the bipyramidal crystal that apparently determines the hysteresis activity of the moderately active AF(G)Ps.

6.4 Factors That Affect the Hysteresis Activity

In the above paragraphs the categorization of AF(G)Ps into moderately active and hyperactive were ascribed to consequences of irreversible adsorption to the ice surface that arises from features of their IBS. In the following, differences in hysteresis activity within each of these categories will be ascribed to the situation that exist prior to the AF(G)Ps becoming irreversibly adsorbed. It will be argued that, while the ice crystal is held at the equilibrium melting temperature, AF(G)Ps acquire an equilibrium distribution between the crystal surface melting region and the surrounding solution. Then, following a cooling event, AF(G)Ps within this surface region freeze onto the solidifying crystal surface and, hence, become irreversibly adsorbed (Kristiansen and Zachariassen 2005). Any change in this distribution pattern prior to the cooling event will result in changes in the surface density of irreversibly adsorbed AF(G)Ps after the cooling event and, hence, to changes in the observed hysteresis activity. Differences in hysteresis activity among hyperactive or among moderately active AF(G)Ps, may be attributed to differences in the solubility of the AF(G)Ps in the solution; a lowered solubility results in a shift in the distribution of the AF(G)Ps toward the ice surface region prior to the cooling event, and hence, to increased hysteresis activity (Kristiansen and Zachariassen 2005; Kristiansen et al. 2008).

6.4.1 The Factors

Several investigators have reported that the size of the AF(G)Ps can have a profound effect on their capacity to cause thermal hysteresis. For structurally similar isoforms, their potency reportedly increases with molecular size for both moderately active AFGPs (Schrag et al. 1982; Chao et al. 1996; Miura et al. 2001; Baardsnes et al. 2003; Nishimiya et al. 2003) and hyperactive AFPs (Leinala et al. 2002; Marshall et al. 2004b; Liu et al. 2005; Mok et al. 2010; Friis et al. 2014). Synthetic oligomers of moderately active AFPs also reportedly have increased potency (Nishimiya et al. 2005; Holland et al. 2008; Can and Holland 2011, 2013; Stevens et al. 2015). In all the cases mentioned above, the increased size is accompanied by an increased IBS or the addition of multiple IBSs. Other investigators have reported that AFPs are potentiated by ligation to, or interaction with, large non-ice binding structures (Deluca et al. 1998; Hakim et al. 2013; Wu and Duman 1991, Wu et al. 1991;

Horwath et al. 1996; Wang and Duman 2005, 2006). In these cases, the IBS is unchanged.

In addition to the effect of molecular size, several authors have reported that the hysteresis activity is also elevated in the presence of various low-mass co-solutes. These low-mass solutes include sugars, polyols, salts, amino acids, salts of polycarboxylates, and NADH. The effect has been reported for both moderately active AF(G)Ps (Kerr et al. 1985; Caple et al. 1986; Evans et al. 2007; Gong et al. 2011) and hyperactive AFPs (Li et al. 1998; Kristiansen et al. 2008; Amornwittawat et al. 2008; Wang et al. 2009a, b; Amornwittawat et al. 2009; Wen et al. 2011; Liu et al. 2015).

There is one thing that variations in molecular size and additives have in common; they change the solubility of proteins in solution. Moreover, they reportedly enhance the hysteresis activity in manners predicted by their general effects on protein solubility. In the following section, the potential importance of the solubility of AF(G)Ps to their antifreeze potency is briefly explored.

6.4.2 *The Solubility of the AF(G)Ps: A General Concept to Explain Variability?*

Several authors have in various ways implicated protein solubility as a relevant factor in antifreeze potency (Kristiansen and Zachariassen 2005; Evans et al. 2007; Kristiansen et al. 2008; Wang et al. 2009a). Solubility of AF(G)Ps have also inadvertently been implicated in the manner the AF(G)Ps are thought to orient toward the ice; these proteins are somewhat amphipathic, were the more hydrophobic side that contains the IBS orient toward the ice (Yang et al. 1988; Sönnichsen et al. 1996; Haymet et al. 1998, 1999). In other words, the less soluble side of the molecule orients toward the ice whereas the more soluble side orients toward the water. The logical extension of this is that a less soluble AFP would have a greater affinity toward the ice surface than a more soluble AFP. In the following paragraphs, a brief examination of the significance of this common denominator, the solubility of the AF(G)Ps, to their potency is presented.

6.4.2.1 The AF(G)P/Ice Interaction Is Temperature Dependent

The ice surface in equilibrium with surrounding liquid water is not distinct but a transition region where the configuration of the water molecules changes from the ordered crystal structure of the ice lattice to the random distribution of the bulk water in the surrounding solution. This change occurs across a 1–2 nm deep region called the interfacial region or the melting/freezing region (Hayward and Haymet 2001).

As stated in the introductory quote by Ramsay (1964), AF(G)Ps act at temperatures below the equilibrium melting temperature of the ice, not at temperatures above it, i.e., the ice crystal does not grow below this temperature but melts above it (but see also next section concerning superheating of ice crystals). This suggests that the AF(G)Ps are irreversibly adsorbed onto the ice crystal surface only at temperatures below the melting temperature. A simple explanation to this is that AF(G)Ps freeze onto the crystal surface as the temperature is lowered to within the hysteresis gap and then melt off the ice when the temperature is raised to the melting temperature (Kristiansen and Zachariassen 2005). Such a temperature-dependent behavior of freezing onto (adsorption) and melting off (desorption) would explain why ice crystals in the presence of AF(G)Ps typically melt at the equilibrium temperature irrespective of any colligative variation in this temperature. It also provides an intuitive and simple explanation to the long-standing conundrum of the origin of the necessary bond strength to achieve irreversible adsorption (Wen and Laursen 1992; Knight et al. 1993; Chao et al. 1995); the bond strength between the irreversibly adsorbed AF(G)P and the ice surface corresponds to those between water molecules in bulk ice at that temperature. Recently, Garnham et al. (2011a) showed that the hydration water of a hyperactive AFP has a clathrate-like configuration and is firmly embedded by extensive H-bonds to the backbone of the protein. Hence, this crystalline-like water at the IBS appears to be prone to fuse together with the solidifying crystalline interface once the temperature is lowered and melt off when the interface disintegrates into chaos on warming to the equilibrium melting temperature. Molecular dynamics studies support this contention (Chakraborty and Jana 2019; Zanetti-Polzi et al. 2019).

Pertaya et al. (2008) reported on the fluorescence associated with an ice crystal in a solution containing fluorescently tagged AFP. When slowly melting a crystal at a temperature just above that of equilibrium the crystal showed no fluorescence, indicating no adsorbed AFPs. When cooled to within the hysteresis gap the crystal surface became fluorescent, indicating irreversible adsorption. Similar results were reported by Pertaya et al. (2007), who used a technique of photo-bleaching of fluorescently tagged AFPs to study the AFP/ice association at the crystal surface at temperatures within, and just above, the hysteresis gap. Bleached AFPs at the surface were not replaced within the hysteresis gap but were replaced at temperatures just above, showing that the AFPs were irreversibly adsorbed within the hysteresis gap and desorbed off the ice at the melting temperature.

While in the desorbed state, at the melting temperature of the crystal surface, there must be a distribution of AF(G)Ps between the melting/freezing region and the bulk solution. It is this distribution pattern that presumably becomes affected by changes in the solubility of the AF(G)Ps; a lowered protein solubility means that the AF(G)P has an increased tendency to move away from the solution and toward the melting/freezing region. This results in more AF(G)P molecules being at the ice/water interfacial region and available to freeze onto the solidifying crystal surface the instant the temperature is lowered. Consequently, lowered solubility of an AF(G)P should result in greater surface density of the AF(G)P below the melting temperature and, hence, to greater hysteresis activity (Kristiansen and Zachariassen 2005).

Superheating of Ice Crystals Several investigators have reported that ice crystals in solutions of AF(G)Ps may superheat slightly (Celik et al. 2010; Cziko et al. 2014). Celik et al. (2010) reported that tiny ice crystals became superheated by 0.04 °C and 0.44 °C in the presence of several hyperactive AFPs. In the case of moderately active AFPs, superheating up to 0.02 °C was reported at high AFP concentrations. The observed superheating reflects the presence of concave surface regions developing between irreversibly adsorbed AF(G)Ps at temperatures above the equilibrium temperature (Knight and DeVries 1989). These observations potentially contradict the notion of an equilibrium distribution of AF(G)Ps developing between the solution and the ice surface region at the equilibrium temperature, as outlined above.

The samples that expressed this superheating also expressed hysteresis activities ranging from 1.7 °C to 4.1 °C. The hysteresis activity increases approximately as a function of the square root of the surface density of AF(G)Ps (Raymond and DeVries 1977; Kristiansen and Zachariassen 2005). Thus, apparently only a small fraction of the AFPs that was originally frozen onto the surface and caused these high hysteresis activities was subsequently involved in the comparatively much lower superheating. That is, most AFPs melted off the ice surface.

The superheating phenomenon requires a cooling event to occur; when Celik et al. (2010) melted out ice in solutions with high concentrations of moderate AFPs or low concentrations of hyperactive AFPs, they observed that the many small crystals decreased uniformly in size. If the melting process was briefly halted, then the crystals began to show slight superheating. This change in melting behavior following a brief cooling event suggests that AFPs in solution do not adsorb irreversibly to the ice surface unless there is a cooling event, i.e., the adsorption is a freezing of the AFPs onto the ice surface. The subsequent desorption as the temperature is raised is for some of the adsorbed AFPs a delayed process.

Why do some of the AF(G)Ps not simply melt off the surface as the temperature is raised to the melting point? The freezing of the AF(G)Ps onto the ice surface imply that the hydration water at the IBS becomes part of the crystal lattice. Above its equilibrium melting temperature, ice melts from its surface, as lattice water molecules are released to the fluid hydrogen-bonding network of the surrounding solution. However, if no liquid water is in contact with the lattice that is to be melted, e.g., in the interior of a crystal, the lattice structure may superheat extensively before a melting nucleation event occurs (Turnbull 1950; Chalmers 1964; Lu and Li 1998). Consequently, if the crystalline water at the IBS of an adsorbed AFP is shielded from the surrounding liquid solution, then the melting process at the IBS is prevented and the AFP will remain adsorbed onto the crystal surface at temperatures above the melting point. The distinct difference in the capacities of moderately and hyperactive AFPs to cause superheating reported by Celik et al. (2010) presumably reflect differences in their respective capacities to shield the crystalline water at the IBS from the surrounding liquid water when adsorbed onto the ice. They observed that in the presence of hyperactive AFPs, crystals sporadically disappeared over time up to 4 h, showing that this situation can be quite stable if it develops. Since the phenomenon is very weak compared to the hysteresis activity, it might be that only those AF

(G)Ps with certain rare orientations at the crystal surface is able to postpone the initiation of the melting process at the IBS.

6.4.3 Basic Concepts in Solubility Theory

The solubility of a protein in water reflects its energetic state in water (Reynolds et al. 1974). Once present in the water, the solubility of a protein is determined by two opposing effects acting on structural features of the protein; favorable attractive forces such as van der Waals- and dipole-type forces lower the energy state of the protein and therefore increase its solubility. This is opposed by an energetic cost associated with occupying a cavity within the water that increases its energetic state and therefore lowers its solubility (Uhlig 1937; Tolls et al. 2002). In the latter case, the presence of the protein in the water effectively adds additional high-energy water surface at the water/protein boundary of the cavity occupied by the solute. The presence of nonpolar surface regions of the protein restricts hydrogen bond formation between water molecules in the surface boundary, and consequently reduces the freedom of these local water molecules to orientate. This structuring of water at the protein/water boundary is known as the hydrophobic effect.

According to Uhlig (1937), the solubility (S) of a dissolved molecule may be expressed as:

$$RT\ \ln(S) = -A\gamma + E \tag{6.2}$$

where R and T are the universal gas constant and the absolute temperature, respectively. The first term on the right side of Eq. (6.2), $A\gamma$, represents the "hydrophobic" effect that lowers the solubility of a molecule. This effect is a function of the nonpolar surface area, A, of the molecule in contact with water, and the energetic state of the water at this surface, expressed as the water surface tension, γ. This hydrophobic effect is opposed by the second term on the right-hand side of Eq. (6.2), the favorable "electrostatic" effect, E, that raises the solubility of the dissolved molecule (Reynolds et al. 1974; Melander and Horváth 1977).

Changing the size of the AF(G)Ps, for instance by adding or removing repetitive peptide segments, inadvertently also changes the nonpolar surface area, A, of the protein and consequently its solubility. Also, for structurally similar isoforms of different size, their nonpolar surface areas, and hence, their solubility, correlate with their size. The small mass solutes that reportedly enhance the hysteresis activity, such as salts, sugars, polyols, and amino acids are known to elevate the surface tension, γ, of water (Washburn 1929; Melander and Horváth 1977; Kaushik and Bhat 1998; Landt 1931; Matubayasi and Nishiyama 2006; Bull and Breese 1974). Thus, their reported enhancement effect may simply be the result of the solutes lowering the solubility of the AF(G)P by elevating γ at the protein/water interface.

The basic framework outlined above may be useful when interpreting natural and induced variations in the antifreeze potency among moderately active and among hyperactive AF(G)Ps. In the following paragraphs, standard solubility theory will be applied to examine some of the reported effects small co-solutes and variations in size have on hysteresis activity.

6.4.4 *Low-Mass Additives, Solubility, and Antifreeze Potency*

The effects of salts on the hysteresis activity in relation to solubility theory will be exemplified by the effects of salts on the hyperactive AFP, RiAFP, from the cerambycid beetle *Rhagium inquisitor* (Kristiansen et al. 2008). Wang et al. (2009a) also had a quite similar approach to this issue. It will be shown that these effects are entirely consistent with being caused by salt-induced lowered solubility of the RiAFP molecules. To support this claim, the nonpolar surface area and the dipole moment of RiAFP is derived from the effects of salts on its antifreeze potency.

6.4.4.1 The Salting-Out Constant, K_s

As mentioned above, salts are known to lower the solubility of proteins. This effect is termed "salting-out." The salting-out effect is qualitatively similar for different kinds of proteins and different kinds of salts in that the solubility of the protein changes in a log-linear manner with the concentration of salt (Cohn 1925; Melander and Horváth 1977):

$$\ln(S) = \beta - K_s m \tag{6.3}$$

where S is the solubility of the protein (mg/ml), β is the solubility of the protein in the absence of salts (mg/ml), m is the concentration of the salt (molal), and K_s is known as the salting-out constant (molal^{-1}). K_s is an expression of the sensitivity of the solubility of a particular protein to the presence of a particular salt. The value of K_s depends on both the salt and the protein and is experimentally determined as the slope of the linear relationship between ln(S) and m.

6.4.4.2 Obtaining Salting-Out Constants from Measurements of Hysteresis Activity

Since the presence of salts increases the hysteresis activity, adding salts is equivalent to increasing the concentration of the AF(G)P. Since solubility is in units of concentration, Eq. (6.3) should describe the salt-induced apparent changes in the concentration of AF(G)P. Thus, the salting-out constant, K_s, in the presence of a particular salt may be obtained from the hysteresis measurements as follows; the actual concentration of AF(G)P in the samples is kept unchanged during the procedure. An "apparent" concentration of AF(G)P in the presence of different concentrations of salts is then obtained by converting the observed enhanced hysteresis activity in the presence of salts to the equivalent concentration of AF(G)P needed to cause this activity in the absence of salt. The value of K_s for that salt is then obtained simply as the slope of the linear relationship obtained by plotting the natural logarithm of the "apparent" concentration of AF(G)P in the samples as a function of the concentration of salt.

Kristiansen et al. (2008) used this method to determine K_s for each of ten different salts from the salt-induced enhancement of the hysteresis activity for RiAFP. As predicted by Eq. (6.3) all "apparent" concentrations were log-linear functions of the concentrations of the different salts tested.

6.4.4.3 The Hofmeister Series and Its Linearity

An experimentally determined salting-out constant, K_s, is an expression of the two opposing effects acting on structural features of the protein, the favorable "electrostatic" effect that increases the solubility of the protein and the unfavorable "hydrophobic" effect that lowers its solubility, as outlined in Eq. (6.2). The net observed salting-out constant, K_s, in Eq. (6.3) is given by (Melander and Horváth):

$$K_s = \Omega\sigma - \Lambda \tag{6.4}$$

where Ω is a protein-specific intrinsic salting-out constant (cm dyn^{-1}), σ is the molal surface tension increment of the salt (10^{-3} dyn g/cm mol), and Λ is a protein-specific intrinsic salting-in constant (molal^{-1}).

By arranging salts according to their ability to lower the solubility of proteins, the so-called Hofmeister series of salts is obtained. The arrangement of different salts in the Hofmeister series may be understood from Eq. (6.4); for a specific protein, the molal surface tension increment, σ, is the only variable in the equation. Thus, for any single protein the arrangement of salts according to their ability to lower the solubility of that protein is similar for all proteins and dictated by the molal surface tension increment, σ, of the different salts. For example, the following eight salts listed in descending order according to their ability to lower protein solubility form the Hofmeister series as (value for σ in parenthesis): $Na_3C_6H_6O_7$ (3.12) > $(NH_4)_2SO_4$ (2.16) > NaCl (1.64) > KCl (1.40) > NH_4Cl (1.39) > $NaNO_3$ (1.20) > NaI, (1.09) > $N(CH_3)_4Cl$ (0.76). Since the value of σ is actually the surface tension increment of the water/air interface and not the protein/water interface, there are slight differences between the predicted and observed Hofmeister series. However, this general arrangement of salts varies little for different proteins. Hence, if the salt-induced enhancement of the hysteresis activity is caused by salt-induced reduction in the solubility of RiAFP, then the enhancement-effect of the different salts should reflect the Hofmeister series. In the case of RiAFP the experimentally determined K_s values were arranged as (value for σ in parenthesis): $Na_3C_6H_6O_7$ (3.12) > $(NH_4)_2SO_4$ (2.16) > NaCl (1.64) > KCl

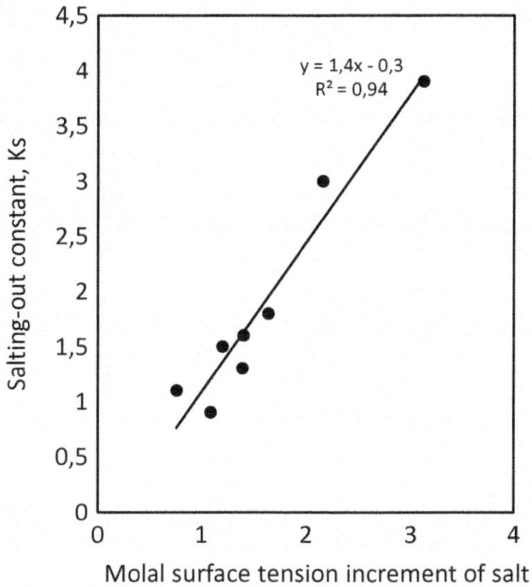

Fig. 6.5 The linearity of the Hofmeister series. Different salting-out constants, K_s, determined from the hysteresis activity in the presence of different kinds of salts, versus the surface tension increment of the salts used. The slope of the linear line is the intrinsic salting-out constant, Ω, for the protein. The intercept value is the intrinsic salting-in constant, Λ, for the protein. These two protein-specific constants may be used to determine the nonpolar surface area and the dipole moment of the protein. Adapted from Kristiansen et al. (2008)

(1.40) > $NaNO_3$ (1.20) > NH_4Cl (1.39) > $N(CH_3)_4Cl$ (0.76) > NaI (1.09). This arrangement is in close agreement with the Hofmeister series. An examination of the results of Wang et al. (2009a) also shows that arrangement of the salts according to their ability to enhance a hyperactive AFP, DAFP, from the beetle *Dendroides canadensis* corresponds well with the Hofmeister series. For those salts where the value for σ is known, they are listed as (value for σ in parenthesis); NaCl (1.64) > KCl (1.40) > KBr (1.31) > NaBr (1.32) > KI (0.84) > NaI (1.09) > $NaClO_4$ (0.55). In their extensive study, Li et al. (1998) found that, among all the different compounds tested, citrate was the strongest enhancer of the antifreeze potency of DAFP. Citrate has among the highest known surface tension increments (σ of 3.12) and was also the strongest enhancer in the case of RiAFP. Evans et al. (2007) did not find differences in the efficacies of LiCl, NaCl, and KCl to enhance the antifreeze activity of different kinds of fish AF(G)Ps. This is consistent with the fact that the molal surface tension increments of these salts are very similar, 1.63, 1.64, and 1.40, respectively.

Since both Ω and Λ of Eq. (6.4) are constant features of the protein, the K_s values obtained for that protein will be a linear function of the molal surface tension increment, σ, of the different salts. Figure 6.5 shows that the salting-out constants of RiAFP, determined from the salt-induced enhancement of the hysteresis activity, vary as a linear function of σ, consistent with this prediction of Eq. (6.4). Thus, the linear relationship depicted in Fig. 6.5 is a quantitative representation of the Hofmeister series for RiAFP.

6.4.4.4 Quantitative Predictions of Protein Properties from Salt-Induced Enhancement

The protein-specific salting-out constant, Ω, and the protein-specific salting-in constant, Λ, of Eq. (6.4) reflect physicochemical properties of the protein (Melander and Horváth 1977). Thus, if the concept of solubility-induced enhancement is correct, then it should be possible to use the information derived from the antifreeze measurements to predict features of the protein that are reflected by these constants.

The protein-specific salting-out constant, Ω, of Eq. (6.4) represents the hydrophobic properties of the protein and is a function of its nonpolar surface area, ϕ. The numeric value of ϕ, in units of square Ångstrøm, may be obtained from Ω as (Melander and Horváth 1977):

$$\phi = 411\Omega - 12 \tag{6.5}$$

According to Eq. (6.4), Ω is given by the slope of the linear relationship depicted in Fig. 6.5. Using the value of 1.4 cm dyn^{-1} for Ω in Eq. (6.5) gives a value for ϕ for RiAFP of 563 Å^2. This is about 20% of the total surface area of the protein (Kristiansen et al. 2008). According to Melander and Horváth (1977), ϕ is typically between 20% and 40% of the total surface area of proteins. The sensitivity of RiAFP to become enhanced by salts therefore seems to correspond well with the expected salt sensitivity of a protein of its size.

The intrinsic protein salting-in constant, Λ, of Eq. (6.4) reflects the favorable electrostatic forces acting to enhance the solubility of the protein and is a function of its dipole moment, μ. The dipole moment, μ, may be numerically obtained in units of Debye from Λ using the formula (Melander and Horváth 1977):

$$\mu = -578\Lambda \tag{6.6}$$

According to Eq. (6.4), the value of Λ is given by the intercept of the linear relationship depicted in Fig. 6.5 and has the value of –0.3 molal^{-1}. This gives a predicted dipole moment for RiAFP of 173 Debye. Since the original study was published (Kristiansen et al. 2008), the crystal structure of RiAFP has become available (Hakim et al. 2013). The structure file (PDB 4DT5) contains two molecules, A and B, which, when submitted to the online Protein Dipole Moments Server (Felder et al. 2007) has predicted dipole moments of 182 Debye and 125 Debye, respectively. It is noteworthy that the dipole moment of RiAFP, derived from its molecular structure, coincides within a few percentage points with the dipole moment derived from the effects of salts on the antifreeze potency of the protein.

Considering the above presented relations, it appears obvious that salts enhance the antifreeze potency by lowering the solubility of AF(G)Ps. Since the other small mass solutes known to enhance the antifreeze potency of AF(G)P, i.e., polyols, amino acids, sugars etc., act on protein solubility in a manner similar to that of salts, they are all likely to operate by the same mechanism.

6.4.5 Molecular Size, Solubility, and Antifreeze Potency

Several explanations are provided for the effect of size on the potency of AF(G)Ps. In those cases where the IBS does not vary with the size of the protein, the size effect is ascribed to the larger AFP–macromolecule complex covering a larger surface area than the AFP alone. This larger coverage effectively reduces the intermolecular adsorbent gap between adsorbed AFPs at the ice surface, thereby displacing the hysteresis freezing point to a lower temperature (Wu et al. 1991). When the variation in molecular size of the protein involves changes in the size of the IBS, then the effect has additionally been ascribed to various aspects of their ice-binding ability (Leinala et al. 2002; Mok et al. 2010; Chao et al. 1996; Liu et al. 2005). The increased potency reported for a natural and several synthetic intramolecular multimers of AFPs is ascribed to an overall greater likelihood of successful adsorption due to the presence of multiple IBSs (Miura et al. 2001; Nishimiya et al. 2003) or to increased overall ice-binding area (Baardsnes et al. 2003).

Although some, or even all, of these explanations may contribute to some extent to the observed effect, there are nevertheless problems associated with their applicability. For instance, Marshall et al. (2004b) pointed out that, explanations relying on differences in interaction energies at the IBS are not likely to be correct, since AF(G)Ps are irreversibly adsorbed onto the ice surface, i.e., it is an all-or-none situation. As alluded to above (Sect. 6.4.2), if the AF(G)Ps become irreversibly adsorbed by freezing onto the interface, then they are as strongly adsorbed to the ice as any piece of ice is to the surface of ice. Thus, changing the size of the IBS, or the like, should not make any difference. In the case of the added surface cover explanation provided by Wu et al. (1991), it is intuitively logical and could well be a satisfactory explanation. However, as pointed out by the original authors, experimentally there is no correlation between the size of the enhancer and the enhancement effect (Wu and Duman 1991). The enhancers, identified by Wu and Duman (1991), range according to efficiency as 70 kDa (endogen enhancer) > 70 kDa (protein ice nucleator) > 800 kDa (lipoprotein ice nucleator) > 150 kDa (antibody) > gelatin (80–375 kDa) > agar (average 120 kDa). The effectiveness of all these enhancers is surpassed by a 28 kDa endogenous enhancer (Wang and Duman 2006). Also, Horwath et al. (1996) reported that an efficient endogenous enhancer from the beetle *Tenebrio molitor* was 12 kDa, about the same size as the AFP. Thus, there seem to be little experimental support for the otherwise logical contention that the enhancement effect of size arises from added surface cover of the adsorbent complex.

Equation (6.2) provides a general explanation to the size effect; variations in size is inevitably accompanied by variations in the nonpolar surface area, *A*, of the protein and probably also variations in the electrostatic forces, *E*, acting between the protein and the solution. Such size-induced differences in solubility is consistent with the gradual increase in antifreeze potency with size that are reported for structurally similar variants of both hyperactive and moderately active AF(G)Ps. This approach also provides an explanation as to why there is no correlation between antifreeze potency and size for macromolecules that are very different; if the structures are different, then differences in their nonpolar surface areas and the strength of the electrostatic forces acting between the structure and the surrounding water do not vary with size. In other words, the solubility of structurally different compounds does not vary with molecular size. This would explain why a 28 kDa protein is a far more efficient enhancer than a protein of 800 kDa; the smaller is simply less soluble.

Some complicated and intriguing findings have been reported that ties in well with the concept of solubility-induced enhancement; Wang and Duman (2005) found that certain of the isoforms of hyperactive AFPs, DAFPs, from *D. canadensis* interact, and the association results in greater activities. This greater activity may be ascribed to a reduced solubility due to the overall larger nonpolar surface area, ϕ, of the complex (Eqs. 6.2, 6.4 and 6.5). But further, they found that the additive glycerol only acted as an enhancer if the isoforms interacted. This may also be understood from Eqs. (6.4) and (6.5); the sensitivity of a protein to some additive increases with increased nonpolar surface area, ϕ. It should be noted that several of the polyols, glycerol included, actually reduces the surface tension of water. Nevertheless, Gekko and Timasheff (1981) found that glycerol lowered the solubility of proteins by the same mechanism as salts, i.e., polyols act differently at the air/water interface than at the protein/water interface.

Amornwittawat et al. (2008) found that many carboxylates enhanced DAFPs and ascribed the effect to aggregation of DAFPs. As in the case with Wang and Duman (2005), such aggregation results in lowered solubility due to increased overall nonpolar surface area, which could explain the increased activity. Wang et al. (2009b) identified the binding sites for these carboxylates to be specific arginine residues in the DAFP structure, since blocking these residues abolished the effect. With intact such residues the monomeric DAFP aggregated in the presence of carboxylates and the complex was more sensitive to other additives. This situation is similar to that of Wang and Duman (2005) described above, i.e., the greater nonpolar surface area, ϕ, of the complex makes the overall complex more sensitive to additives than the monomers alone (Eqs. 6.2, 6.4 and 6.5).

As have been outlined above, ascribing variations in antifreeze potency to variations in protein solubility explain many aspects of hysteresis activity, including the significance of size and how additives enhance AF(G)Ps. This approach also explains why interactions between isoforms cause enhancement and the increased sensitivity to additives when isoforms interact. Ascribing variability of antifreeze potency to variations in protein solubility give a plausible explanation to the natural variability reported among AF(G)Ps that are either hyperactive or moderately active. Both qualitative and quantitative agreements with predictions based on established theory support this approach.

The presence of a 1 molal solution of sodium citrate has the effect on the hysteresis activity of RiAFP equivalent to elevating its concentration 50-fold (Kristiansen et al. 2008). Thus, aside from the categorization into hyperactive and moderately active, which are consequences of structural aspects of their IBS, the physicochemical property of solubility is probably the most dominant determinant of AF(G)P potency.

6.5 AF(G)Ps and Ice Nucleation

The convex surface zones that grow out at the ice surface within the hysteresis gap are developing ice nuclei with their critical radius at that temperature. At the hysteresis freezing point, the phenomenon is terminated by a surface nucleation event, as one of these surface nuclei initiates nucleation. Apart from causing thermal hysteresis by controlling the development of nuclei at the ice surface, AFPs also interact with structures in the body fluids that can trigger an ice nucleation event. Such a structure is known as an ice-nucleating agent, INA, and the nucleation process triggered by INAs is referred to as heterogeneous nucleation. This is to distinguish this kind of nucleation from that which occur by spontaneous ordering of water, so-called homogenous nucleation. Evidence suggest that the ice nucleation sites of INAs are structurally related to the IBSs of AFPs. Thus, the mechanism of ice nucleation by INAs may be very similar to the mechanism of adsorption of AFPs to ice.

6.5.1 Biological Relevance of INAs

Freeze-avoiding species die if their body fluids freeze out. Consequently, they rely on extensive supercooling of their body fluids to survive subfreezing temperatures. Any incidental INAs in the body fluids of such an organism would therefore be potentially lethal. Freeze-avoiding insects are known to remove or reduce the amount of such incidental INAs that could pose a threat (Neven et al. 1986; Olsen and Duman 1997a, b). In addition, AFPs prevent incidental INAs from initiating freezing by physically interacting with such structures (Olsen and Duman 1997a, b; Duman 2002). By removal of INAs from their body fluids and by producing high concentrations of AFPs, the supercooling points of freeze avoiding larvae of the pychroid beetle *Dendroides canadensis* changes from about −7 °C in the summer to below −30 °C during winter (Olsen and Duman 1997a, b).

Freeze-tolerant species, that adaptively allow their body fluids to freeze out, often produce INAs and allocate them to the extracellular fluid. The principal function of such adaptive INAs in freeze tolerance is to prevent harmful cellular freezing by initiating a preemptive nucleation event outside the cells at a temperature above the nucleation temperature of any incidental harmful cellular INAs (Zachariassen and Hammel 1976). Since solutes are excluded from the growing ice mass, the extracellular freezing event causes the remaining unfrozen extracellular fluid fraction to become increasingly concentrated. This in turn initiates a concomitant osmotic efflux of water out from the cells. The extracellular freezing process and consequent efflux of cell water continues until the melting point of the remaining unfrozen fluid fraction is colligatively depressed to the environmental temperature, at which point the danger of harmful cellular freezing is eliminated.

6.5.2 Overall Structural Aspects of INAs

It is vital for the functionality of adaptive INAs found in freeze-tolerant species that their nucleation temperature is above that of any incidental harmful cellular INAs. The efficiency of INAs to initiate nucleation depends on their size. This may be understood from Eq. (6.1); the larger the diameter of the INA, d, the less supercooling, ΔT, is required to initiate nucleation. Consistent with this, adaptive INAs found in freeze tolerant species are very large structures. It is likely that the great potency of biologically adaptive INAs results from association between monomeric INA molecules; it has been shown that an adaptive 800 kDa INA from freeze-tolerant larvae of the cranefly, *Tipula trivittata*, form long chains of 800 kDa monomers, akin to pearls-on-a-string, and that two such chains align side by side into extended dimers (Yeung et al. 1991). This association apparently relies on the presence of phosphatidylinositol, PI, at the surface of the INA, as enzymatic removal of PI depressed the nucleation temperature (Neven et al. 1989). PI has also been shown to anchor highly active bacterial INAs to the bacterial membrane (Kozloff et al. 1991) and thereby possibly causing them to cooperate. The proposed structure of a large repetitive segment of the 123 kDa INA from *Pseudonomas borealis* suggests that the operating INA consists of at least two monomers (Garnham et al. 2011b).

AFPs are known to physically interact with INA molecules (Wu and Duman 1991). A simple explanation to how AFPs depress the nucleation temperature of INAs would be if they act by preventing them from forming larger associations, analogous to the effect of reducing the diameter of growing surface nuclei at the ice surface (Eq. 6.1). A peculiar aspect of this AFP/INA association is that it apparently does not involve the IBS of the AFP (Duman 2001). This is evident from the fact that the hysteresis activity, which requires the IBS to be free to adsorb onto the ice surface, is enhanced by the AFP/INA interaction (Wu and Duman 1991). It remains unclear if AFPs contain some secondary functional surface-site outside the IBS dedicated to the interaction of structures other than ice (Duman 2001).

Although the details of how INAs trigger freezing is not entirely identified, it is likely that they do so by structuring their hydration water to mimic that of ice. It has been shown that the hydration water at the IBS of β-helical hyperactive AFPs are clathrate-like, and this structured water has been implicated in the process of adsorption (Garnham et al. 2011a). Large internal repetitive parts of several bacterial INAs have been modeled to fold into β-helixes (Graether and Jia 2001; Garnham et al. 2011b). The structural similarity between the IBS of the β-helical AFPs and the suspected nucleation sites of the INAs suggest they share a similar mode of operation. Supporting this contention, Kobashigawa et al. (2005) reported that a recombinant protein corresponding to an internal part of one of these bacterial INAs shape ice crystals into hexagonal bipyramids. Similar results were also reported by Xu et al. (1998), who found that a 164 kDa molecule with INA activity shaped ice crystals into hexagonal bipyramids. Apparently, these bacterial INAs have some kind of internal IBS. It is not clear if the part of the INA responsible for the observed structuring of the ice, an IBS, corresponds to the site that causes nucleation. Another aspect is the shape of the ice crystals in the presence of the INAs reported by Kobashigawa et al. (2005); these INAs are those of the species *Pseudonomas syringae*, the same INA modeled as a β-helix by Graether and Jia (2001). The ice crystals in the presence of all known β-helical AFPs express multiple ice crystal planes, e.g., in the form of hexagonal discs. These INAs, on the other hand, shape ice into hexagonal bipyramids, as seen in the presence of the monoplane-specific AF(G) Ps of fish.

If the IBS of AF(G)Ps is structurally comparable to the nucleation sites of INAs, then why are AF(G)Ps not INAs? The explanation may in part rely on differences in the structure of the hydration water at the IBS/nucleation site and in part be due to the large difference in size of AFPs and INAs. What is clear is that β-helical insect AFPs do not act as INAs within the supercooling range of the freeze avoiding insects, i.e., down to about −30 °C, or even below.

6.6 Conclusions

This chapter has dealt with the *modus operandi* of AF(G)Ps. The characteristic prevention of ice growth within the hysteresis gap is explained by ice/water vapor pressure equilibrium being maintained by the Kelvin effect as the ice surface grows out as microscopic curvatures between adsorbed AF(G)Ps. The different potencies of moderately and hyperactive AF(G)Ps are ascribed to differences in their adsorption habits, whereas variations in antifreeze potencies within each of these categories are ascribed to variations in their solubilities. In the latter case, experimental proof of concept is discussed in the context of basic solubility theory. Some characteristics of ice-nucleating agents (INAs) in relation to AF(G)Ps and their relevance in cold tolerance was also briefly examined.

AF(G)Ps as a group are defined by their shared capacity to prevent ice in solution from growing at temperatures below the melting point. However, another widespread trait observed for many of these proteins when at very low concentrations occurs at the melting temperature; they inhibit the spontaneous process by which larger ice crystals grow at the expense of smaller crystals. This trait is not an exclusive property of AF(G)Ps but are also found among non-antifreeze proteins and organic solutes. This fascinating phenomenon of recrystallisation inhibition is both biologically and commercially important and is the topic of the next chapter.

References

Amornwittawat N, Wang S, Duman JG, Wen X (2008) Polycarboxylates enhance beetle antifreeze protein activity. Biochim Biophys Acta 1784:1942–1948

Amornwittawat N, Wang S, Banatlao J, Chung M, Velasco E, Duman JG, Wen X (2009) Effects of polyhydroxy compounds on beetle antifreeze protein activity. Biochim Biophys Acta 1794:341–346

Baardsnes J, Kuiper MJ, Davies PL (2003) Antifreeze protein dimer. When two ice binding faces are better than one. J Biol Chem 278:38942–38947

Bull HB, Breese K (1974) Surface tension of amino acid solutions: a hydrophobicity scale of amino acid residues. Arch Biochem Biophys 161:665–670

Can O, Holland NB (2011) Conjugation of type I antifreeze protein to polyallylamine increases thermal hysteresis activity. Bioconjug Chem 22:2166–2171

Can O, Holland NB (2013) Utilizing avidity to improve antifreeze protein activity: a type III antifreeze protein trimer exhibits increased thermal hysteresis activity. Biochemistry 52:8745–8752

Caple G, Kerr WL, Burcham TS, Osuga DT, Yeh Y, Feeney RE (1986) Superadditive effects in mixtures of fish antifreeze glycoproteins and polyalcohols or surfactants. J Colloid Interface Sci 111:299–304

Celik Y, Graham LA, Mok Y-F, Bar M, Davies PL, Braslavsky I (2010) Superheating of ice crystals in antifreeze protein solutions. Proc Natl Acad Sci 107:5423–5428

Celik Y, Drori R, Pertaya-Braun N, Altan A, Barton T, Bar-Dolev M, Groisman A, Davies PL, Braslavsky I (2013) Microfluidic experiments reveal that antifreeze proteins bound to ice crystals suffice to prevent their growth. Proc Natl Acad Sci 110:1309–1314

Chakraborty S, Jana B (2019) Ordered hydration layer mediated ice adsorption of a globular antifreeze protein: mechanistic insight. Phys Chem Chem Phys 21:19298–19310

Chalmers B (1964) Principles of solidification. Wiley, New York

Chao H, DeLuca CL, Davies PL (1995) Mixing antifreeze protein types changes ice crystal morphology without affecting antifreeze activity. FEBS Lett 357:183–186

Chao H, Hodges RS, Kay CM, Gauthier SY, Davies PL (1996) A natural variant of Type I antifreeze protein with four ice-binding repeats is a particularly potent antifreeze. Protein Sci 5:1150–1155

Cohn EJ (1925) The physical chemistry of the proteins. Physiol Rev 5:349–437

Cziko PA, DeVries AL, Evans CW, Cheng C-HC (2014) Antifreeze protein-induced superheating of ice inside Antarctic notothenioid fishes inhibits melting during summer warming. Proc Natl Acad Sci 111:14583–14588

DeLuca CI, Comley R, Davies PL (1998) Antifreeze proteins bind independently to ice. Biophys J 74:1502–1508

DeVries AL (1971) Glycoproteins as biological antifreeze agents in Antarctic fishes. Science 172:1152–1155

DeVries AL (1982) Biological antifreeze agents in Coldwater fishes. Comp Biochem Physiol A 73:627–640

Drori R, Davies PL, Braslavsky I (2015) Experimental correlation between thermal hysteresis activity and the distance between antifreeze proteins on an ice surface. RSC Adv 5:7848–7853

Duman JG (2001) Antifreeze and ice nucleator proteins in terrestrial arthropods. Annu Rev Physiol 63:327–357

Duman JG (2002) The inhibition of ice nucleators by insect antifreeze proteins is enhanced by glycerol and citrate. J Comp Physiol B 172:163–168

Duman JG, Bennett V, Sformo T, Hochstrasser R, Barnes BM (2004) Antifreeze proteins in Alaskan insects and spiders. J Insect Physiol 50:259–266

Evans PE, Hobbs RS, Goddard SV, Fletcher GL (2007) The importance of dissolved salts to the *in vivo* efficacy of antifreeze proteins. Comp Biochem Physiol A 148:556–561

Felder CE, Prilusky J, Silman I, Sussman JL (2007) A server and database for dipole moments of proteins. Nucleic Acids Res 35:W512–W521. http://dipole.weizmann.ac.il/

Fletcher GL, Hew CL, Davies PL (2001) Antifreeze proteins in teleost fishes. Annu Rev Physiol 63:359–390

Friis DS, Kristiansen E, von Solms N, Ramløv H (2014) Antifreeze activity enhancement by site directed mutagenesis on an antifreeze protein from the beetle *Rhagium mordax*. FEBS Lett 588:1767–1772

Garnham CP, Campbell RL, Davies PL (2011a) Anchored clathrate waters bind antifreeze proteins to ice. Proc Natl Acad Sci 108:7363–7367

Garnham CP, Campbell RL, Walker VK, Davies PL (2011b) Novel dimeric β-helical model of an ice nucleation protein with bridged active sites. BMC Struct Biol 11:36

Gekko K, Timasheff SN (1981) Mechanism of protein stabilization by glycerol: preferential hydration in glycerol-water mixtures. Biochemistry 20:4667–4676

Gong HS, Croft K, Driedzic WR, Ewart VK (2011) Chemical chaperoning action of glycerol on the antifreeze protein of rainbow smelt. J Therm Biol 36:78–83

Graether SP, Jia Z (2001) Modeling *Pseudomonas syringae* ice-nucleation protein as a β-helical protein. Biophys J 80:1169–1173

Graether SP, Kuiper MJ, Gagné SM, Walker VK, Jia Z, Sykes BD, Davies PL (2000) β-helix structure and ice-binding properties of a hyperactive antifreeze protein from an insect. Nature 406:325–328

Graham LA, Davies PL (2005) Glycine-rich antifreeze proteins from snow fleas. Science 310:461

Grandum S, Yabe A, Nakagomi K, Tanaka M, Takemura F, Kobayashi Y, Frivik P-E (1999) Analysis of ice crystal growth for a crystal surface containing adsorbed antifreeze proteins. J Cryst Growth 205:382–390

Hakim A, Nguyen JB, Basu K, Zhu DF, Thakral D, Davies PL, Isaacs FJ, Modis Y, Meng W (2013) Crystal structure of an insect antifreeze protein and its implications for ice binding. J Biol Chem 288:12295–12304

Hansen TN, Baust JG (1988) Serial dilution of *Tenebrio molitor* haemolymph: analysis of antifreeze activity by differential scanning calorimetry. Cryo-Letters 9:386–391

Hansen TN, DeVries AL, Baust JG (1991) Calorimetric analysis of antifreeze glycoproteins of the polar fish, *Dissostichus mawsoni*. Biochim Biophys Acta 1079:169–173

Haymet ADJ, Ward LG, Harding MM, Knight CA (1998) Valine substituted winter flounder 'antifreeze': preservation of ice growth hysteresis. FEBS Lett 430:301–306

Haymet ADJ, Ward LG, Harding MM (1999) Winter flounder "antifreeze" proteins: synthesis and ice growth inhibition of analogues that probe the relative importance of hydrophobic and hydrogen-bonding interactions. J Am Chem Soc 121:941–948

Hayward JA, Haymet ADJ (2001) The ice/water interface: molecular dynamics simulations of the basal, prism, 2021 and 2110 interfaces of ice Ih. J Chem Phys 114:3713–3726

Holland NB, Nishimiya Y, Tsuda S, Sönnichsen FD (2008) Two domains of RD3 antifreeze protein diffuse independently. Biochemistry 47:3955–3941

Horwath KL, Easton CM, Poggioli GJ Jr, Myers K, Schnorr IL (1996) Tracking the profile of a specific antifreeze protein and its contribution to the thermal hysteresis activity in cold hardy insects. Eur J Entomol 93:419–433

Jia Z, Davies PL (2002) Antifreeze proteins: an unusual receptor-ligand interaction. Trends Biochem Sci 27:101–106

Kaushik JK, Bhat R (1998) Thermal stability of proteins in aqueous polyol solutions: role of the surface tension of water in the stabilizing effect of polyols. J Phys Chem B 102:7058–7066

Kerr WL, Burcham TS, Osuga DT, Yeh Y, Feeney RE (1985) Synergistic depression of the freezing temperature in solutions of polyhydroxy compounds and antifreeze glycoproteins. Cryo-Letters 6:107–114

Knight CA, DeVries AL (1988) The prevention of ice crystal growth from water by "antifreeze proteins". In: Wagner PE, Vali G (eds) Atmospheric aerosols and nucleation. Lecture notes in physics 309. Springer, Berlin

Knight CA, DeVries AL (1989) Melting inhibition and superheating of ice by an antifreeze glycopeptide. Science 245:505–507

Knight CA, Cheng CC, DeVries AL (1991) Adsorption of α-helical antifreeze peptides on specific ice crystal surface planes. Biophys J 59:409–418

Knight CA, Driggers E, DeVries AL (1993) Adsorption to ice of fish antifreeze glycopeptides 7 and 8. Biophys J 64:252–259

Kobashigawa Y, Nishimiya Y, Miura K, Ohgiya S, Miura A, Tsuda S (2005) A part of ice nucleation protein exhibits the ice-binding ability. FEBS Lett 579:1493–1497

Kozloff LM, Turner MA, Arellano F, Lute M (1991) Phosphatidylinositol, a phospholipid of ice-nucleating bacteria. J Bacteriol 173:2053–2060

Kristiansen E, Zachariassen KE (2005) The mechanism by which fish antifreeze proteins cause thermal hysteresis. Cryobiology 51:262–280

Kristiansen E, Pedersen SA, Zachariassen KE (2008) Salt-induced enhancement of antifreeze protein activity: a salting-out effect. Cryobiology 57:122–129

Landt E (1931) The surface tension of solutions of various sugars. Z Ver Dtsch Zucher-Ind 81:119–124

Laursen RA, Wen D, Knight CA (1994) Enantioselective adsorption of the D- and L-forms of an α-helical antifreeze polypeptide to the {2021} planes of ice. J Am Chem Soc 116:12057–12058

Leinala EK, Davies PL, Doucet D, Tyshenko MG, Walker VK, Jia Z (2002) β-Helical antifreeze protein isoform with increased activity. J Biol Chem 277:33349–33352

Li N, Andorfer C, Duman JG (1998) Enhancement of insect antifreeze protein activity by solutes of low molecular mass. J Exp Biol 201:2243–2251

Liou YC, Tocilj A, Davies PL, Jia Z (2000) Mimicry of ice structure by surface hydroxyls and water of a beta-helix antifreeze protein. Nature 406:322–324

Liu K, Jia Z, Chen G, Tung C, Liu R (2005) Systematic size study of an insect antifreeze protein and its interaction with ice. Biophys J 88:953–958

Liu Z, Li H, Pang H, Me J, Mao X (2015) Enhancement effect of solutes of low molecular mass on the insect antifreeze protein ApAFP752 from *Anatolica polita*. J Therm Anal Calorim 120:307–315

Lu K, Li Y (1998) Homogeneous nucleation catastrophe as a kinetic stability limit for superheated crystal. Phys Rev Lett 80:4474–4477

Marshall CB, Tomczak MM, Gauthier SY, Kuiper MJ, Lankin C, Walker VK, Davies PL (2004a) Partitioning of fish and insect antifreeze proteins into ice suggests they bind with comparable affinity. Biochemistry 43:148–154

Marshall CB, Daley ME, Sykes BD, Davies PL (2004b) Enhancing the activity of a β-helical antifreeze protein by the engineered addition of coils. Biochemistry 43:11637–11646

Matubayasi N, Nishiyama A (2006) Thermodynamic quantities of surface formation of aqueous electrolyte solutions VI. Comparison with typical nonelectrolytes, sucrose and glucose. J Colloid Interface Sci 298:910–913

Melander W, Horváth C (1977) Salt effects on hydrophobic interactions in precipitation and chromatography of proteins: an interpretation of the lyotropic series. Arch Biochem Biophys 183:200–215

Miura K, Ohgiya S, Hoshino T, Nemoto N, Suetake T, Miura A, Spyracopoulos L, Kondo H, Tsuda S (2001) NMR analysis of Type III antifreeze protein intramolecular dimer. Structural basis for enhanced activity. J Biol Chem 276:1304–1310

Mok Y-F, Lin F-H, Graham LA, Celik Y, Braslavsky I, Davies PL (2010) Structural basis for the superior activity of the large isoform of snow flea antifreeze protein. Biochemistry 49:2593–2603

Neven LG, Duman JG, Beals JM, Castellino FJ (1986) Overwintering adaptations of the stag beetle, *Ceruchus piceus*: removal of ice nucleators in winter to promote supercooling. J Comp Physiol B 156:707–716

Neven L, Duman JG, Low MG, Sehl LC, Castellino FJ (1989) Purification and characterization of an insect hemolymph lipoprotein ice nucleator: evidence for the importance of phosphatidylinositol and apolipoprotein in the ice nucleator activity. J Comp Physiol B 159:71–82

Nishimiya Y, Ohgiya S, Tsuda S (2003) Artificial multimers of the type III antifreeze protein. Effects on thermal hysteresis and ice crystal morphology. J Biol Chem 278:32307–32312

Nishimiya Y, Sato R, Takamichi M, Miura A, Tsuda S (2005) Co-operative effect of the isoforms of type III antifreeze protein expressed in Notched-fin eelpout, *Zoarces elongatus Kner*. FEBS J 272:482–492

Olsen TM, Duman JG (1997a) Maintenance of the supercooled state in overwintering pyrochroid beetle larvae, *Dendroides canadensis*: role of hemolymph ice nucleators and antifreeze proteins. J Comp Physiol B 167:105–113

Olsen TM, Duman JG (1997b) Maintenance of the supercooled state in the gut fluid of overwintering pyrochroid beetle larvae, *Dendroides canadensis*: role of ice nucleators and antifreeze proteins. J Comp Physiol B 167:114–122

Olsen TM, Sass SJ, Li N, Duman JG (1998) Factors contributing to seasonal increases in inoculative freezing resistance in overwintering fire-colored beetle larvae *Dendroides canandensis* (Pyrochroidae). J Exp Biol 201:1585–1594

Pertaya N, Marshall CB, DiPrinzio CL, Wilen L, Thomson ES, Wettlaufer JS, Davies PL, Braslavsky I (2007) Fluorescence microscopy evidence for quasi-permanent attachment of antifreeze proteins to ice surfaces. Biophys J 92:3663–3673

Pertaya N, Marshall CB, Celik Y, Davies PL, Braslavsky I (2008) Direct visualization of spruce budworm antifreeze protein interacting with ice crystals: basal plane affinity confers hyperactivity. Biophys J 95:333–341

Poynting JH (1881) Change of state: solid–liquid. Philos Mag 5th series 12:32–48

Ramsay JA (1964) The rectal complex of the mealworm *Tenebrio molitor* L. (Coleoptera, Tenebrionidae). Philos Trans R Soc B 348:279–314

Raymond JA, DeVries AL (1977) Adsorption inhibition as a mechanism of freezing resistance in polar fishes. Proc Natl Acad Sci 74:2589–2593

Reynolds JA, Gilbert DB, Tanford C (1974) Empirical correlation between hydrophobic free energy and aqueous cavity surface area. Proc Natl Acad Sci 71:2925–2927

Schrag JD, O'Grady SM, DeVries AL (1982) Relationship of amino acid composition and molecular weight of antifreeze glycopeptides to non-colligative freezing point depression. Biochim Biophys Acta 717:322–326

Sönnichsen FD, DeLuca CI, Davies PL, Sykes BD (1996) Refined solution structure of type III antifreeze protein: hydrophobic groups may be involved in the energetics of the protein-ice interaction. Structure 4:1325–1337

Sørensen TF, Ramløv H (2001) Variations in antifreeze activity and serum inorganic ions in the eelpout *Zoarces viviparus*: antifreeze activity in the embryonic state. Comp Biochem Physiol A 30:123–132

Stevens CA, Drori R, Zalis S, Braslavsky I, Davies PL (2015) Dendrimer-linked antifreeze proteins have superior activity and thermal recovery. Bioconjug Chem 26:1908–1915

Thomson W (1871) On the equilibrium of vapour at a curved surface of liquid. Philos Mag 42:448–452

Tolls J, van Dijk J, Verbruggen EJM, Hermens JLM, Loeprecht B, Schüürmann G (2002) Aqueous solubility-molecular size relationships: a mechanistic case study using C10- to C19-alkanes. J Phys Chem A 106:2760–2765

Turnbull D (1950) Kinetics of heterogenous nucleation. J Chem Phys 18:198–203

Uhlig HH (1937) The solubilities of gases and surface tension. J Phys Chem 41:1215–1225

Wang L, Duman JG (2005) Antifreeze proteins of the beetle *Dendroides canadensis* enhance one another's activities. Biochemistry 44:10305–10312

Wang L, Duman JG (2006) A thaumatin-like protein from larvae of the beetle *Dendroides canadensis* enhances the activity of antifreeze proteins. Biochemistry 45:1278–1284

Wang S, Amornwittawat N, Banatlao J, Chung M, Kao Y, Wen X (2009a) Hofmeister effects of common monovalent salts on the beetle antifreeze protein activity. J Phys Chem B 113:13891–13894

Wang S, Amornwittawat N, Juwita V, Kao Y, Duman JG, Pascal TA, Goddard WA, Wen X (2009b) Arginine, a key residue for the enhancing ability of an antifreeze protein of the beetle *Dendroides canadensis*. Biochemistry 48:9696–9703

Washburn EW (1929) International critical tables of numerical data, physics, chemistry and technology, vol 4. McGraw-Hill, New York

Wen D, Laursen RA (1992) A model for binding of an antifreeze polypeptide to ice. Biophys J 63:1659–1662

Wen X, Wang S, Amornwittawat N, Houghton EA, Sacco MA (2011) Interaction of reduced nicotinamide adenine dinucleotide with an antifreeze protein from *Dendroides canadensis*: mechanistic implication of antifreeze activity enhancement. J Mol Recognit 24:1025–1032

Westh HP, Ramløv H, Wilson PW, DeVries AL (1997) Vapor pressure of aqueous antifreeze glycopeptide solutions. Cryo-Letters 18:277–282

Wilson PW (1993) Explaining thermal hysteresis by the Kelvin effect. Cryo-Letters 14:31–36

Wilson PW, Beaglehole D, DeVries AL (1993) Antifreeze glycopeptide adsorption on single crystal ice surfaces using ellipsometry. Biophys J 64:1878–1884

Wöhrmann APA (1996) Antifreeze glycopeptides and peptides in Antarctic fish species from the Weddell Sea and the Lazarev Sea. Mar Ecol Prog Ser 130:47–59

Wu DW, Duman JG (1991) Activation of antifreeze proteins from larvae of the beetle *Dendroides canadensis*. J Comp Physiol B 161:279–283

Wu DW, Duman JG, Xu L (1991) Enhancement of antifreeze protein activity by antibodies. Biochim Biophys Acta 1076:416–420

Xu H, Griffith M, Patten CL, Glick BR (1998) Isolation and characterization of an antifreeze protein with ice nucleation activity from the plant growth promoting rhizobacterium *Pseudomonas putida* GR12-2. Can J Microbiol 44:64–73

Yang DSC, Sax M, Chakrabartty A, Hew CL (1988) Crystal structure of an antifreeze polypeptide and its mechanistic implications. Nature 333:232–237

Yeung KL, Wolf EE, Duman JG (1991) A scanning tunneling microscopy study of an insect lipoprotein ice nucleator. J Vac Sci Technol B 9:1197–1201

Zachariassen KE, Hammel HT (1976) Nucleating agents in the haemolymph of insects tolerant to freezing. Nature 262:285–287

Zachariassen KE, Husby JA (1982) Antifreeze effect of thermal hysteresis agents protects highly supercooled insects. Nature 298:865–867

Zachariassen KE, DeVries AL, Hunt B, Kristiansen E (2002) Effect of ice fraction and dilution factor on the antifreeze activity in the hemolymph of the cerambycid beetle *Rhagium inquisitor*. Cryobiology 44:132–141

Zanetti-Polzi L, Biswas AD, Del Galdo S, Barone V, Daidone I (2019) Hydration shell of antifreeze proteins: unveiling the role of non-ice-binding surfaces. J Phys Chem B 123:6474–6480

Zepeda S, Yokoyama E, Uda Y, Katagiri C, Furukawa Y (2008) *In situ* observation of antifreeze glycoprotein kinetics at the ice interface reveals a two-step reversible adsorption mechanism. Cryst Growth Des 8:3666–3672

Chapter 7
Inhibition of Recrystallization

Carsten Budke and Thomas Koop

7.1 Introduction

The term antifreeze protein (AFP) has developed from the observation that the blood serums of Antarctic fishes freeze at lower temperatures than would be expected from the colligative ice melting point depression (DeVries 1971). This effect, observed as the separation of the ice melting point and ice freezing point, is termed thermal hysteresis (TH). In the TH temperature range the growth of ice crystals is stopped entirely. Although microscopic ice formation may occur, the growth of ice crystals to macroscopic sizes is inhibited. The discovery of AFPs in freeze-avoiding marine teleost fish together with observation of TH activity led to the suggestion that freeze avoidance and TH are linked. Marine fish, insects and other terrestrial arthropods are assigned to this group and the primary AFP function here is supposedly the prevention from freezing (Duman 2015). However, another aspect of AFPs may be controlling the size of smaller ice crystals, a protection mechanism that would belong to the class of freeze tolerance (Storey and Storey 2017). Inhibition of ice recrystallization by AFP, explained in depth in Sect. 7.4.5, may lower the risk that larger ice crystals may be clogging small blood vessels (Cziko et al. 2014). Indeed, the very high ice recrystallization inhibition (IRI) efficacy of antifreeze glycoproteins (AFGPs) from the Antarctic toothfish, *Dissostichus mawsoni*, which reduce the recrystallization rate by half at a concentration of only 22 ng mL^{-1} (1.0 nmol L^{-1}), appears to lend support for this notion (Budke et al. 2014; Olijve et al. 2016b).

Already more than 30 years ago, a correlation was found between recrystallization of intracellular ice and cell death (Mazur 1984). It was observed experimentally that high cooling rates lead to the formation of very small crystals. If such small

C. Budke · T. Koop (✉)
Faculty of Chemistry, Atmospheric and Physical Chemistry, Bielefeld University, Bielefeld, Germany
e-mail: thomas.koop@uni-bielefeld.de

H. Ramløv, D. S. Friis (eds.), *Antifreeze Proteins Volume 2*,
https://doi.org/10.1007/978-3-030-41948-6_7

crystals are melted upon fast heating, the cells survived. In contrast, slow warming damaged the cells most likely due to formation of larger ice crystals by way of recrystallization during the longer time period of warming. These observations support the supposition that IRI may be an important survival strategy for psychrophilic organisms living at sub-zero temperatures (Bar Dolev et al. 2016; Duman 2015).

7.2 Ice Recrystallization Inhibition in Nature

Generally, ice recrystallization processes can occur in various types of settings: apart from dedicated experiments in the laboratory, it is important in familiar everyday products, for example in ice cream and frozen food, in technological applications, for example in the production of porous materials, and it also occurs in nature, often with life-threatening consequences for the affected species. Next, we briefly summarize natural ice recrystallization in plant and animal species and their strategies for protecting against it.

7.2.1 Plant Species

Freeze tolerance is predominantly prevalent in plants (Ouellet and Charron 2013). These organisms can often survive extracellular freezing. AFPs in plants do typically show only low TH activity, and they are thought to inhibit ice recrystallization in the extracellular matrix, while intracellular AFPs might deactivate intracellular ice nucleators (Gupta and Deswal 2014). One exception from low TH activity of plant AFPs is an antifreeze glycolipid from the bittersweet nightshade, *Solanum dulcamara*, which shows a TH value of 3.1 °C (Walters et al. 2011). AFPs have been found in many freeze-tolerant plants (Griffith and Yaish 2004; Gupta and Deswal 2014): about 60 species have been assigned to this group. Doucet et al. showed that about 25% of overwintering species from the United Kingdom and all investigated Antarctic species exhibited antifreeze activity after cold acclimation (Doucet et al. 2000). Proteinaceous AFP molecules were isolated from 11 plants, mostly in the apoplast, and their molecular masses are in the range of 1.32–70 kDa. AFGPs were isolated from five plants with molecular masses from 29 to 200 kDa.

The presence of AFPs in the apoplast of a plant might prevent lethal propagation of extracellular ice into the cytoplasm, and another supposedly important effect is keeping the apoplastic water transport route functioning (Fig. 7.1a). The porous cell walls and the extracellular spaces forming the apoplast enable a high water transport velocity and a transport of high amounts to the xylem. Water transport through the apoplast is one of the two main pathways in plants and the important one in the cortex. When this path is blocked by the Casparian strip and water is forced to enter the symplast of endodermal cells at the junction between the cortex and the stele, the

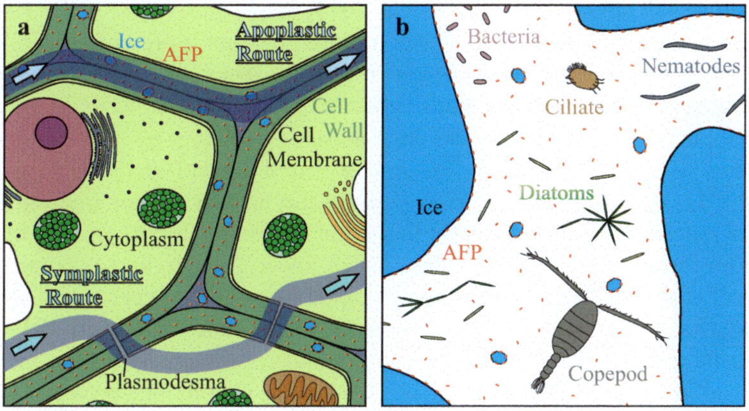

Fig. 7.1 Recrystallization in biological environments. (**a**) AFPs allow water transport in plants along the apoplastic route. (**b**) The lower part of sea ice is a habitat for several organism that use AFPs to keep water channels open within the ice

slower symplastic route through the cytoplasm and the plasmodesmata becomes dominant (Chen et al. 2011).

High IRI activity might be the key to control ice crystal growth, especially in cooperation with ice-nucleating proteins, which nucleate ice in the extracellular space at low supercooling, and with membrane-associated AFPs that prevent inoculation of the cytoplasm (Duman 2015). For example, cooperation of ice-nucleating proteins and antifreeze proteins is supposed to be important for bacteria or pollen (Dreischmeier et al. 2017; Lorv et al. 2014).

7.2.2 Animals

Freeze tolerance also occurs in other organisms, supporting the role of AFPs to function by recrystallization inhibition there, too. For example, non-proteinaceous antifreeze glycolipids showing both TH and IRI were found in the freeze-tolerant adult darkling beetle, *Upis ceramboides*, and other winter-acclimated species (Walters et al. 2009, 2011). Not all antifreeze are proteins or glycoproteins. For example, Xylomannan antifreezes exist that contain a β-mannopyranosyl-(1 $\rightarrow$ 4) β-xylopyranose saccharide backbone and a fatty acid. Such Xylomannan antifreezes were isolated from three freeze-tolerant insect species (darkling beetle *Upis ceramboides*, crane fly *Tipula trivittata*, and stonefly *Nemoura arctica*), from three freeze-avoiding insects (flat bark beetle *Cucujus clavipes puniceus*, fire-coloured beetle *Dendroides canadensis*, and stag beetle *Ceruchus piceus*), from the bittersweet night shade plant, *Solanum dulcamara*, and from muscles of the freeze-tolerant frog *Rana lessonae* (Walters et al. 2011). The freeze-tolerant Alaskan wood frog *Rana sylvatica* can survive being frozen for several months at temperatures well below zero (Larson et al. 2014; Storey and Storey 2017). Based on experiments, the authors proposed that antifreeze glycolipids present in muscles and internal organs contribute to its freeze tolerance. They act together with high concentrations of colligative solutes: glucose concentrations were found to be about 300 μmol g^{-1} wet mass in the thigh muscles (~0.050 g g^{-1}), ~600 μmol g^{-1} in the heart (~0.100 g g^{-1}), and ~790 μmol g^{-1} in the liver (~0.140 g g^{-1}).

High TH values (2–13 °C) for AFPs from the hemolymph of freeze-avoiding terrestrial arthropods indicate clearly that these AFPs are able to prevent freezing. AFPs from freeze-tolerant species show much lower TH activities thus raising the question of efficacy. TH in the hemolymph of the freeze-tolerant centipede *Lithobius forticatus,* for example, is only occasionally measurable. However, IRI can be observed throughout the winter suggesting another AFP function than in freeze-avoiding animals here (Tursman et al. 1994). An ice-binding protein from the Antarctic nematode *Panagrolaimus davidi* shows ice crystal shaping and IRI but very little or no TH (Wharton et al. 2005). Plant AFPs typically show high IRI activity but low TH. It can therefore be concluded that the main function of AFPs in plants is the inhibition of recrystallization (Gupta and Deswal 2014). Because TH measurements require AFP concentrations of several mg mL^{-1} (mmol L^{-1}) for significant values and IRI can be observed in the range of μg mL^{-1} (μmol L^{-1}), usually at least some IRI activity is observable for most ice-binding molecules. We note that in some exceptional cases of bacterial AFPs, low IRI activity is observed in spite of high TH activity. For example, two AFPs produced by the Arctic bacterium *Pseudomonas ficuserectae* revealed strong and very similar TH activity of about 2 °C at concentrations of ~2 mg mL^{-1}. However, one of them showed only very moderate IRI at 0.1 mg mL^{-1}, while the other one did show very high IRI activity already at 0.05 mg mL^{-1} when compared to AFPs from other bacterial strains that show only moderate activity even at concentrations of 0.25 mg mL^{-1} (Singh et al. 2014).

Apart from organisms that use AFPs in their body fluids and in their extracellular matrices, IRI may also be considered in influencing the habitat of a species. For example, several microorganisms such as diatoms, algae, fungi, yeasts, and bacteria apparently secrete AFPs for maintaining their environment in a liquid state, see e.g. Fig. 7.1b (Davies 2014; Raymond 2011). Raymond et al. discovered the presence of AFPs in the platelet layer of sea ice in McMurdo Sound, Antarctica (Raymond et al. 1994). Ice-pitting activity was found in extracts from diatoms and in the interstitial water from diatom-rich ice, while diatom-free seawater did not show activity. All sea ice diatom species investigated so far do produce AFPs, in contrast to their mesophilic counterparts (Raymond and Kim 2012). The extracellular concentrations in the range of μg mL^{-1} of the secreted AFPs (~25 kDa) do not allow for neither a significant freezing point depression nor TH, thus indicating the potential importance of IRI. It is interesting to note in this context that marine planktonic diatoms *Thalassiosira pseudonana* have been reported to show moderate ice nucleation activity (Alpert et al. 2011; Knopf et al. 2010).

Another example for sea ice inhabitants are bacteria belonging to the Gram-negative genus *Colwellia* (Raymond et al. 2007), from which an extracellular

~25 kDa protein was purified using its affinity to ice. IRI experiments indicate that the protein may protect membranes by acting as an ice recrystallization inhibitor. However, later studies also showed pronounced TH of 3.8 °C at 3.3 mg mL^{-1} (Hanada et al. 2014). Singh et al. found that 8 out of 14 investigated Arctic bacterial strains inhabiting cryoconites produce AFPs, too (Singh et al. 2014), and their measurements showed that TH and IRI were not correlated. The same behaviour was observed for an AFP from an Antarctic psychrophilic bacterium that forms a biological consortium with a ciliate. The AFP showed high IRI efficacy with an observable effect even at a concentration of 2.5 nmol L^{-1} and a moderate TH of 0.53 °C at 50 μmol L^{-1} (Mangiagalli et al. 2016). The authors hypothesized that this AFP either is located at the cell surface or is secreted to the surrounding in order to protect the entire cell consortium.

Another example is an Antarctic plankton, the calanoid copepod *Stephos longipes*, which inhabits brine channels within sea ice and expresses high amounts of AFPs in all of its cells (Kiko 2010). This ability could possibly explain why this copepod is capable of colonizing the ice surface layer from spring to autumn even though strong temperature fluctuation of up to 5 °C per day can occur during this period. In contrast, the copepod *Paralabidocera antarctica*, which is also a dominant calanoid copepod of Antarctic sea ice, does not express AFPs and inhabits only during wintertime the bottom layer of fast ice (i.e. sea ice that is fast along the coast and is not moving with winds or currents).

In summary, AFPs showing low TH but pronounced IRI activity from freeze-tolerant organisms are supposed to function through the inhibition of ice recrystallization. However, the presence of the same type of AFP in both freeze-avoiding and freeze-tolerant species might indicate that the simple classification system for AFP function, i.e. TH in freeze-avoiding and IRI in freeze-tolerant species, may not apply universally. To provide a deeper insight into the different efficacies of AFPs in terms of TH and IRI, the following paragraphs will focus on the underlying mechanisms.

7.3 Ice-Binding Molecules and the Gibbs–Thomson Effect

Whether TH or IRI is the main function of ice-binding AFPs in biota, the underlying molecular mechanism is their adsorption to specific faces of ice crystals. The adsorption inhibition mechanism originally proposed by Raymond and DeVries is now well accepted (Raymond and DeVries 1977). For example, experiments investigating fluorescently labelled AFP have shown that adsorption of AFPs onto ice is irreversible (Celik et al. 2013; Haleva et al. 2016; Pertaya et al. 2007). TH, IRI, and the third effect of ice binding, namely ice crystal habit modification, can be explained by the development of a local ice surface curvature between adsorbed AFP during cooling-induced ice growth.

The formation of a curved crystal surface is accompanied by a decrease in the local melting temperature. The melting temperature $T_m(r_i)$ of a crystal with a higher curvature, corresponding to a smaller 'particle' radius r_i, is below the bulk melting temperature $T_m(r_\infty)$, and in a liquid environment it can be described by the Gibbs–Thomson equation in the following form:

$$T_m(r_i) = T_m(r_\infty) - \frac{2 \cdot \sigma_{sl} \cdot T_m(r_\infty)}{\rho_s \cdot \Delta H_m} \cdot \frac{1}{r_i} \quad (7.1)$$

Here, σ_{sl} is the interface energy between the solid and the liquid phase, ρ_s the solid phase density, and ΔH_m the heat of fusion. In a solution containing AFPs, these can adsorb to the ice surface. In that case, not the actual crystal size is important, but instead the radius of the local convexities that develop in between adsorbed AFP molecules during growth. Within the TH gap, one can envision the convexities as virtual ice particles with diameters that are larger than the distances between adjacent AFPs. Hence, the macroscopic crystal growth is stopped as long as the ambient temperature is equal to or higher than the melting temperatures of the maximum convexities (corresponding to virtual ice particles that just fit in between the adsorbed AFP). At temperatures below the TH gap, the spaces between adsorbed AFPs are too large for the even smaller virtual ice particles and are unable to lead to a sufficient local melting point depression. Thus, ice growth commences, often visible in form of a burst event when growth starts explosively.

All three effects of AFPs such as TH, crystal habit modification, and IRI are due to ice binding. Hence, it might be intuitive to expect similar efficacy for a particular AFP regarding the three effects. Moreover, one may expect a similar rank-order for each effect when comparing different AFPs. However, several studies have shown that TH and IRI activity are not necessarily correlated (Capicciotti et al. 2015; Kim et al. 2017; Olijve et al. 2016a; Yu et al. 2010). Observation of TH usually requires relatively high AFP concentrations, e.g. in the range of mg mL^{-1} (mmol L^{-1}). In comparison, crystal habit modification and IRI do not require that ice crystal growth is stopped entirely and often growth inhibition is sufficient. Hence, crystal habit modification and IRI can be observed already at much lower concentrations down to µg mL^{-1} or even ng mL^{-1} (µmol L^{-1} and nmol L^{-1}, respectively) (Budke et al. 2014; Olijve et al. 2016b). This concentration dependence of activity can be understood, if we take a closer look at the fundamental processes that are at work during recrystallization.

7.4 Fundamentals of Ice Recrystallization

The process of ice recrystallization belongs to a class of so-called Ostwald ripening processes. These may involve different physical states and can occur in different kinds of chemical systems such as polycrystalline mixtures, solid solutions, or colloidal emulsions and suspensions. As we will see below, it is important for the rate of ice recrystallization whether it occurs in a solid or liquid matrix: this is not only true in terms of a theoretical description of the involved processes and their particular kinetic behaviour, but it can also affect our experimental ability to distinguish active from non-active molecular IRI inhibitors. In the following, we will discuss first the thermodynamic driving force behind ice recrystallization, before we have a closer look at its kinetics.

7.4.1 Ostwald Ripening

Ice recrystallization is an Ostwald ripening process of polycrystalline ice in a liquid matrix in which larger crystals grow at the expense of smaller ones at constant temperature. This process leads to an increase of the mean crystal size while the number of crystals and the total crystal surface area decrease with time. Figure 7.2a shows example microphotographs of ice crystals grown from an aqueous 45 wt% sucrose solution at −8 °C after certain time periods. The overall driving force of this process is the reduction of high energy surfaces between the solid and the liquid phase. This reduction in overall crystal surface area can be inferred from the increase in the mean crystal radius (Fig. 7.2b) while at the same time the solid-to-liquid ratio stays constant after an initial period as indicated by the ice volume fraction (Fig. 7.2c). The change in ice volume fraction during the first minutes of recrystallization probably indicates the melting point depression of the initial, very small crystals as given by the Gibbs–Thomson Eq. (7.1). When these initial crystals reach

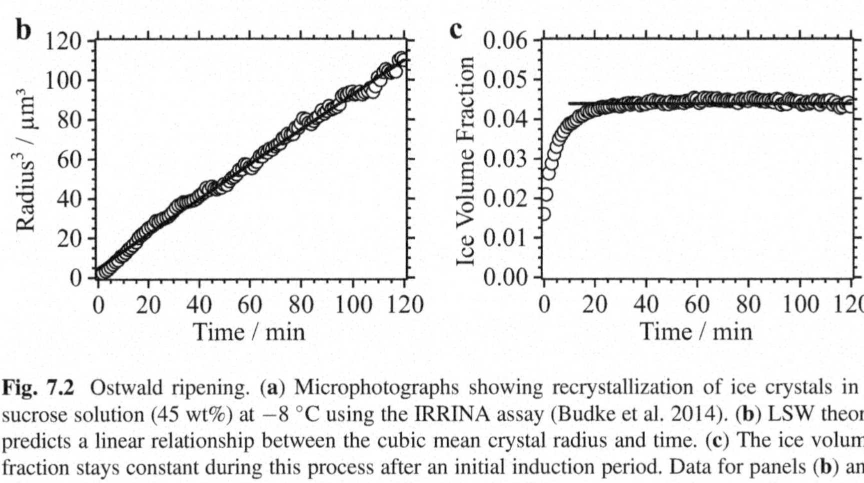

Fig. 7.2 Ostwald ripening. (**a**) Microphotographs showing recrystallization of ice crystals in a sucrose solution (45 wt%) at −8 °C using the IRRINA assay (Budke et al. 2014). (**b**) LSW theory predicts a linear relationship between the cubic mean crystal radius and time. (**c**) The ice volume fraction stays constant during this process after an initial induction period. Data for panels (**b**) and (**c**) were taken from a previous publication (Budke et al. 2009)

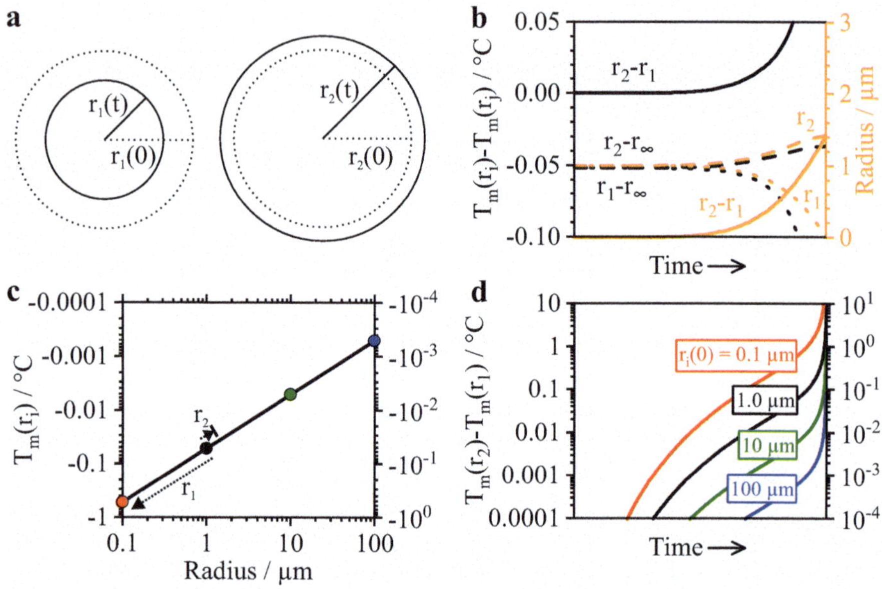

Fig. 7.3 Gibbs–Thomson effect. (**a**) Two identical particles 1 and 2 (dotted lines) are allowed to grow (r_2) and shrink (r_1) in a recrystallization process resulting in different sizes after time t. (**b**) An example with a starting radius of 1.0 µm. The melting points of both crystals are lower than 0 °C due to their small particle sizes and deviate from each other when recrystallization occurs. (**c**) If the radius of a crystal is larger, the absolute melting point depression is smaller, see the Gibbs–Thomson Eq. (7.1). (**d**) Smaller differences between the particle melting points result for larger initial radii (also indicated in **c**)

a certain size, the difference to a bulk crystal becomes negligible and the volume fraction reaches a steady-state value, at about 20–30 min in Fig. 7.2c. Although the amount of ice does not change any more, growth of larger crystals and shrinkage of smaller crystals still occurs, in accordance with the Gibbs–Thomson effect. Due to minute size differences, the larger crystals have a lower curvature thus leading to a slightly higher melting temperature. This difference between small and large crystals results in a net flux of water molecules from the smaller crystals to the larger ones even though the total amount of ice stays constant at constant temperature. For example, in Fig. 7.3 we show the melting temperature difference as a function of time for two crystals with radius $r_1(t)$ and $r_2(t)$, respectively. It is assumed that both crystals have the same initial size, i.e. $r_1(0) = r_2(0) = 1$ µm, see dotted circles in Fig. 7.3a. Once their sizes become different, for example due to small temperature fluctuations, the crystals' radii begin to further deviate from each other and, hence, also their melting temperatures until the smaller crystal vanishes completely (Fig. 7.3b).

The absolute melting temperature T_m is given as a function of ice particle radius in Fig. 7.3c. Several particle sizes are shown exemplarily as symbols with different colours. The arrows indicate the recrystallization process described above, i.e. beginning at 1 μm as shown in Fig. 7.3b. For example, such a 1 μm radius crystal possesses a melting temperature of −0.05 °C. Obviously, the melting temperature difference in a recrystallization process of two such crystals starting at this size must be small, too. Indeed, Fig. 7.3d shows that the difference is smaller than 0.1 °C most of the time during recrystallization for starting crystal sizes that are larger than 1 μm (corresponding to the black, green, and blue lines). This analysis implies that for such larger crystals with their relatively small driving force only minute changes in local curvature are needed in order to stop the recrystallization process. Therefore, it is sufficient to slightly increase the melting temperature of the small crystals and decrease it for the larger ones for compensating their melting point difference via the adsorption of AFPs at comparatively low surface coverage. In turn, such low surface coverage requires only relatively small AFP solution concentrations when compared to the high AFP concentrations required for establishing significant TH values of several tens of a degree or even several degrees.

7.4.2 Recrystallization Kinetics

As described in the previous section, AFPs can stop ice recrystallization by changing the local ice melting temperature via adsorption to the ice surface. It is sufficient to slightly increase the melting temperature of the small crystals and decrease it for the larger ones in order to compensate their melting point difference: when all surface curvatures are equal, recrystallization finally stops. Experiments show, however, that also nonbinding substances that do not induce TH or ice crystal shaping can inhibit recrystallization (Tam et al. 2008), i.e. slow it down or even stop it entirely. A natural protection strategy of organisms living at sub-zero temperatures is the increased production of cryoprotectants, e.g. polyols or sugars such as glycerol or trehalose. Solutes reduce the equilibrium melting point in a colligative manner and, therefore, they are typically assigned to the protection mechanism of freeze avoidance. At higher solute concentrations, nucleation of ice is shifted to lower temperatures and an increased solution viscosity may additionally impede ice formation (Koop and Zobrist 2009). However, elevated solute concentrations also reduce recrystallization. In order to understand this behaviour we need to have a closer look at the detailed kinetics of ice recrystallization.

Figure 7.2b shows a linear relationship between the cubic mean crystal radius r^3 and time t, which is in perfect agreement with predictions of LSW theory developed by Lifshitz, Slyozov and Wagner (Lifshitz and Slyozov 1961; Wagner 1961).

$$r^3(t) = r_0^3 + k_d \cdot t \tag{7.2}$$

Here, r_0 is the initial mean radius, and the constant k_d is the rate constant of diffusion-controlled recrystallization. This equation is valid, if the diffusion of water molecules through the matrix between ice crystals is the rate-limiting step of the recrystallization process. In the presence of AFPs, however, the liquid-to-ice interface transfer processes may become important and even rate limiting. In this case, the exponent in Eq. (7.2) is supposed to change to 2, i.e. the value for reaction-controlled recrystallization (Wagner 1961). In contrast, the consideration of coalescence in the encounter-modified theory by Lifshitz and Slyozov yields a similar linear $r^3(t)$ dependence as the original LSW approach, but with the difference of a higher recrystallization rate constant in case of coalescence (Davies et al. 1980).

7.4.3 LSW Theory at Idealized Conditions

The diffusion-controlled recrystallization rate constant is determined from LSW theory using the Gibbs–Thomson equation in terms of the concentration c_r of dissolved water molecules in local equilibrium with an ice crystal of radius r_i relative to the concentration c_∞ in equilibrium with bulk ice crystals.

$$c_r = c_\infty \exp\left(\frac{2\sigma_{sl}\Omega}{r_i RT}\right) \tag{7.3}$$

Here, σ_{sl} is the ice/liquid interface energy, Ω is the molar volume of ice, R is the gas constant, and T is absolute temperature. We note that c_∞ is the bulk thermodynamic equilibrium concentration and is usually obtained from the ice/solution liquidus curve of the phase diagram (see e.g. Fig. 7.4a). Together with Fick's law of diffusion and using the assumption of a negligible ice volume fraction Q (i.e. $Q \rightarrow 0$) the rate constant k_d can be calculated from LSW theory.

$$\lim_{Q\rightarrow 0} k_d = k_{\text{LSW}} = \frac{8\sigma_{sl}\Omega^2 D_w c_\infty}{9RT} \tag{7.4}$$

Equation (7.4) indicates the primary dependence of recrystallization on the diffusion coefficient of water D_w in the liquid matrix, which can vary by orders of magnitude due to a similar variation in the solution viscosity. The other parameters typically do not vary by the same magnitude when different mixtures are evaluated. For example, the ice/water interface energy decreases by less than 10% from 0 °C to −15 °C (Ickes et al. 2015; Koop and Murray 2016). At present, there are only very limited data available for ice/solution interface energies, but these values as well as those for interfacial energies between air and solution (i.e. surface tension) do change by a similar amount or less (e.g. Horibe et al. 1996; Hrubý et al. 2014; Lide 2004; Van Oss et al. 1992). The bulk equilibrium water concentration c_∞ does of course decrease when solutes are added. In the presence of ice, the solute concentration follows the ice melting point curve, as shown in Fig. 7.4a for aqueous sucrose solution. Cooling leads to growth of ice until the remaining solution reaches the equilibrium concentration. At −15 °C the water concentration in the liquid

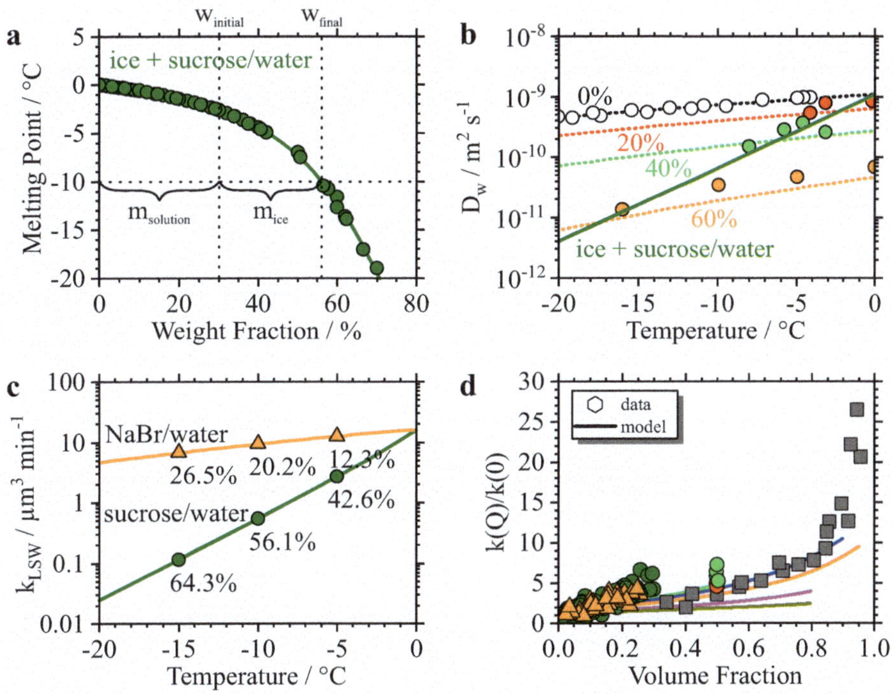

Fig. 7.4 LSW theory and volume fraction dependence of IRI. (**a**) The phase diagram and the initial solute concentration determine the amount of ice at a particular temperature (Lide 2004; Washburn 1928; Young and Jones 1949). (**b**) Change of the water diffusion coefficient with sucrose concentration and temperature (Gillen et al. 1972; Girlich et al. 1994; Hagiwara et al. 2006; He et al. 2006; Martin et al. 1999; Price et al. 1999). The dark green line corresponds to the ice melting curve shown in panel a. (**c**) Calculated LSW recrystallization rate constants based on Eq. (7.4) for ice crystals in aqueous sodium bromide and aqueous sucrose solutions. (**d**) Experimental data and model results show an increase in the recrystallization rate with increasing solid volume fraction $k(Q)$ relative to the LSW prediction at zero volume fraction $k(0)$ (Asimow 1963; Budke 2010; Davies et al. 1980; Hagiwara et al. 2006; Hardy and Voorhees 1988; Jayanth and Nash 1989; Voorhees and Glicksman 1984)

matrix is about 36 wt% (26 mol L^{-1}); for comparison the water concentration in pure water is about 55 mol L^{-1}. Other solutes do show different dependencies of the water mass fraction, e.g. solutions of electrolytes such as sodium bromide typically show lower ice melting temperatures than sucrose, thus leading to much higher water concentrations. At very low temperatures, two other processes may occur that can significantly affect the kinetics of water diffusion in the matrix: solute precipitation or glass formation. If one of them occurs, water concentrations probably never fall below about 10 mol L^{-1}, even at low temperature.

In Eq. (7.4) σ_{sl} and c_∞ in the numerator are both supposed to decrease with decreasing temperature, as does temperature itself in the denominator. As a result, their quotient only changes slightly with decreasing temperature, depending on the phase diagram of the particular solute/water mixture as mentioned above. For example, assuming the water diffusion coefficient D_w was constant, k_{LSW} would decrease by about 50% for a temperature decrease from 0 to −15 °C in sucrose solutions, and by only 5% in sodium bromide solutions. However, D_w sometimes strongly depends on temperature and solute concentration. Figure 7.4b shows experimental D_w data for pure water and sucrose solutions, as well as a parametrization used to indicate the temperature dependence of D_w at several sucrose weight fractions, i.e. 20, 40, and 60% (He et al. 2006). D_w decreases with decreasing temperature, by a factor of 2 for pure water and by a factor of 4 for a 60% sucrose solution. However, in the presence of ice the solute concentration decreases with temperature following the ice melting curve shown in panel a (dark green line), further enhancing the temperature dependence. Adding the same line to panel b shows that D_w decreases by two orders of magnitude when going from 0 °C to −15 °C. Figure 7.4c shows calculations of k_{LSW} for ice in aqueous sodium bromide and sucrose solutions, where the solute concentrations at −5 °C, −10 °C, and − 15 °C have been indicated. When the decrease in D_w along the ice melting curve is taken into account k_{LSW} decreases from about 16 μm^3 min^{-1} at 0 °C in a pure ice/water sample to about 7 μm^3 min^{-1} for sodium bromide solutions and only about 0.1 μm^3 min^{-1} in aqueous sucrose solutions at −15 °C, respectively. These calculations clearly indicate the importance of D_w for k_{LSW} and the fact that water diffusion is the rate-limiting process in ice recrystallization at negligible ice volume fractions.

7.4.4 Modifications to LSW Theory for Real Systems

In principle, LSW theory is applicable for predicting ice recrystallization rate constants only for negligible ice volume fraction, i.e. in the limit of Q approaching a value of zero. In real systems, however, the volume fraction of ice is never close enough to zero for Eq. (7.4) to apply strictly (compare Fig. 7.2c). At finite values of Q, the diffusion fields surrounding adjacent ice crystals may overlap, thus increasing diffusion rates and, hence, the IRI rate constant k_d. In general, for systems with a known phase diagram, one can calculate Q using the relationship shown in Fig. 7.4a, sometimes referred to as the lever rule. For a sucrose solution with an initial solute weight fraction of $w_{\mathrm{initial}} = 30\%$ (vertical dotted line in Fig. 7.4a) ice can form at the ice melting point at about −2.6 °C. This implies that at −2.6 °C this value of 30 wt% is the equilibrium solution concentration in the presence of minute amounts of ice. If the sample is cooled further, the amount of ice increases whereas the equilibrium sucrose concentration follows the ice melting point curve (line with green symbols). As an example, the mass ratio of ice (m_{ice}) and solution (m_{solution}) is shown in Fig. 7.4a at −10 °C, where the equilibrium sucrose weight fraction is about $w_{\mathrm{final}} = 56\%$. Using this relationship the ice mass fraction W can be calculated according to the lever rule:

$$W = \frac{m_{\text{ice}}}{m_{\text{ice}} + m_{\text{solution}}} = \frac{w_{\text{final}} - w_{\text{initial}}}{w_{\text{final}}} \tag{7.5}$$

With the additional knowledge of the densities of ice, ρ_s, and of the liquid phase, $\rho_l(w)$, which itself depends on solute concentration, the ice volume fraction Q is calculated as follows:

$$Q = \frac{V_{\text{ice}}}{V_{\text{ice}} + V_{\text{solution}}} = \left(1 + \frac{\rho_s}{\rho_l}\left(\frac{w_{\text{final}}}{w_{\text{final}} - w_{\text{initial}}} - 1\right)\right)^{-1} \tag{7.6}$$

Here, $V_{\text{ice}} = m_{\text{ice}} \cdot \rho_s$ is the volume of ice and $V_{\text{solution}} = m_{\text{solution}} \cdot \rho_l(w)$ is the volume of the aqueous solution matrix. Figure 7.4a indicates that lowering the temperature automatically increases both the solute concentration in the matrix as well as the ice volume fraction. This behaviour implies a lower water concentration c_∞ and a higher viscosity and, thus, a lower water diffusivity D_w in the matrix due to a decrease in temperature and increase in solute concentration. According to Eq. (7.4), the value of D_w is typically the most important parameter for the IRI rate constant as mentioned above.

In order to study the effect of Q on the IRI rate constant, one has to vary Q while keeping all other parameters in Eq. (7.4) constant. The latter can be achieved by performing experiments always at the same temperature, but with varying initial solute weight fractions w_{initial}. According to Fig. 7.4a the composition of the matrix w_{final} always remains the same and so c_∞, σ_{sl}, D_w, and, thus also k_{LSW} remain constant, since the properties of the matrix do not change. The only parameter that varies is the amount of ice, and hence Q. Especially in the high ice volume fraction regime where agglomeration and subsequent coalescence are dominant factors, an increase in Q will increase the recrystallization rate constant dramatically (see Fig. 7.4d). In contrast, the rate of recrystallization can be inhibited of the rate by the addition of solute, which enlarges w_{initial} and, thereby, reduces Q. The magnitude of this effect depends on the particular phase diagram and, therefore, depends on the type of solute. Figure 7.4d shows the relative dependence of the recrystallization rate constant on the volume fraction of the solid phase $k(Q)$. The rate constants are scaled relative to the value at $Q = 0$, i.e. $k(0) \equiv k_{\text{LSW}}$ to allow for a meaningful comparison by compensating the dependencies related to Eq. (7.4) (The absolute values may differ by up to several orders of magnitude as shown in Fig. 7.4c above for sucrose and sodium bromide solutions at -15 °C). All experimental data in Fig. 7.4d from alloys or from ice in fructose/water, sucrose/water or sodium bromide/water solutions show an increase of the recrystallization rate constant up to a volume fraction Q of about ~0.8, and this behaviour is also well described by several models (Budke 2010; Hagiwara et al. 2006; Hardy and Voorhees 1988; Jayanth and Nash 1989). At higher volume fraction $Q \geq 0.8$ the models fail to describe the observed steeper increase, predominantly because the assumption of spherical particles becomes incorrect when the sample begins to resemble grains in a solid matrix where recrystallization occurs by migration of grain boundaries. Additionally, the presented models do account for an overlap of the diffusion fields surrounding the particles, but most of them do not take into account the effects of coalescence.

Nevertheless, the data of Fig. 7.4d suggest that at high ice volume fraction the addition of solutes may usually lead to a decrease in the recrystallization rate, irrespective of whether the solutes actually bind to ice or not. However, in this regime it is also important to think about a potential solidification of the matrix. As outlined above, the solute concentration in the liquid matrix increases during cooling in the presence of ice. For the case of solutes such as sucrose, this increase in concentration leads to an increase in the matrix viscosity and may finally produce a glassy, i.e. amorphous solid, matrix at very low temperatures. As a result, recrystallization practically completely ceases at this point. Other solutes may reach their solubility limit when the concentration rises during cooling below the eutectic temperature and, hence, they may actually precipitate, thus also very strongly inhibiting diffusion through the now crystalline matrix between ice crystals. If only partial precipitation occurs, or if for instance polymer molecules aggregate, the resulting solid particles may trigger Zener pinning at high volume fractions. This effect describes the inhibition of grain boundary movement by second-phase particles (Humphreys and Hatherly 2004). The grain boundary area decreases when the particle intersects with it, which leads to a drag on the boundary. The production of a new surface when the boundary is moving away from the particle is unfavourable and therefore recrystallization is inhibited.

7.4.5 Recrystallization Kinetics in the Presence of Ice-Binding Molecules

As discussed above, many organisms produce AFPs or AFGPs to inhibit ice recrystallization. Here we show how the presence of an ice-binding AFGP affects ice recrystallization and its analysis by means of the LSW approach. As an example we show how AFGP8, the smallest fraction of AFGPs obtained from the Antarctic toothfish, *Dissostichus mawsoni*, inhibits ice recrystallization (DeVries et al. 1970; Nagel et al. 2011). The experimental data were obtained with the IRRINA (Ice Recrystallization Rate INhibition Analysis) assay (Budke et al. 2009), and Fig. 7.5a shows microphotographs of polycrystalline ice samples very similar to those discussed in Fig. 7.2. The two left images show how ice recrystallization proceeds over 120 min in a 45 wt% aqueous sucrose solution at −8 °C. Under such conditions, diffusion through the liquid sucrose matrix is the rate-limiting step, and hence recrystallization proceeds according to LSW theory, see Fig. 7.5b.

In two other experiments, AFGP8 was added to the 45 wt% sucrose solution in minute concentrations of 0.5 μg mL^{-1} and 1.0 μg mL^{-1}, see middle and right panel of Fig. 7.5a, respectively. In both cases, there are more and smaller ice crystals observed after 120 min when compared to pure sucrose. This behaviour is supported by LSW analysis, which shows that the IRI rate constant k_d is reduced after about

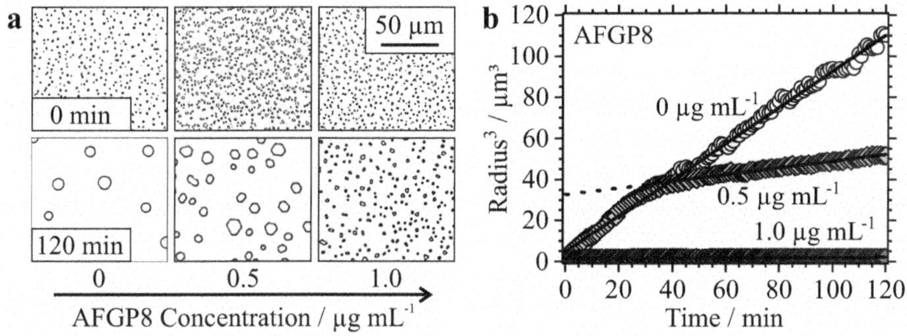

Fig. 7.5 Recrystallization inhibition by ice-binding molecules. (**a**) Microphotographs of polycrystalline ice samples formed in 45 wt% aqueous sucrose solutions at the start and after annealing at −8 °C for 120 min, without and in the presence of 0.5 or 1.0 μg mL^{-1} AFGP8, respectively (Budke et al. 2009; Nagel et al. 2011). (**b**) Temporal development of the cubic mean crystal radius for the three experiments shown in panel (**a**)

30–40 min at 0.5 μg mL^{-1}, and throughout the experiment at 1.0 μg mL^{-1}. Note that the very small concentration of AFGP8 does not alter the properties of the liquid matrix such as ice volume fraction or viscosity and water diffusion coefficient in any significant way, as these are determined by the very high sucrose concentration of 52 wt% at these conditions. This is also supported by the fact that for the experiment at 0.5 μg mL^{-1} the initial growth of ice during the first 30 min occurs at the same rate constant as for the control solution without AFGP8. At 0.5 μg mL^{-1}, the adsorption of AFGP8 molecules to the ice crystals is also evident from the emerging hexagonal crystal habits. The crystals' growth becomes significantly inhibited from about 30 min onwards: now the rate-limiting process is the release of water molecules from the melting smaller ice crystals and the incorporation of water molecules into the growing larger ice crystals. Both processes are retarded due to the ice surface curvature change by the adsorbing AFGP molecules leading to a lower melting point difference between small and large crystals, as mentioned in Sect. 7.4.1. At the higher AFGP8 concentration of 1.0 μg mL^{-1}, the turnover happens already in the first few minutes of the experiment and, hence, is not visible in Fig. 7.5b. Additionally, the surface coverage becomes high enough here to reach a complete compensation of the melting point difference between the crystals of different size. As a result, ice recrystallization is entirely stopped.

7.4.6 Summary

Ice recrystallization can be inhibited by the addition of solutes in several very different physical processes. IRI may indeed be caused by ice binding via adsorption of molecules to the growing ice crystal faces. However, IRI can also be caused by a reduced accretion and coalescence or another change in the recrystallization mechanism at high ice volume fraction, for example by an increase in viscosity of the liquid matrix, by the formation of a glassy solid or even a crystalline matrix, or by the formation of precipitate particles. Therefore, one cannot conclude based on observed recrystallization inhibition alone, whether a substance is ice binding or not.

We note, however, that the amount of substance needed to obtain a significant IRI effect may sometimes give a hint. Only very low amounts in the range of $\mu g\ mL^{-1}$ of ice-binding molecules are usually required for recrystallization inhibition, because the melting point difference between crystals of varying size can be quite small and so only a slight increase or decrease of the crystal surface curvature is sufficient to stop recrystallization, see discussion related to Fig. 7.3. In contrast, concentrations in the percent range ($>10\ mg\ mL^{-1}$) are typically required for significantly reducing the amount of ice or increasing the solution viscosity. However, the latter statement is true only in the range of lower ice volume fraction, e.g. Q below 0.5 or so. If experiments are performed at high ice volume fraction, e.g. Q larger than ~0.9, then relatively small amounts of solute can reduce the recrystallization rate even for solutes without ice-binding ability by further reducing Q and, thus, reducing accretion processes, see discussion of Fig. 7.4d. We further note that addition of very small solute amounts at very high ice volume fractions of Q close to 1 may actually lead to very high local solute concentrations in the matrix between the grain boundaries, because of the ice melting point curve. These high concentrations may induce high local viscosity or even precipitates in the matrix between the grain boundaries, and may result in a significant IRI effect even without ice binding.

7.5 Experimental Determination of Ice Recrystallization

After we have described above the underlying theoretical framework for the kinetics of ice recrystallization, we now turn to different experimental procedures for its investigation. This discussion involves different experimental setups, alternative types of sample preparation, and also various ways for the analysis of experimental data. Some of the different procedures come about because experimenters have different goals, for example to screen a large variety of molecules for IRI activity, or to quantify and compare the IRI efficacy of different molecules or mutants.

7.5.1 Methods

Ice recrystallization is investigated experimentally by way of several—sometimes just slightly different—methods (see also Chap. 9 on the measurement of antifreeze activity). To the best of our knowledge, all of them involve studying optically the behaviour of a polycrystalline ice sample over time at constant temperature. In most of these cases, some type of optical microscopy is employed such that individual ice crystals can be identified. For the same reasons, thin samples are used, i.e. slices or wavers of polycrystalline ice with typical thicknesses between ~10 μm and up to a few hundred micrometres. Exceptions are assays involving the study of polycrystalline ice in capillary or nuclear magnetic resonance (NMR) tubes (Tomczak et al. 2003; Wharton et al. 2007). It has been discussed, however, whether the observed process in capillary tubes should indeed be called ice recrystallization or not (Yu et al. 2010).

The two most common preparation methods for polycrystalline ice samples were both originally developed by Knight et al. (Knight and Duman 1986; Knight et al. 1988), and since then they have been modified at different levels of detail. The first method is usually referred to as 'sandwich splat cooling'. It involves the rapid cooling of a thin solution film between two glass slides until ice nucleation occurs. Rapid cooling at rates of tens of degrees per minute are required in order to obtain high ice nucleation rates, which then leads to a polycrystalline ice sample consisting of many individual ice crystals. Thereafter, the film is heated to a temperature somewhere below the ice melting point of the sample, typically between −8 °C and − 4 °C.

In the first study, a 300 μm thick film of 5–30 μL of the hemolymph of several insect species as well as of a solution containing the purified AFP from *Dendroides canadensis* were examined (Knight and Duman 1986). As a control an aqueous solution containing 0.9% sodium chloride, 500 mmol L^{-1} glycerol, and 100 mmol L^{-1} sorbitol was used, which are the main ingredients of the winter hemolymph of *D. canadensis*. The sandwich was cooled first to a temperature where nucleation occurs in the range between −6 °C and − 8 °C, then heated to −4 °C, and annealed at this temperature for 24 h. Photographs were taken, which reveal that recrystallization inhibition is associated with the AFP. The hemolymph of *D. canadensis* showed IRI activity at dilutions of up to 10^4 and the pure AFP was able to inhibit recrystallization at 20 μg mL^{-1} (1.3 μmol L^{-1}). For comparison, hemolymph fractions that did not contain AFPs and the hemolymph of other species that do not produce AFPs did not show any recrystallization inhibition.

The second technique is usually termed 'splat cooling', which is still commonly used with some modifications. In this assay a 20–50 μm thin ice wafer is produced by dropping a ~10 μL sample volume from a height of 1 to 3 m onto a metal block, often aluminium or copper, which is precooled to some low temperature, often about −80 °C. The formed ice wafer is then transferred onto a pre-cooled microscope stage, where recrystallization is observed at constant annealing temperature for some time.

In the pioneer study by Knight et al., annealing of an ice wafer obtained from pure water produced maximum linear ice grain dimensions of ~250 μm after 18 hours at −8 °C (Knight et al. 1988). In the presence of 100 ng mL^{-1} AFGP1–5 from the Antarctic fish *Dissostichus mawsoni*, the crystals reached a largest grain dimension of only ~100 μm, and a significant deviation from the pure water sample was found already at a concentration of 1 ng mL^{-1}.

One advantage of the sandwich splat cooling method is the possibility to produce and analyse the polycrystalline ice sample in the same apparatus, and an assay allowing for the study of up to ten samples on the same slide simultaneous has recently been introduced (Graham et al. 2018). In the splat cooling method, the polycrystalline ice sample is produced outside of the analysing apparatus and, hence, it has to be transferred to a cold stage mounted on a microscope subsequently, bearing the risk that water from laboratory air may condense onto the sample.

Both sample preparation methods are employed nowadays in slightly modified versions. In both the splat cooling and the sandwich splat cooling method often some colligative solute is added as a background, for the reasons discussed above. This goes back to a study of Knight et al. on ice-binding and non-ice binding synthetic derivatives of AFP from the winter flounder (Knight et al. 1995). At a concentration of 2 μg mL^{-1} in pure water, all investigated analogues showed IRI activity after 6 h at −6 °C, independently of whether or not they were able to induce TH. However, when analogues were investigated in 100 μg mL^{-1} aqueous sodium chloride solutions, only those peptides with TH activity were able to inhibit recrystallization. At present, sucrose is the most common solute employed for such purposes. When used in a thin film between two glass slides, it is now typically referred to as the 'sucrose sandwich splat cooling' method (Griffith et al. 2005; Pudney et al. 2003; Smallwood et al. 1999).

There are also different ways to analyse IRI experiments. One commonly used method is to compare images of a control ice samples with ice samples containing potential ice-binding substances after a certain annealing time, typically after between 30 min and 2 h. When a substance has been already identified as ice binding, then a concentration series allows the determination of the lowest concentration at which IRI activity exists by just visually comparing the different images to that of a control without the ice-binding substance, often a buffer sample (e.g. Knight et al. 1995; Tomczak et al. 2003; Yu et al. 2010). Similarly, when a larger number of substances, such as synthetic analogues is to be investigated for their IRI activity, they can be compared directly when they are added at the same concentration to the sample film (Lauersen et al. 2013). This kind of assay is very easy and efficient as it allows comparing many different substances with moderate effort. It has the disadvantage that substances with an IRI activity below the investigated concentration may actually be missed. Moreover, often the visual inspection is somewhat subjective. Hence, a more objective way to analyse such images is to determine the size of the ice crystals in the image, thus allowing for the determination of, for example, the mean ice crystal size of a sample (Sutton et al. 1994). The exact determination of the individual grain boundaries can by quite difficult, however, particularly at high ice volume fraction where some ice crystals in the film may actually overlap in the image, making the use of an automated image analysis problematic. In that case, the addition of fluorescent dyes or the use of polarized microscopy has shown to be helpful in the identification of individual ice crystals (Knight et al. 1995). But even then, the smallest crystals are often missed. Hence, one way around this problem is to focus on the larger ice crystals in an image, e.g. by just determining the size of the largest or a few of the largest ice crystals (Eniade et al. 2003; Wu et al. 2017), although this implies accepting some uncertainty when dealing with small amounts of data.

Addition of solutes such as sucrose does help in that respect. For example, adding solutes allows for determining the average ice crystal size at certain time intervals (Budke et al. 2009; Gaukel et al. 2014; Hagiwara et al. 2006, 2009; Olijve et al. 2016b; Sutton et al. 1994, 1996). Apart from the advantage of adding solute for avoiding accretion processes by reducing the ice volume fraction as described above, the separation of the individual ice crystals allows for a much more accurate ice crystal size analysis by optical methods. Overlap and contact of crystals is especially a problem if automated image processing software is used, which allows for the fast analysis of high amounts of crystals (Buch and Ramløv 2016; Budke et al. 2009; Jackman et al. 2007; Leiter et al. 2016; Olijve et al. 2016b). Finally, the most advanced way of IRI analysis is the determination of kinetic growth parameters, for example following the LSW approach discussed in detail above (Abraham et al. 2015; Budke et al. 2009; Hagiwara et al. 2006; Olijve et al. 2016b; Sutton et al. 1996).

7.5.2 Results for Different Compounds

The different IRI assays have been applied for various purposes, for example the identification of natural ice-binding substances, the quantification of their IRI efficacy, the study of the effects of mutations, and the development of synthetic ice-binding analogues.

For example, in a recent study a large number of natural AFPs and AFGPs as well as synthetic analogues have been compared regarding their IRI efficacy (Budke et al. 2014). The recrystallization rate constants were determined according to the LSW approach and as an indicator for the individual efficacy, the authors used that concentration c_i of a substance, at which the IRI rate constant was reduced to 50% of the diffusion-limited value in sucrose control solutions. The comparison yielded a wide range of efficacies with the least active substance being a 1.5 kDa peptide with a chemical structure similar to natural AFGPs and an efficacy of about 3.2 mg mL^{-1} (1.9 mmol L^{-1}). The natural AFGP1–5 fraction with a mean molecular mass of 22 kDa was found to be the most efficient substance. It significantly reduces the ice recrystallization rate at 22 ng mL^{-1} (1.0 nmol L^{-1}), which is consistent with an earlier assay showing significant reduction after 18 hours at about 1 ng mL^{-1} (Knight et al. 1988).

Very significant progress has been made in the elucidation of important chemical motifs in AFGPs. This became possible due to the development of synthetic routes that enable the reconstruction of natural AFGPs as well as modifications at several positions (Heggemann et al. 2010; Peltier et al. 2010; Tachibana et al. 2002). In terms of IRI efficacy the effects of chain length, sugar moiety modification, and amino acid sequence variation of various AFGP analogues were studied (Budke et al. 2014). The most important findings are that the natural AFGP fraction 1–5 shows high efficacy because of its length, the disaccharide group, and the alanine–alanine–threonine repeating unit. For instance, the exchange of the disaccharide by a monosaccharide in the tetramer leads to an increase of c_i by two orders of magnitude. However, activity of the backbone without any sugar moiety was observed, suggesting that the sugar moieties are not essential, even though the lower solubility of the pure backbone restricts its inhibitory effect to a limited concentration range.

A non-toxic, biodegradable, and already widely used synthetic substance that possesses an even simpler structure than the AFGP backbone and that nevertheless shows IRI activity is poly(vinyl alcohol). The inhibitory effect was first presented with the remark that TH was not measurable (Knight et al. 1995). However, poly (vinyl alcohol) showed strong IRI even in the presence of 0.5% of sodium chloride. A later study focussed on a sample of poly(vinyl alcohol) with a molecular mass of 13–23 kDa and a degree of hydrolysis of 0.98. For this sample, a high concentration of 50 mg mL^{-1} was required in order to observe a minute TH of just 0.04 °C, which is about the same value as the corresponding colligative melting point depression (Inada and Lu 2004). In contrast, efficacious IRI was determined for polymers in that size range even at concentrations of only about 50 μg mL^{-1} (Budke et al. 2014; Congdon et al. 2013). IRI activity is observable for chain lengths down to 19 repeating units (1.7 kDa) at 250 μg mL^{-1}, and probably even smaller variants may be active at higher concentrations (Congdon et al. 2013). The observation of ice crystal shaping demonstrates that this effect is caused by ice binding, and a proposed adsorption model based on the poly(vinyl alcohol) structure supports this hypothesis (Budke and Koop 2006).

Ice recrystallization inhibition is also used as a marker for ice-binding activity in studies of carbon-linked AFGP analogues, which are not susceptible to hydrolysis (Eniade et al. 2003; Liu and Ben 2005). Despite the rather pronounced modifications in the chemical structure of the synthetic analogues when compared to natural AFGPs, some of them did show IRI and TH activity as well as ice crystal shaping, which in combination indicates their ice-binding ability. In vitro studies revealed also promising results showing that an analogue, which possesses IRI activity, did not show cytotoxicity even at high concentrations in contrast to the natural AFGP 8 (Liu et al. 2007). Starting from these observations, small molecules like diols, mono-, and disaccharides were identified that do also inhibit recrystallization (Capicciotti et al. 2013; Deller et al. 2013; Trant et al. 2013). The absence of TH and ice shaping indicate, however, that in these cases the recrystallization inhibitory effect may not be due to ice binding (Tam et al. 2008). In contrast, the reduction in the mean largest grain size after 30 min at about −6 °C correlated with their molecular hydration number. We speculate that substances with higher hydration number may interact more strongly with water molecules and with each other, thus resulting in a decrease of the water diffusion coefficient. If this were indeed the case, a reduction of the ice recrystallization rate constant would result according to LSW theory (see Eq. 7.4). Moreover, significant IRI activity was observed at concentrations of 11 to 44 mmol L^{-1}, i.e. at concentrations large enough such that the corresponding melting point depression may have led to a decrease of the ice volume fraction. As these experiments were performed at ice volume fractions near one, the reduction of ice volume fraction very likely also led to a decreasing rate constant. In the presence of natural ice-binding AFPs that efficiently inhibit recrystallization at concentrations in the range of a few μmol L^{-1}, however, the colligative melting point depression is too low to have a measurable effect (Budke et al. 2014).

Recently, the synthetic dye safranine O chloride was found to inhibit ice growth in a manner similar to antifreeze glycoproteins (Drori et al. 2016). Despite its low molecular weight of about 350 g mol^{-1}, ice crystals growing in solutions of this particular dye show hexagonal shapes at concentrations below 3 mmol L^{-1} and bipyramidal needles at higher concentrations. Reduced ice recrystallization accompanied by hexagonal crystal shaping is also observed in the range from 2 to 3 mmol L^{-1}, complete inhibition at 4.2 mmol L^{-1}. TH activity in molar concentrations is similar to AFGP8 for growth along the *a*-direction. A TH of 0.2 °C is found for example at 10 mmol L^{-1}. The ice growth rate along the *c*-axis was, however, not reduced much. A dependence on the crystal orientation was also found for the inorganic electrolytes zirconium (IV) acetate and zirconium (IV) acetate hydroxide, which have been shown to inhibit recrystallization and to possess ice-shaping ability, too (Mizrahy et al. 2013). On first glance, one may attribute the IRI activity of safranine O and the zirconium electrolytes to the same effects of other small molecules, i.e. ice volume fraction reduction and water diffusion reduction. However, ice binding is indicated by the observed ice crystal habit modification. The apparent reason for the ice-binding activity of the zirconium (IV) salts is suggested to be the formation of hydroxyl-bridged oligomers of zirconium acetate. In a certain pH range these oligomers are large enough to interact strongly with ice surfaces and even produce ice shaping, but at the same time they are small enough to remain soluble in water (Deville et al. 2011, 2012). For safranine O chloride a similar mechanism is proposed suggesting the formation of large aggregates that have a periodic structure with flat surfaces (Drori et al. 2016). We note that the heavy metal zirconium salts and the aromatic safranine dyes are likely to be far less suitable for some applications than AFP or AFGP because of the limited biocompatibility and biodegradability.

7.6 Conclusions

Ice recrystallization inhibition is an important biological strategy for organisms living at subfreezing temperatures. The combined effect of ice-binding AFPs and non-binding colligative antifreezes might be a key to inhabit extreme environments. In this chapter, we have discussed various fundamental processes that can lead to ice recrystallization inhibition. These processes do not only include the adsorption of ice-binding solutes, which is traditionally viewed as the predominant cause for IRI activity. Instead, we have used LSW theory to show that non-ice binding 'colligative' solutes can also cause ice recrystallization inhibition by way of different processes, in particular by increasing solution viscosity and by reducing the volume fraction of ice. The latter processes have to be taken into account when studying IRI activity of natural or synthetic compounds in laboratory experiments, in order to avoid falsely attributing observed IRI activity to the process of ice binding. On the other hand, both ice binding as well as colligative effects may contribute to IRI processes in natural organisms. However, the very low concentrations at which ice-binding molecules can completely inhibit ice recrystallization clearly emphasize their very important role in natural cryoprotection.

References

Abraham S, Keillor K, Capicciotti CJ, Perley-Robertson GE, Keillor JW, Ben RN (2015) Quantitative analysis of the efficacy and potency of novel small molecule ice recrystallization inhibitors. Cryst Growth Des 15:5034–5039

Alpert PA, Aller JY, Knopf DA (2011) Ice nucleation from aqueous NaCl droplets with and without marine diatoms. Atmos Chem Phys 11:5539–5555

Asimow R (1963) Clustering kinetics in binary alloys. Acta Metall 11:72–73

Bar Dolev M, Braslavsky I, Davies PL (2016) Ice-binding proteins and their function. Annu Rev Biochem 85:515–542

Buch JL, Ramløv H (2016) An open source cryostage and software analysis method for detection of antifreeze activity. Cryobiology 72:251–257

Budke C (2010) Hemmung der Eisrekristallisation in wässrigen Lösungen durch Antigefrierglykopeptide. Dissertation

Budke C, Koop T (2006) Ice recrystallization inhibition and molecular recognition of ice faces by poly(vinyl alcohol). ChemPhysChem 7:2601–2606

Budke C, Heggemann C, Koch M, Sewald N, Koop T (2009) Ice recrystallization kinetics in the presence of synthetic antifreeze glycoprotein analogues using the framework of LSW theory. J Phys Chem B 113:2865–2873

Budke C, Dreyer A, Jaeger J, Gimpel K, Berkemeier T, Bonin AS, Nagel L, Plattner C, DeVries AL, Sewald N et al (2014) Quantitative efficacy classification of ice recrystallization inhibition agents. Cryst Growth Des 14:4285–4294

Capicciotti CJ, Doshi M, Ben RN (2013) Ice recrystallization inhibitors: from biological antifreezes to small molecules. In: Wilson P (ed) Recent developments in the study of recrystallization. InTech, New York, pp 177–224

Capicciotti CJ, Poisson JS, Boddy CN, Ben RN (2015) Modulation of antifreeze activity and the effect upon post-thaw HepG2 cell viability after cryopreservation. Cryobiology 70:79–89

Celik Y, Drori R, Pertaya-Braun N, Altan A, Barton T, Bar-Dolev M, Groisman A, Davies PL, Braslavsky I (2013) Microfluidic experiments reveal that antifreeze proteins bound to ice crystals suffice to prevent their growth. Proc Natl Acad Sci USA 110:1309–1314

Chen T, Cai X, Wu X, Karahara I, Schreiber L, Lin J (2011) Casparian strip development and its potential function in salt tolerance. Plant Signal Behav 6:1499–1502

Congdon TC, Notman R, Gibson MI (2013) Antifreeze (glyco)protein mimetic behaviour of poly (vinyl alcohol): Detailed structure-ice recrystallisation inhibition activity study. Biomacromolecules 14:1578–1586

Cziko PA, DeVries AL, Evans CW, Cheng C-HC (2014) Antifreeze protein-induced superheating of ice inside Antarctic notothenioid fishes inhibits melting during summer warming. Proc Natl Acad Sci 111:14583–14588

Davies PL (2014) Ice-binding proteins: a remarkable diversity of structures for stopping and starting ice growth. Trends Biochem Sci 39:548–555

Davies CKL, Nash P, Stevens RN (1980) Precipitation in Ni-Co-AI alloys Part 1 Continuous precipitation. J Mater Sci 15:1521–1532

Deller RC, Congdon T, Sahid MA, Morgan M, Vatish M, Mitchell DA, Notman R, Gibson MI (2013) Ice recrystallisation inhibition by polyols: comparison of molecular and macromolecular inhibitors and role of hydrophobic units. Biomater Sci 1:478

Deville S, Viazzi C, Leloup J, Lasalle A, Guizard C, Maire E, Adrien J, Gremillard L (2011) Ice shaping properties, similar to that of antifreeze proteins, of a zirconium acetate complex. PLoS One 6:e26474

Deville S, Viazzi C, Guizard C (2012) Ice-structuring mechanism for zirconium acetate. Langmuir 28:14892–14898

DeVries AL (1971) Glycoproteins as biological antifreeze agents in Antarctic fishes. Science 172:1152–1155

DeVries AL, Komatsu SK, Feeney RE (1970) Chemical and physical properties of freezing point-depressing glycoproteins from Antarctic fishes. J Biol Chem 245:2901–2908

Doucet CJ, Byass L, Elias L, Worrall D, Smallwood M, Bowles DJ (2000) Distribution and characterization of recrystallization inhibitor activity in plant and lichen species from the UK and maritime Antarctic. Cryobiology 40:218–227

Dreischmeier K, Budke C, Wiehemeier L, Kottke T, Koop T (2017) Boreal pollen contain ice-nucleating as well as ice-binding 'antifreeze' polysaccharides. Sci Rep 7:41890

Drori R, Li C, Hu C, Raiteri P, Rohl AL, Ward MD, Kahr B (2016) A supramolecular ice growth inhibitor. J Am Chem Soc 138:13396–13401

Duman JG (2015) Animal ice-binding (antifreeze) proteins and glycolipids: an overview with emphasis on physiological function. J Exp Biol 218:1846–1855

Eniade A, Purushotham M, Ben RN, Wang JB, Horwath K (2003) A serendipitous discovery of antifreeze protein-specific activity in C-linked antifreeze glycoprotein analogs. Cell Biochem Biophys 38:115–124

Gaukel V, Leiter A, Spieß WEL (2014) Synergism of different fish antifreeze proteins and hydrocolloids on recrystallization inhibition of ice in sucrose solutions. J Food Eng 141:44–50

Gillen KT, Douglass DC, Hoch MJR (1972) Self-diffusion in liquid water to −31 °C. J Chem Phys 57:5117–5119

Girlich D, Lüdemann H-D, Buttersack C, Buchholz K (1994) c, T-dependence of the self diffusion in concentrated aqueous sucrose solutions. Zeitschrift für Naturforschung C 49:258–264

Graham LA, Agrawal P, Oleschuk RD, Davies PL (2018) High-capacity ice-recrystallization endpoint assay employing superhydrophobic coatings that is equivalent to the 'splat' assay. Cryobiology 81:138–144

Griffith M, Yaish MWF (2004) Antifreeze proteins in overwintering plants: a tale of two activities. Trends Plant Sci 9:399–405

Griffith M, Lumb C, Wiseman SB, Wisniewski M, Johnson RW, Marangoni AG (2005) Antifreeze proteins modify the freezing process in planta. Plant Physiol 138:330–340

Gupta R, Deswal R (2014) Antifreeze proteins enable plants to survive in freezing conditions. J Biosci 39:931–944

Hagiwara T, Hartel RW, Matsukawa S (2006) Relationship between recrystallization rate of ice crystals in sugar solutions and water mobility in freeze-concentrated matrix. Food Biophys 1:74–82

Hagiwara T, Sakiyama T, Watanabe H (2009) Estimation of water diffusion coefficients in freeze-concentrated matrices of sugar solutions using molecular dynamics: correlation between estimated diffusion coefficients and measured ice-crystal recrystallization rates. Food Biophys 4:340–346

Haleva L, Celik Y, Bar-Dolev M, Pertaya-Braun N, Kaner A, Davies PL, Braslavsky I (2016) Microfluidic cold-finger device for the investigation of ice-binding proteins. Biophys J 111:1143–1150

Hanada Y, Nishimiya Y, Miura A, Tsuda S, Kondo H (2014) Hyperactive antifreeze protein from an Antarctic sea ice bacterium *Colwellia* sp. has a compound ice-binding site without repetitive sequences. FEBS J 281:3576–3590

Hardy SC, Voorhees PW (1988) Ostwald ripening in a system with a high volume fraction of coarsening phase. Metall Trans A 19:2713–2721

He X, Fowler A, Toner M (2006) Water activity and mobility in solutions of glycerol and small molecular weight sugars: implication for cryo- and lyopreservation. J Appl Phys 100:74702

Heggemann C, Budke C, Schomburg B, Majer Z, Wißbrock M, Koop T, Sewald N (2010) Antifreeze glycopeptide analogues: microwave-enhanced synthesis and functional studies. Amino Acids 38:213–222

Horibe A, Fukusako S, Yamada M (1996) Surface tension of low-temperature aqueous solutions. Int J Thermophys 17:483–493

Hrubý J, Vinš V, Mareš R, Hykl J, Kalová J (2014) Surface tension of supercooled water: no inflection point down to −25 °C. J Phys Chem Lett 5:425–428

Humphreys FJ, Hatherly M (2004) Recrystallization and related annealing phenomena. Elsevier, Boston

Ickes L, Welti A, Hoose C, Lohmann U (2015) Classical nucleation theory of homogeneous freezing of water: thermodynamic and kinetic parameters. Phys Chem Chem Phys 17:5514–5537

Inada T, Lu S-S (2004) Thermal hysteresis caused by non-equilibrium antifreeze activity of poly (vinyl alcohol). Chem Phys Lett 394:361–365

Jackman J, Noestheden M, Moffat D, Pezacki JP, Findlay S, Ben RN (2007) Assessing antifreeze activity of AFGP 8 using domain recognition software. Biochem Biophys Res Commun 354:340–344

Jayanth CS, Nash P (1989) Factors affecting particle-coarsening kinetics and size distribution. J Mater Sci 24:3041–3052

Kiko R (2010) Acquisition of freeze protection in a sea-ice crustacean through horizontal gene transfer? Polar Biol 33:543–556

Kim H, Lee J, Hur Y, Lee C, Park S-H, Koo B-W (2017) Marine antifreeze proteins: structure, function, and application to cryopreservation as a potential cryoprotectant. Mar Drugs 15:27

Knight CA, Duman JG (1986) Inhibition of recrystallization of ice by insect thermal hysteresis proteins: a possible cryoprotective role. Cryobiology 23:256–262

Knight CA, Hallett J, DeVries AL (1988) Solute effects on ice recrystallization: an assessment technique. Cryobiology 25:55–60

Knight CA, Wen DY, Laursen RA (1995) Nonequilibrium antifreeze peptides and the recrystallization of ice. Cryobiology 32:23–34

Knopf DA, Alpert PA, Wang B, Aller JY (2010) Stimulation of ice nucleation by marine diatoms. Nat Geosci 4:88–90

Koop T, Murray BJ (2016) A physically constrained classical description of the homogeneous nucleation of ice in water. J Chem Phys 145:211915

Koop T, Zobrist B (2009) Parameterizations for ice nucleation in biological and atmospheric systems. Phys Chem Chem Phys 11:10839–10850

Larson DJ, Middle L, Vu H, Zhang W, Serianni AS, Duman J, Barnes BM (2014) Wood frog adaptations to overwintering in Alaska: new limits to freezing tolerance. J Exp Biol 217:2193–2200

Lauersen KJ, Vanderveer TL, Berger H, Kaluza I, Mussgnug JH, Walker VK, Kruse O (2013) Ice recrystallization inhibition mediated by a nuclear-expressed and -secreted recombinant ice-binding protein in the microalga *Chlamydomonas reinhardtii*. Appl Microbiol Biotechnol 97:9763–9772

Leiter A, Rau S, Winger S, Muhle-Goll C, Luy B, Gaukel V (2016) Influence of heating temperature, pressure and pH on recrystallization inhibition activity of antifreeze protein type III. J Food Eng 187:53–61

Lide DR (ed) (2004) CRC handbook of chemistry and physics, 85th edn. CRC Press, Boca Raton, FL

Lifshitz IM, Slyozov VV (1961) The kinetics of precipitation from supersaturated solid solutions. J Phys Chem Solids 19:35–50

Liu S, Ben RN (2005) C-linked galactosyl serine AFGP analogues as potent recrystallization inhibitors. Org Lett 7:2385–2388

Liu S, Wang W, Von Moos E, Jackman J, Mealing G, Monette R, Ben RN (2007) In vitro studies of antifreeze glycoprotein (AFGP) and a C-linked AFGP analogue. Biomacromolecules 8:1456–1462

Lorv JSH, Rose DR, Glick BR (2014) Bacterial ice crystal controlling proteins. Scientifica (Cairo) 2014:1–20

Mangiagalli M, Bar-Dolev M, Tedesco P, Natalello A, Kaleda A, Brocca S, de Pascale D, Pucciarelli S, Miceli C, Bravslavsky I et al (2016) Cryo-protective effect of an ice-binding protein derived from Antarctic bacteria. FEBS J 284:163–177

Martin DR, Ablett S, Darke A, Sutton RL, Sahagian M (1999) Diffusion of aqueous sugar solutions as affected by Locust Bean Gum studied by NMR. J Food Sci 64:46–49

Mazur P (1984) Freezing of living cells: mechanisms and implications. Am J Phys 247:C125–C142

Mizrahy O, Bar-Dolev M, Guy S, Braslavsky I (2013) Inhibition of ice growth and recrystallization by zirconium acetate and zirconium acetate hydroxide. PLoS One 8:e59540

Nagel L, Plattner C, Budke C, Majer Z, DeVries AL, Berkemeier T, Koop T, Sewald N (2011) Synthesis and characterization of natural and modified antifreeze glycopeptides: glycosylated foldamers. Amino Acids 41:719–732

Olijve LLC, Meister K, DeVries AL, Duman JG, Guo S, Bakker HJ, Voets IK (2016a) Blocking rapid ice crystal growth through nonbasal plane adsorption of antifreeze proteins. Proc Natl Acad Sci 113:3740–3745

Olijve LLC, Oude Vrielink AS, Voets IK (2016b) A simple and quantitative method to evaluate ice recrystallization kinetics using the circle hough transform algorithm. Cryst Growth Des 16:4190–4195

Ouellet F, Charron J-B (2013) Cold acclimation and freezing tolerance in plants. In: eLS. Wiley, Chichester

Peltier R, Brimble MA, Wojnar JM, Williams DE, Evans CW, DeVries AL (2010) Synthesis and antifreeze activity of fish antifreeze glycoproteins and their analogues. Chem Sci 1:538

Pertaya N, Marshall CB, DiPrinzio CL, Wilen L, Thomson ES, Wettlaufer JS, Davies PL, Braslavsky I (2007) Fluorescence microscopy evidence for quasi-permanent attachment of antifreeze proteins to ice surfaces. Biophys J 92:3663–3673

Price WS, Ide H, Arata Y (1999) Self-diffusion of supercooled water to 238 K using PGSE NMR diffusion measurements. J Phys Chem A 103:448–450

Pudney PDA, Buckley SL, Sidebottom CM, Twigg SN, Sevilla M-P, Holt CB, Roper D, Telford JH, McArthur AJ, Lillford PJ (2003) The physico-chemical characterization of a boiling stable antifreeze protein from a perennial grass (*Lolium perenne*). Arch Biochem Biophys 410:238–245

Raymond JA (2011) Algal ice-binding proteins change the structure of sea ice. Proc Natl Acad Sci 108:E198–E198

Raymond JA, DeVries AL (1977) Adsorption inhibition as a mechanism of freezing resistance in polar fishes. Proc Natl Acad Sci 74:2589–2593

Raymond JA, Kim HJ (2012) Possible role of horizontal gene transfer in the colonization of sea ice by algae. PLoS One 7:e35968

Raymond JA, Sullivan CW, Devries AL (1994) Release of an ice-active substrance by Antarctic sea ice diatoms. Polar Biol 14:71–75

Raymond JA, Fritsen C, Shen K (2007) An ice-binding protein from an Antarctic sea ice bacterium. FEMS Microbiol Ecol 61:214–221

Singh P, Hanada Y, Singh SM, Tsuda S (2014) Antifreeze protein activity in Arctic cryoconite bacteria. FEMS Microbiol Lett 351:14–22

Smallwood M, Worrall D, Byass L, Elias L, Ashford D, Doucet CJ, Holt C, Telford J, Lillford P, Bowles DJ (1999) Isolation and characterization of a novel antifreeze protein from carrot (*Daucus carota*). Biochem J 340:385

Storey KB, Storey JM (2017) Molecular physiology of freeze tolerance in vertebrates. Physiol Rev 97:623–665

Sutton RL, Evans ID, Crilly JF (1994) Modeling ice crystal coarsening in concentrated disperse food systems. J Food Sci 59:1227–1233

Sutton RL, Lips A, Piccirillo G, Sztehlo A (1996) Kinetics of ice recrystallization in aqueous fructose solutions. J Food Sci 61:741–745

Tachibana Y, Matsubara N, Nakajima F, Tsuda T, Tsuda S, Monde K, Nishimura SI (2002) Efficient and versatile synthesis of mucin-like glycoprotein mimics. Tetrahedron 58:10213–10224

Tam RY, Ferreira SS, Czechura P, Chaytor JL, Ben RN (2008) Hydration index – a better parameter for explaining small molecule hydration in inhibition of ice recrystallization. J Am Chem Soc 130:17494–17501

Tomczak MM, Marshall CB, Gilbert JA, Davies PL (2003) A facile method for determining ice recrystallization inhibition by antifreeze proteins. Biochem Biophys Res Commun 311:1041–1046

Trant JF, Biggs RA, Capicciotti CJ, Ben RN (2013) Developing highly active small molecule ice recrystallization inhibitors based upon C-linked antifreeze glycoprotein analogues. RSC Adv 3:26005

Tursman D, Duman JG, Knight CA (1994) Freeze tolerance adaptations in the centipede, Lithobius forficatus. J Exp Zool 268:347–353

Van Oss CJ, Giese RF, Wentzek R, Norris J, Chuvilin EM (1992) Surface tension parameters of ice obtained from contact angle data and from positive and negative particle adhesion to advancing freezing fronts. J Adhes Sci Technol 6:503–516

Voorhees PW, Glicksman ME (1984) Ostwald ripening during liquid phase sintering – effect of volume fraction on coarsening kinetics. Metal Trans Sect A 15:1081–1088

Wagner C (1961) Theorie der Alterung von Niederschlägen durch Umlösen (Ostwald-Reifung). Zeitschrift für Elektrochemie, Berichte der Bunsengesellschaft für Phys. Chemie 65:581–591

Walters KR, Serianni AS, Sformo T, Barnes BM, Duman JG (2009) A nonprotein thermal hysteresis-producing xylomannan antifreeze in the freeze-tolerant Alaskan beetle Upis ceramboides. Proc Natl Acad Sci USA 106:20210–20215

Walters KR, Serianni AS, Voituron Y, Sformo T, Barnes BM, Duman JG (2011) A thermal hysteresis-producing xylomannan glycolipid antifreeze associated with cold tolerance is found in diverse taxa. J Comp Physiol B Biochem Syst Environ Physiol 181:631–640

Washburn EW (ed) (1928) International critical tables of numerical data, physics, chemistry and technology, 1st edn. McGraw-Hill, New York

Wharton DA, Barrett J, Goodall G, Marshall CJ, Ramløv H (2005) Ice-active proteins from the Antarctic nematode *Panagrolaimus davidi*. Cryobiology 51:198–207

Wharton DA, Wilson PW, Mutch JS, Marshall CJ, Lim M (2007) Recrystallization inhibition assessed by splat cooling and optical recrystallometry. Cryo Letters 28:61–68

Wu S, Zhu C, He Z, Xue H, Fan Q, Song Y, Francisco JS, Zeng XC, Wang J (2017) Ion-specific ice recrystallization provides a facile approach for the fabrication of porous materials. Nat Commun 8:15154

Young FE, Jones FT (1949) Sucrose hydrates. The sucrose-water phase diagram. J Phys Colloid Chem 53:1334–1350

Yu SO, Brown A, Middleton AJ, Tomczak MM, Walker VK, Davies PL (2010) Ice restructuring inhibition activities in antifreeze proteins with distinct differences in thermal hysteresis. Cryobiology 61:327–334

Chapter 8
Other Protective Measures of Antifreeze Proteins

Hans Ramløv and Dennis Steven Friis

8.1 Introduction

Antifreeze proteins are in general considered being amphiphilic molecules with the ice-binding side being the most hydrophobic part of the molecule (Kristiansen et al. 2012; Pentelute et al. 2008). However, it has been proposed that for fish antifreeze proteins, which also can be considered amphiphilic, it is the hydrophilic side which is binding to the ice (Cheng and DeVries 1991).

Quite early after the discovery of the fish antifreeze glycoproteins (AFGPs) in the 1960s, eight weight classes of these proteins (AFGP 1–AFGP 8) ranging from 34,000 kDa to 2600 kDa, respectively, were separated on the basis of their relative migration on gel-electrophoresis gels (DeVries 1971; Feeney 1974; Feeney and Yeh 1978). When determining the antifreeze activity of the eight weight classes it was discovered that AFGP 1–6 had strong antifreeze activity, whereas 7 and 8 had only very weak activity (Feeney 1974). However, AFGP 6 only appears in the fish blood in trace quantities (Rubinsky et al. 1990). These results have later been confirmed several times, see for example (Ramløv et al. 2005). The presence of isoforms of AFGPs or AFPs with very low activity in the blood of antifreeze protein bearing fishes has been a topic of discussion and research for many years, for example, it has been proposed that they might work in concert with the more active AFGPs and hereby enhance the antifreeze activity of these. Some studies have not been able to show any enhancement or synergy whereas others have (Baardsnes et al. 2003; Nishimiya et al. 2005; Ramlov et al. 2005). When measuring the antifreeze activity of AFGPs 1–5 and 7+8 using differential scanning calorimetry (DSC) Ramløv et al.

H. Ramløv (✉)
Department of Natural Sciences, Roskilde University, Roskilde, Denmark
e-mail: hr@ruc.dk

D. S. Friis
Copenhagen, Denmark

H. Ramløv, D. S. Friis (eds.), *Antifreeze Proteins Volume 2*,
https://doi.org/10.1007/978-3-030-41948-6_8

found that the combination of AFGPs 1–5 and 7+8 did not result in a higher activity than AFGP 1–5 alone although the activity in the combination was much higher than when AFGP 7+8 was measured alone (Ramløv et al. 2005). Later results on isoforms of both AFGPs and Type III AFPs have showed that synergistic and additive effects indeed exist between isoforms of various AFPs. In AFP type III it is well established that the rather passive SP isoforms, that are found in much higher abundance in AFP type III-bearing fishes than the QAE isoforms, are enhanced by the QAE isoforms (Nishimiya et al. 2005; Takamichi et al. 2009). However, until now the exact mechanism of the synergistic effect resulting from the combination of one or more isoforms of AFPs has not been resolved. One explanation may be that there is a sequential binding of the isoforms to the ice surface due to differences in adsorption rates and in the strength of the binding to the ice. This would mean that the isoform with the highest adsorption rate binds first, slowing down the ice growth which facilitates the binding of the slower and more passive isoform. In a recent paper Berger et al. studied the combinations effects on enhancement or reduction of antifreeze activity synergistic effects on a combination of AFPI, AFPIII-QAE, AFPIII-SP as well as AF(G)Ps isoforms from fish and plants using nanoliter osmometer, microfluidics and fluorescence microscopy. The authors showed that in experiments with various AF(G)Ps even in experiments which included cross-organismal AFPs, there was no direct correlation between the structure of the protein or its biological origin in its effect of the antifreeze activity. The three-dimensional structure of the AFP determines to which ice crystal plane it binds and that in turn determines the effect on the antifreeze activity in the mixture. From these experiments the authors were able to put forward a kinetic model that explains the synergistic or additive effects on the antifreeze activity in binary mixtures of isoforms. The model is limited to AFPs that cannot bind to basal planes and thus combinatory effects of insect AFPs are not explained by the model. In the proposed model the active isoform of AF(G)P shows high adsorption rate as well as binding to the prism plane. The passive isoform binds to the pyramidal planes and at a slower rate. This way a so-called complex consisting of ice and the active isoform bound to the prism plane and the passive isoform bound to the pyramidal plane. Thus the inhibition of ice growth by the AF(G)Ps is a result of solution concentration (the passive form having the highest concentration) and their intrinsic adsorption rates (Berger et al. 2019). Future experiments on insect AFPs using a similar approach may indeed yield very interesting results concerning the interplay of various isoforms in the same organism.

Because of the amphiphilic nature of the antifreeze proteins and the low activity in some, other protective measures than the inhibition of the growth of ice have been proposed and some properties have been discovered. One such protective property could be the stabilization of membranes when the organisms are exposed to cold, another could be the stabilization of the organisms ion balance during cold exposure and it seem that some antifreeze proteins have some anti-virulence effects, which may or may not be adaptive to cold-tolerant organisms.

8.2 Stabilization of Membranes

When the temperature decreases, animals may be subjected to direct chilling injury (Bayley et al. 2018; Morris 1987) which may include phase transitions in the cell membranes. During cooling membrane phospholipids may undergo phase transitions from the liquid crystalline phase to the gel phase (Hazel 1995) leading to different diffusion rates—the membranes become transiently leaky during the lipid phase transition, leading to loss of intracellular metabolites and leakage of various potentially damaging substances into the cytosol (Clerc and Thompson 1995; Hazel and Williams 1990) increased hydrophobic thickness of the lipid membrane affecting the activity of integral proteins (Lee 2004), aggregation of integral proteins (Lee 2004) as a consequence of exclusion from domains of gel-phase lipids (Hazel 1995) and pressure/stretch forces within the membrane (Anishkin et al. 2014) affecting the function of integral proteins. During cooling the membranes may consist of coexisting domains of liquid crystalline lipids and domains of gel phase lipids. Along the phase boundaries packing defects may exist, which leads to leakage of nutrients and ions through the membrane (Morris 1987; Morris and Clarke 1987). However, these changes in membrane structure and function may be counteracted via homeoviscous adaptation where saturated membrane lipids are exchanged by unsaturated ones as a response to the cooling environment (Hazel and Williams 1990), a process which may be called acclimatory (short term) where changes in membrane lipid composition occurs within days (Snyder et al. 2012). Homeoviscous adaptation can also be a consequence of evolutionary (long time scale) adaptation to a stenothermal environment because the organisms do not experience large fluctuations in temperature and therefore do not need a fast exchange of membrane lipids (Hazel 1995). Malekar et al. have shown that membrane lipid compositions were consistent across three Antarctic notothenioid fish species which were significantly different from a related New Zealand fish. They also showed that when exposing two of the Antarctic species to 6 °C for 7 days one of them showed acclimatory homeoviscous adaptation by increasing the amount of saturated fatty acids and decreasing the amount of unsaturated fatty acids in their membranes (Malekar et al. 2018).

The AFGPs 7 and 8 with molecular weights of 3500 Da and 2600 Da, respectively, are present in the fish serum in high concentrations although they do not show very high antifreeze activity. Thus, one could speculate that these molecules might have other roles than inhibiting the growth of ice crystals. Such roles could be interactions with and stabilization of macromolecules or cell membranes. Hays et al. in fact showed that liposomes composed of the phospholipid DEPC were completely protected against leakage by AFGP 8 at 10 mg/ml when cooled through the phase transition temperature. However, much more of the small AFGP 8 was required to achieve the same protection against leakage compared to the larger AFGPs 2–6 (Hays et al. 1996).

In a study of the membrane potential in pig oocytes Rubinsky et al. found that oocytes measured at 22 °C in a phosphate buffered saline for 4 h (PBS) and in a PBS solution containing 40 mg/ml AFGPs 1–8 had a mean membrane potential of −31 mV with an SD of 4.5 mV, which is within the normal range of membrane potentials of fresh immature pig oocytes. This showed that the AFGPs 1–8 did not affect the membrane potentials of the oocytes (Rubinsky et al. 1990). When pig oocytes were exposed to 4 °C for 4 h and reheated to 22 °C almost 80% showed a membrane potential which had dropped below two standard deviations from the mean in the non-treated oocytes. On the basis of this, Rubinsky and co-workers concluded that significant ion-leakage had occurred from the cold exposed oocytes. However, when oocytes were exposed to 4 °C for 4 h but in a PBS containing 40 mg/ml AFGPs 1–8 and returned to 22 °C around 83% showed a membrane potential within two standard deviations from that of untreated oocytes leading to the hypothesis that the AFGPs had afforded some kind of protection against leakage of ions across the cell membrane. Rubinsky et al. (1990) varied the concentration of the AFGPs and within the range of 1–40 mg/ml protection against leakage was observed. When the oocytes were exposed to 4 °C for 24 h the effect of the AFGPs 1–8 was even more obvious. Rubinsky and co-workers further showed that it was only PBS with the integrated composition of AFGPs 1–8 that conveyed the protection to the oocytes. Significant fewer oocytes exposed to 4 °C in solutions with either 40 mg/ml AFGPs 1–5 or AFGPs 7, 8 retained a membrane potential within two standard deviations from that of the nonexposed oocytes when returned to 22 °C (Rubinsky et al. 1990). Rubinsky and co-workers do not offer any explanation of the observed results and point out that one may suggest several explanations, one of these being an interaction between that AFGPs and the hydrophilic part of membrane proteins. To investigate if antifreeze proteins in general had other properties than those related to inhibiting ice crystal growth by recognizing and binding to ice crystal surfaces, Rubinsky and co-workers investigated a number of various types of antifreeze proteins for their ability to convey resistance to hypothermic conditions in immature bovine oocytes (Rubinsky et al. 1991).

In an attempt to elucidate by which mechanism antifreeze proteins confer protection to mammalian cells during hypothermic exposure, Rubinsky and co-workers used porcine granulosa cells and antifreeze proteins from the winter flounder (*Pseudopleuronectes americanus*), wfAFP, a fish antifreeze protein type I, and patch clamp technique (Rubinsky et al. 1992). Rubinsky and co-workers showed that the wfAFP in a concentration of 0.5 mg/ml and at a temperature of 22 °C was able to completely suppress both the Ca^{2+} and K^{+} currents at 40 and 200 s, respectively, after infusion. They also showed that higher concentrations of the antifreeze protein resulted in a faster decay of both currents after infusion. Because there was no effect of the wfAFP on the currents at 0.1 mg/ml, a nonlinear relationship between wfAFP concentration and ion current suppression as well as saturation at a concentration of 0.5 mg/ml the authors argue that the suppression of the ion currents may be via an AFP–protein interaction (Rubinsky et al. 1992).

Hays et al. studied liposomes prepared from dielaidoylphophatidylcholine (DEPC) loaded with carboxyflourescein (CF) and cooled through the phase transition temperature (T_m) in the presence of AFGPs from the Antarctic fishes *D. mawsoni* and *Pagothenia borchgrevinci* (Hays et al. 1996). The results showed that liposomes, which were only composed of phospholipids, leaked 50% of their contents when cooled through T_m when no AFGPs were present in the solvent. However, at concentrations as low as 1 mg/ml AFGP from *D. mawsoni* leakage of the DEPC liposomes was prevented up to 100% when run through the same cooling cycle. When liposomes of DEPC were warmed through T_m after first having been cooled to 2 °C and held at this temperature for 20 min (so all lipid molecules were in the gel phase), it was observed that the AFGPs had the same effect on leakage of CF as was observed during cooling alone. Thus the authors conclude that the AFGPs decrease the permeability of the liposomes during the phase transition to the same extent during cooling and warming. It is however interesting that when similar temperature regimes were used on DEPC liposomes using differential scanning calorimetry (DSC), where the phase transition can be observed as an exo- or endotherm, the authors found no depression of the phase transition temperature by the AFGPs (Hays et al. 1996). According to the authors this may be due to difficulties of achieving a similar ratio of phospholipid to AFGP in the DSC experiments as in the CF leakage experiments.

In addition to the experiments performed on DEPC liposomes Hays et al. investigated the effects of AFGPs on the leakage of liposomes consisting of dipalmitoylphophatidylcholine (DPPC) and a dimyristoylphosphatidylcholine (DMPC)/egg phosphatidylcholine mixture (1:1 ratio), these liposomes had T_m's of 41 °C and 15 °C, respectively. Leakage of CF from these liposomes, like from the DEPC liposomes, corresponded with T_m. The results showed that the AFGPs also in these cases inhibited the leakage of CF when the liposomes were cooled through T_m. Hays et al. conclude that the effects of AFGPs on the leakage of liposomes composed of various phospholipids going through the phase transition temperature is a general effect.

On the basis of their results Hays et al. suggest that the stabilizing effect on intact mammalian cells observed by Rubinsky et al. (1992) may rather be due to a nonspecific effect on the lipid components in the membranes than a direct interaction with specific ion channels (Hays et al. 1996).

However, experiments performed by Hincha and co-workers showed that AFGPs and AFPs did not protect spinach thylakoid membranes during freezing (Hincha et al. 1993, 1996). Hincha et al. measured the release of the luminal protein plastocyanin from thylakoid vesicles. The antifreeze proteins used were AFGP in a physiological composition from Antarctic fish *Dissostichus mawsoni*, AFPs were from the arctic starry flounder *Platichthys stellatus* and the Antarctic eel pout *Austrolyichtys brachycephalus*. The thylakoids were stored for up to 6 h at either −20 or 0 °C in solutions with AFGP or AFP concentrations ranging from 0 to 1 mg/ml. In all cases freeze–thaw cycles of the thylakoids to −20 °C resulted in a concentration-dependent release of plastocyanin. It was found that AFGP concentrations as low as 50 μg/ml resulted in freeze–thaw induced plastocyanin release. The rate of plastocyanin release almost doubled in the presence of 1 mg/ml AFGP compared to controls and all the proteins investigated resulted in an increase of the rapid component of plastocyanin release which was also seen in solutions not containing AF(G)Ps. The authors concluded that as with animal cells AF(G)Ps interact with the thylakoid membranes. However, in contrast to the animals cells where the membranes are stabilized with the interaction the thylakoids are damaged. This difference may rely on the fact that thylakoids contain approximately 80% galactolipids and only 10% phospholipids whereas animal cell membranes mainly contain phospholipids (Hincha et al. 1993).

The mechanism by which the AF(G)Ps interact with membranes and exert the protective effect on various cell types has been a subject of study by several authors (Inglis et al. 2006; Kar et al. 2016; Tomczak et al. 2002a, b; Wu and Fletcher 2000). Wu and Fletcher investigated the interactions of four different antifreeze proteins; AFP I, II, III and AFGP with three kinds of liposome types: dielaidoylphosphaicylcholine (DEPC), dielaidoylphosphatidylethanolamine (DEPE), and dielaidoylphosphatidylglycerol (DEPG) (Wu and Fletcher 2000). Liposomes of the three phospholids were cooled from 20 to 0 °C through their phase transition temperature (T_m) and the leakage of carboxyflourescein (CF) was measured spectrometrically. The results showed that all four AF(G)Ps suppressed leakage from DEPC liposomes but also the control protein BSA inhibited leakage to the same extent as the antifreeze proteins and all of them in a concentration-dependent manner with the highest leakage of around 55% occurring with AFP type III at 25 μg/ml. The leakage from control liposomes, where no AFGP was present in the solution, was approximately 80%. On the basis of their study and on that of Hays et al. (1996) Wu and Fletcher propose that AFGP in general bind to zwitterionic liposomes preventing them from leaking when cooled through their phase transition temperature. They further argue that this may rely on that AFGPs are amphiphilic but with a very hydrophilic component wherefore the binding of the AFGPs to zwitterionic and negatively charged liposomes may be confined to the surface of the lipid bilayers. Thus, the carbohydrate moiety of the AFGPs may interact with the polar head groups via charged and/or hydrogen bond interactions. They also propose an alternative explanation that maybe the hydrophobic site of the AFGPs partly immerses into the lipid bilayer. One could however, argue that in the last case AFGPs should then interact with all kinds of liposomes. In the case of the AFPs and BSA they all induced some leakage when added to the DEPC/DEPG liposomes at room temperature. This is interpreted as the AFPs penetrate the bilayer and because of the negatively charged head groups of the DEPG the ordered structure of the membranes was disrupted. The overall interpretation of the results is that protection of DEPC liposomes is achieved by the investigated proteins and possibly by other proteins too by binding of these to the membrane and thereby maintaining the ordered structure during the phase transition (Wu and Fletcher 2000). In the case of the liposomes composed of DEPE and DEPG the charge of the polar head groups differ from that of DEPC and possibly this means that the proteins, although they are bound to the lipid bilayer, are not able to maintain an organized state of the membrane when the liposomes are cooled through the phase transition temperature or that they may gain entry into and disrupt the bilayer and this way induce leakage. Studies by Tomczak et al. further underline the complexity of understanding the interactions of antifreeze proteins with membranes (Tomczak et al. 2001). Tomczak et al. produced liposomes with a mixture of galactolipids; either monogalactosyldiacylglycerol (MGDG) or digalacosyldiacylglycerol

(DGDG) and phospholipids egg phophatidylcholine (EPC). The compositions of these liposomes were 50% DGDG/50% EPC, 15% MGDG/85% EPC or 100% EPC. The antifreeze proteins used were AFGP fractions (fr. 1–5, the large AFGPs, fr. 3,4 middle sized AFGPs and fr. 8, a small AFGP) and AFP I and the proteins were added only to the outside of the liposomes. The control was leakage of CF from unfrozen samples without protein and this was used as 0% leakage. Leakage form the liposomes was assessed by measuring fluorescence of CF in the medium after the liposomes had been cooled rapidly to −18 °C incubated at that temperature for 3 h and returned to room temperature. Tomczak et al. found that AFP I at a concentration of 1 mg/ml caused a significant increase in CF leakage compared to the control in all three kinds of liposomes when frozen to −18 °C. Results also showed that a 10 mg/ml AFP I caused a loss of 30% from the EPC liposomes compared to the control and the DGDG and MGDG liposomes lost 40% more than the control. Tomczak et al. also states that AFP I led to a concentration-dependent increase in CF leakage from unfrozen EPC liposomes. Investigation of the fusion of liposomes in treatments with AFP I during freezing showed that there was a linear correlation between fusion of the liposomes and the protein concentration and that the effect was almost equal for liposomes consisting of pure EPC and DGDG but that significantly more fusion occurred in the MGDG liposomes. Tomczak et al. explain this observation by the fact that MGDG liposomes are non-bilayers which promote membrane fusion (Tomczak et al. 2001). As a result of these experiments Tomczak et al. suggest that the observed leakage and fusion are related events and that they are caused by the protein during freezing. The experiments performed with the AFGPs showed that fr. 1–5 and fr. 3,4 had almost the same effect on liposome stability so here they were taken as one "larger fractions." They increased the CF leakage from the liposomes although less than AFP I. The largest effect of increasing AFGP concentration was on the DGDG-containing liposomes and the smallest was on the MGDG-containing vesicles. Liposome fusion showed the same trend with AFGP (larger fractions) as AFP I. However, the liposome fusion was almost 10 times less in the experiments with AFGPs than with AFP I and as seen in the previous experiments with AFP I, MGDG-containing liposomes showed the highest degree of fusion.

When DGDG:EPC, MGDG:EPC, and pure EPC liposomes were frozen in the presence of AFGP fr. 8 there was an initial decrease in CF leakage at low protein concentrations and the leakage did not increase with increasing protein concentration. A small increase in fusion was observed with AFGP fr. 8, but this was significantly smaller than with AFGP (larger fractions) at 10 mg/ml or with AFP I. Tomczak et al. conclude that AFGP fr. 8 had a small protective effect on DGDG-containing liposomes and hardly any effect on MGDG or pure EPC liposomes during freezing. In contrast AFP I was cryotoxic to liposomes and the effect of AFGP (large fraction) is dependent on the lipid composition of the liposomes most noticeably between MGDG- and DGDG-containing liposomes. Tomczak et al. also conclude that antifreeze proteins can induce membrane fusion and that fusion is highest in liposomes containing the non-bilayer lipid MGDG (Tomczak et al. 2001).

In a later study Tomczak et al. investigated if the interactions of AFP I with model membranes were dependent on lipid unsaturation during thermotropic phase transitions (Tomczak et al. 2002b). Tomczak et al. had earlier shown that AFP I inhibits leakage form 1:1 DGDG:DMPC liposomes when they were cooled through their phase transition temperature (Tomczak et al. 2002a). As mentioned above DGDG is highly unsaturated whereas DMPC is purely saturated. On the basis of the earlier study Tomczak et al. hypothesize that DGDG is necessary for the interaction of AFP I with membranes (Tomczak et al. 2002a). In the present study Tomczak et al. hypothesize that it is the unsaturation of the acyl chains and/or the digalactose head group that stabilize the membranes during the thermotropic phase transition from the liquid crystalline state to the gel state. Using a hydrogenated DGDG in a 1:1 ratio to DMPC Tomczak et al. found that the AFP I does not affect lipid order and that T_m was unchanged in the presence of AFP I when the DGDG acyl chains were saturated (Tomczak et al. 2002b). They also found that AFP I was not associated with pelleted liposomes when all lipids are saturated. This is in contrast to the results of the earlier study where Tomczak et al. using Fourier-transform infrared spectroscopy (FTIR) showed that AFP I affected the packing order of the bilayer in liposomes containing unsaturated DGDG and DMPC (Tomczak et al. 2002a). The finding that AFP I was not associated with pelleted liposomes of the saturated lipids after chilling through T_m suggests that unsaturation is required for the hydrophobic interaction of AFP I with the liposomes which had previously been observed (Tomczak et al. 2002a). An interesting finding by Tomczak et al. is that the folding of AFP I is inhibited in the presence of liposomes containing unsaturated or hydrogenated DGDG. This observation indicates that the protein is still interacting with the liposomes and that temperature-dependent folding of AFP I is inhibited (Tomczak et al. 2002b).

The mechanism of interaction between AFP I and the membrane is indicated by the above mentioned results. AFP type I interacts directly with the membranes, shifting T_m upwards in DGDG:DMPC liposomes, inducing a significant change in the packing order resulting in reduced membrane permeability presumably through hydrophobic interaction between the protein and the lipid bilayer (Tomczak and Crowe 2002; Tomczak et al. 2002a). It was suggested by Tomczak and Crowe that the interaction of AFGPs with membranes is different from the one observed with AFP I. Tomczak and Crowe suggest that AFGPs form a monolayer covering the liposomes which is preventing the transient leakage, which would occur if the bilayer was cooled through T_m without the presence of the AFGP but that both types of interactions mentioned here are dependent on the composition of the bilayer (Tomczak and Crowe 2002).

The interaction with membranes has further been investigated by Kun et al. who used differential scanning calorimetry (DSC) and fluorescence anisotropy to study interactions of multilamellar and small unilamellar vesicles of dimyristoylphosphatidylcholine (DMPC) with AFP I and short segments of this protein (Kun et al. 2008). Kun and Mastai showed earlier that short segments of AFP I can possess up to 60% of the antifreeze activity of the native AFP I (Kun and Mastai 2007). The DSC parameters investigated on the multilamellar vesicles in the study by Kun et al. were changes in T_m, the enthalpy and the shape of the DMPC phase transition in the thermograms generated by the DSC. The native AFP I (HPLC

6) shifted T_m downward by 0.32 °C and decreased the enthalpy of the phase transition by 3.66 kJ/mol, the smallest of the prepared AFP I segments (HK 3) showed a similar decrease of the enthalpy and a downshift of the phase transition temperature of 0.45 °C. The largest of the AFP I (HK 1) segments decreased the phase transition temperature by 0.33 °C and decreased the enthalpy by 1.15 kJ/mol. The middle sized segment of AFP I (HK 2) had almost no effect on the phase transition temperature (0.02 °C) but increased the enthalpy by 0.81 kJ/mol. All the AFPs increased the DSC peak width at half height by between 0.19 and 0.31 °C. The authors suggest that the decrease in phase transition temperature indicates that the liquid crystalline phase is stabilized through the interactions with the AFPs. However, from an evolutionary viewpoint these values seem quite small and probably of limited adaptive value. The observed decrease in enthalpy indicates that the AFPs facilitated the transition from the liquid crystalline to the gel state by ordering the acyl chains in the membranes above the phase transition temperature (Kun et al. 2008). The fluorescence anisotropy experiments showed that above the phase transition temperature, $T > T_{m,}$ the order of acyl chain region (the core) increased in the presence of AFP with HK2 as the most active. Below the phase transition temperature $T < T_m$ the order of this region also increased and relatively more than was observed in the control, where AFPs were not added to the vesicles. In this case all the AFPs had an ordering effect on the acyl chains where HK2 and HPLC6 had the same effect and HK1 and HK3 also induced order, but to a lesser extent. At a high weight ratio of HK2 to DMPC it was observed that HK2 showed a higher polarization than HPLC6, which may indicate that HK2 induces a higher degree of ordering of the acyl chains (Kun et al. 2008). From their fluorescence results Kun et al. calculated possible changes in microviscosity of the lipid bilayers in the presence of the AFPs. These calculations showed that microviscosity was increased both above and below T_m when AFPs were present compared to the control. The effect on the microviscosity of the AFPs was larger below T_m than above. Kun et al. compare the effects on T_m and membrane fluidity to cholesterol and saturated fatty acids, which as the AFPs gets incorporated into the lipid composition of the membrane. The AFPs investigated in the study by Kun et al. affect the order of the membrane core, which is comparable to cholesterol but unlike cholesterol they also change T_m which is comparable to incorporation of saturated fatty acids. It should be noted that the results found by Kun et al. are in disagreement with the results shown by Tomczak et al. referred above where AFP I (which is equal to HPLC6 in Kun et al.) shifts T_m (Tomczak et al. 2002b). On the basis of their results Kun et al. suggest that the AFPs incorporated into the membrane core induce order in isolated domains which decrease phase transition cooperativity. They arrive at this conclusion due to the observation that order is induced in the membrane as observed with polarized fluorescence and at the same time the phase transition peak observed with DSC is broadened. They also observed that there is an inverse relationship between the AFP/lipid ratio and the order of the membrane. At higher AFP concentrations more ordered domains are formed eventually leading to a "defective" membrane indicating a disruption of the order of the membrane (Kun et al. 2008).

Thus, a balance between the concentration of AFP and the lipid must be maintained for the AFPs to induce order in the membrane without disrupting it.

To further elucidate the interactions between the native AFP I and the truncated segments mentioned above Kun and Mastai performed an Isothermal Calorimetric (ITC) study of these AFPs with DMPC large unicellular vesicles (LUV) above and below the phase transition temperature (Kun and Mastai 2010). According to Kun and Mastai "lipid-peptide interactions are proposed to consist of three molecular steps: (1) conformational changes of the peptides prior to interactions with the membrane; (2) electrostatic adsorption-electrostatic attraction of the peptide to the membrane surface; (3) penetration of the peptide into the membrane" (Kun and Mastai 2010). Above it can be seen that AFP I is not folded when it interacts with the membranes (Tomczak et al. 2002b) and this is in concordance with the first point in the quote from Kun and Mastai above. Further it has also been shown that AFP I gets incorporated into the lipid core of the membrane (Kun et al. 2008). In their ITC experiments with AFP I and DMPC vesicles Kun and Mastai inject the AFP into the solution containing the DMPC vesicles in a concentration so that the lipid-to-protein ratio is very large. This leads the authors to the assumption that all peptides in the sample are bound to the DMPC membranes (Kun and Mastai 2010). In their ITC experiments Kun and Mastai found that the heat of interaction increased with increasing temperature so that below the phase transition heat release was linear with temperature whereas above the phase transition the heat release remained almost constant at an enthalpy $\Delta H = -1.47$ kcal/mol. When performing similar experiments with the truncated AFP I called HK2, Kun and Mastai found that the heat of interaction of HK2 with the DMPC vesicles was lower above the phase transition temperature than the heat of interaction below the phase transition temperature. Kun and Mastai interpret their results in the context of two defined temperature zones: above the phase transition temperature and below the phase transition temperature. When considering earlier results with polarized fluorescence (Kun et al. 2008) as well as the results found using ITC Kun and Mastai suggest that above the phase transition temperature the found enthalpy of interaction is a result of the interaction with AFP I and HK2 with the membrane where these peptides induce very little modification of the membrane structure—this may be interpreted as the enthalpy of binding of the peptides to the membranes (Kun and Mastai 2010). This assumption is substantiated by a study by Garner et al. who via solid state nuclear magnetic resonance (NMR) showed that AFP did not affect the ordering of DMPC membranes above the phase transition temperature (Garner et al. 2008). The negative enthalpy leads Kun and Mastai to the assumption that the enthalpy, although small, is favorable for the interaction between the peptides and the membrane (Kun and Mastai 2010). Below the phase transition temperature Kun and Mastai state that the value of heat release is very small and this implicates that it may be accounted for induction of order in the lipid matrix when the AFP is inserted into this. Kun and Mastai thus conclude that their measurements indicate that the following sequence when native AFP I and the HK2 truncate associate with the DMPC membranes: above the phase transition temperature the interaction is mainly the binding of the AFPs to the membrane with a minimal perturbation of the membrane structure, below the phase transition temperature the AFPs partition into the membrane and induce some order of the lipids (Kun and Mastai 2010).

In an attempt to further elucidate the interactions of antifreeze proteins with membranes Kar et al. employed high and low resolution of biophysical characterization techniques such as circular dichroism (CD) and NMR spectroscopy as well as molecular dynamics simulations to study interactions between AFPs and micelle models mimicking natural membrane systems (Kar et al. 2016). Kar et al. used Winter flounder (*Pleuronectes americanus*) (AFP I) *wf*AFP and presumably fabricated recombinantly, though not stated, and an antifreeze peptide derived from an Antarctic yeast (*Glaciozyma antarctica*) called Peptide lm, both purchased commercially. The micelle models studied were made from dodecylphosphocholine (DPC) and the anionic detergent sodium dodecyl sulfate (SDS) also perdeuterated SDS and DPC as well as spin-labeled lipids 5-doxyl-stearic acid and 16-doxyl-stearic acid were used in the experiments. When assessing the structure of the AFPs in the presence of the DPC and SDS micelles with far-UV CD spectroscopy the *wf*AFP showed α-helical conformation indicating that this AFP is not perturbed in the presence of the micelles. Similar results were obtained with Peptide lm (Kar et al. 2016). When analyzing the results further Kar et al. showed that both *wf*AFP and Peptide lm were dominated by α-helical structures in the presence of DPC and SDS micelles. When analyzing the far-UV CD spectroscopy results for *wf*AFP and Peptide lm in water as a control these also showed a large component of α-helical structure, indicating little difference between the conformation of the AFPs in the presence of the micelles and in water. However, some variations existed in the secondary structure of the AFPs in different micelle conditions compared to when the AFPs were only exposed to water, indicating an interaction between the micelles and the AFPs (Kar et al. 2016).

The paramagnetic relaxation experiments employing the spin-labeled fatty acids indicated an interaction of Peptide lm with both the DPC and the SDS micelles. Kar et al. suggest that the Peptide lm is forming salt bridges or interacting electrostatically with the phosphate/anionic groups of the micelles and positively charged amino acids on the peptide or that there exists a hydrophobic interaction between the acyl chains and the aliphatic/aromatic residues of Peptide lm. On the basis of their experiments and on the molecular modelling Kar et al. concludes that both AFP's studied interact with zwitterionic as well as negatively charged model membranes and that type I AFPs retain their helical conformation in the presence of both these types of membranes. Concerning the interaction between the model membranes and the AFPs Kar et al. find that the AFPs are found in the interface of the micelle—solvent systems and that the interaction between Peptide lm is electrostatic as well as hydrophobic. Such hydrophobic interaction is also indicated between certain residues in the *wf*AFP and the micelles. The hydrophobic interaction between *wf*AFP and the micelles also perturbs the secondary structure of this AFP to a certain extent and induce some unordered structure (Kar et al. 2016).

The study of the interaction between AF(G)Ps and membranes were initiated because of the amphiphilic nature of these proteins which could indicate that they might interact with membranes. After having discovered that some kind of interaction was occurring some authors argued that this interaction is a second fundamental property of antifreeze proteins and that this preceded sea level glaciation, as a response to low water temperatures and that the inhibition of ice crystal growth is a secondary property conveying protection to fish living in ice laden water (Rubinsky et al. 1991, 1992). However, no evolutionary evidence was provided in the papers. In studies of the evolution of fish antifreeze proteins arguments has been advanced that the evolution of the AFGPs about 5–14 million years ago is consistent with the time frame during which the Antarctic ocean cooled to freezing temperatures (Chen et al. 1997; Zhuang et al. 2019). Further, as mentioned earlier, organisms adapted to both stenothermal and eurythermal conditions show homeoviscous adaptations of their membranes (Hazel 1995; Hazel and Williams 1990; Malekar et al. 2018; Snyder et al. 2012). Thus, the adaptive value of the interaction of antifreeze proteins with membranes could therefore be debated and it may be only coincidental. However, it still remains to be elucidated if this interaction is of adaptive value to cold-tolerant organisms synthesizing antifreeze proteins.

In conclusion, several authors have studied AF(G)Ps in the context of membrane interactions and in all cases interaction has been observed. However, the results from different groups of researchers are not always in agreement. Some researchers find that the secondary structure of the AFPs is very disturbed in the presence of model membranes compared to the secondary structures of the same AFP in water (Tomczak et al. 2002b). Others find that the membranes to a large extent do not perturb the secondary structures although they do change some parts of the structure from helical to unordered (Kar et al. 2016). There also seem to be somewhat opposing viewpoints as to the location of the interacting AF(G)Ps while they are interacting with the model membranes. Some authors argue that some AF(G)Ps are inserted into the core of the membranes (Kun et al. 2008) and other AF(G)Ps are surrounding the micelles interacting with the phospholipid heads on the surface of the membranes. Other authors do not find that the AF(G)Ps are inserted into the core of the membranes but that they rather are found in the interface between the membranes and the solvent (Kar et al. 2016). The above differences may rely on the use of different membrane lipids or micelle forming molecules and some consensus seems not to have been reached. Thus further studies are needed to fully elucidate the AF(G)P interactions with membranes employing different techniques on the same kind of membranes across laboratories. Seen from an adaptational viewpoint it should also be pointed out that the studies referred to above are studies of fairly simple membrane systems and they only to a limited extent emulate the membranes found in cold-tolerant ectothermic organisms. Future studies would therefore benefit from applying the available knowledge about the composition of the naturally occurring membranes in cold-tolerant ectothermic organisms. This would possibly give surprising results and maybe an understanding of the adaptive value of the interactions would be more obvious. It should further be noted that very few studies have been performed on the interaction of non-fish AF(G)Ps with membranes and studies employing insect AFPs or AFPs from other organisms may lead to more insights into the nature of membrane—AFP interactions and the adaptive value of this.

8.3 Antifreeze Proteins in the Context of Bacteria

From a small number of studies the interaction of antifreeze proteins and bacteria leading to anti-virulence or improved tolerance to changes in the environment of the bacteria has been reported.

Mueller et al. produced an artificial chimeric protein from the DNA sequence encoding one of the Winter flounder (*Pleuronectes americanus*) antifreeze proteins and a portion of the *spa* gene which encodes for the staphylococcal protein A gene. This chimeric protein retained its recrystallization inhibiting properties already known from the winter flounder AFPs (Mueller et al. 1991). Holmberg et al. produced similar chimeric protein A/winter flounder AFPs in *E. coli* to study osmoprotecting properties of AFP analogues in bacteria (Holmberg et al. 1994). In the experiments Holmberg et al. grew *E.coli* containing various analogues of the Winter flounder AFPs fused with the staphylococcal protein A. *E.coli* were grown in broth containing various concentrations of NaCl of up to 0.74 M and the optical density of the broth was monitored spectrophotometrically. Bacteria expressing the chimeric antifreeze proteins showed only a slighter increase in generation time at 0.5 M NaCl than in 0.17 M NaCl. In the control where the *E.coli* bacteria contained a truncated version of protein A the generation time was substantially higher in bacteria grown in 0.5 M than in those grown in 0.17 M. The lag phase times for the chimeric protein A/AFP proteins were approximately 30% shorter in the medium containing 0.5 M NaCl than in the control. Holmberg et al. also investigated if the improved NaCl tolerance of the bacteria containing the chimeric proteins conferred a protection from damage by freezing. Bacteria were grown to the late exponential phase and harvested by centrifugation and resuspended in dH_2O. 1 ml samples were frozen at −3 °C for 1 h and slowly thawed for 1 h. This procedure was repeated six times over a 12 h period. Survival (the fraction of non-lysed cells) was assessed by determining the conductivity of the cell suspensions. The conductivity of completely integrated cells was obtained by boiling a sample for 15 min. The results exhibited that cells with the chimeric protein containing four 11 amino acid repeats had an improved freezing tolerance compared to the control (Holmberg et al. 1994).

A tick antifreeze glycoprotein (IAFGP) from the tick *Ixodes scapularis* was described by Neelakanta et al. (2010). The presence of the pathogen, *Anaplasma phagocytophilum* which gives rise to human granulytic anaplasmosis, in *I. scapularis* increases the survival to cold exposure of this tick. IAFGP in *I. scapularis* is upregulated both as a response to cold and upon infestation with *A. phagocytophilum* (Neelakanta et al. 2010). The mechanism by which IAFGP protects *I. scapularis* during cold is not elucidated. However, it may be through the inhibition of ice crystal growth inside the ticks but a second mechanism may also be suggested as Neelakanta et al. observed that membrane disruption of *A. phagocytophilum*-infected tick cells that expressed more *iafgp* was reduced. Neelakanta et al. also showed that yeast cells which had been transfected with the *iafgp* gene showed higher viability when subjected to cold (Neelakanta et al. 2010).

Heisig et al. studied the anti-virulence properties of the IAFGP on infections with *Staphylycoccus aureus* and other bacteria in vitro and in vivo (Heisig et al. 2014). They showed that GST-tagged IAFGP bound to various bacteria and that purified peptidoglycan was sufficient for the IAGP binding. In vitro IAFGP interfered with biofilm formation in static *S. aureus* cultures, where the GST-IAFGP treated cultures exhibited at significantly reduced biofilm formation, whereas GST-IAFGP did not change planktonic growth of the microbes. Further Heisig et al. showed that in bacteria exposed to IAFGP the exopolysaccharide poly-*N*-acetylglucosamine (PNAG) decreased significantly. Heisig et al. produced at a peptide (P1), which consisted of a domain Pro-Ala-Arg-Lys-Ala-Arg (PAKAR) and six AAT triplets and investigated this peptides affinity for bacteria and its antimicrobial activity. They tested both the full length IAFGP and the P1 peptide on *P. aureus* as well as *Listeria monocytogenes*, *Pseudomonas entomophila*, and *Serratia marcescens* so that both Gram-negative and Gram-positive bacteria were included in the study. P1 showed the same properties as full length IAFGP on binding to the various bacteria species. P1 also decreased biofilm formation in *S. aureus*, but did not influence the bacterial viability. Heisig et al. therefore arrive at the conclusion that IAFGP and P1 directly bind Gram-negative as well as Gram-positive pathogens and that they have antibiofilm properties (Heisig et al. 2014).

Heisig et al. further investigated the effect of IAFGP on *Drosophila melanogaster* expressing the *iafgp* gene (Neelakanta et al. 2012) and transgenic mice expressing the *iafgp* gene (Heisig et al. 2014). *Drosophila melanogaster* expressing the *iafgp* gene that was microinjected with *S. aureus*, prick-infected with *L. monocytogenes* or oral-infected with *S. marcescens* or *P. entomophila* all exhibited a higher survival than controls. The pathogen load was quantified to elucidate whether the anti-infective effect was due to tolerance or resistance mechanisms. The results showed that the bacterial burden in the control flies increased for 4 days whereas in the *iafgp*-expressing flies the pathogen load remained stable. The stable pathogen load could be explained either as a consequence of reduced replication of the pathogens or by growth homeostasis with host-induced killing. Heisig et al. conclude that the difference between transgenic and control flies in bacterial titers after infection rely on the IAFGP conferring infection resistance.

The next question addressed by Heisig et al. was whether the observed resistance in the *iafgp*-expressing flies was due to an increased host response to infection (an immune response) or a direct effect of IAFGP on the bacteria. According to Heisig et al. the increased resistance was not mediated through any of the three effector mechanisms comprising *D. melanogaster* immunity; (1) hemocytes equivalent to macrophages that engulf pathogens, (2) antimicrobial peptides that have direct bactericidal effects, and (3) melanin deposition which encapsulates microbes. Heisig et al. blocked phagocytosis and melanization in both transgenic and control flies with similar effects in both groups. The authors were not able to measure any increase in antimicrobial peptides in the *iafgp*-expressing flies compared to controls. Thus the authors conclude that the resistance is not mediated through the immune system. However, using immunohistochemical detection of *S. aureus* and PNAG the authors found that bacterial colonization was deceased and as well was the PNAG level. When infecting *iafgp*-expressing flies with a strain of *S. aureus* not expressing PNAG the protective effect of IAFGP totally disappeared leading to the conclusion that the anti-infective effect of IAFGP correlates with the bacterial PNAG synthesis (Heisig et al. 2014).

Heisig et al. investigated if the observed effect of IAFGP in the invertebrate model (the *iafgp*-expressing *D. melanogaster*) could be extended to a vertebrate model. They produced a transgenic mouse expressing *iafgp* and by cecal ligation and puncture which induces polymicrobial sepsis they showed that the *iafgp*-expressing mice had a 35% longer average survival time and serum cytokine release of MCP-1 (which is a marker of sepsis) was reduced, when compared to control animals. These results indicate that IAFGP also in mice leads to some protection against bacterial infection.

When investigating if IAFGP had any effect on biofilm formation by *S. aureus* on intra venous catheters Heisig et al. incubated intra venous catheters in solutions of IAFGP or P1. Both molecules associated with the catheters and reduced biofilm formation. It was found that both IAFGP and P1 decrease associated bacteria 9- and 40-fold, respectively, compared to controls. In vivo in mice where P1 coated catheters were implanted subcutaneously and left for 72 h upon examination of the catheters it was found that the number of bacteria was 40-fold lower than controls and that PNAG was reduced in *S. aureus* biofilms.

Heisig et al. conclude that bacterial binding of IAFGP or P1 inhibits biofilm formation in *S. aureus* which leads to an increased host survival upon bacterial challenge and that "IAFGP or P1 interferes with envelope structure or associated proteins, affecting cellular signaling or PNAG secretion and assembly" (Heisig et al. 2014). Based upon the results shown in their paper and upon some unpublished data the authors also conclude that although IAFGP does not lead to an altered immune response in the host it leads to a greater susceptibility of the pathogen to the host immune response. Thus, the prevention of biofilm formation leads to an increased susceptibility of the bacteria to clearance by the immune response.

Finally Heisig et al. have some interesting remarks on the evolutionary significance of the observed anti-virulence of IAFGP. The authors argue that ectothermic animals become less effective at combatting pathogens when subjected to low temperature both due to a decrease in nutrient uptake and to an overall decrease of the metabolic activity. This claim is based on findings by Triggs and Knell (2012) as well as a large amount of literature arguing that insects with a higher temperature better survive viral, microbial, and microsporial infections (Triggs and Knell 2012). The authors also speculate that upon warming there is time lag of the immune response of the insects to bacterial infections due to the time it takes before they have regained their temperature and optimal metabolic activity (Heisig et al. 2014). Thus the anti-virulence activity of IAFGP is adaptive in *I. scapularis* whereby the AFGP serves two functions in this animal.

In addition to the properties of IAFGP and P1 shown above, these peptides have been shown to enhance the efficacy of several antibiotics (daptomycin, ciprofloxacin, and gentamycin) against *S. aureus*. This combined effect is apparently achieved by improving the permeability of the peptidoglycan layer to small molecules and thereby the permeation of the antibiotic into the bacterial cell (Abraham et al. 2017).

8.4 Antifreeze Proteins and Plant Pathogens

Antifreeze proteins have been reported in plants since 1992. These proteins often do not show any marked thermal hysteresis but are efficient ice recrystallization inhibitors (see Chap. 7 on plant antifreeze proteins). They have been found in many overwintering vascular plants occurring as a response to low temperatures and they are only found in plants tolerant to ice formation in their tissues (Griffith and Yaish 2004). Interestingly, most plant antifreeze proteins are homologous to pathogenesis-related proteins (PR proteins) and they are found also to provide protection against psychrophilic pathogens active at temperatures below 20 °C (Griffith and Yaish 2004). Psychrophilic pathogens are active under snow cover, where the temperature is relatively constant and humidity is high. At the low temperatures the plants have difficulties raising a de novo defense against the pathogens. But it has been shown that many over-wintering grasses become more resistant to fungal diseases during cold acclimation. Hon et al. purified various AFPs from winter rye (*Secala cereale* L. cv Musketeer) which are similar to three types of PR proteins; β-1,3-glucanases, chitinases, and thaumatin-like proteins (TL proteins). Hon et al. propose that the AFPs have evolved on the basis of small structural changes in the PR proteins, which accumulate in plants as a response to cold and in this way conferring ice-binding activity to these. PR proteins all inhibit fungal growth (Hon et al. 1995). The purified AFPs from winter rye have been shown to exhibit antifungal and hydrolytic activity as well as ice-binding activity (Griffith and Yaish 2004). Thus, by synthesizing AFPs similar to PR proteins the plants achieve both protection against recrystallization of ice as well as against psychrophilic pathogens.

8.5 Conclusions

In conclusion we have seen that AF(G)Ps may have secondary functions such as interactions with membranes, anti-virulence in arthropods and pathogen-inhibiting effects in plants to the ice growth or ice recrystallization inhibition which is their primary function. These properties may be of adaptive value to the organisms, for example, stabilization of membranes may be important in some cases although homeoviscous adaptation in the membranes of intact organisms is likely to take place both on an evolutionary time scale (long term) and on an acclimatory time scale (short term). In the case of anti-virulence it can be argued that if antifreeze proteins are antibacterial/antifungal then this may be adaptive as both animals and plants are not able to raise a de novo defense against pathogens when low temperatures prevail. Thus if antifreeze proteins have been synthesized as response to decreasing temperature these are then conferring resistance to infective organisms during the cold period or on warming before the organisms have mobilized a full metabolism. There is quite some evidence that the plant AFPs have developed from pathogen-related proteins so the evolution of these is rather well understood.

The studies of these "other protective measures" are rather few and sporadic (except for the plants). Thus, further studies are needed to elucidate if these "measures" only pertain to few organisms or if it is a more general trait.

Membrane–AFP interactions need to be studied more in the context of naturally occurring membrane compositions. The mechanisms of anti-virulence properties of AFPs need to be investigated further; for example how does the IAFGP inhibit biofilm formation and reduce the amount of PNA.

References

Abraham NM, Liu L, Jutras BL, Murfin K, Acar A, Yarovinsky TO, Sutton E, Heisig M, Jacobs-Wagner C, Fikrig E (2017) A tick antivirulence protein potentiates antibiotics against *Staphylococcus aureus*. Antimicrob Agents Chemother 61:e00113-17

Anishkin A, Loukin SH, Teng J, Kung C (2014) Feeling the hidden mechanical forces in lipid bilayer is an original sense. Proc Natl Acad Sci U S A 111:7898–7905

Baardsnes J, Kuiper MJ, Davies PL (2003) Antifreeze protein dimer: when two ice-binding faces are better than one. J Biol Chem 278:38942–38947

Bayley JS, Winther CB, Andersen MK, Gronkjaer C, Nielsen OB, Pedersen TH, Overgaard J (2018) Cold exposure causes cell death by depolarization-mediated Ca(2+) overload in a chill-susceptible insect. Proc Natl Acad Sci U S A 115:E9737–E9744

Berger T, Meister K, DeVries AL, Eves R, Davies PL, Drori R (2019) Synergy between antifreeze proteins is driven by complementary ice-binding. J Am Chem Soc 141(48):19144–19150

Chen L, DeVries AL, Cheng CH (1997) Evolution of antifreeze glycoprotein gene from a trypsinogen gene in Antarctic notothenioid fish. Proc Natl Acad Sci U S A 94:3811–3816

Cheng CC, DeVries AL (1991) The role of antifreeze glycopeptides and peptides in the freezing avoidance of cold-water fish. In: di Prisco G (ed) Life under extreme conditions. Springer, Berlin, pp 1–14

Clerc SG, Thompson TE (1995) Permeability of dimyristoyl phosphatidylcholine/dipalmitoyl phosphatidylcholine bilayer membranes with coexisting gel and liquid-crystalline phases. Biophys J 68:2333–2341

DeVries AL (1971) Glycoproteins as biological antifreeze agents in antarctic fishes. Science 172:1152–1155

Feeney RE (1974) A biological antifreeze. Am Sci 62:712–719

Feeney RE, Yeh Y (1978) Antifreeze proteins from fish bloods. In: Anfinsen CB, Edsall JT, Richards FM (eds) Advances in protein chemistry, vol 32. Academic Press, New York, pp 191–282

Garner J, Inglis SR, Hook J, Separovic F, Harding MM (2008) A solid-state NMR study of the interaction of fish antifreeze proteins with phospholipid membranes. Eur Biophys J 37:1031–1038

Griffith M, Yaish MW (2004) Antifreeze proteins in overwintering plants: a tale of two activities. Trends Plant Sci 9:399–405

Hays LM, Feeney RE, Crowe LM, Crowe JH, Oliver AE (1996) Antifreeze glycoproteins inhibit leakage from liposomes during thermotropic phase transitions. Proc Natl Acad Sci U S A 93:6835–6840

Hazel JR (1995) Thermal adaptation in biological membranes: is homeoviscous adaptation the explanation? Annu Rev Physiol 57:19–42

Hazel JR, Williams EE (1990) The role of alterations in membrane lipid composition in enabling physiological adaptation of organisms to their physical environment. Prog Lipid Res 29:167–227

Heisig M, Abraham NM, Liu L, Neelakanta G, Mattessich S, Sultana H, Shang Z, Ansari JM, Killiam C, Walker W et al (2014) Antivirulence properties of an antifreeze protein. Cell Rep 9:417–424

Hincha DK, DeVries AL, Schmitt JM (1993) Cryotoxicity of antifreeze proteins and glycoproteins to spinach thylakoid membranes—comparison with cryotoxic sugar acids. Biochim Biophys Acta 1146:258–264

Hincha DK, Sieg F, Bakaltcheva I, Köth H, Schmitt JM (1996) Freeze-thaw damage to thylakoid membranes: specific protection by sugars and proteins. In: Steponkus PL (ed) Advances in low-temperature biology, vol 3. JAI Press, London, pp 141–183

Holmberg N, Lilius G, Bulow L (1994) Artificial antifreeze proteins can improve NaCl tolerance when expressed in E. coli. FEBS Lett 349:354–358

Hon WC, Griffith M, Mlynarz A, Kwok YC, Yang DSC (1995) Antifreeze proteins in winter rye are similar to pathogenesis-related proteins. Plant Physiol 109:879

Inglis SR, Turner JJ, Harding MM (2006) Applications of type I antifreeze proteins: studies with model membranes & cryoprotectant properties. Curr Protein Pept Sci 7:509–522

Kar RK, Mroue KH, Kumar D, Tejo BA, Bhunia A (2016) Structure and dynamics of antifreeze protein—model membrane interactions: a combined spectroscopic and molecular dynamics study. J Phys Chem B 120:902–914

Kristiansen E, Wilkens C, Vincents B, Friis D, Lorentzen AB, Jenssen H, Lobner-Olesen A, Ramlov H (2012) Hyperactive antifreeze proteins from longhorn beetles: some structural insights. J Insect Physiol 58:1502–1510

Kun H, Mastai Y (2007) Activity of short segments of Type I antifreeze protein. Pept Sci 88:807–814

Kun H, Mastai Y (2010) Isothermal calorimetry study of the interactions of type I antifreeze proteins with a lipid model membrane. Protein Pept Lett 17:739–743

Kun H, Minnes R, Mastai Y (2008) Effects antifreeze peptides on the thermotropic properties of a model membrane. J Bioenerg Biomembr 40:389

Lee AG (2004) How lipids affect the activities of integral membrane proteins. Biochim Biophys Acta 1666:62–87

Malekar VC, Morton JD, Hider RN, Cruickshank RH, Hodge S, Metcalf VJ (2018) Effect of elevated temperature on membrane lipid saturation in Antarctic notothenioid fish. Peer J 6: e4765

Morris GJ (1987) Direct chilling injury. In: Grout BWW, Morris GJ (eds) The effects of low temperature on biological system. Edward Arnold, London, pp 120–146

Morris GJ, Clarke A (1987) Cells at low temperature. In: Grout BWW, Morris GJ (eds) The effects of low temperatures on biological systems. Edward Arnold, London, pp 72–119

Mueller GM, McKown RL, Corotto LV, Hague C, Warren GJ (1991) Inhibition of recrystallization in ice by chimeric proteins containing antifreeze domains. J Biol Chem 266:7339–7344

Neelakanta G, Sultana H, Fish D, Anderson JF, Fikrig E (2010) Anaplasma phagocytophilum induces Ixodes scapularis ticks to express an antifreeze glycoprotein gene that enhances their survival in the cold. J Clin Invest 120:3179–3190

Neelakanta G, Hudson AM, Sultana H, Cooley L, Fikrig E (2012) Expression of Ixodes scapularis antifreeze glycoprotein enhances cold tolerance in Drosophila melanogaster. PLoS One 7: e33447–e33447

Nishimiya Y, Sato R, Takamichi M, Miura A, Tsuda S (2005) Co-operative effect of the isoforms of type III antifreeze protein expressed in Notched-fin eelpout, Zoarces elongatus Kner. FEBS J 272:482–492

Pentelute BL, Gates ZP, Tereshko V, Dashnau JL, Vanderkooi JM, Kossiakoff AA, Kent SBH (2008) X-ray structure of snow flea antifreeze protein determined by racemic crystallization of synthetic protein enantiomers. J Am Chem Soc 130:9695–9701

Ramlov H, DeVries AL, Wilson PW (2005) Antifreeze glycoproteins from the antartic fish Dissostichus mawsoni studied by differential scanning calorimetry (DSC) in combination with nanolitre osmometry. CryoLetters 26:73–84

Ramløv H, DeVries AL, Wilson PW (2005) Antifreeze glycoproteins from the antarctic fish Dissostichus mawsoni studied by differential scanning calorimetry (DSC) in combination with nanolitre osmometry. CryoLetters 26:73–84

Rubinsky B, Arav A, Mattioli M, Devries AL (1990) The effect of antifreeze glycopeptides on membrane potential changes at hypothermic temperatures. Biochem Biophys Res Commun 173:1369–1374

Rubinsky B, Arav A, Fletcher GL (1991) Hypothermic protection—a fundamental property of "Antifreeze" proteins. Biochem Biophys Res Commun 180:566–571

Rubinsky B, Mattioli M, Arav A, Barboni B, Fletcher GL (1992) Inhibition of Ca2+ and K+ currents by "antifreeze" proteins. Am J Phys Regul Integr Comp Phys 262:R542–R545

Snyder RJ, Schregel WD, Wei Y (2012) Effects of thermal acclimation on tissue fatty acid composition of freshwater alewives (Alosa pseudoharengus). Fish Physiol Biochem 38:363–373

Takamichi M, Nishimiya Y, Miura A, Tsuda S (2009) Fully active QAE isoform confers thermal hysteresis activity on a defective SP isoform of type III antifreeze protein. FEBS J 276:1471–1479

Tomczak MM, Crowe JH (2002) The interaction of antifreeze proteins with model membranes and cells. In: Fish antifreeze proteins, vol 1. World Scientific, London, pp 187–212

Tomczak MM, Hincha DK, Estrada SD, Feeney RE, Crowe JH (2001) Antifreeze proteins differentially affect model membranes during freezing. Biochim Biophys Acta 1511:255–263

Tomczak MM, Hincha DK, Estrada SD, Wolkers WF, Crowe LM, Feeney RE, Tablin F, Crowe JH (2002a) A mechanism for stabilization of membranes at low temperatures by an antifreeze protein. Biophys J 82:874–881

Tomczak MM, Vigh L, Meyer JD, Manning MC, Hincha DK, Crowe JH (2002b) Lipid unsaturation determines the interaction of AFP type I with model membranes during thermotropic phase transitions. Cryobiology 45:135–142

Triggs A, Knell RJ (2012) Interactions between environmental variables determine immunity in the Indian meal moth Plodia interpunctella. J Anim Ecol 81:386–394

Wu Y, Fletcher GL (2000) Efficacy of antifreeze protein types in protecting liposome membrane integrity depends on phospholipid class. Biochim Biophys Acta Gen Subj 1524:11–16

Zhuang X, Yang C, Murphy KR, Cheng CHC (2019) Molecular mechanism and history of non-sense to sense evolution of antifreeze glycoprotein gene in northern gadids. Proc Natl Acad Sci U S A 116:4400

Chapter 9
Measuring Antifreeze Protein Activity

Johannes Lørup Buch

9.1 Introduction

The methods most commonly used today to quantify AFP are the very same methods that were originally used to simply detect the presence of AFP in a sample of interest. From a quantitative point of view, much progress has been made. Unlike enzymes, which catalyze chemical reactions at predictable rates, the interactions between AFPs and their surroundings are purely physical. For this reason, detection and quantification of AFP has historically been a discipline requiring significant amounts of creativity. In AFP research, there has always been a strong reliance on microscopy. This relates to the physical nature of the AFP mechanism, and the fact that changes to ice crystal morphology is at the very core of said mechanism. But manual observation can potentially lead to subjectivity and possible confirmation bias in a data set. In recent years, much effort has gone toward development of methods that automate these steps, improving objectivity. In this chapter we shall look closer at some of the most commonly used techniques for detecting and quantifying AFP, as well as some of the newest and most novel approaches within the field.

9.2 Quantification of AFP Activity

In the biological context, the strong selective pressure for AFPs is a direct consequence of their ability to improve survival through cold or freezing tolerance. This relates to thermal hysteresis (T_h) and recrystallization inhibition, respectively. There are generally two approaches to the methods that rely on thermal hysteresis or recrystallization inhibition, and these are the microscopic observation techniques

---

J. L. Buch (✉)
Department of Biology, University of Southern Denmark, Odense, Denmark

H. Ramløv, D. S. Friis (eds.), *Antifreeze Proteins Volume 2*,
https://doi.org/10.1007/978-3-030-41948-6_9

and the calorimetric techniques. Both have strengths and weaknesses, and the choice of method often comes down to available hardware. Many studies have dealt with the downstream effects of AFP, e.g., tissue freezing viability (Liu et al. 2014; Boonsupthip and Lee 2003; Heisig et al. 2015), AFP-coated ice-free surfaces (Lv et al. 2014; Esser-Kahn et al. 2010; Gwak et al. 2015), and the effects on various frozen foods (Zhang et al. 2007a; Panadero et al. 2005; Kontogiorgos et al. 2008; Yeh et al. 2009; Baier-Schenk et al. 2005). Interesting as that may be, the start of this chapter focuses on the methods used to directly measure the effects of AFP at the ice–water interface. For an overview of some of the indirect methods used in AFP research, see the second part of this chapter.

The AFP-induced thermal hysteresis seen in the blood of Arctic fishes is also the historical origin of the term "Antifreeze protein." With this in mind, it makes sense to start this chapter with the quantification methods that rely on T_h.

9.2.1 Thermal Hysteresis

Thermal hysteresis is the separation of melting (T_m) and freezing temperatures (T_f), which would in most cases otherwise be one and the same. For pure water at atmospheric pressure, these two temperatures are both 0 °C. In the absence of AFP, T_m and T_f are determined by osmolality (mOsm/kg), following Blagden's Law of freezing point depression in a solution, see Eq. (9.1).

$$\Delta T_f = K_f \times b \tag{9.1}$$

where

K_f: cryoscopic constant (1.853 K · kg/mol)
b: osmolality of the solution

In the presence of AFP, T_m and T_f are not affected equally as increasing osmolality enhances the antifreeze activity (Wang et al. 2009a). T_m decreases following (9.1), but T_f relates both to AFP concentration and solution osmolality. See Chap. 6 for an outline of the T_h concept. Comparative studies of AFP and T_h should therefore always discus the biological origin of both AFP and chosen sample osmolality (Evans et al. 2007). From an experimental point of view, low osmolality samples are not easy to work because the freezing process occurs rapidly in pure water, leaving no unfrozen fraction. Samples which are essentially pure water with added AFP are not biologically relevant, but they may however provide the researcher with a baseline effect of AFP for comparative studies (Kristiansen et al. 2008). The introduction of osmolytes increases the functional temperature range where partially frozen samples can exist. Equation (9.2) below introduces the concept of ice fraction (F_i):

$$F_i = 1 - \frac{T_0 - T_m}{T_0 - T_i} \tag{9.2}$$

where

F_i: ice fraction
T_m: equilibrium melting point of solution
T_i: temperature of partially frozen solution
T_0: equilibrium melting point of pure water, 273.15 K

In Eq. (9.2), T_m is determined by sample osmolality. As T_i decreases, the frozen fraction of the sample increases. For a sample with $T_m = 0.5$ °C, held at -5 °C, 90% of the sample is frozen. Because the freezing process in most cases is directed from an ice nucleus, producing replicable T_h measurements requires a stable ice nucleus in equilibrium with the surrounding solution. This is especially important for cryomicroscopy, but has until recently also been the case for calorimetric measurements of T_h.

9.2.1.1 Cryomicroscopy

The most well-known and proven technique for detecting AFP in solution is microfluidic cryoscopy performed under a microscope. It relies on manual observation of a single ice crystal, and very fine temperature manipulation. The technique requires a light microscope/binocular and a temperature-controlled stage, or cryostage. To this end, the Clifton Nanoliter Osmometer has for decades been the gold standard. But the instrument is not in production any longer, and functional units are becoming scarce. As a result of this, researchers across the globe have come up with new, innovative solutions to the problem of ageing hardware. At its very core, a cryostage is simply a sample holder, equipped with subzero temperature capabilities. Temperature control is made possible with the use of a thermoelectric component (Peltier element), the vapor from liquid nitrogen or a combination of the two. This sample holder must be outfitted in such a way, that single ice crystals can be observed with the use of a microscope. At the same time, the ice/water interface should not be disturbed by heterogeneous nucleators, such as metals, glass, or plastics. Cryoscopy is often performed in a low humidity environment, in order to decrease condensation on optics, which would otherwise interfere with the experiment. Therefore, evaporation can easily become an issue for small sample sizes, if a standing droplet was used to contain the ice crystal. These practical challenges can be overcome by suspending a small nanoliter droplet of aqueous AFP solution inside a viscous oil, which is suspended within a drilled hole in a thin metal plate. This methodology requires training and specialized equipment, but most, if not all of it, can be manufactured at a relatively low price. Figure 9.1 shows a generalized cross section of the metal plate used to perform cryoscopic measurements on a nanoliter osmometer. The principle behind AFP cryoscopy is that thermal hysteresis by definition is only observable in a certain temperature range, the hysteresis gap. It

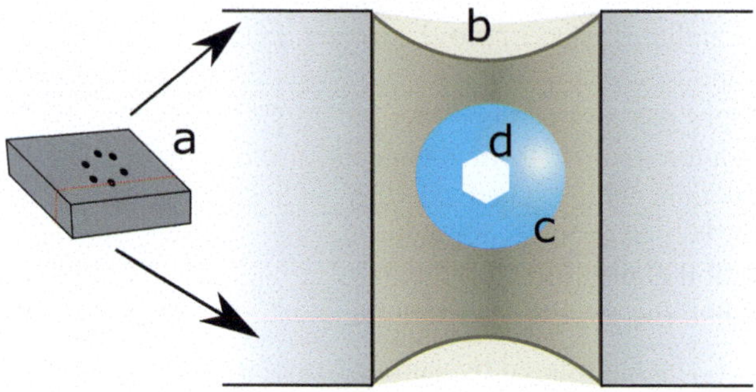

Fig. 9.1 Nanoliter osmometry sample cross section. A cross section of the experimental cell in a nanoliter osmometer setup (**a**). The cylindrical hole through the metal plate is filled with viscous oil (**b**), and a droplet of water (**c**) is suspended within. Inside the droplet of water, an ice crystal (**d**) is visible through the microscope. The droplet is surrounded by oil, which hinders evaporation of solvent, which would otherwise interfere with the sample osmolality

is therefore possible to completely freeze a sample containing AFP, just not at the T_f determined by osmolality. A crucial step in the cryoscopic method is the initial freezing of the sample, at a very low temperature. By flash freezing the AFP-containing solution, the resulting ice will consist of many ice crystals of different sizes and orientations. The idea is then to slowly increase temperature toward T_m. The ice will gradually melt away, and when only a single crystal remains, the temperature increase is stopped and the sample is kept at a constant temperature. This procedure requires extremely fine temperature control, and this is exactly why the Clifton Nanoliter Osmometer has been popular for so many years. The practical temperature range in which the crystals slowly melt is affected by sample osmolality, with higher osmolalities leading to a broader range of temperatures that are practically viable. To determine T_h, the temperature is decreased at a low cooling rate. The single ice crystal remains stable within the hysteresis gap, but will grow explosively below the hysteresis freezing point as described by the adsorption–inhibition theory (Kristiansen and Zachariassen 2005; Raymond and DeVries 1977). This approach to thermal hysteresis measurement produces replicable results, and has been used for decades (Chakrabartty and Hew 1991; Kim et al. 2017). Sample osmolality is usually between 150 mOsm/kg (Marshall et al. 2002; Chakrabartty and Hew 1991) and 500 mOsm/kg (Ramløv et al. 2005) depending on the originating organism, as discussed earlier in this chapter. Figure 9.2 shows a schematic overview of the cryoscopy process in AFP research. To load samples of such minuscule volumes, two special loading syringes are required. First, to fill the metal plate drill holes with the viscous oil, a thin glass capillary can be drawn out using a flame. This glass capillary can be fixed to an oil-filled syringe using glue, lacker or another air-tight sealant. Second, to inject and suspend a nanoliter droplet of sample inside a metal drilled hole filed with oil (see Fig. 9.1), a very thin loading instrument is needed. Another syringe is filled with a nonviscous paraffin oil, and a new glass capillary can be drawn out using a flame and the same procedure as the viscous oil syringe. However, the tip of the drawn out glass must have a smaller diameter than the metal

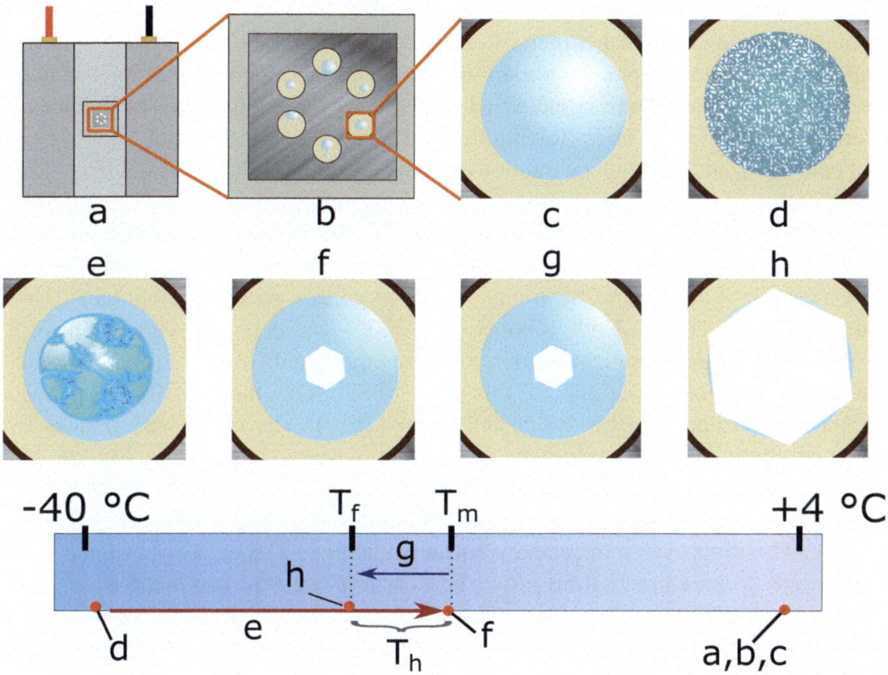

Fig. 9.2 Nanoliter osmometry. Top panels: (**a**) Top view of a cryostage with the sample holder over the light path in the middle (red square); (**b**) Zoomed view of the sample holder plate with several wells visible (red square); (**c**) Zoomed view of a single well, containing a suspended droplet of water; (**d**) Flash frozen water droplet consisting of tiny ice crystals; (**e**) The ice is heated to near-melting temperature, and the ice starts melting; (**f**) The ice is heated until a minute single crystal remains; (**g**) The solution is cooled down within the hysteresis gap; (**h**) The crystal reaches the hysteresis freezing point and starts growing rapidly. Bottom panel: Temperature scale from −40 °C to 4 °C, with letters corresponding to the view in the top panels. Blue arrow indicates cooling, red arrow indicates heating

drilled holes, for it must be able to pierce the surface of the oil and deposit a small droplet inside of the oil phase. The loading capillary often features a small bend near the tip, that allows for easier access to the metal drill holes. The thinness and angle of the capillary makes it unfit for sample extraction. However, a short extractor capillary is often used to as an intermediary to transfer sample from a container to the loading syringe. A small volume of sample can be drawn into a 10 μL capillary tube by capillary force, and the tube is then cut off right above the liquid level. By inserting the sample loading syringe into the extractor tube, the sample can be drawn into the fine glass tip with little risk of breaking. Once the injector syringe is filled with sample, it is time to load the metal sample holder. As this procedure requires a very steady hand, most sample holders feature several drilled holes to add some redundancy, as depicted in Fig. 9.2b. When droplets have been deposited into the sample chambers to a satisfactory degree, the initial freezing can take place. It is important that there is no contact between the sides of the metal plate and the water droplet, as this could potentially nucleate ice growth in the droplet. The objective is to obtain a single ice crystal, which can be used to observe the melting and growing temperatures of the sample. To obtain this single crystal, the entire droplet is flash frozen, and the temperature is increased to just below T_m, which can be calculated beforehand using Eq. (9.1). Following Eq. (9.2), only a tiny ice fraction should exist right below T_m, and this tiny ice fraction often results in a single ice crystal (Fig. 9.2f) in at least one of the suspended droplets. When the single crystal is determined to be in equilibrium, the temperature can be decreased to determine the hysteresis freezing point T_f and increased to determine the experimental melting point T_m, thus providing a measure of T_h (Fig. 9.2h). Many variations of this general system exist (Hon et al. 1994; Haymet et al. 1998; Ramløv et al. 2005), and in recent years much work has gone into the development of more modern equivalents of the Clifton Nanoliter Osmometer. A good example of this type of work is the LabVIEW-operated cryostage presented by Braslavsky and Drori (2013). The cryostage operates by the same thermoelectric principles that earlier designs have made use of, but everything is controlled through a laboratory computer. Because temperature control is handled by a microprocessor, real-time temperature feedback and logging is made possible. This improves transparency and repeatability. The system has been used several times to measure T_h from spruce budworm (*Choristoneura fumiferana*) AFP (Pertaya et al. 2007, 2008), ice crystal growth rates using inorganic ice inhibitors (Mizrahy et al. 2013) and ice morphology with T_h from Antarctic bacterial symbionts (Mangiagalli et al. 2017). Many of the microscopic methods rely on the growing and melting temperatures of single ice crystals. This provides the researcher with insights on ice crystal morphology, and visual growth patterns; something which may be just as relevant as the thermal hysteresis itself. Many of these techniques are however of somewhat qualitative character, and will therefore be presented later in this chapter. But there are also quantitative techniques that rely on calorimetric feedback, rather than visual interpretation.

9.2.1.2 Calorimetry

The visual change during crystallization of water is accompanied by thermodynamic event that is just as dramatic. The latent heat of fusion of water is 333.55 J/g, and the freezing event is therefore easily observable in a calorimeter. The use of differential scanning calorimetry (DSC) as a method for determining T_h was pioneered by Hansen and Baust (1988) and has been improved and modernized ever since (Ramløv et al. 2005; Baier-Schenk et al. 2005; Hassa-Roudsari and Goff 2012). DSC works by exposing a sample holder capsule and an empty reference capsule to the same steep temperature gradient, created either by liquid nitrogen or a compressor/refrigerator system. These two sample holders capsules are constantly heated, which counters the low temperature exposure, enabling precise temperature control. The result of a DSC experiment is the difference in heat flow to the two sample holder capsules, while performing a temperature scan. A generalized outline of such a thermogram can be seen in Fig. 9.3, where the checkered curve represents a sample

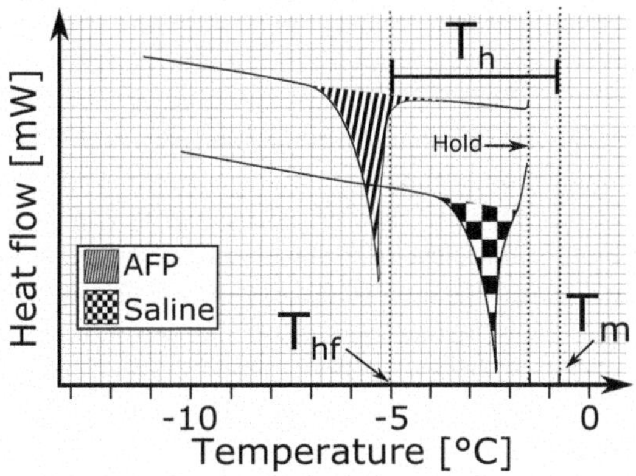

Fig. 9.3 Calorimetric determination of thermal hysteresis. A generalized thermogram showing two freezing experiments performed on a DSC. The exotherms signify a freezing event, which is the same order of magnitude in saline (checkered) and saline + AFP (striped) experiments. The two graphs are shifted on the *y*-axis, so that they are easier to tell apart. Once the samples are cooled below the holding temperature, crystallization takes place immediately in the saline, but is delayed in the AFP-containing solution. The delay corresponds to the hysteresis gap and is thus THA for that AFP

containing only buffer and the striped curve is a sample containing the same buffer with added AFP, giving it T_h capabilities. As water freezes, the enthalpy of fusion is released in an exothermic event, and the heat flowing toward the sample holder in order to keep the set temperature gradient decreases, as can be seen in Fig. 9.3. The concept of ice fraction makes an appearance yet again, as the calorimetric approaches to T_h also rely on a stable ice crystal as the nucleus for crystallization. But without the visual aspect of cryomicroscopy, measuring T_h with DSC relies on the instruments ability to reach a set temperature precisely below the calculated T_m. A small temperature offset or overshoot may lead to a complete melting of the sample solution, which is something to keep in mind when performing DSC measurements of T_h. The sample is allowed to stabilize at the calculated temperature for a short while, before cooling at a low rate begins. A sample without AFP will immediately begin freezing when the temperature scan starts. An ice crystal that is interacting with AFP will remain stable within the hysteresis gap, which can be observed as a straight horizontal line in a thermogram. DSC samples are always weighed, so the latent heat released during crystallization creates an exotherm of a predictable magnitude, represented as the peak area in the thermogram. By performing several freezing scans in the DSC, T_m can be calculated from the peak areas and an accurate determination of T_h can be performed (Ding et al. 2014). A calorimetric determination of T_h can be performed in many different ways, and it is often done in addition to cryoscopy, as a verification procedure. The two methods rely on similar temperature profiles, as samples must be flash-frozen, heated to a hysteresis gap temperature just below T_m and then slowly cooled. Ramløv et al. (2005) presented several temperature profiles that were used to quantify T_h of different AFGPs from Antarctic fish. The temperature scan profiles were performed on a Perkin Elmer DSC 7, but may be easily transferred to newer equipment (Romero et al. 2008; Cao et al. 2016). As a last note about calorimetry, it is clear that exact measures of T_h are obtainable through this approach. But it is also possible to measure ice recrystallization inhibition (IRI) using a calorimetric method (Hassa-Roudsari and Goff 2012). The heat flow to a frozen sample of water should be more or less constant over time. Crystals will change size through Ostwald ripening, but the net ice fraction does not change. However, Hassa-Roudsari and Goff (2012) observe an exothermic event lasting up to 4 h in frozen samples containing winter wheat (*Triticum aestivum cv. Vienna*) AFP and type I AFP. This exothermic event is believed to be AFPs moving around the boundary layer between ice crystals. The slope of the exothermic event on the thermogram correlates well with AFP concentration, but only for winter wheat AFP. Plant AFPs generally show strong IRI characteristics, and this approach may therefore not be suited for fish AFP. Insect AFP has yet to be tested with this method.

Measuring T_h using either cryomicroscopy or DSC requires crystallization taking place from a single ice nucleus. This is not possible to observe in the DSC, and must therefore be accounted for by calculation. Ding et al. (2014) shows this calculation, and also notes that apparent T_h is dependent upon ice-fraction, F_i. The DSC method has therefore historically involved some uncertainties. A relatively new technique that has become very popular in recent years is sonocrystallization of a supercooled sample (Gaede-Koehler et al. 2012). Instead of relying on a crystallization nucleus consisting of ice crystals, the freezing process can be initiated by supersonic mechanical excitation. By supercooling a large (~1 mL) volume of AFP-containing solution and initiating nucleation using sound waves, it is possible to accurately determine the freezing point of said solution. This is achieved by data-logging the temperature of the solution, and observing the temperature plateau associated with the freezing process of water, due to the release of latent heat. Continuous heating of the frozen sample will eventually lead to melting, which is again accompanied by a melting plateau. The temperature difference between these two plateaus is T_h of the sample. This technique produces reliable results (Gaede-Koehler et al. 2012; Olijve et al. 2016), but in its current form requires relatively large volumes of AFP solution to work.

9.3 Recrystallization Inhibition

Another hallmark feature of AFP is the ability to inhibit the recrystallization of ice. As we know from T_h experiments, the addition of AFP to a liquid only shifts the freezing point a few degrees, in most cases. From a biological point of view, IRI is often important in freeze-tolerant organisms (Ramløv et al. 1996), whereas thermal hysteresis activity (THA) is often related to cold tolerance through freeze avoidance (Olsen et al. 1998). Measuring IRI for a given AFP is generally less of a challenge than THA. There are several reasons for this, but the most important is probably that there is no requirement for obtaining single ice crystals. In fact, IRI experiments rely on the Ostwald ripening process that takes place amongst ice crystals in physical contact with one another. Due to the mechanism by which AFPs cause thermal hysteresis (adsorption–inhibition described in Chaps. 5 and 6), ice crystals with adsorbed AFP are unable to equilibrate with their surroundings within the hysteresis gap. At temperatures lower than the hysteresis freezing point, AFPs still interact with the ice, even if the ice fraction is drastically increased. Water molecules in the ice are not static, and there is a net flux of water from small ice crystals to larger ones, due to differences in surface tension. The flux is conducted through the liquid phase, and it is right at this interface that AFP is able to have an effect on systems that are essentially frozen solid. By masking the differences in surface tension between ice crystals, the molecular flux and equilibration is halted or slowed significantly. There are several ways of observing this phenomenon, and like the techniques for measuring T_h, they generally rely on microscopy or calorimetry.

9.3.1 Microscopy

As anyone who has ever found a long forgotten popsicle in the freezer can tell you, recrystallization of ice eventually becomes a macroscopic phenomenon. So while it is possible to observe the effects of AFP on ice recrystallization macroscopically, it is highly impractical in the laboratory setting due to the time and volume required to do so. By freezing a small volume of AFP-containing liquid and observing the ice crystals under a microscope, recrystallization can be observed and quantified within minutes. Measuring T_h requires precise temperature control and the use of temperature gradients or scans. Recrystallization, however, is observable at a static temperature. So while the temperature profile of a T_h experiment resembles that of an IRI-experiment, there is essentially only need for two set temperatures. One very low (~−40 °C) to ensure complete and fast freezing of the sample in question, and one below T_m to observe recrystallization. The temperature used is often calculated from sample osmolality and a chosen F_i, see below. This reduces the hardware requirements of IRI analysis, and IRI is generally considered much simpler to measure than T_h. The experimental samples used for IRI assays do not have to be loaded in to an oily suspension, but can simply be held between two coverslips (Budke et al. 2009).

Figure 9.4 shows micrographs from two experiments at different points in time. There is a striking difference in crystal structure and size after 30 min at $F_i = 0.9$, which is a direct consequence of recrystallization. The imagery seen in Fig. 9.4 shows clear borders between the growing ice crystals, but in many cases this can be enhanced further by the use of crossed polarizers. Polarized imagery, such as the photographs shown by Ramløv et al. (1996), not only emphasizes the boundary layers between ice crystals, it can also provide insights to possible patterns within the overall orientation of each individual ice crystal. For automated analysis of crystal growth, the use of cross polarizers essentially blocks out half of a micrograph and is therefore often left out (Budke et al. 2009, 2014; Buch and Ramløv 2016). Modern

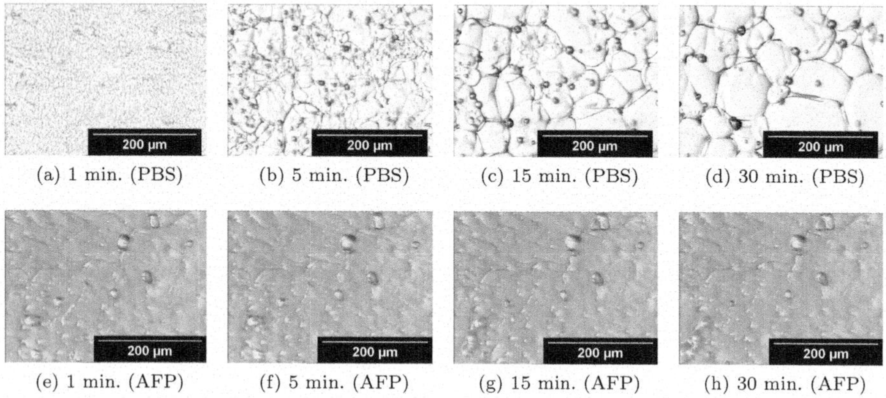

Fig. 9.4 Recrystallization of ice. Ice recrystallization at various time points in saline (top) and saline + AFP (bottom). Post flash-freeze the samples were kept at $F_i = 0.9$ and photographed after 1, 5, 15 and 30 min. Notice how the shape and size of the saline ice changes over time, whereas ice crystals formed with AFP do not. Data from Buch and Ramløv (2016)

cryostages can flash freeze small samples of experimental solution to a satisfying degree (Halwani et al. 2014; Buch and Ramløv 2016), but when IRI was introduced as an aspect of AFP research, one of the most popular techniques was the "splat-cooling" method (Knight et al. 1988). A small droplet (10 μL) was dropped from a height of 3 m on to an aluminum plate, which was cooled down to −76 °C by dry ice. The droplet would freeze as a very thin disc, which could then be transferred to the view of a microscope and photographed. The analysis of IRI was based on the captured photographs, and by measuring the average crystal diameter, growth rates could be established. However, due to the binary nature of IRI, the results were often evaluated on a yes/no basis. Such analysis was used to detect IRI in the Antarctic nematode *Panagrolaimus davidii* and refute the hypothesis that the Alpine Weta *Hemideina maori* contained AFP (Ramløv et al. 1996). As new AFPs were discovered and hypothesized, a demand for a faster and broader IRI assay arose. The capillary assay presented by Tomczak et al. (2003) allows for simultaneous observation of 10–15 recrystallizing samples at any given time during an experiment. Instead of preparing a single, thin wafer of ice by splat-cooling or flash freezing between two cover slips, the capillary IRI assay instead relies on the freezing of several liquid-filled 10 μL capillary tubes. The experiment is meant to be run over night, and the capillary tubes, once prepared, can be, frozen, thawed, or archived a number of times. This is a convenience that none of the classic methods offer. This allows for several IRI assays to be performed at the same time, either as replicates or and at different conditions, such as AFP concentration or type. The capillary analysis allows for accurate determination of the lowest concentration at which an AFP can cause IRI: the IRI efficacy or C_i (Olijve et al. 2016).

As time progressed and desktop computers became more and more powerful, there was a need for new quantification techniques that made use of these new possibilities. The idea of software-assisted quantification through digital image analysis was introduced by Jackman et al. (2007). The researchers had written custom C software, that allowed for recognition of recrystallization within a thin ice wafer, produced with the splat-cooling method, described earlier. Through software analysis, it was possible to obtain a much greater amount of data points throughout an experiment. Some of the subjectivity inherent to microscopy can also be circumvented by letting a piece of software randomly choose which ice crystals to measure. Following this trend, a new method was presented by Budke et al. (2009) which also makes use of automated image analysis through software. This new analysis method is called ice recrystallization rate inhibition assay (IRRINA) (Budke et al. 2014), and it is a very sensitive quantification technique. Rather than relying on average crystal diameter, which might as well be measured from single still images, the end result is a crystal growth rate at low F_{i}, modeled after Eq. (9.3) (Budke et al. 2009).

$$r^3(t) = {r_0}^3 + k_{\mathrm{d}}t \tag{9.3}$$

where

$r^3(t)$ [μm^3]: mean ice crystal radius at given time, t
${r_0}^3$ [μm^3]: mean crystal radius at $t = 0$
k_{d} [μm^3 t^{-1}]: crystal growth rate

C_{i} is generally much lower (nM–μM) than the AFP concentration at which T_{h} can be observed (μM–mM), and IRI is therefore a good endpoint for detecting AFP activity, even if T_{h} has historically been the method of choice. The requirements for specialized equipment are also lower for recrystallization experiments, than those of T_{h}-based experiments, mainly because measuring T_{h} requires a cryoscopic approach with fine-tuned temperature control. However, this may be changing due to the popularity of the sonocrystallization method presented above (Gaede-Koehler et al. 2012). The need for relatively large volumes of AFP-containing solution in order to observe recrystallization (Tomczak et al. 2003) has also dissipated in recent years, and antifreeze activity can now be detected with sub-microliter quantities (Buch and Ramløv 2016).

9.4 Detection of AFP

The need for quantitative methods in AFP research came from the wonder and interest generated by some of the first micrographs showing AFP activity. See Haymet et al. (1998, 1999), Takamichi et al. (2009), and Raymond et al. (1989) for a few examples of such micrographs. The existence of AFPs could explain many of the extreme survival strategies seen in both polar fish species (Raymond and DeVries 1977) and cold-hardy insects (Duman 1980). In the previous pages we have discussed some of the most common quantification techniques used in AFP research, and all of these can off course be used for qualitative studies just as well. But there is a whole subgroup of techniques that primarily see use in qualitative AFP research, and some of these will be presented here. Most techniques are based on microscopy of ice, but a few are of a more indirect nature and some are even macroscopic.

9.4.1 Crystal Structure

The core mechanism of action for AFP relates to the growth of ice crystals, or rather, the interference of ice crystal growth. As is evident from the structure of snow flakes, ice crystals grow in a hexagonal fashion. AFPs truncate this growth, but not necessarily equally along all crystal planes. Figure 9.5 shows a hexagonal crystal with growth planes visible. Unequal binding to crystal growth planes could explain the pyramidal shape of ice crystals grown with fish AFP, as observed by Raymond et al. (1989). To illustrate the differences in AFP-binding specificity, the so called ice etching technique was employed by Knight et al. (1991, 1993). Because crystal growth in macroscopic volumes of water often takes place from several nucleation points, it can be difficult to obtain a large, single ice crystal. With very few modifications, the method still used today relies on large volumes of water at very low degrees of supercooling (Knight et al. 1991; Basu et al. 2014). By allowing the volume of water to slowly freeze over several days, the risk of "overnucleating" the crystal growth is reduced. After several days of freezing, an upper layer of ice is likely to have formed. The layer may be one large crystal or several smaller. Because single ice crystals interact very little with the light passing through them, crystal orientation and purity can be checked using a crossed polarizer, much like recrystallization experiments. It may therefore be apparent to the naked eye where the large, single crystals have formed, but orientation of the crystals must also be

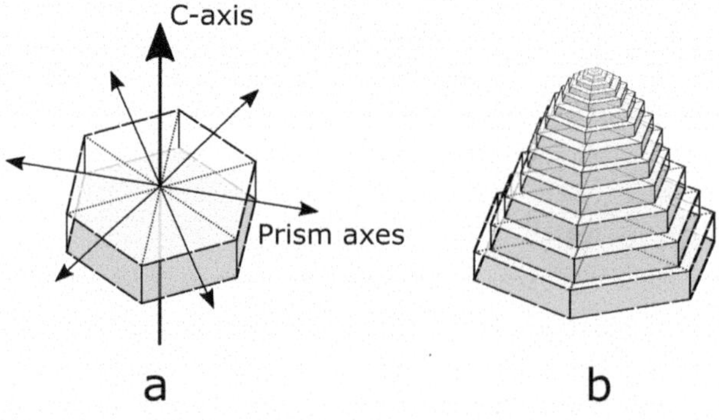

Fig. 9.5 Growth planes of an ice crystal. (**a**) Drawing of a hexagonal ice crystal showing C- and prism-axes, and (**b**): ice crystal with truncated growth along the prism axes, commonly seen in the presence of fish AFPs. The thin, truncated crystals stack on top of each other and form a new pyramidal plane. The pyramid also extends downward, creating the well-known bipyramidal ice spicule shape

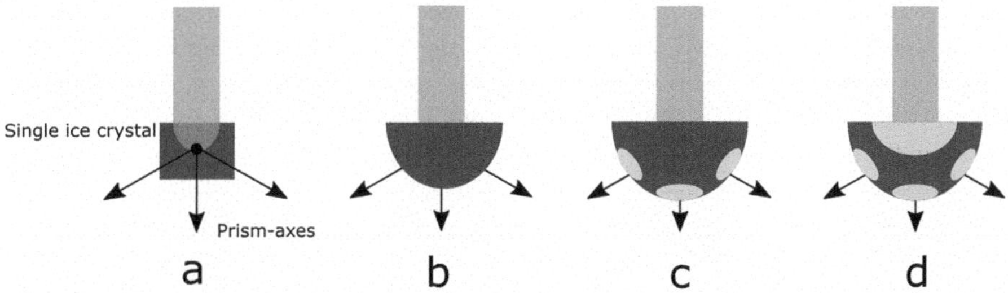

Fig. 9.6 Ice etching technique. Overview of ice etching performed on a single, large ice crystal. The prism-axes are shown as arrows throughout and the *c*-axis is normal to the image, i.e., it points toward the reader. (**a**) A single ice crystal is prepared and frozen to the cold-finger; (**b**) The crystal is allowed to grow in a dilute solution of AFP, which results in a large hemisphere of ice; (**c**) The ice is scraped clean and allowed to evaporate, which reveals an etched pattern where AFP have adsorbed to the crystal surface. In this case, only the prism-axes, which indicates a fish type AFP; (**d**) A hyperactive AFP from an insect would create a etched pattern on the *c*-axis as well, which can be seen as the additional white spot

determined. To this end, the crystal may be transferred to a colder atmosphere and observation of frost forming on the surface can reveal the approximate orientation of the prism faces (Knight et al. 1991). A more accurate determination of the crystal planes involves a small puncture on the single ice crystal, which is then transferred to a freeze dryer. This process is known as ice pitting (Knight 1966). Evaporation happens faster along the prism-axes than the *c*-axis, and a clear hexagonal pattern will form within 20 min (Basu et al. 2014). Knowing crystal orientation is important for interpreting the AFP-binding patterns that eventually form. The single ice crystal is often sawed along the *c*-axis and a small hole is melted in the middle of the crystal, parallel to the *c*-plane. The melting is often performed using a brass cold-finger, with the coolant turned off. Once the cold-finger has melted its way into the ice, the cooling may be turned on and the ice freezes to the cold-finger. Figure 9.6a shows the single ice crystal attached to the cold-finger. Prism-axes are shown on the figure and the *c*-axis is normal to the plane of the figure. The ice crystal is then grown in a dilute AFP solution, so that the irreversible binding does happen, but ice growth continuously covers the adsorbed AFPs. The ice crystal grows as a hemisphere, which is also the way it is presented in Fig. 9.6b. The classic approach to ice etching (Haymet et al. 1998, 1999) is to then scrape the outermost layer of AFP-containing film off of the ice hemisphere and transfer the ice crystal to a cold room with low humidity. The low humidity leads to a sublimation reaction that leaves the ice hemisphere completely mirror clean and shiny. But the patches with adsorbed AFP (i.e., prism faces for fish AFP) will remain unclear with a scratched or *etched* appearance. Due to the circular growth pattern of the ice, these patches will appear as ellipsoids on the surface of the hemisphere, such as the white patches seen in Fig. 9.6c. The ice etching method was first used to investigate binding patters of fish AFP, which generally only bind the prism faces of the growing ice crystal, leading to the pattern seen in Fig. 9.6c. However, the hyperactive AFPs found in insects are also able to bind the basal plane (Scotter et al. 2006), leading to more etched areas on the growing hemisphere (Graether et al. 2000), see Fig. 9.6d. The ice etching technique has seen much use over the years, but the approach suffers somewhat from the visual inspection aspect of an experiment. It may not always be completely clear what is actual bound protein, and what is just impurities in the ice. With the developments in molecular biology and production platforms for fusion proteins with fluorescent tags, it has now become viable to extend the original ice etching technique to also include fluorescence (Garnham et al. 2010; Park et al. 2012b; Kondo et al. 2012; Bar-Dolev et al. 2012). The technique is called fluorescence-based ice plane affinity (FIPA) and is carried out in more or less the same fashion as traditional ice etching (Basu et al. 2014). The AFP binding patches are still visible to the naked eye, which is a good experimental indicator throughout the experiment. But since the AFPs used in FIPA are fusion proteins with GFP attached, UV excitation of the bound AFP reveals binding patterns much clearer than traditional ice etching (Basu et al. 2016). Because of the clear high-contrast binding patterns one can obtain through FIPA analysis, observing AFP bound to other planes than prism and basal is also possible. The pyramidal plane (Wen and Laursen 1992), which forms in ice spicules grown with fish AFP (Raymond et al. 1989), is a good example of one such plane (Garnham et al. 2010). Since AFPs exert their effect at the boundary between ice and water, a cryostage setup that can observe this boundary layer can be very useful. By constructing the cryostage in such a way that a temperature gradient exists over the light path, a movable slide glass holder can be used to place a liquid sample on top of the gradient. The boundary layer between water and ice can then be observed, and even manipulated (Namperumal and Coger 1998). To look even closer at the ice crystals formed in the presence of AFP, electron microscopy can be used. By freezing AFP solution in a capillary and essentially breaking it at the middle, the exposed ice crystals can be photographed using transmitted electron microscopy (TEM) (Wilson et al. 2002). This can provide insight to the ice-plane affinity of an AFP, much like the FIPA technique. To go even further down the size scale, scanning tunnelling microscopy (STM) can be used to map the surface of an ice crystal, down to a resolution in picometers. Using STM, Grandum et al. (1999) was able to observe grooves on the surface of an ice crystal growing in the presence of AFP. These grooves were 2–10 nm deep and corresponded to points of adsorption by AFP. The curvature between adsorbed AFP agrees with the adsorption–inhibition hypothesis, in that thermal hysteresis is, in part, caused by local convexity and thus higher interfacial tension. By observing ice crystals grown in the presence of AFP, much can be deduced about the structure and function of AFPs. But just as AFP can influence the structure of a growing ice crystal, their presence can also alter the "structure" of the hydrogen bonded network of water molecules surrounding them.

9.4.2 Terahertz Spectroscopy and the Hydration Shell

Water molecules in bulk water, on average, take on a tetrahedral structure, something which is disrupted by impurities (Ickes et al. 2015; Bulavin et al. 2008). These impurities may include proteins, which often have both hydrophilic and hydrophobic residues. Proteins generally fold to expose the hydrophilic residues while creating a hydrophobic core. Around the protein is a dynamic hydration shell which has water molecules residing near the protein in the picosecond time scale. A strong hydration shell discourages freezing, as water molecules strongly bound in the shell are less likely to break their network of hydrogen bonds (Ickes et al. 2015), compared to bulk water. Ebbinghaus et al. (2007) showed that it was possible to probe this hydration shell using terahertz (THz) spectroscopy, and see the effects of a protein on the residence time and size of the dynamic hydration shell. Using the same Thz spectroscopy technique, Ebbinghaus et al. (2010) showed that the hydroxyl groups on AFGPs were able to perturb water molecules in the hydration shell for up to 35 Å at 5 °C. Using site-directed mutagenesis Ebbinghaus et al. (2012) showed that type I winter flounder AFP was also able to induce an extended hydration shell in solution. But the mechanism was not through excess of hydroxyl groups like in AFGPs, but rather a few very specifically placed threonine residues. The same was found for *Dendroides canadensis* DAFP-1 (Meister et al. 2013) and Antarctic eelpout (*Pachycara brachycephalum*) type III AFP (Xu et al. 2016). The THA of AFPs can be enhanced by various small molecules, such as trehalose, glycerol, or citrate (Li et al. 1998; Wang et al. 2009b). Most of these enhancers can interact with an arginine residue (Wang et al. 2009b) and on DAFP-1 it is hypothesized to be Arg9, the only Arg-residue that is not directly localized in the ice-binding site. It is not completely understood how these low molecular weight enhancers function, but Meister et al. (2013) show that addition of 0.5 M sodium citrate to DAFP-1 will extend the dynamic hydration shell from 20 Å to 27 Å. The hydration shell may also assist in the binding between ice and AFP, as shown by Meister et al. (2014b) using sum frequency generation spectroscopy. An ice-like structure of water could be observed near the ice-binding surface of a type III AFP, which is likely involved in the initial docking between AFP and the growing ice crystal. Likewise, the lifetime of hydrogen bonds in the hydration shell surrounding DAFP-1 was shown to increase with the addition of sodium sulfate, but the increase was relatively stronger on the ice binding site, from 72 to 90 ps (Meister et al. 2014a). The lifetime of a hydrogen bond in bulk water is ~23 ps. It seems to be the case that AFPs have evolved to induce the highly structured hydration shell near the nonbinding side of the protein, which halts the freezing process of those water molecules. On the ice binding site, however, the presence of an ice-like hydration shell likely assists the docking between AFP and ice. These phenomena can be detected using THz spectroscopy and sum frequency generation spectroscopy, respectively. These spectroscopic methods have already lead to extended versions of the original adsorption–inhibition hypothesis (Xu et al. 2016; Meister et al. 2013), and may be a valuable tool in the future, especially for explaining new, novel antifreeze agents or inorganic AFP analogues.

9.4.3 Detection in Tissue

So far, the detection methods presented in this chapter have worked on the basis of pure AFP in somewhat nonphysiological conditions. But as the name would imply, AFPs are indeed proteins and are therefore immunologically active. This off course enables the use of Western blotting for qualification of AFP, either from serum (Wu and Duman 1991) or recombinant proteins produced through molecular biology techniques (Qiu et al. 2009). But perhaps more interesting in the biological context, are the possibilities for immunohistochemistry and immunofluorescence. The ELISA technique has also been employed for AFP detection (Tong et al. 2000), but remains somewhat novel within this field of research. Thus far we have been able to ask the question of *how* AFPs work, but with the use of immunolocalization we can ask *where* they work. By capturing live specimen, e.g., insect larvae, and fixing them in a preservative solution it is possible to embed a whole organism in a hard medium, which allows for very fine sectioning on a microtome. Cross sections only 3–5 μm thick can then be stretched on a flotation bath and ligated to a glass slide. The section on the glass then contains the entire cross sectional tissue, but essentially none of the water. The embedding process replaces all liquid water with the hard medium of choice. By using polysine slide glass, the section sticks to the slide easier, and the embedding compound can be removed in a serious of washes with xylene, ethanol, and water. The tissue cross section is thus slowly returned to a condition resembling that in the living organism, where water is the main solvent. Once the tissue is returned to this condition, it can be exposed to primary antibodies specific to one or several AFPs found in that organism. The primary antibodies are allowed to react and bind, and the whole slide is then washed in a blocking solution, often consisting of BSA or another well-defined protein. The blocking solution reduces the amount of unspecific binding, which would lead to false positive fluorescence. Once the tissue-bound AFPs are coated with primary antibodies, the second immunological reaction can take place. Primary antibodies bind the protein of interest and the secondary antibodies bind the primary antibodies, creating a tandem of antibodies. But the secondary antibodies can feature either a fluorescent tag (e.g., fluoresceine or rhodamine) for immunofluorescence (Olsen et al. 1998; Buch and Ramløv 2017) or an enzyme (e.g., horse radish peroxidase) for immunohistochemistry (Easton and Horwath 1994). Using immunohistochemistry Horwath et al. (1996) was able to detect a high concentration of an AFP in the primary culture of fat body cells from the yellow meal worm beetle *Tenebrio molitor* which varied with ambient temperature. According to the study by Horwath et al. (1996) *T. molitor* accumulates AFP at higher temperatures, which may be part of a rapid-response mechanism to protect against sudden cold spells. The same pattern was observed in cuticle of *Rhagium mordax* larvae using immunofluorescence (Buch and Ramløv 2017). In

D. canadensis the pattern seems to be the opposite, as shown by Olsen et al. (1998). Here AFP could be detected in the cuticle of larvae only when ambient temperature was low. Immunofluorescence and immunohistochemistry are qualitative techniques, and they can provide a great deal of biologically relevant information about the AFP in question. Attempts have been made toward quantifying the fluorescence levels of bound antibodies, and thereby the amount of tissue AFP (Buch and Ramløv 2017). While not without its flaws, automated high resolution image analysis may have an important place in future qualitative AFP research, just as it has become for tracking recrystallization inhibition from AFP (Budke et al. 2014).

9.4.4 Macroscopic and Indirect Methods

To wrap up this chapter on detection and quantification techniques used in AFP research, we will here present several approaches to macroscopic or indirect detection of AFP. The use of AFP as an additive in food and other consumables has expanded the catalogue of methodologies used in AFP research. This part of the chapter may therefore serve as a source of inspiration, when developing new methods for future research.

When a solution freezes, relatively pure ice crystals will form, which exclude nearly all solutes. These solutes are concentrated in boundary layers surrounding the ice crystals, and it is in these layers that the recrystallization water flux takes place. This means that particulate matter may aggregate and even precipitate. In the biological context, the steep gradients of solutes can lead to membrane and protein damage. The reduced space for solutes and particles to occupy with subsequent aggregation is exploited in a novel antifreeze assay, which uses the colorimetric shift of gold nanoparticles (AuNPs) to quantify AFP activity (Park et al. 2012a). Whether or not the AFPs in solution interact directly with the AuNPs or only interact with ice crystals is not fully known, but aggregation does happen with freezing, and AFPs are able to slow this process down. The AuNP suspension has an absorption maximum at 520 nm and will appear bright red. As AuNPs aggregate over time, the particles become larger and the absorption maximum shifts toward 650, which will make the suspension appear blue. By filling a microtiter plate with AuNP suspension and varying concentrations of AFP, the colorimetric shift from red to blue following freezing and thawing correlates well with AFP concentration (Park et al. 2012a). A similar technique was used by Mitchell et al. (2015) to quantify the effects of recrystallization inhibition of several non-AFP compounds. Here the authors measure absorbance at 520 nm in suspensions of AuNPs after a freeze/thaw cycle. At the optimal AuNP concentration of 80 $\mu g \times mL^{-1}$ the absorbance at 520 nm in a 50 μL well will be 0.15 A if any IRI activity is present (Mitchell et al. 2015). The use of a colorimetric assay is very useful if the end goal is a completely automated screening process for IRI activity. Even though these techniques are fairly novel at the time of writing, they may see much use in the future. The somewhat strong correlation between AFP concentration and absorbance shift seen by Park et al. (2012a) makes it a strong candidate for a quantitative method of the future.

As mentioned in the beginning of this chapter, the involvement of AFP in food science has led to a number of interesting methods for determining the downstream consequences of adding AFP to a complex matrix. These indirect methods all assume a specific effect of AFP, and it cannot be ruled out that the observations are not due to any of the classic AFP mechanisms, such as thermal hysteresis or IRI. Nevertheless, there seems to be some consensus that especially the IRI capabilities of AFP may play an important role as a food additive in the future. Zhang et al. (2007a, b) did experiments with AFP from carrots (*Daucus carota* AFP—DcAFP) as additives in frozen bread dough. The hypothesis was that DcAFP would have cryopreservative properties that would ensure a lower mortality rate in the yeast cells present in the dough. They measured the volume expansion of the dough when it was allowed to rise after thawing. AFP was shown to have a positive effect on dough volume, which would indicate a cryopreservative effect on the yeast. The same effect was observed in frozen dough using a recombinant type I AFP, which was produced by a food-grade *Lactococcus lactis* strain (Yeh et al. 2009). Like dough containing yeast, the complex matrix of meat is also damaged during freezing and thawing (Leygonie et al. 2012). Thawed meat should however not ferment, and the expansion volume approach is thus useless. A sign of freezing damage to meat is a liquid run off, which is essentially water that has failed to be reabsorbed. It is generally associated with a decrease in quality (Leygonie et al. 2012). By measuring drip loss after increasing periods in the freezer Yeh et al. (2009) determined that freezer damage could be significantly decreased by infusing meat with AFP prior to freezing. The same effect was observed in meat from lambs that were injected with AFGP prior to slaughter (Payne and Young 1995). In a similar study by Payne et al. (1994) it was shown by scanning electron microscopy that the radius of ice crystals within frozen meat could be significantly reduced by soaking the meat in an AFP solution prior to freezing. But not all food types are ideally stored frozen, for instance fish. The seafood product known as surimi is a gel formed by fish meat, or rather the actomyosin fibers within the muscles. The gel forming capabilities decrease rapidly with storage, both frozen and chilled, and is correlated to the activity of the actomyosin Ca^{2+}ATPase. Unlike AFPs, the Ca^{2+}ATPase is an enzyme which predictably catalyses a chemical reaction, and is therefore easily quantified. Boonsupthip and Lee (2003) showed that addition of AFP to fish muscle homogenate in both chilled (6 °C) and frozen (−12 °C to −18 °C) improved Ca^{2+}ATPase activity significantly.

9.5 Conclusions

Here we have described some of the most important techniques used in AFP research, a field which is clearly influenced by many scientific disciplines. It is clear that a significant amount of creativity has shaped this branch of research into what it is today. From the first observations of insects and fish apparently able to avoid death by freezing to quantification of the hydration shell surrounding ice crystals interacting with AFPs, a great deal of innovation and cross-disciplinary work has been involved. With these scientific advances in mind, the reader can now continue exploring the industrial and technological applications of AFPs. And it is the author's hope that this chapter has served as source of inspiration and perhaps a starting point for new scientific endeavors.

References

Baier-Schenk A, Handschin S, Conde-Petit B (2005) Ice in prefermented frozen bread dough - an investigation based on calorimetry and microscopy. Cereal Chem 82(3):251–255. https://doi.org/10.1094/CC-82-0251, wOS:000229284300004

Bar-Dolev M, Celik Y, Wettlaufer JS, Davies PL, Braslavsky I (2012) New insights into ice growth and melting modifications by antifreeze proteins. J R Soc Interface 9(77):3249–3259. https://doi.org/10.1098/rsif.2012.0388

Basu K, Garnham CP, Nishimiya Y, Tsuda S, Braslavsky I, Davies P (2014) Determining the ice-binding planes of antifreeze proteins by fluorescence-based ice plane affinity. J Vis Exp 83: e51185. https://doi.org/10.3791/51185

Basu K, Wasserman SS, Jeronimo PS, Graham LA, Davies PL (2016) Intermediate activity of midge antifreeze protein is due to a tyrosine-rich ice-binding site and atypical ice plane affinity. FEBS J 283(8):1504–1515. https://doi.org/10.1111/febs.13687

Boonsupthip W, Lee TC (2003) Application of antifreeze protein for food preservation: effect of type III antifreeze protein for preservation of gel-forming of frozen and chilled actomyosin. J Food Sci 68(5):1804–1809. https://doi.org/10.1111/j.1365-2621.2003.tb12333.x, wOS:000183947000043

Braslavsky I, Drori R (2013) LabVIEW-operated novel nanoliter osmometer for ice binding protein investigations. J Vis Exp 72:e4189. https://doi.org/10.3791/4189

Buch JL, Ramløv H (2016) An open source cryostage and software analysis method for detection of antifreeze activity. Cryobiology 72(3):251–257. https://doi.org/10.1016/j.cryobiol.2016.03.010

Buch J, Ramløv H (2017) Detecting seasonal variation of antifreeze protein distribution in *Rhagium mordax* using immunofluorescence and high resolution microscopy. Cryobiology 74:132–140. https://doi.org/10.1016/j.cryobiol.2016.11.003

Budke C, Heggemann C, Koch M, Sewald N, Koop T (2009) Ice recrystallization kinetics in the presence of synthetic antifreeze glycoprotein analogues using the framework of LSW theory. J Phys Chem B 113(9):2865–2873. https://doi.org/10.1021/jp805726e

Budke C, Dreyer A, Jaeger J, Gimpel K, Berkemeier T, Bonin AS, Nagel L, Plattner C, DeVries AL, Sewald N, Koop T (2014) Quantitative efficacy classification of ice recrystallization inhibition agents. Cryst Growth Des 14(9):4285–4294. https://doi.org/10.1021/cg5003308

Bulavin LA, Lokotosh TV, Malomuzh NP (2008) Role of the collective self-diffusion in water and other liquids. J Mol Liq 137(1):1–24. https://doi.org/10.1016/j.molliq.2007.05.003

Cao H, Zhao Y, Zhu YB, Xu F, Yu JS, Yuan M (2016) Antifreeze and cryoprotective activities of ice-binding collagen peptides from pig skin. Food Chem 194:1245–1253. https://doi.org/10.1016/j.foodchem.2015.08.102

Chakrabartty A, Hew CL (1991) The effect of enhanced alpha-helicity on the activity of a winter flounder antifreeze polypeptide. Eur J Biochem 202(3):1057–1063. https://doi.org/10.1111/j.1432-1033.1991.tb16470.x

Ding X, Zhang H, Liu W, Wang L, Qian H, Qi X (2014) Extraction of carrot (*Daucus carota*) antifreeze proteins and evaluation of their effects on frozen white salted noodles. Food Bioprocess Technol 7(3):842–852. https://doi.org/10.1007/s11947-013-1101-0

Duman JG (1980) Factors involved in overwintering survival of the freeze tolerant beetle, *Dendroides canadensis*. J Comp Physiol B 136(1):52–59. https://doi.org/10.1007/BF00688622

Easton CM, Horwath KL (1994) Characterization of primary cell cultures derived from fat body of the beetle, *Tenebrio molitor*, and the immunolocalization of a thermal hysteresis protein in vitro. J Insect Physiol 40(6):537–547. https://doi.org/10.1016/0022-1910(94)90127-9

Ebbinghaus S, Kim SJ, Heyden M, Yu X, Heugen U, Gruebele M, Leitner DM, Havenith M (2007) An extended dynamical hydration shell around proteins. PNAS 104(52):20749–20752. https://doi.org/10.1073/pnas.0709207104

Ebbinghaus S, Meister K, Born B, DeVries AL, Gruebele M, Havenith M (2010) Antifreeze glycoprotein activity correlates with long-range protein-water dynamics. J Am Chem Soc 132 (35):12210–12211. https://doi.org/10.1021/ja1051632

Ebbinghaus S, Meister K, Prigozhin MB, DeVries AL, Havenith M, Dzubiella J, Grue-bele M (2012) Functional importance of short-range binding and long-range solvent interactions in helical antifreeze peptides. Biophys J 103(2):L20–L22. https://doi.org/10.1016/j.bpj.2012.06.013

Esser-Kahn AP, Trang V, Francis MB (2010) Incorporation of antifreeze proteins into polymer coatings using site-selective bioconjugation. J Am Chem Soc 132(38):13264–13269. https://doi.org/10.1021/ja103038p

Evans RP, Hobbs RS, Goddard SV, Fletcher GL (2007) The importance of dissolved salts to the in vivo efficacy of antifreeze proteins. Comp Biochem Physiol A 148(3):556–561. https://doi.org/10.1016/j.cbpa.2007.07.005

Gaede-Koehler A, Kreider A, Canfield P, Kleemeier M, Grunwald I (2012) Direct measurement of the thermal hysteresis of antifreeze proteins (AFPs) using sonocrystallization. Anal Chem 84 (23):10229–10235. https://doi.org/10.1021/ac301946w, wOS:000311815300014

Garnham CP, Natarajan A, Middleton AJ, Kuiper MJ, Braslavsky I, Davies PL (2010) Compound ice-binding site of an antifreeze protein revealed by mutagenesis and fluorescent tagging. Biochemistry 49(42):9063–9071. https://doi.org/10.1021/bi100516e

Graether SP, Kuiper MJ, Gagne SM, Walker VK, Jia Z, Sykes BD, Davies PL (2000) Beta-helix structure and ice-binding properties of a hyperactive antifreeze protein from an insect. Nature 406(6793):325–328

Grandum S, Yabe A, Nakagomi K, Tanaka M, Takemura F, Kobayashi Y, Frivik PE (1999) Analysis of ice crystal growth for a crystal surface containing adsorbed antifreeze proteins. J Cryst Growth 205(3):382–390. https://doi.org/10.1016/S0022-0248(99)00267-5, wOS:000082217200018

Gwak Y, Ji P, Kim M, Kim HS, Kwon MJ, Oh SJ, Kim YP, Jin E (2015) Creating anti-icing surfaces via the direct immobilization of antifreeze proteins on aluminum. Sci Rep 5:12019. https://doi.org/10.1038/srep12019

Halwani DO, Brockbank KGM, Duman JG, Campbell LH (2014) Recombinant *Dendroides canadensis* antifreeze proteins as potential ingredients in cryopreservation solutions. Cryobiology 68(3):411–418. https://doi.org/10.1016/j.cryobiol.2014.03.006

Hansen TN, Baust JG (1988) Differential scanning calorimetric analysis of antifreeze protein activity in the common mealworm, *Tenebrio molitor*. BBA-Protein Struct M 957(2):217–221. https://doi.org/10.1016/0167-4838(88)90275-0

Hassa-Roudsari M, Goff HD (2012) A new quantitative method to measure activity of ice structuring proteins using differential scanning calorimetry. Cryoletters 33(2):118–125

Haymet ADJ, Ward LG, Harding MM, Knight CA (1998) Valine substituted winter flounder 'antifreeze': preservation of ice growth hysteresis. FEBS Lett 430(3):301–306. https://doi.org/10.1016/S0014-5793(98)00652-8, wOS:000074797600034

Haymet ADJ, Ward LG, Harding MM (1999) Winter flounder "antifreeze" proteins: synthesis and ice growth inhibition of analogues that probe the relative importance of hydrophobic and hydrogen-bonding interactions. J Am Chem Soc 121(5):941–948. https://doi.org/10.1021/ja9801341

Heisig M, Mattessich S, Rembisz A, Acar A, Shapiro M, Booth CJ, Neelakanta G, Fikrig E (2015) Frostbite protection in mice expressing an antifreeze glycoprotein. PLoS One 10(2):e0116562. https://doi.org/10.1371/journal.pone.0116562

Hon WC, Griffith M, Chong P, Yang DS (1994) Extraction and isolation of antifreeze proteins from winter rye (*Secale cereale* L.) leaves. Plant Physiol 104(3):971–980

Horwath KL, Easton CM, Poggioli GJ, Myers K, Schnorr IL (1996) Tracking the profile of a specific antifreeze protein and its contribution to the thermal hysteresis activity in cold hardy insects. Eur J Entomol 93(3):419–433, wOS:A1996VN62700013

Ickes L, Welti A, Hoose C, Lohmann U (2015) Classical nucleation theory of homogeneous freezing of water: thermodynamic and kinetic parameters. Phys Chem Chem Phys 17 (8):5514–5537. https://doi.org/10.1039/C4CP04184D

Jackman J, Noestheden M, Moffat D, Pezacki JP, Findlay S, Ben RN (2007) Assessing antifreeze activity of AFGP 8 using domain recognition software. Biochem Biophys Res Commun 354 (2):340–344. https://doi.org/10.1016/j.bbrc.2006.12.225

Kim EJ, Lee JH, Lee SG, Han SJ (2017) Improving thermal hysteresis activity of antifreeze protein from recombinant *Pichia pastoris* by removal of N-glycosylation. Prep Biochem Biotechnol 47 (3):299–304. https://doi.org/10.1080/10826068.2016.1244682

Knight CA (1966) Formation of crystallographic etch pits on ice, and its application to the study of hailstones. J Appl Meteorol 5(5):710–714

Knight CA, Hallett J, DeVries AL (1988) Solute effects on ice recrystallization: an assessment technique. Cryobiology 25(1):55–60. https://doi.org/10.1016/0011-2240(88)90020-X

Knight C, Cheng C, DeVries A (1991) Adsorption of alpha-helical antifreeze peptides on specific ice crystal surface planes. Biophys J 59(2):409–418. https://doi.org/10.1016/S0006-3495(91)82234-2

Knight C, Driggers E, DeVries A (1993) Adsorption to ice of fish antifreeze glycopeptides 7 and 8. Biophys J 64(1):252–259. https://doi.org/10.1016/S0006-3495(93)81361-4

Kondo H, Hanada Y, Sugimoto H, Hoshino T, Garnham CP, Davies PL, Tsuda S (2012) Ice-binding site of snow mold fungus antifreeze protein deviates from structural regularity and high conservation. PNAS 109(24):9360–9365. https://doi.org/10.1073/pnas.1121607109

Kontogiorgos V, Goff HD, Kasapis S (2008) Effect of aging and ice-structuring proteins on the physical properties of frozen flour–water mixtures. Food Hydrocoll 22(6):1135–1147. https://doi.org/10.1016/j.foodhyd.2007.06.005

Kristiansen E, Zachariassen KE (2005) The mechanism by which fish antifreeze proteins cause thermal hysteresis. Cryobiology 51(3):262–280. https://doi.org/10.1016/j.cryobiol.2005.07.007

Kristiansen E, Pedersen SA, Zachariassen KE (2008) Salt-induced enhancement of antifreeze protein activity: a salting-out effect. Cryobiology 57(2):122–129. https://doi.org/10.1016/j.cryobiol.2008.07.001

Leygonie C, Britz TJ, Hoffman LC (2012) Impact of freezing and thawing on the quality of meat: review. Meat Sci 91(2):93–98. https://doi.org/10.1016/j.meatsci.2012.01.013

Li N, Andorfer CA, Duman JG (1998) Enhancement of insect antifreeze protein activity by solutes of low molecular mass. J Exp Biol 201(15):2243–2251

Liu J, Xu X, Xu Q, Wang S, Xu J (2014) Transgenic tobacco plants expressing PicW. Plant Cell Tiss Org Cult 118(3):391–400. https://doi.org/10.1007/s11240-014-0491-7

Lv J, Song Y, Jiang L, Wang J (2014) Bio-inspired strategies for anti-icing. ACS Nano 8 (4):3152–3169. https://doi.org/10.1021/nn406522n

Mangiagalli M, Bar-Dolev M, Tedesco P, Natalello A, Kaleda A, Brocca S, de Pascale D, Pucciarelli S, Miceli C, Braslavsky I, Lotti M (2017) Cryo-protective effect of an ice-binding protein derived from Antarctic bacteria. FEBS J 284(1):163–177. https://doi.org/10.1111/febs.13965

Marshall CB, Daley ME, Graham LA, Sykes BD, Davies PL (2002) Identification of the icebinding face of antifreeze protein from *Tenebrio molitor*. FEBS Lett 529(2–3):261–267. https://doi.org/10.1016/S0014-5793(02)03355-0

Meister K, Ebbinghaus S, Xu Y, Duman JG, DeVries A, Gruebele M, Leitner DM, Havenith M (2013) Long-range protein–water dynamics in hyperactive insect antifreeze proteins. PNAS 110 (5):1617–1622. https://doi.org/10.1073/pnas.1214911110

Meister K, Duman JG, Xu Y, DeVries AL, Leitner DM, Havenith M (2014a) The role of sulfates on antifreeze protein activity. J Phys Chem B 118(28):7920–7924. https://doi.org/10.1021/jp5006742

Meister K, Strazdaite S, DeVries AL, Lotze S, Olijve LLC, Voets IK, Bakker HJ (2014b) Observation of ice-like water layers at an aqueous protein surface. PNAS 111(50):17732–17736

Mitchell DE, Congdon T, Rodger A, Gibson MI (2015) Gold nanoparticle aggregation as a probe of antifreeze (glyco) protein-inspired ice recrystallization inhibition and identification of new IRI active macromolecules. Sci Rep 5:15716. https://doi.org/10.1038/srep15716

Mizrahy O, Bar-Dolev M, Guy S, Braslavsky I (2013) Inhibition of ice growth and recrystallization by zirconium acetate and zirconium acetate hydroxide. PLoS One 8(3):e59540. https://doi.org/10.1371/journal.pone.0059540

Namperumal R, Coger R (1998) A new cryostage design for cryomicroscopy. J Microsc (Oxford) 192(2):202–211. https://doi.org/10.1046/j.1365-2818.1998.00416.x

Olijve LLC, Meister K, DeVries AL, Duman JG, Guo S, Bakker HJ, Voets IK (2016) Blocking rapid ice crystal growth through nonbasal plane adsorption of antifreeze proteins. PNAS 113 (14):3740–3745. https://doi.org/10.1073/pnas.1524109113

Olsen TM, Sass SJ, Li N, Duman JG (1998) Factors contributing to seasonal increases in inoculative freezing resistance in overwintering fire-colored beetle larvae *Dendroides canadensis* (Pyrochroidae). J Exp Biol 201(10):1585–1594, wOS:000074211300007

Panadero J, Randez-Gil F, Prieto JA (2005) Heterologous expression of type I antifreeze peptide GS-5 in Baker's yeast increases freeze tolerance and provides enhanced gas production in frozen dough. J Agric Food Chem 53(26):9966–9970. https://doi.org/10.1021/jf0515577

Park JI, Lee JH, Gwak Y, Kim HJ, Jin E, Kim YP (2012a) Frozen assembly of gold nanoparticles for rapid analysis of antifreeze protein activity. Biosens Bioelectron 41:752–757. https://doi.org/10.1016/j.bios.2012.09.052

Park KS, Do H, Lee JH, Park SI, Jung Kim E, Kim SJ, Kang SH, Kim HJ (2012b) Characterization of the ice-binding protein from Arctic yeast *Leucosporidium sp. AY30*. Cryobiology 64 (3):286–296. https://doi.org/10.1016/j.cryobiol.2012.02.014

Payne SR, Young OA (1995) Effects of pre-slaughter administration of antifreeze proteins on frozen meat quality. Meat Sci 41(2):147–155. https://doi.org/10.1016/0309-1740(94)00073-G

Payne SR, Sandford D, Harris A, Young OA (1994) The effects of antifreeze proteins on chilled and frozen meat. Meat Sci 37(3):429–438. https://doi.org/10.1016/0309-1740(94)90058-2

Pertaya N, Celik Y, DiPrinzio CL, Wettlaufer JS, Davies PL, Braslavsky I (2007) Growth–melt asymmetry in ice crystals under the influence of spruce budworm antifreeze protein. J Phys Condens Matter 19(41):412101. https://doi.org/10.1088/0953–8984/19/41/412101

Pertaya N, Marshall CB, Celik Y, Davies PL, Braslavsky I (2008) Direct visualization of spruce budworm antifreeze protein interacting with ice crystals: basal plane affinity confers hyperactivity. Biophys J 95(1):333–341. https://doi.org/10.1529/biophysj.107.125328, wOS:000256668200033

Qiu L, Wang Y, Wang J, Zhang F, Ma J (2009) Expression of biologically active recombinant antifreeze protein His-MpAFP149 from the desert beetle *Microdera punctipennis dzungarica* in *Escherichia coli*. Mol Biol Rep 37(4):1725–1732. https://doi.org/10.1007/s11033-009-9594-3

Ramløv H, Wharton DA, Wilson PW (1996) Recrystallization in a freezing tolerant Antarctic nematode, *Panagrolaimus davidi*, and an Alpine Weta, *Hemideina maori* (Orthoptera; Stenopelmatidae). Cryobiology 33(6):607–613. https://doi.org/10.1006/cryo.1996.0064

Ramløv H, DeVries A, Wilson P (2005) Antifreeze glycoproteins from the Antarctic fish *Dissostichus mawsoni* studied by differential scanning calorimetry (DSC) in combination with nanolitre osmometry. Cryoletters 26(2):73–84

Raymond JA, DeVries AL (1977) Adsorption inhibition as a mechanism of freezing resistance in polar fishes. PNAS 74(6):2589–2593

Raymond JA, Wilson P, DeVries AL (1989) Inhibition of growth of nonbasal planes in ice by fish antifreezes. PNAS 86(3):881–885

Romero I, Fernandez-Caballero C, Goñi O, Escribano MI, Merodio C, Sanchez-Ballesta MT (2008) Functionality of a class I beta-1,3-glucanase from skin of table grapes berries. Plant Sci 174 (6):641–648. https://doi.org/10.1016/j.plantsci.2008.03.019

Scotter AJ, Marshall CB, Graham LA, Gilbert JA, Garnham CP, Davies PL (2006) The basis for hyperactivity of antifreeze proteins. Cryobiology 53(2):229–239. https://doi.org/10.1016/j.cryobiol.2006.06.006, wOS:000241145300008

Takamichi M, Nishimiya Y, Miura A, Tsuda S (2009) Fully active QAE isoform confers thermal hysteresis activity on a defective SP isoform of type III antifreeze protein. FEBS J 276 (5):1471–1479. https://doi.org/10.1111/j.1742-4658.2009.06887.x, wOS:000263451600027

Tomczak MM, Marshall CB, Gilbert JA, Davies PL (2003) A facile method for determining ice recrystallization inhibition by antifreeze proteins. Biochem Biophys Res Commun 311 (4):1041–1046. https://doi.org/10.1016/j.bbrc.2003.10.106

Tong L, Lin Q, Wong WKR, Ali A, Lim D, Sung WL, Hew CL, Yang DSC (2000) Extracellular expression, purification, and characterization of a winter flounder antifreeze polypeptide from *Escherichia coli*. Protein Expr Purif 18(2):175–181. https://doi.org/10.1006/prep.1999.1176

Wang S, Amornwittawat N, Banatlao J, Chung M, Kao Y, Wen X (2009a) Hofmeister effects of common monovalent salts on the beetle antifreeze protein activity. J Phys Chem B 113 (42):13891–13894. https://doi.org/10.1021/jp907762u, wOS:000270670800029

Wang S, Amornwittawat N, Juwita V, Kao Y, Duman JG, Pascal TA, Goddard WA, Wen X (2009b) Arginine, a key residue for the enhancing ability of an antifreeze protein of the beetle *Dendroides canadensis*. Biochemistry 48(40):9696–9703. https://doi.org/10.1021/bi901283p

Wen D, Laursen R (1992) A model for binding of an antifreeze polypeptide to ice. Biophys J 63 (6):1659–1662, wOS:A1992KF55100025

Wilson P, Gould M, DeVries A (2002) Hexagonal shaped ice spicules in frozen antifreeze protein solutions. Cryobiology 44(3):240–250. https://doi.org/10.1016/S0011-2240(02)00028-7

Wu DW, Duman JG (1991) Activation of antifreeze proteins from larvae of the beetle *Dendroides canadensis*. J Comp Physiol B 161(3):279–283. https://doi.org/10.1007/BF00262309

Xu Y, Bäumer A, Meister K, Bischak CG, DeVries AL, Leitner DM, Havenith M (2016) Protein-water dynamics in antifreeze protein III activity. Chem Phys Lett 647:1–6. https://doi.org/10.1016/j.cplett.2015.11.030

Yeh CM, Kao BY, Peng HJ (2009) Production of a recombinant type 1 antifreeze protein analogue by *L. lactis* and its applications on frozen meat and frozen dough. J Agric Food Chem 57 (14):6216–6223. https://doi.org/10.1021/jf900924f

Zhang C, Zhang H, Wang L (2007a) Effect of carrot (*Daucus carota*) antifreeze proteins on the fermentation capacity of frozen dough. Food Res Int 40(6):763–769. https://doi.org/10.1016/j.foodres.2007.01.006, wOS:000247326000014

Zhang C, Zhang H, Wang L, Gao H, Guo XN, Yao HY (2007b) Improvement of texture properties and flavor of frozen dough by carrot (*Daucus carota*) antifreeze protein supplementation. J Agric Food Chem 55(23):9620–9626. https://doi.org/10.1021/jf0717034

Part III
Applications of Antifreeze Proteins

Chapter 10
Antifreeze Proteins in Foods

Nebahat Sule Ustun and Sadettin Turhan

10.1 Introduction

The distribution of living things on Earth is influenced by many factors such as climate, geographical formations, soil characteristics, population growth, technology and industrial development, hunting, destruction of forests, opening of new settlements, dam construction, climate change, and continental drift. For this reason, different plants, animals, and microorganism species live in different regions. Climate is very important within these factors, and ice is an important factor influencing the distribution of living organisms in cold regions. When the temperature falls below the freezing point, ice crystals may form and the cells are damaged by the physical effects of growing ice crystals. However, since many organisms have developed ways of controlling the lethal growth of ice crystals, these organisms can, to a certain extent, counteract the effect of climate on their distribution.

Adaptation of living things to subzero conditions is explained through two basic strategies. In the first of these, the "freeze avoidance" strategy, the living organisms prevent their intracellular fluids from freezing by maintaining them either as a supercooled solution or in a vitreous state, by the accumulation of small organic antifreeze compounds having cryoprotective effects (e.g., sugars, glycerol, amino acids) in high concentrations, which help maintain the cellular components and proteins in a functional state. In the latter, the "freeze tolerance" strategy, the organisms can use the same components to avoid osmotic freezing and to balance the osmotic force that leads to an increase in solute concentration in the extracellular fluid. In addition, living organisms can prevent sudden catastrophic freezing by

N. S. Ustun (✉) · S. Turhan
Engineering Faculty, Department of Food Engineering, Ondokuz Mayis University, Samsun, Turkey
e-mail: sustun@omu.edu.tr; sturhan@omu.edu.tr

H. Ramløv, D. S. Friis (eds.), *Antifreeze Proteins Volume 2*,
https://doi.org/10.1007/978-3-030-41948-6_10

promoting the formation of extracellularly controlled ice through the use of ice nucleating compounds or structures (Venketesh and Dayananda 2008).

Many organisms synthesize specialized antifreeze proteins (AFPs) to protect from the effects of ice. These proteins are produced by plants, animals, or microorganisms and are from the group of polypeptides that allow them to survive at subzero temperatures (Jin-Yao et al. 2005; Kontogiorgos et al. 2007; Petzold and Aguilera 2009; Ustun and Turhan 2015). These compounds are also known as ice-structuring proteins (ISP) because they bind to ice crystals and control crystal growth (Crevel et al. 2002; Venketesh and Dayananda 2008). One of the protective mechanisms of AFPs is recrystallization inhibition. In this process, ice-bound AFPs prevent the growth of large ice crystals on the expense of smaller ones—a process that would otherwise damage the cells and tissues. The recrystallization inhibition mechanism is described in depth in Chap. 7.

After the discovery of the first AFPs, the proteins with these properties were identified not only in fish living in ice-laden seas but also in many plants and insects (Crevel et al. 2002; Jin-Yao et al. 2005; Hassas-Roudsari and Goff 2012). The ability of AFPs to influence the growth of ice crystals has resulted in the use of these proteins in a wide range of areas, such as improving the properties of frozen foods, preserving transplant organs and cells, developing cryosurgery, and aquaculture (Crevel et al. 2002; Harding et al. 2003; Jin-Yao et al. 2005).

AFPs are a family of proteins with thermal hysteresis activity that can lower the freezing point of solutions noncolligatively. They can modify the growth of ice crystals, increase the stabilization of ice crystals and inhibit the recrystallization of ice during temperature fluctuations. These distinct properties make AFPs promising additives for frozen food in recent years (Ding et al. 2015). Initially the work was limited to ice cream products (Warren et al. 1992; Clarke et al. 2003) and meat (Payne et al. 1994; Payne and Young 1995), but today frozen dough (Jia et al. 2012; Ding et al. 2014, 2015; Zhang et al. 2015) and fruit and vegetables (Kong et al. 2016, 2017) are also among the investigated products. There have been many studies of successful application of AFPs in freezing and thawing food (Payne et al. 1994; Payne and Young 1995; Yeh et al. 2009; Ding et al. 2015). In this section, more detailed information is given about the studies on the use of AFPs in food. It also focuses on the dietary sources of these proteins, their use in foods, their toxicity, and the factors that influence their use.

10.2 Dietary Sources of Antifreeze Proteins

AFPs are found in a wide variety of organisms, including bacteria, fungi, insects, animals, and plants, which must protect themselves against freeze damage in nature (Griffith and Ewart 1995; Feeney and Yeh 1998; Crevel et al. 2002; Bouvet and Ben 2003). Fish such as cod, herring, and ocean pout are among the earliest identified sources of such proteins (Griffith and Ewart 1995; Crevel et al. 2002). Antifreeze glycoproteins (AFGPs) have been isolated from southern pole notothenioid fish as well as from the codfish *Gadus agac* and from the other high-latitude northern codfish of the *Gadidae* family. Most studied AFGPs are those isolated from the Antarctic fish *Pagothenia borchrevinki* and *Dissostichus mawsoni* and northern fish *Boreogadus saida* (Harding et al. 2003). In addition, type I AFP was isolated from winter flounder (*Pleuronectes americanus*), yellow tail flounder (*P. ferrugineus*), shorthorn sculpin (*Myoxocephalus scorpius*), and grubby sculpin (*M. aenarus*), type II AFP from sea raven (*Hemitripterus american*), rainbow smelt (*Osmerus mordax*), and Atlantic herring (*Clupea harengus haren*), and type III AFP from ocean pout (*Macrozoarces americanus*), Atlantic wolffish (*Anarhichas lupus*), radiated shanny (*Ulvaria subbifurcata*), rock gunnel (*Pholis gunnellus*) and eelpout (*Lycoes lavalaei*) (Griffith and Ewart 1995) as well as the eelpout (*Zoarces viviparus*) (Sørensen et al. 2006).

Antifreeze activity has been reported in higher plants, as well as in more primitive plants such as ferns and algae. Antifreeze protein expression has only been observed when plants are acclimated to low temperatures (Griffith and Ewart 1995). AFPs have also been isolated from widely consumed foods such as oats (*Avena sativa*), rye (*Secale cereale*), barley (*Hordeum vulgare*), wheat (*Triticum aestivum*), carrots (*Daucus carota*), and potatoes (*Solanum tuberosum*). Different parts of the foodstuffs have different AFP contents and some are found only in the inedible parts of some plants, while others are in the edible parts of many plants such as carrot pile root, potato lobster, and Brussels cabbage (Griffith and Ewart 1995; Crevel et al. 2002).

It has also been reported that AFPs are found in fungi, bacteria, and plants (Griffith and Ewart 1995; Crevel et al. 2002). Thermal hysteresis has been observed in the extracts obtained from oyster mushroom *Pleurotus ostreatus*, winter mushroom *Flammulina velupites*, *Stereum* sp. and *Coriolus versicolor*. Many invertebrate animals, mostly insects as well as terrestrial arthropods such as spiders and mites, contain AFPs (Griffith and Ewart 1995; Duman et al. 2004). Up to 50 insect species including *Coleoptera*, *Collembola*, *Plecoptera*, *Orthoptera*, *Hemiptera*, *Mecoptera*, *Lepidoptera*, *Diptera*, and *Neuropteran* have been reported to produce AFPs (Duman et al. 2004, 2010; Jin-Yao et al. 2005). Antifreeze activity was also observed in *Micrococcus cryophilus*, a well-known psychrophile, and *Rhodococcus erythropolis* and *Pseudomonas putida* cultures, which are common soil bacteria (Griffith and Ewart 1995; Singh et al. 2014).

The AFP content of organisms is influenced by many factors. In most cases, organisms exposed to cold weather have higher levels of AFP in their tissues. Differences in AFP levels in fish occur among species with different distributions, among different populations within species, and within the individuals according to developmental stage and time of year (Griffith and Ewart 1995; Crevel et al. 2002). For example, fish that are constantly in cold Arctic waters may have high AFP levels throughout the year. In August, the shorthorn sculpin in the Arctic Region has AFP levels equal to the maximum winter AFP levels in Newfoundland populations. On the other hand, AFP levels in fishes living in northern temperate environments rise to maximum levels in the winter following the seasonal cycles. AFGP production coincides with the temperature in the Atlantic cod (Griffith and Ewart 1995), while

AFP production in winter flounder follows a seasonal cycle according to the light period, reaching the maximum level in January and disappearing in May (Griffith and Ewart 1995). In addition to seasonal cycles, AFP expression varies significantly with species and age. While the content of Type III AFP in the ocean pout blood is around 30 mg/mL, Atlantic cod AFGP, value ranges from about 7 mg/mL in adult fish to 14 mg/mL in seed fish (Crevel et al. 2002).

There are also AFPs that accumulate in crops such as carrots, cabbage, and Brussels sprouts after being harvested in autumn. Since antifreeze activity is not observed in temperate climates in summer, it is possible for plants such as carrots and cabbage to synthesize AFP in cold storage for final sale in the fresh produce market (Griffith and Ewart 1995), since AFP synthesis can be stimulated at low temperatures (Galindo et al. 2005). Crevel et al. reported that there was 0.307 mg/g AFP w/wt in winter rye leaves (Crevel et al. 2002). In both *Pagothenia borchggrevinki* and *Dissostichus mawsoni*, the total AFGP concentration is 25 mg/mL, of which about 25% is AFGP 1–5 and the remaining 75% contains smaller AFGP 6–8 (Harding et al. 2003).

10.3 Basic Functions of Antifreeze Proteins in Food

The primary role of AFPs is not to prevent the nucleation of extracellular ice but to decrease the rate of further crystal growth, i.e., recrystallization. The functions of AFPs include the reduction of the freezing point of food without affecting the melting point (thermal hysteresis) (Boonsupthip and Lee 2003; Kontogiorgos et al. 2007; Zhang et al. 2008), the modification of the ice crystal morphology (Boonsupthip and Lee 2003; Kontogiorgos et al. 2007), inhibiting the recrystallization of ice crystals (Boonsupthip and Lee 2003; Kontogiorgos et al. 2007; Zhang et al. 2008), increasing cell integrity and reducing microbial growth (Boonsupthip and Lee 2003). The ability of AFPs to reduce freezing temperature, to prevent recrystallization during freezing–thawing and to neutralize the effects of ice nuclei reveals the potential to be used as natural ice modulators during freeze storage (Griffith and Ewart 1995). Thus, they can protect the food texture by decreasing cell damage and they can reduce nutrient loss to a minimum by decreasing drip loss. AFPs are found in many foods consumed as part of human nutrition. One of its potential applications is the prevention of recrystallization resulting in poor quality textures in the event of temperature fluctuations during storage and transportation (Venketesh and Dayananda 2008). Besides, the main functions of applying AFPs to food can be summarized under two main headings.

10.3.1 Reducing the Freezing Temperature of Food

Foodstuffs can undergo microbial contamination from different sources during production. Different applications are used to prevent the growing of contaminating microorganisms or to reduce the microbial load before processing. One of these applications is to prevent the food from spoilage by slowing down or completely stopping the microorganism activity by lowering the ambient temperature. By adding AFP, it is possible to store food at lower temperatures without freeze, so the AFPs added to food slow down microbial activity, thus extending the shelf life of food (Griffith and Ewart 1995). In several cases, the addition of AFPs directly into plants and animals influences their freezing characteristics, because AFPs are 200–300 times more effective than ideal solutions in reducing the freezing point. The lethal freezing temperature of the rainbow trout is reduced in directly proportion to the amount of type I AFP injected into the fish. The response in the plant tissues is a somewhat more variable. When the leaves were subjected to vacuum infiltration with 1 mg/mL winter flounder type I AFP, the lower limit of nucleation was lowered by 1.8 °C for canola (*Brassica napus*) and 4 °C for *Arabidopsis thaliana* (Griffith and Yaish 2004). In contrast, there was an increase from −7.2 to −0.8 °C in the threshold ice nucleation temperature for potato leaves (*Solanum tuberosum*) vacuum infiltrated with AFP, a result that suggests that not all plant tissues respond in the same way to the addition of AFPs. The freezing characteristics of bromegrass cell cultures suspended in 20 mg/mL type I AFP by NMR was also examined and it was determined that there was a decrease in the amount of freezable water that froze at a given temperature, as well as a decrease in the rate at which water froze (Griffith and Ewart 1995).

In addition, low temperature applications reduce the rate of deterioration by creating resistance to diseases in some fruits and vegetables such as apples, pears, carrots, citrus fruits, green peppers, peaches, and tomatoes (Wilson et al. 1994). Low-temperature induced pathogenesis-related (PR) proteins in winter grains increase disease resistance by inhibiting the growth of fungi in plant tissue. Therefore, the presence of cold-induced PR proteins in food during harvest may act to reduce microbial contamination of foodstuffs prior to freezing and to prevent microbial contamination during thawing (Griffith and Ewart 1995). AFPs from plants are homologous to PR proteins and also provide protection against psychrophilic pathogens. Most of the plant AFPs (both peptide and corresponding genes) are homologous to PR proteins (Hon et al. 1995). PR proteins are released into the apoplast in response to pathogen infection and act together to degrade fungal cell walls enzymatically and inhibit fungal enzymes. PR proteins with and without antifreeze activity are likely to be different members of gene families that are differentially regulated by cold and by pathogens. By accumulating PR proteins during cold acclimation, overwintering grasses and cereals acquire a systemic, nonspecific, preemptive defense against pathogens and exhibit greater disease resistance (Griffith and Yaish 2004). In winter rye, the AFPs exhibit antifungal, hydrolytic activities and ice-binding activity. Therefore, cold-acclimated plants are more resistant to injury caused either by snowmolds or by freezing (Hiilovaara-Teijo et al. 1999).

10.3.2 Control or Prevent Recrystallization During Freeze Storage

Ice crystals may grow due to temperature gradients that occur during freezing or thawing or temperature fluctuations during melting cycles, storage, or during transportation of food. Recrystallization of ice occurs within the subzero range through different mechanisms such as disappearance of smaller ice crystals and the growth of larger ice crystals. The resulting large ice crystals are more potent to cause physical damage to tissues and cells than smaller ones. There are numerous types of recrystallization processes described in the literature. Isomass, migratory and accretive recrystallization are the types that are most common. The reduction in surface free energy is the thermodynamic driving force of all these mechanisms (Hassas-Roudsari and Goff 2012). Ice recrystallization causes an increment at the amount of free water that has negative sensory implications for frozen foods (Ramlov and Johnsen 2014). According to Zhang et al. (2007), frozen dough with concentrated carrot AFP had less free water than the control dough. Recrystallization creates gradients of salt and pH that can lyse the cells of yeast in the dough as they are thawed. Carrot AFP addition reduced recrystallization and caused more yeast cells viable upon thawing (Panadero et al. 2005; Zhang et al. 2007). More viable cells produced higher CO_2, so the dough rose more than the control dough (Panadero et al. 2005). The gluten protein matrix in the dough can also be damaged by salt and pH gradients, and gluten is better protected when recrystallization is inhibited (Ramlov and Johnsen 2014). Large volumes of free water are not reabsorbed well, and cause drip loss in frozen meat after thawing. The addition of recombinant type I AFP to frozen meat showed a decrease in the amount of drip loss after thawing (Yeh et al. 2009).

In addition to lowering the freezing temperature, AFPs are attracting interest as potential food additives, as they delay the recrystallization or the formation of large ice crystals associated with recrystallization during freezing and storage. Recrystallization is faster in temperatures just below freezing and during glassy heating. Glass transition occurs when a supercooled, malleable liquid/rubbery material is changed into a disordered solid glass upon cooling, or conversely when a brittle glass is changed upon heating into a supercooled liquid/rubbery material (Balasubramanian et al. 2016). Glass transition is a time, temperature, and water activity dependent transition, which is characterized by a discontinuity in physical, mechanical, electrical, thermal, and other properties of a material. Semicrystalline solids are having both amorphous and crystalline regions. The temperature at which the transition in the amorphous regions of polymers between the glassy and rubbery state occurred is called the glass transition temperature (Tg) (Karaoglu et al. 2009). The phenomenon of glass transition could be applied as an integrated approach along with water activity, physical and chemical changes in food during processing and storage to determine food stability. Agglomeration, crystallization, caking, sticking, collapse, oxidation reactions, and nonenzymatic browning of food systems also depend on Tg (Balasubramanian et al. 2016). Carbohydrates and proteins in food exhibit glass transition. When water activity is high in foods, deterioration reactions can be explained by water activity. However, in foods with low water activity containing amorphous components, Tg becomes a factor determining stability. When these types of foods are stored at temperatures lower than the Tg, rates of deterioration reactions can be minimized and by this way, the stability can be maximized. Glass transition has been related with physical and some chemical deterioration reactions in foods. Microbial deteriorations have not been related to glass transition as they occur at high water activity values (Kilic and Evranuz 2006). The development of milk powder and ice cream manufacturing showed challenges in the stabilization of amorphous lactose at high humidities and partially frozen food systems. These food materials have high solute concentrations, which are required for glass formation in foods and may be achieved by freezing of water to concentrate solutes in an unfrozen solute phase or by dehydration. The key processes requiring understanding of the amorphous state and glass transitions of food systems are those occurring at limited water contents. Hence, these cover cereal systems and processing, food freezing and frozen foods, confectionary and candies, dehydration processes and dehydrated foods, extrusion, and extruded foods, among others. Rapid freezing decreases ice crystal size, and provides a better retention of food quality after frozen storage and thawing of frozen foods. The main changes of frozen foods during storage are chemical and biochemical reactions such as enzymatic changes and oxidation, affecting product quality as well as crystallization of solutes, e.g., lactose in ice cream and recrystallization of ice. Knowledge of the glass transition behavior of frozen food systems is important in understanding the freezing process and frozen state stability of foods (Roos 2010).

Temperature fluctuations during storage and handling of frozen foods promote ice crystal growth. The crystal growth rate is very slow at lower storage temperatures, especially when the product is stored below its glass transition temperature. Above the glass transition temperature, the greater molecular mobility of water leads to faster growth of ice crystals. Because the typical average storage temperature in household freezers is well above −20 °C and fluctuates because of automatic defrost cycles, formation of large ice crystals and deterioration of textural qualities of frozen foods is a common occurrence under household conditions. Thus, one of the major challenges faced by frozen foods manufacturers is developing appropriate technological conditions and ingredient formulations that can inhibit ice crystal growth during storage and handling (Damodaran 2007).

In various studies, the addition of various AFPs in ice cream, dough, and white salted noodles increased the glass transition temperature of food and then reduced the difference between the storage temperature and the glass transition temperature (Ding et al. 2014, 2015; Zhang et al. 2016). So, the addition of AFPs can increase the glass transition temperature of frozen foods, thus inhibiting its crystallization and recrystallization process and reducing the damage caused by frozen and temperature fluctuation, thus resulting in superior properties of frozen food.

When the ambient temperatures fluctuate at subzero temperatures, ice recrystallizes. Inhibition of ice recrystallization is an important factor in the formation of textures of frozen foods such as ice cream and iced sugar (Griffith and Ewart 1995). Studies have shown that the most successful results are obtained in frozen dairy products and that they inhibit or slow down recrystallization in these products (Warren et al. 1992; Feeney and Yeh 1998).

Prevention of recrystallization is also important in foods consumed after they are thawed. Large ice crystals can damage membranes when intracellularly formed in foods such as meat and fish and can cause increased drip loss during thawing (Muela et al. 2010). Decreased water holding capacity and loss of nutrients in tissues can result in a poor quality frozen product (Rahman et al. 2014). Most fruits and vegetables are over 90% water of total weight. The water and dissolved solutes inside the rigid plant cell walls give support to the plant structure, and texture to the fruit or vegetable tissue. In the process of freezing, when water in the cells freezes, an expansion occurs and ice crystals cause the cell walls to rupture. Consequently, the texture of the product is generally much softer after thawing when compared to non-frozen produce. This textural difference is especially noticeable in products normally consumed raw (Barbosa-Cánovas et al. 2005). Prevention of recrystallization is important for foods such as strawberries, raspberries, and tomatoes in which cellular structure is sensitive to freezing. If the cellular structures of these foods are damaged because of freezing, undesirable texture and taste changes may occur. One of the most common forms of quality degradation due to moisture migration in frozen foods is freezer burn, a condition defined as the glassy appearance in some frozen products produced by ice crystals evaporating on the surface area of a product. The grainy, brownish spots occurring on the product cause the tissue to become dry and tough and to develop off-flavors (Barbosa-Cánovas et al. 2005). Recrystallization is thus involved in impairing foodstuff quality in many stages. As described thoroughly in Chap. 7, the AFPs inhibit this recrystallization. Where and how this can be exploited in the food industry is described in the following sections.

10.4 Application of Antifreeze Proteins to Food

While the use of antifreeze proteins as a food additive has been limited to liquid foods such as frozen milk and ice cream products that are badly affected by recrystallization, meat, fish, frozen dough, and fruit vegetables are now among the investigated products today (Ustun and Turhan 2015). Isolation, characterization, and cloning of a number of different AFPs have made it possible to obtain large quantities of AFPs and to incorporate these proteins into some foods (Griffith and Ewart 1995; Ökkes and Nalbantoğlu 2003; Zhang et al. 2015, 2016). But mainly due to cost and many other factors such as isolation and purification, thermal stability, chemical synthesis, and development in molecular biology, studies on the use potential of AFPs in foods are still in a limited number of products and at laboratory scale, so commercial use is not yet in question (Ustun and Turhan 2015). AFPs can be added to foods by physical means such as mixing (Zhang et al. 2007, 2015, 2016; Yeh et al. 2009; Xu et al. 2009a, b; Ding et al. 2015), injection (Payne et al. 1994; Payne and Young 1995), dipping (Yeh et al. 2009; Kong et al. 2016, 2017), or vacuum infiltration (Griffith and Ewart 1995). In addition, AFPs can be applied to food by gene transfer. Introducing AFPs through gene transfer into plants and animals has resulted in lower freezing temperatures and more ice recrystallization inhibition in organisms and the synthesis of AFPs in *Arabidopsis thaliana* plants (Huang et al. 2002), tobacco (Wang et al. 2008), salmon (Fletcher et al. 1988; Hew et al. 1992), and tomato (Hightower et al. 1991; Kumar et al. 2017). Below is a list of studies conducted on the application of AFPs to different food items.

10.4.1 Ice-Cream Products

Demand for quality and quantity in ice cream production is increasing day by day. Smooth and creamy texture is the most important quality parameter of the ice cream. *Creaminess of the ice cream depends on the size of the ice crystals* that form during freezing—the smaller the crystals, the creamier the texture. Rapid chilling and constant churning encourage the water in the ice cream mixture to form lots of minuscule "seed" crystals; this process is known as propagation. Ice cream containing lots of tiny ice crystals feels smoother and creamier than ice cream that is equally thick but with fewer, larger crystals. Freezing point depression of a solution is associated with the number of dissolved molecules. *The lower the molecular weight, the greater the ability of a molecule to depress the freezing point* because there will be more molecules present. Thus monosaccharides such as fructose or glucose produce a much softer ice cream than disaccharides such as sucrose (Bahramparvar and Tehrani 2011).

Fat is one of the main components that provide smoothness to ice cream. Longer heating times will result in smoother ice cream because the water-binding capacity of a denatured protein is greater than that of the native protein. The original ice cream emulsifier is egg yolk. Emulsifiers keep the ice cream smooth and aid the distribution of the fat molecules throughout the colloid. Today, two emulsifiers frequently show up on the ingredients label of many ice cream formulations, namely, mono- and di-glycerides. Stabilizers help hold the air bubble structure together and give the ice cream a better texture. Stabilizers are used in ice creams to modify water-binding capacity, freezing rates, ice crystal formation, and rheological properties. Stabilizers achieve these modifications by altering the behavior of water. Although gelatin was originally used as a stabilizer, today polysaccharide stabilizers that are used commonly to control ice crystal growth in ice cream include locust bean gum, sodium carboxymethylcellulose, alginate, carrageenan, and xanthan gum (Adapa et al. 2000; Bahramparvar and Tehrani 2011).

The use of AFPs in ice cream production can improve the texture and flavor of frozen ice by inhibiting the growth of ice crystals in the mixture (Kong et al. 2016). The fact that ice cream products have a good ice crystal structure is important in order to preserve its smooth and creamy texture. Antifreeze proteins inhibit ice crystallization in freezing, thus ensuring the smooth structure of ice cream (Feeney and Yeh 1998). One of the problems with ice cream production is the recrystallization caused by temperature fluctuations during storage where small ice crystals are melted and refrozen into large ice crystals. Abusive-storage conditions, particularly high and fluctuating temperatures, cause rapid recrystallization as evidenced by an increase in mean size and width of the crystal size distribution. The recrystallization process primarily involves small crystals melting, large crystals growing and many crystals fusing together, resulting in fewer and larger crystals for a given ice phase volume. A rounding process is also observed, where crystals with rougher surfaces become rounder through a thermodynamic ripening process. While these processes occur at constant temperature, rates of recrystallization are especially enhanced when temperature fluctuates (Hartel 1998).

Preserving the textural and sensory characteristics of frozen food to ensure product quality is a challenge faced by the food industry. The use of cryoprotectants such as ethylene glycol, glycerol, and dimethyl sulfoxide has been well studied for the cryopreservation of cells in the medical field as well as for biological systems. However, they are normally used in moderately high concentrations such as 0.5 M due to their colligative antifreeze properties. The use of cryoprotectants in high concentrations in food products raises toxicity concerns as well as concerns for changes to the natural color, taste, and texture of the food product caused by cryoprotectant use. The use of antifreeze pretreatment is of major interest in this regard (Kong et al. 2016). The destruction of texture and flavor sense of ice cream by large ice crystals is the reason for AFPs to be recommended as natural ice growth inhibitors in such products. In order to prevent ice recrystallization and improve the final product texture, AFPs can be added to the ice cream mix in half-frozen stage. AFPs can be added to eliminate the need for rapid freezing of the mix, and ice cream can be gradually hardened between −18 and −30 °C without formation of large ice crystals (Kong et al. 2016). The use of AFPs from different sources in ice cream products and the results obtained are summarized in Table 10.1.

In small quantities, AFP was added to the samples of commercial foodstuffs and then samples were frozen at −80 °C and stored at different periods at −6 and −8 °C. Changes in recrystallization during storage were examined microscopically and large crystals were visible after 1 h in the control sample stored at −6 and −8 °C, whereas samples containing AFP did not show any crystal growth. In the same study, vanilla ice cream was also studied, and the AFP-containing specimens showed very small ice crystal growth, while a significant increase in the ice crystal size was observed in the control ice cream without AFP, after 1 h of storage (Warren et al. 1992).

In one study, Clarke et al. (2003) have stored ice cream samples containing Type III antifreeze protein between −10 and −20 °C for 3 weeks and examined the structure images by scanning electron microscopy. The study showed that there

Table 10.1 Potential uses of AFPs in ice-cream products

Source of AFP	Effect on properties of product
Fish AFP (Warren et al. 1992)	– Prevented growth of ice crystals
Type III AFP (Clarke et al. 2003)	– Prevented the growth of ice crystals
Winter wheat grass ice-structuring protein (Regand and Goff 2006)	– Reduced the rate of recrystallization – Produced smoother textures
AP-3 ice-binding pig collagen peptide (Cao et al. 2016)	– Decreased recrystallization – Improved resistance to melting – Reduced the transition temperature
Oats (*Avena sativa* L.) (Zhang et al. 2016)	– Prevented recrystallization – Improved melting resistance – Elevated transition temperature

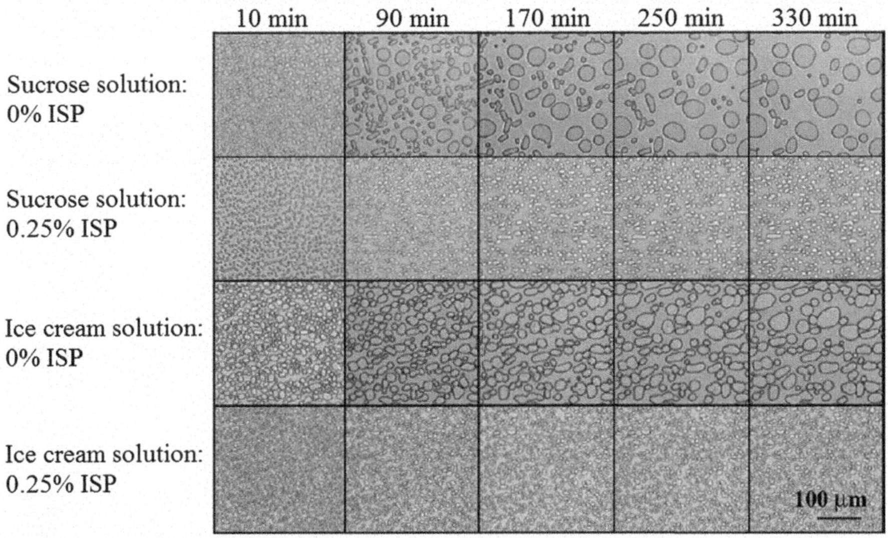

Fig. 10.1 Bright field images acquired every 80 min (starting at 10 min) at −5 °C from sucrose (Formulation A) and ice cream (Formulation D) solutions containing 0 and 0.25% total protein from cold-acclimated winter wheat grass extract (Regand and Goff 2006)

was little difference in ice crystal size compared to the samples before storage and at the end of storage, the size of the ice crystals of the antifreeze protein-containing samples was much smaller than that of the control samples.

In a study of the effect of ice-structuring proteins (ISPs) from winter wheat on the inhibition of ice recrystallization in ice cream, significant ISP activity was observed in sucrose and ice cream solutions containing 0.13% total protein at the delay of ice crystal growth (Fig. 10.1). As shown in Fig. 10.1, adding ISP to sucrose and ice cream solutions prevented the growth of ice crystals and this inhibitory effect

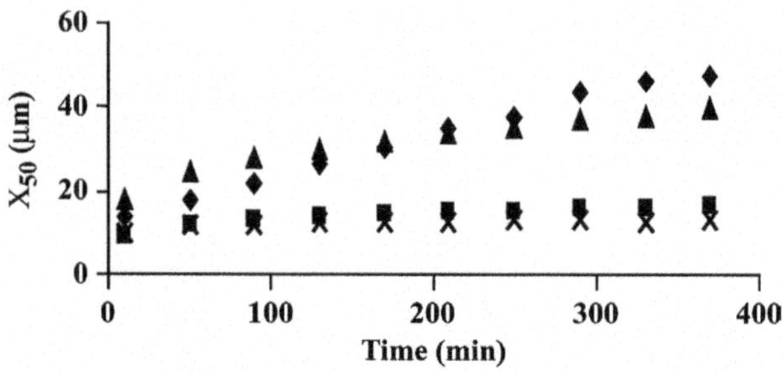

Fig. 10.2 Mean ice crystal diameter from acquired images at −5 °C of sucrose solution; filled diamond, sucrose solution containing 0.25% total protein (TP) from cold-acclimated winter wheat grass extract (AWWE); filled square, ice cream solution; filled triangle and ice cream solution containing 0.25% TP from AWWE; X (Regand and Goff 2006)

appeared more prominent over time. The investigators also determined mean ice crystal diameters (Fig. 10.2). The ice crystal growth in the ice cream solution is mainly controlled by the diffusion of water molecules to the ice and the counter-diffusion of the other solids from the ice surroundings. With the addition of 0.25% total protein from winter wheat, the ice crystal growth changed radically and as seen from Fig. 10.2, once the accretion has reached its maximum, the ice recrystallization is practically stopped and no Oswald ripening was observed in either of the two samples. Total protein additions of 0.025% and 0.0037% from winter wheat germ antifreeze significantly reduced the rate of ice recrystallization by 40% and 46% in heat-shocked freezing. A synergistic effect was observed between the ISP and the stabilizer (locust bean gum), and ISP activity decreased when there was no stabilizer. ISP-containing ice creams have been confirmed by sensory analysis to have significantly smoothened uniform texture after heat-shocked storage (Regand and Goff 2006).

Hyperactive AP-3 ice-binding collagen peptide purified from porcine collagen using enzyme hydrolysis and a series of column chromatographic separation techniques showed antifreeze and cryoprotective properties in ice cream. Due to their much higher TH at low concentrations, these AFPs are termed "hyperactive." The AP-3 ice-binding collagen peptide addition at a concentration of 0.2% (w/v) improved the recrystallization and thawing resistance of ice cream similar to that of natural antifreeze proteins. Also, the addition of AP-3 ice-binding collagen peptide to the ice cream reduced the glass transition temperature (Tg) to −17.6 °C. Investigators have suggested that AP-3 ice-binding collagen peptide may be added as a beneficial additive novel antifreeze protein to probiotics and to the other processed foods that require low temperature storage (Cao et al. 2016).

A distinction between moderate and hyperactive AFPs, aside from the order of magnitude difference in TH, is the way in which they direct ice crystal growth. For moderately active AFPs, ice grows parallel to the *c*-axis, leaving the basal planes unprotected and vulnerable to growth. This is evidenced by "bursts" at the tips of hexagonal bipyramidal ice crystals. For hyperactive AFPs, ice appears to grow along the *a*-axis; the rate of growth along the *a*-axis is ~100 times faster than along the *c*-axis, resulting in a circular disk-like morphology. Clearly, hyperactive AFPs combined with the basal plane give better protection from ice growth by providing more complete ice surface coverage than non-hyperactive AFPs. The ability of hyperactive AFPs to retard recrystallization and depress the freezing temperature below the melting point suggests potential for utilization as natural ice modulators in the storage of frozen products. However, the limited availability of these natural hyperactive AFPs has restricted their application in the food industry. Thus, there is a need for alternative ingredients to control ice crystal growth during storage of frozen foods (Cao et al. 2016).

A slow melting rate and good shape retention are generally considered desirable qualities in ice cream. Ice cream has three main structural components: air cells, ice crystals, and fat globules, which are dispersed throughout a continuous phase of unfrozen solution (Muse and Hartel 2004). These components affect the melting rate. Structural attributes include properties of the air phase (overrun and air cell size distribution), fat phase (total fat content, fat globule size distribution and extent of fat destabilization), ice phase (ice phase volume and ice crystal size distribution), and the continuous phase (viscosity) (Hartel et al. 2003).

Zhang et al. (2016) who examined cryoprotectant activity in ice cream of AFPs obtained from cold-acclimated oats (*Avena sativa* L.) reported that 0.1% oat AFP addition improves recrystallization (Fig. 10.3) and the melt resistance of ice cream, as well as raises its glass transition temperature from −29.1 to −27.74 °C. As seen from Fig. 10.3, addition of 0.1% oat AFP to the ice cream reduced the growth of ice crystal after seven cycles at −4 and −6 °C, compared with the control group.

As can be seen from the results of studies, AFPs obtained from different sources can be added to ice cream and similar products to inhibit ice crystallization in freezing, to prevent or retard ice recrystallization or crystal growth, to reduce the glass transition temperature and to depress the freezing temperature below the melting point, to prevent damage caused by frost and temperature fluctuations in frozen foods and thus to obtain frozen food in superior quality. Depending on these abilities, the use of AFPs in ice cream production can improve the texture and flavor of frozen ice, ensuring the smooth structure of ice cream, and improving the final product texture.

10.4.2 Chilled and Frozen Meat and Fish

The recrystallization inhibition ability of antifreeze proteins can also be used for the frozen storage of meat and fish. Ice recrystallization seen during frozen storage of meat may limit the effective storage period. In ice recrystallization and the average crystal diameter increases as the number of ice crystals decreases. Large ice crystals damage membranes and other muscle structures leading to protein denaturation and drip loss (Payne and Young 1995). The use of AFPs from different sources in meat and fish products and the results obtained are summarized in Table 10.2.

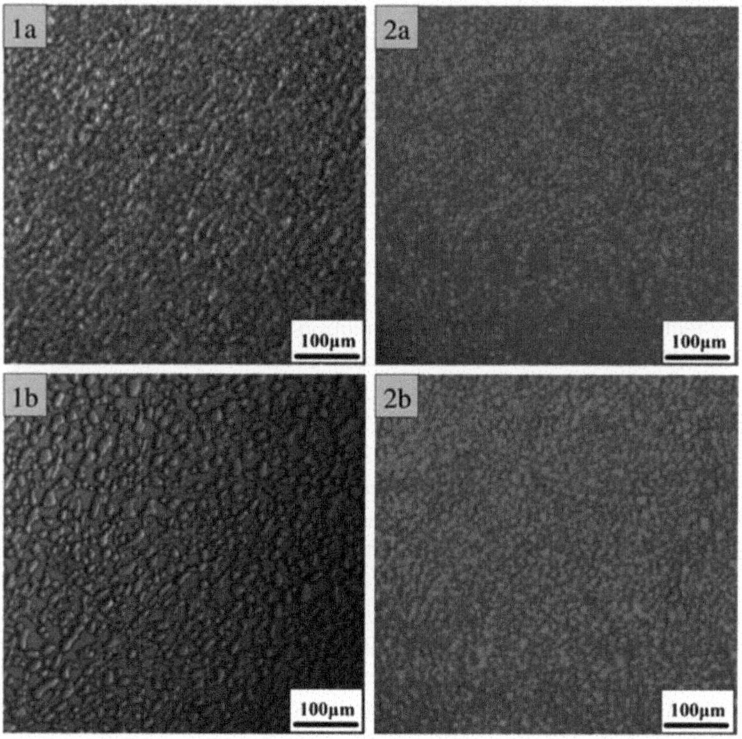

Fig. 10.3 Bright field images of ice cream containing (1) Control, (2) Oat AFP (0.1%); (a) ice cream before seven cycles at −4 and −6 °C, (b) ice cream after seven cycles at −4 and −6 °C (Zhang et al. 2016)

Table 10.2 Potential uses of AFPs in meat and fish products

Source of AFP	Effect on properties of product
Fish AFP (Payne et al. 1994)	– Protected the ice crystal size
AFGP obtained from Antarctic cod fish (Payne and Young 1995)	– Reduced drip loss – Reduced the ice crystal size
Type III fish AFP (Boonsupthip and Lee 2003)	– Maintained the ability of muscle proteins to form gels
Recombinant Type I AFP analogue produced by *Lactobacillus lactis* (Yeh et al. 2009)	– Reduced drip loss – Reduced protein loss – Developed sensory properties (juiciness)

In the study by Payne et al. (1994), small pieces of beef were soaked in concentrated AFP solution, the water was removed and they were dried. With frozen samples, AFPs reduced the size of ice crystals, compared to the control. However, the duration of the retention in the AFP solution was long (2 weeks) and during this period, the breakdown of the meat caused problems.

In another study, AFGPs obtained from the Antarctic Cod (*Dissostichus mawsoni*) were given intravenously to sheep at various concentrations and at various times before slaughtering and the effects on freezing and thawing quality of the meat were investigated. The meat samples were vacuum packaged and checked after storage at −20 °C during 2–16 weeks. Antifreeze injection intravenously into lambs, 1 or 24 h before slaughtering reduced drip loss and ice crystal size. The smallest ice crystals were obtained in the meat of the sheep injected with the final concentration of 0.01 mg/kg AFGP 24 h before slaughter. Investigators have indicated that AFGP addition can reduce losses during freezing storage of meat, if these proteins can be introduced at an affordable cost and in a consumer-acceptable way before freezing (Payne and Young 1995).

In another study investigating the potential usefulness of Type III AFP in preserving the gel-forming properties of the fish muscle under frozen and cold storage conditions, it was determined that AFP preserves its ability to form gels in both cold and freeze storage much better than conventional cryoprotectants (a sucrose sorbitol mixture). In addition, in this line of research, it has been considered as a commercial benefit that the relatively small concentrations of AFP for protection are sufficient and that it does not add a sweet taste to the product (Boonsupthip and Lee 2003).

Protein loss by drip loss in frozen meat is a main problem of meat processing industry. In a study, a novel recombinant type I antifreeze protein analogue (rAFP) was produced and secreted by *Lactococcus lactis*. A food-grade expression and fermentation system for the production and secretion of high levels of rAFP was developed. Lyophilized, crude rAFP produced by *L. lactis* was tested in a frozen pork fillet processing model. The fillets were immersed in rAFP solution for varying time intervals. The drip loss and protein loss of frozen fillets were analyzed after 15 and 30 days of storage at −20 °C. The frozen meat treated with rAFP showed less drip loss, less protein loss, and a high score on juiciness by sensory evaluation. The rAFP provides solution to protein and drip loss problem (Yeh et al. 2009). The researchers compared their results with the previous reports which used AFP from natural resources and concluded that the recombinant AFP analogue was not only economically easy to obtain but also provided an efficient way to apply in frozen meat or other forms of refrigerated processed meats.

To summarize, the AFPs have shown various positive effects in chilled and frozen meat and fish; AFPs has been shown to protect the ice crystal size; AFP injection intravenously into lambs before slaughtering reduced drip loss and ice crystal size; AFGP addition reduced losses during freezing storage of meat; AFP preserved fish muscle's ability to form gels in both cold and freeze storage much better than the conventional cryoprotectants; the frozen meat treated with rAFP showed less drip loss, less protein loss, and a high score on juiciness by sensory evaluation and the rAFP provided solution to protein and drip loss problem.

10.4.3 Frozen Dough

Due to the short shelf life of traditional dough, the technique of freezing dough has attracted a great deal of attention since the 1960s. However, the technique of freezing dough has disadvantages such as weakening the dough structure, lowering the CO_2 retention capacity, prolonging the fermentation period, and impairing the final product texture (Zhang et al. 2007, 2008). In addition, freezing during storage is often lethal to yeast cells, as it has many negative effects caused by water loss such as denaturation of macromolecules, disruption of cell membranes and contraction of cells and has negative impact on bread quality. To counteract this effect, the amount of yeast is increased by 4–10% of the normal dose. However, this increases production costs and negatively affects flavor and texture (Jørgensen et al. 2008). The use of AFPs obtained from different sources in frozen dough and the results obtained are summarized in Table 10.3.

A method for using AFPs in frozen dough products using an expression carrier containing an type I AFP coding sequence (GS-5) has been studied. The sequence was obtained from the grubby sculpin (*Myoxocephalus aenaeus*) and recombined with *Saccharomyces cerevisiae*. Recombinant cells were inoculated in a flourless liquid dough model system and cooled to −20 °C and then thawed several times to determine functional cell numbers. In the control group without the AFP gene, cells rapidly lost their function even after being frozen for only 1 day, while half of the cells containing the recombinant AFP gene remained alive for 13 days at the same temperature. Researchers also studied the ability of the yeast cell to generate gas and determined that after 3 h of freezing at −20 °C, both groups gave almost identical results (Panadero et al. 2005).

The effects of carrot concentrate protein (CCP) additions containing 18.3% (w/w) carrot antifreeze protein on frozen dough fermentation capacity were evaluated by Zhang et al. (2007). CCP was concentrated from cold-acclimated carrot tap roots and stock solutions containing bovine serum albumin (BSA), soy protein isolate (SPI) and CCP, each containing 6.2 g of protein, were prepared. All the dough obtained were fermented; the resulted dough was molded, covered with polymer film and immediately stored at −30 °C. The molded dough was thawed and baked in an oven. The fermentation capacity of the frozen dough was improved by the CCP supplementation through raising the yeast survival and retarding the ice crystal forming. Rheological properties of the CCP group showed stronger fermentation capacity than those of other groups. Retention capacity of the CCP group was also higher than the other groups. The bread quality of the CCP group was similar to that of other groups according to the Chinese National Standard of bread sensory evaluation. So the investigators have informed that the CCP addition to frozen dough has a promising future as an improver for frozen dough fermentation.

Zhang et al. (2008) also studied the effects of CCP containing 15.4% (w/w) carrot (*Daucus carota*) antifreeze protein on textural properties of frozen dough and volatile compounds of bread. The data of hardness, adhesiveness, springiness, chewiness, gumminess, cohesiveness, and resilience were determined by a

Table 10.3 Potential uses of AFPs in dough products

Source of AFP	Effect on properties of product
The type I AFP isolated from the grubby sculpin and named GS-5 was introduced into *Saccharomyces cerevisiae* (Panadero et al. 2005)	– Increased number of living cells
Carrot (*Daucus carota*) (Zhang et al. 2007)	– Increased CO_2 capture capacity of dough – Reduced mortality rate of yeast – Increased fermentation capacity – Decreased freezing rate – Delayed crystal formation in dough
Carrot (*Daucus carota*) (Zhang et al. 2008)	– Developed a soft and stable texture in dough – Stronger capacity for holding loaf volume – Brought a pleasant aroma felt like Michelia alba DC by trans-caryophyllene to crumb
Winter wheat (Xu et al. 2009a)	– Increased water-binding capacity of dough – Improved bread making features of dough – Had a protective effect against the destructive effect of frozen storage and freeze–thaw cycling
Winter wheat (Xu et al. 2009b)	– Reduction in the amount of freezable water content and freezing point depression of dough – Affected the specific heat and effective thermal conductivity properties of dough
A recombinant Type I AFP analogue produced by *Lactobacillus lactis* (Yeh et al. 2009)	– Increased dough fermentation capacity – Bread likability increased positively
Privet (*Ligustrum vulgare*) leaf (Jia et al. 2012)	– Reduced free water molecules and ice melting enthalpy – Improved cooking properties of frozen dough – Increased fermentation capacity – Improved micro structure of dough – Increased residual gluten fibril content – Reduced exposed starch granules content – Enriched gas production and holding capacity of frozen dough – Reduced crumb hardness – Increased specific volume of bread
Carrot (*Daucus carota*) (Ding et al. 2014)	– Decreased the initial and end freezing point – Reduced freezable water content – Improved cooking and textural properties of frozen noodles – Protected gluten matrix from damage caused by freezing and temperature fluctuations
Barley (*Hordeum vulgare* L.) (Ding et al. 2015)	– Increased melting temperature and transition temperature – Reduced freezable water content and melting enthalpy of fresh dough – Slowed down the decrease in moisture content – Decreased the effect of the freeze–thaw cycle on water mobility

(continued)

Table 10.3 (continued)

Source of AFP	Effect on properties of product
	– Influenced the water distribution on frozen dough
Oat (*Avena sativa* L.) (Zhang et al. 2015)	– Reduced the freezable water content of dough – Increased fermentation capacity – Prevented degradation of gluten matrix – Provided bread with superior textural properties

texturometer, and the sensorial quality of final bread was ranked by the panelists according to the National Sensory Evaluation (China). The texture properties of dough and changes of the volatile compounds of crumb were investigated after adding CCP. The sensory quality and the texture profile analysis results of bread of CCP group were similar to those of the control. Meanwhile, CCP supplementation brought a beneficial ability in holding loaf volume. Therewith the loaf volume changes of CCP group were less than those of other groups. Results of the texture property analysis showed that the hardness of CCP group was softer and steadier than that of other groups, due to the low freezable water content during frozen storage. CCP supplementation did not give negative influence on volatile compounds of crumb and brought a pleasant aroma felt like *Michelia alba* DC by trans-caryophyllene. In conclusion, researchers have reported that CCP could be used as a beneficial additive to frozen dough.

In another study evaluating water-binding capacity and bread characteristics of frozen dough containing winter wheat ice-structuring proteins (ISPs), it was determined that the addition of ISPs affected frozen dough water-binding capacity and baking properties positively. ISPs showed a dough protecting effect against the detrimental effects of frozen storage and freeze–thaw cycling. In addition, a strong correlation between water-binding capacity and proofing time and bread specific volume of frozen dough was determined in this study (Xu et al. 2009a).

AFPs or ISPs can improve frozen dough quality by stabilizing ice crystals and preserving texture. For this purpose, the addition of 0.6% of ice-structuring proteins obtained from winter wheat to dough has resulted in a maximum freezing point depression of 0.23 °C and a decrease of 8% in the amount of freezable water. A relatively low effect was observed on the freezable water fraction but the AFPs significantly affected the apparent specific heat and effective thermal conductivity properties in the frozen stage. These information may be helpful for simple and rapid engineering calculations and for complex mathematical modelling of heat transfer (Xu et al. 2009b).

Antifreeze proteins are powerful cryoprotectants used to protect food products from cold. Production of a recombinant type I antifreeze protein analogue by *Lactobacillus lactis* (rAFP) and its application on frozen dough were studied by Yeh et al. (2009). Frozen dough treated with the rAFP showed better fermentation capacity than untreated frozen dough. Breads baked from frozen dough treated with rAFP acquired the same consumer acceptance with regard to the volume, shape, crust color, crumb color, grain, texture, odor, and overall acceptance as fresh bread.

Jia et al. (2012) have used thermostable ISP extracts from leaves of privet (*Ligustrum vulgare*) in frozen dough and have reported that the addition of these proteins to frozen dough caused a reduction in free water molecules and ice melting enthalpy, which improves frozen dough cooking properties, fermentation capacity, and micro structure. In addition, by the addition of ISPs, residual gluten fibril was increased, the exposed starch granules were decreased, and the frozen dough gas production and retention were enriched. These effects led to a decrease in the hardness and an increase in the specific volume of the bread crumb of baked frozen dough.

Extraction of carrot antifreeze proteins and the effect of these extracts on the thermophysical, textural, cooking and microstructural properties of frozen white salted noodles were analyzed by Ding et al. (2014). AFP addition not only reduced the initial and final freezing points, but also reduced the content of freezable water and improved the cooking and textural properties of frozen noodles. The results show that the AFP addition can protect the gluten matrix from damage caused by freezing and temperature fluctuations and thus white salted noodles with superior texturing and cooking properties can be obtained, even after freezing.

Ding et al. (2015) examined the effects of barley (*Hordeum vulgare* L.) antifreeze protein (BaAFP-1) on the thermal properties and water state of dough during freezing and freeze–thaw cycles. The thermal properties of treated and untreated fresh dough, including the apparent specific heat, freezing temperature, melting temperature, freezable water content, and glass transition temperature were determined and compared. For frozen dough samples, the change in melting performance, freezable water content, pasting properties, moisture content, water mobility, and water distribution during freezing and freeze–thaw cycles were analyzed. The results demonstrated that the addition of BaAFP-1 increased the apparent specific heat of dough after freezing, increased the temperature range of the melting and glass transition temperatures, and decreased the melting enthalpy and freezable water content of fresh dough. The addition of BaAFP-1 also influenced the melting performance and gelation property of frozen dough after freeze–thaw cycles. It slowed the decrease in moisture content, weakened the influence of the freeze–thaw treatment on water mobility, and influenced the water distribution in frozen dough.

In another study, the effect of oat (*Avena sativa* L.) antifreeze proteins on the physicochemical, rheological, and fermentation properties of frozen dough and on the textural characteristics of bread baked with previously frozen dough were investigated. The oat antifreeze protein supplement provided some beneficial effects on the final bread quality by reducing the content of frozen water. The oat AFP-treated group showed higher fermentation capacity than the control group and also prevented the degradation of the gluten matrix (Fig. 10.4), ensuring that the bread was of superior textural quality. The dominant features such as gluten matrix (G), the starch granules (S), and the voids formed in the gluten matrix (shown by arrows) are indicated in Fig. 10.4. According to the figure, the dough stored

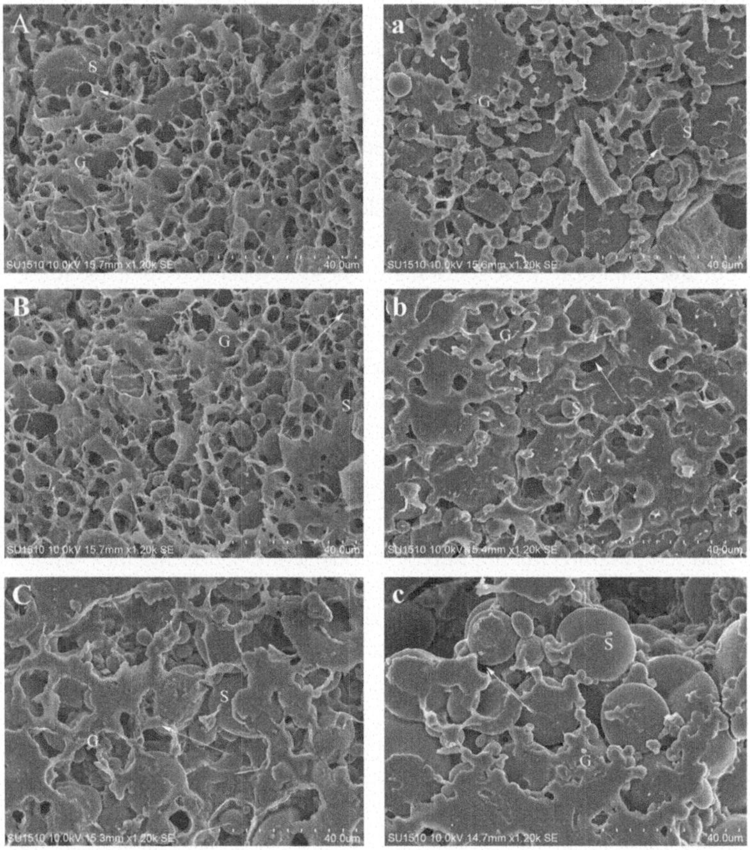

Fig. 10.4 Effects of oat AFPs on the microstructure of frozen dough (**A**, **B**, **C** represent the AFP added dough stored at −18 °C for 1, 2, 3 weeks; **a**, **b**, **c** represent control dough stored at −18 °C for 1, 2 and 3 weeks) (Zhang et al. 2015)

1 week at −20 °C had a very coherent structure, with starch granules firmly embedded in a gluten matrix. After storing for 2 weeks, the microstructure of the control sample appeared less cohesive, the angular voids were less uniform, and large gas voids were present. After storage for 3 weeks, the control sample showed further disruption. As a result, it has been shown that oat AFPs can be used as a useful additive (Zhang et al. 2015).

To summarize the studies of AFPs on preservation of frozen dough, half of the *Saccharomyces cerevisiae* cells containing the recombinant AFP gene remained alive for 13 days at −20 °C; the fermentation capacity of the frozen dough was improved by the CCP supplementation through raising the yeast survival and retarding the ice crystal forming; the rheological properties of the CCP group showed stronger fermentation capacity than those of other groups and the retention capacity of the CCP group was also higher than the other groups; CCP supplementation brought a beneficial ability in holding loaf volume; results of the texture property analysis showed that the hardness of CCP group was softer and steadier than that of other groups, due to the low freezable water content during frozen storage; the addition of ISPs affected frozen dough water-binding capacity and baking properties positively; ISPs showed a dough protecting effect against the detrimental effects of frozen storage and freeze–thaw cycling; ice-structuring proteins addition to dough has resulted in a maximum freezing point depression of 0.23 °C and a decrease of 8% in the amount of freezable water; frozen dough treated with the rAFP showed better fermentation capacity than untreated frozen dough; the addition of ISP extracts to frozen dough caused a reduction in free water molecules and ice melting enthalpy, that improved frozen dough cooking properties, fermentation capacity and micro structure, the exposed starch granules were decreased and the frozen dough gas production and retention were enriched; AFP addition protected the gluten matrix from damage caused by freezing and temperature fluctuations and thus white salted noodles with superior texturing and cooking properties could be obtained, even after freezing; the AFP-treated group showed higher fermentation capacity than the control group and also prevented the degradation of the gluten matrix, ensuring that the bread was of superior textural quality.

10.4.4 Fruits and Vegetables

In the freezing of foods consisting of tissue, formation of large ice crystals which are mostly extracellular, results in significant damages to the tissue. On the other hand, formation of fine crystals that are evenly distributed both inside and outside the cells, leads to the quality of the product to be better preserved due to less damages to the tissue (Kiani and Sun 2011). Freezing rate and fluctuations in temperature are two major factors in ice formation and recrystallization in fruits and vegetables. When the vegetables are frozen at low speed, randomly scattered large ice crystals fill the extracellular spaces (De Ancos et al. 2006). If there is a fluctuation at subzero temperatures, recrystallization will occur, resulting in tissue damage as well as cellular damage. In the temperature range between −7 and 0 (generally called the "zone of maximum ice crystal formation") the recrystallization rate is at its highest, and so is tissue damage, and is thus not a preferable temperature range for freezing storage. Conventionally, this problem was overcome using deep freezers that operate at −60 to −80 °C or liquid nitrogen (LN2, −196 °C), as they reduce the exposure time of the materials to temperatures that promote ice crystal formation thereby minimizing crystal formation. The minute ice crystals that form at these lower temperatures do not effectively destroy the internal structure and allow preservation of quality and/or activity of these materials in the deep freezer or LN2. Special refrigerators that mechanically inhibit ice crystal growth have also been developed. Although these techniques are useful in sample preservation, they generate CO_2. Global reduction of CO_2 is a priority in addressing global warming. In seeking an alternative cryopreservation strategy, AFP was considered. Use of extremely low temperature (−196 to −60 °C) can effectively inhibit ice crystal growth, but it needs high-energy cost (Nishimiya et al. 2008), so AFP participation preventing ice growing and recrystallization in vegetables can significantly reduce costs and maintain food quality and nutrients at subzero temperatures (Jørgensen et al. 2008).

Plant antifreeze activity is cold-induced and is located in the extracellular compartment where ice crystallizes in freeze-tolerant species (Smallwood et al. 1999). Like other plants, carrots increase their freezing tolerance in response to a period of exposure to low, but nonfreezing temperature, a process known as cold acclimation. Some of the low temperature-responsive genes are predicted to encode AFPs. Accumulation and activity of AFPs are always cold-induced and it has been suggested they could be used as a biological marker for crop improvement programs. When cultivated at low temperature, carrot storage roots accumulate large amounts of AFPs in their cell walls (Galindo et al. 2005).

The effects of antifreeze peptides on crystal formation were evaluated during freezing and thawing of carrots. Three synthetic analogues from natural antifreeze peptide were used in this study. By modifying ice crystal morphology, AFPs have proven antifreeze activities in vitro. The ability of synthetic AFP analogues to minimize drip loss and their ability to retain color, structure, texture, and volatile components of frozen carrot were evaluated using Scanning Electron Microscopy (SEM), Gas Chromatography Mass Spectrometry (GC-MS), and texturing analyses. It has been shown that the immersion of carrot samples in the AFP solutions prior to freezing storage helped to protect the frozen carrot closely to the taste of the fresh carrot, particularly the terpenoids that give the carrot's unique flavor were protected, reduced drip loss, and minimized freezing damage. These beneficial effects were attributed to the ability of these AFPs to reduce the ice crystal size and prevent the formation of large intracellular ice crystals, resulting in preservation of the cellular structure of the carrot during freezing process (Fig. 10.5). The prolongation of the immersion time allowed the AFPs to be better diffused into the carrot specimens significantly increasing the beneficial effects of the AFP application. However, it should not be forgotten that this may increase food oxidation, degradation, and bacterial growth risk due to the long immersion time (Kong et al. 2016).

The effects on the quality of frozen cherries which were pretreated with different antifreeze peptides (AFPs) were investigated (Kong et al. 2017) by measuring their physical characteristics and biochemical composition in controlled experiments. Prior to freezing, cherry samples were pretreated by soaking in solutions of three synthetic AFPs, which are based on antifreeze proteins found in nature. Their results showed that AFP pretreatment decreased the drip loss and preserved the color, structure, texture, total anthocyanin, phenolic and antioxidant content of frozen cherries.

The exception that AFPs are located in extracellular tissues of freeze-tolerant organisms suggests that these proteins can be physically added to food by means of mixing (Zhang et al. 2007, 2015, 2016; Yeh et al. 2009; Xu et al. 2009a, b; Ding et al. 2015), injection (Payne et al. 1994; Payne and Young 1995), immersion (Yeh et al. 2009; Kong et al. 2016, 2017), vacuum filtration (Griffith and Ewart 1995), and gene transfer (Fletcher et al. 1988; Hightower et al. 1991; Hew et al. 1992; Feeney

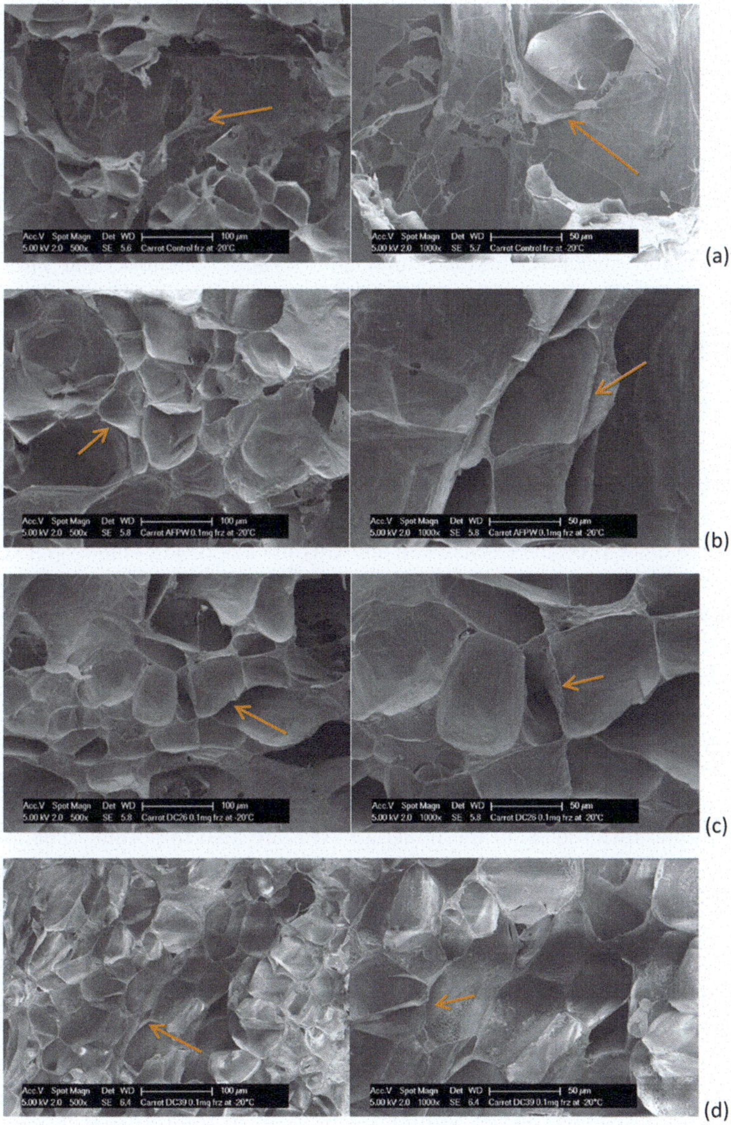

Fig. 10.5 Cryo-scanning electron microscopy images of frozen carrot samples: (**a**) control; (**b**) AFPW 0.1 mg/mL; (**c**) DCR26 0.1 mg/mL; (**d**) DCR39 0.1 mg/mL. Cell walls were arrowed (Kong et al. 2016)

and Yeh 1998; Huang et al. 2002; Wang et al. 2008; Kumar et al. 2017). Among them, gene transfer is probably the most appropriate application for fruits and vegetables. The reason of this is the inheritance of this new feature to the offspring of transgenic plants and thus the cost is reduced compared to other methods. Studies have shown that chimeric AFP gene insertion into tomato plants has been achieved and AFPs generally show promise for application in fruits and vegetables (Jørgensen et al. 2008).

When summarizing the limited number of research results on the use of AFP in fruits and vegetables, the immersion of carrot samples in the AFP solutions prior to freezing storage helped to protect the frozen carrot closely to the taste of the fresh carrot; AFPs reduced the ice crystal size and prevented the formation of large intracellular ice crystals, resulting in preservation of the cellular structure of the carrot during freezing process. AFP pretreatment decreased the drip loss and preserved the color, structure, texture, total anthocyanin, phenolic and antioxidant content of frozen cherries.

10.5 AFP Intake and Reliability

AFPs must be reliable to be used as a food additive. Because of the involvement of these new proteins in food, there is a risk of inducing an allergic reaction in people susceptible to these proteins (Venketesh and Dayananda 2008). However, AFPs can be taken as direct consumption from dietary sources. Fish and plants are the most important sources of dietary AFPs. Amounts of intake vary according to acute or long-term exposure. One portion (200 g) of oceanic pout contains 180–420 mg of AFP type II, and one portion of cod contains 42–196 mg AFGP. AFP intake is also influenced by the consumption frequency of dietary sources. In the United States, which is a moderately fish-consuming country, about 449–4483 mg of AFP per person per year is consumed, while Iceland has a very high consumption of fish, receives between 19,110 and 191,100 mg of AFP per person per year (Crevel et al. 2002). Regional differences in the intake of AFP are also important. The intake of AFP from the diet is thought to be considerable in the northern and temperate regions. This intake is met by edible plants, which are an important part of most diets. In some regions, however, intake from fish may also be important (Crevel et al. 2002; Venketesh and Dayananda 2008).

The history of AFP consumption indicates that these proteins are not acutely toxic, their functional properties do not add any visible reverse toxicological properties to proteins, and AFPs are not associated with any known toxic effect of fish consumption in the short or long term (Crevel et al. 2002). However, analysis of possible allergic reactions in the addition of proteins to foodstuffs is a prerequisite for food marketing. It is not possible to determine whether a protein is allergen or not based solely on its physical or chemical properties. To determine this, the sensitivity of a protein to pepsin digestion, lack of amino acid sequence similarity to known allergens, IgE binding from protein-allergic patients from the source or related proteins, histamine release from basophils should be determined experimentally (Venketesh and Dayananda 2008).

Protein toxicity generally appears as acute toxicity, endocrine modulation, and/or allergic. Fish allergy is a known adverse health effect of fish and allergic proteins have been investigated by various investigators. As a result, it has been determined that AFPs are not among the allergen proteins (Crevel et al. 2002; Venketesh and Dayananda 2008). It has also been shown that AFPs have no adverse effect on general health and are not immunogenic (Venketesh and Dayananda 2008).

10.6 Factors Affecting the Use of AFPs in Foods

Many studies conducted to this date have revealed that antifreeze proteins have potent utility in various fields such as industrial, commercial, and medical applications (Ustun and Turhan 2015). However, many of the factors described below will most likely be effective in future commercial use of these compounds.

Isolation and Purification AFPs are commonly found in a wide variety of dietary and non-dietary organisms including bacteria, fungi, insects, animals, and plants (Griffith and Ewart 1995; Feeney and Yeh 1998; Crevel et al. 2002; Bouvet and Ben 2003). In order for AFPs to be used as pure ingredients in food, the extracts from different sources must be subjected to time-consuming and expensive isolation and purification processes. Since this is not appropriate for mass production and commercial applications, chemical synthesis of AFGP and AFGP analogues should be considered as an important alternative in the future (Bouvet and Ben 2003).

Thermal Stability The use of AFP extracts obtained from plant sources directly in food without any treatment is problematic due to the content of natural enzymes. Since application of vegetable extracts containing proteolytic and lipolytic enzymes may be detrimental to sensory (bitter peptides, oil hydrolysis) or textural (microstructural product degradation) characteristics of the final product, they must be removed from the crude extract by heat application. However, heat application leads to loss of activity by causing loss of secondary structure of AFPs. For this reason, the thermal stability of plant-derived extracts to be used in food applications appears to be a prerequisite (Kontogiorgos et al. 2007).

Price One of the most important factors that will affect the use of AFPs as food additives in future foods is probably the price. The price of AFPs is still high for commercial uses. The price of type III antifreeze obtained from Cold Ocean Teleost Fish is about 10 dollars/mg depending on the purity of the product. However, in the future, chemical synthesis of these substances will probably cause prices to fall (Ustun and Turhan 2015).

Chemical Synthesis of AFPs AFPs can still be produced in limited quantities from plants and other natural sources, with the vast majority being fish. Because the AFP content of natural resources is quite low. For this reason, too much natural resource is required for commercial production of AFPs. In addition, time-consuming and costly isolation methods require the development of alternative sources and methods for the synthesis of these compounds (Harding et al. 2003; Zhang et al. 2016). The price of synthetic materials will probably fall in the coming years, as chemical and biochemical researches on the synthesis of biologically important substances have been rapidly and successfully developed in recent years. However, synthetic food additives are considered to be unreliable due to possible toxicity and their use is gradually diminishing. For this reason, producers are concentrating more on the use of natural additives because they are reliable in the stabilization of foods. In addition to this problem, whether or not the AFPs interact with the other ingredients as well as with the components of the food they used in, should be investigated and elaborated in detail (Ustun and Turhan 2015).

Advances in Molecular Biology In recent years there has been considerable progress in the introduction of appropriate genetic material for protein synthesis and gene transfer in agricultural products (Feeney and Yeh 1998). These rapid developments may allow the production of antifreeze proteins at much lower cost (Ustun and Turhan 2015). Today, antifreeze proteins are already produced in trangenic *E. coli* in many laboratories around the world. However, the use of gene transfer to produce AFP-producing food organisms requires careful selection of the most appropriate AFP strain for the desired concentrations and the organism used. Concerns about the acceptance of transgenic crops by consumers should also not to be ignored (Griffith and Ewart 1995; Ustun and Turhan 2015). It has been proven that the transformation of tomatoes with fish-derived AFP gene is unacceptable for some consumers. Thus, consumer acceptance may be due to the transfer of genes from close relative organisms (Griffith and Ewart 1995). A major food additive company has already recombinant fish AFP in several of their products.

10.7 Conclusions

Initially, studies about the use of AFPs in food concentrated on ice cream and meat, but recently frozen dough and fruits have been included among the investigated foods. In the studies it has been proven that AFPs improve texture and flavor by inhibiting the growth of ice crystals in ice cream, minimize the nutrient loss by preventing drip loss and recrystallization in meat and increase the gas holding ability and reduce the yeast mortality rate in frozen dough. It has also been shown that AFPs have significant potential in industrial, commercial, and medical applications. However, the commercial usability of AFPs depends primarily on their availability. Currently, the most important sources to obtain natural AFPs are plants, insects, fish, and other organisms living in cold regions. However, these natural resources are not sufficient for mass production and commercial applications. Even if AFPs are synthetically produced, it is likely that these forms are less accepted by consumers due to health concerns. One of the important issues is new allergies that may arise from new proteins contained in transgenic products. For these reasons, commercial production, usability, and availability of AFPs are unfortunately limited today. In addition, many tests with different foodstuffs are required to show that AFPs do not react with food additives as well as with food ingredients, which likely would be a time-consuming process. Moreover, the fact that these proteins are expensive and can be produced biotechnologically on a commercial scale limits the potential for commercial use in foods. In spite of all these negativities, the fact that they can fulfill important functions in food and can provide these functions even in very low concentrations suggests that these proteins have the possibility of using as potential additives for foods in the future.

References

Adapa S, Schmidt KA, Jeon IJ, Herald TJ, Flores RA (2000) Mechanisms of ice crystallization and recrystallization in ice cream: a review. Food Rev Int 16:259–271

Bahramparvar M, Tehrani MM (2011) Application and functions of stabilizers in ice cream. Food Rev Int 27:389–407

Balasubramanian S, Devi A, Singh KK, Bosco SJD, Mohite AM (2016) Application of glass transition in food processing. Crit Rev Food Sci Nutr 56:919–936

Barbosa-Cánovas GV, Altunakar B, Mejía-Lorío DJ (2005) Freezing of fruits and vegetables: an agribusiness alternative for rural and semi-rural areas. FAO Agricultural Services Bulletin 158. Food and Agriculture Organization of the United Nations Rome, 76 pp

Boonsupthip W, Lee TC (2003) Application of antifreeze protein for food preservation: effect of type III antifreeze protein for preservation of gel-forming of frozen and chilled actomyosin. J Food Sci 68:1804–1809

Bouvet V, Ben RB (2003) Antifreeze proteins: structure, conformation and biological applications. Cell Biochem Biophys 39:133–144

Cao H, Zhao Y, Zhu YB, Xu F, Yu JS, Yuan M (2016) Antifreeze and cryoprotective activities of ice-binding collagen peptides from pig skin. Food Chem 194:1245–1253

Clarke CJ, Buckley S, Lindner N (2003) Ice-structuring proteins in ice cream. In: Goff HD, Tharp BW (eds) International dairy federation. Proceedings of the second IDF international symposium on ice cream, Thessaloniki, 14–16 May, pp 33–34

Crevel RWR, Fedyk JK, Spurgeon MJ (2002) Antifreeze proteins: occurrence and human exposure. Food Chem Toxicol 40:899–903

Damodaran S (2007) Inhibition of ice crystal growth in ice cream mix by gelatin hydrolysate. J Agric Food Chem 55:10918–10923

De Ancos B, Sanchez-Moreno C, De Pascual-Teresa S, Cano MP (2006) Fruit freezing principles. In: Hui YH (ed) Handbook of fruits and fruit processing. Blackwell, Iowa, pp 59–80

Ding X, Zhang H, Liu W, Wang L, Qian H, Qi X (2014) Extraction of carrot (Daucus carota) antifreeze proteins and evaluation of their effects on frozen white salted noodles. Food Bioprocess Technol 7:842–852

Ding XL, Zhang H, Wang L, Qian HF, Qi XG, Xiao JH (2015) Effect of barley antifreeze protein on thermal properties and water state of dough during freezing and freeze-thaw cycles. Food Hydrocoll 47:32–40

Duman JG, Bennett V, Sformo T, Hochstrasser R, Barnes BM (2004) Antifreeze proteins in Alaskan insects and spiders. J Insect Physiol 50:259–266

Duman JG, Walters KR, Sformo T, Carrasco MA, Nickell PK, Lin X, Barnes BM (2010) Antifreeze and ice-nucleator proteins. In: Denlinger DL, Lee RE Jr (eds) Low temperature biology of insects. Cambridge University Press, Cambridge, pp 59–90

Feeney RE, Yeh Y (1998) Antifreeze proteins: current status and possible food uses. Trends Food Sci Technol 9:102–106

Fletcher GL, Shears MA, King MJ, Davies PL, Hew CL (1988) Evidence for antifreeze protein gene transfer in Atlantic Salmon (*Salmo salar*). Can J Fish Aquat Sci 45:352–357

Galindo FG, Elias L, Gekas V, Herppich WB, Smallwood M, Sommarin M, Worrall D, Sjoholm I (2005) On the induction of cold acclimation in carrots (*Daucus carota* L.) and its influence on storage performance. Food Res Int 38:29–36
Griffith M, Ewart KV (1995) Antifreeze proteins and their potential use in frozen foods. Biotechnol Adv 13:375–402
Griffith M, Yaish MWF (2004) Antifreeze proteins in overwintering plants: a tale of two activities. Trends Plant Sci 9:399–405
Harding MM, Anderberg PI, Haymet ADJ (2003) Antifreeze glycoproteins from polar fish. Eur J Biochem 270:1381–1392
Hartel RW (1998) Mechanisms and kinetics of recrystallization in ice cream. In: Raid RW (ed) The properties of water in foods ISOPOW 6. Springer, London, pp 287–319
Hartel RW, Muse M, Sofjan R (2003) Effects of structural attributes on hardness and melting rate of ice cream. Proceedings of the 2nd IDF ice cream symposium, Thessoliniki, Greece
Hassas-Roudsari M, Goff HD (2012) Ice structuring proteins from plants: mechanism of action and food application. Food Res Int 46:425–436
Hew CL, Davies PL, Fletcher G (1992) Antifreeze protein gene transfer in Atlantic salmon. Mol Mar Biol Biotechnol 1:309–317
Hightower R, Baden C, Penzes E, Lund P, Dunsmuir P (1991) Expression of antifreeze proteins in transgenic plants. Plant Mol Biol 17:1013–1021
Hiilovaara-Teijo M, Hannukkala A, Griffith M, Yu X-M, Pihakaski-Maunsbach K (1999) Snow-mold-induced apoplastic proteins in winter rye leaves lack antifreeze activity. Plant Physiol 121:665–673
Hon WC, Griffith M, Mlynarz A, Kwok YC, Yang DSC (1995) Antifreeze proteins in winter rye are similar to pathogenesis-related proteins. Plant Physiol 109:879–889
Huang T, Nicodemus J, Zarka DG, Thomashow MF, Wisniewski M, Duman JG (2002) Expression of an insect (*Dendroides canadensis*) antifreeze protein in *Arabidopsis thaliana* results in a decrease in plant freezing temperature. Plant Mol Biol 50:333–344
Jia C, Huang W, Wu C, Zhong J, Rayas-Duarte P, Guo C (2012) Frozen bread dough properties modified by thermostable ice structuring proteins extract from Chinese privet (ligustrum vulgare) leaves. Cereal Chem 89:162–167
Jin-Yao L, Ji M, Fu-Chun Z (2005) Recent advances in research of antifreeze proteins. Chin J Biochem Mol Biol 21:717–722
Jørgensen SKK, Keskin S, Kitsios D, Commerou PW, Nilsson B (2008) Antifreeze proteins: the applications of antifreeze proteins in the food industries. Roskilde University, p 54
Karaoglu MM, Boz H, Kotancilar HG, Gercekaslan KE (2009) Glass transition temparature and it's importance in cereal products. Atatürk Üniv Ziraat Fak Derg 40:93–99
Kiani H, Sun D-W (2011) Water crystallization and its importance to freezing of foods: a review. Trends Food Sci Technol 22:407–426
Kilic M, Evranuz O (2006) The relatipnship between glass transition and quality in foods. Gida 31:253–257
Kong CHZ, Hamid N, Liu T, Sarojini V (2016) Effect of antifreeze peptide pretreatment on ıce crystal size, drip loss, texture, and volatile compounds of frozen carrots. J Agric Food Chem 64:4327–4335
Kong CHZ, Hamid N, Ma Q, Lu J, Wang B-G, Sarojini V (2017) Antifreeze peptide pretreatment minimizes freeze-thaw damage to cherries: an in-depth investigation. LWT Food Sci Technol 84:441–448
Kontogiorgos V, Regand A, Yada RY, Goff HD (2007) Isolation and characterization of ice structuring proteins from cold-acclimated winter grass extract for recrystallization inhibition in frozen foods. J Biochem 31:139–160
Kumar SR, Kiruba R, Balamurugan S, Cardoso HG, Helia G, Birgit AS, Zakwan A, Sathishkumar R (2017) Carrot antifreeze protein enhances chilling tolerance in transgenic tomato. Acta Physiol Plant 36:21–27

Muela E, Sañudo C, Campo MM, Medel I, Beltrán JA (2010) Effect of freezing method and frozen storage duration on instrumental quality of lamb throughout display. Meat Sci 84:662–669
Muse MR, Hartel RW (2004) Ice cream structural elements that affect melting rate and hardness. J Dairy Sci 87:1–10
Nishimiya Y, Mie Y, Hirano Y, Kondo H, Miura A, Tsuda S (2008) Mass preparation and technological development of an antifreeze protein—toward the practical use of biomolecules. Synthesiology (English edn) 1:7–14
Ökkes A, Nalbantoğlu B (2003) Antifreeze proteins in higher plants. Phytochemistry 64:1187–1196
Panadero J, Randez-Gil F, Prieto JA (2005) Heterologous expression of type I antifreeze peptide GS-5 in Baker's yeast increases freeze tolerance and provides enhanced gas production in frozen dough. J Agric Food Chem 53:9966–9970
Payne SR, Young OA (1995) Effects of pre-slaughter administration of antifreeze proteins on frozen meat quality. Meat Sci 41:147–155
Payne SR, Sandford D, Harris A, Young OA (1994) The effects of antifreeze proteins on chilled and frozen meat. Meat Sci 37:429–438
Petzold G, Aguilera JM (2009) Ice morphology: fundamentals and technological applications in foods. Food Biophys 4:378–396
Rahman MH, Hossain MM, Rahman SME, Hashem MA, Oh D-E (2014) Effect of repeated freeze-thaw cycles on beef quality and safety. Korean J Food Sci Anim Resour 34:482–495
Ramlov H, Johnsen JL (2014) Controlling the freezing process with antifreeze proteins. In: Sun D-W (ed) Emerging technologies for food processing, 2nd edn. Academic Press, San Diego, pp 539–562
Regand A, Goff HD (2006) Ice recrystallization inhibition in ice cream as affected by ice structuring proteins from winter wheat grass. J Dairy Sci 89:49–57
Roos YH (2010) Glass transition temperature and its relevance in food processing. Annu Rev Food Sci Technol 1:469–496
Singh P, Hanada Y, Singh SM, Tsuda S (2014) Antifreeze protein activity in Arctic cryoconite bacteria. FEMS Microbiol Lett 351:14–22
Smallwood M, Worrall D, Byass L, Elias L, Ashford D, Doucet CJ, Holt C, Telford J, Lillford P, Bowles DJ (1999) Isolation and characterization of a novel antifreeze protein from carrot (*Daucus carota*). Biochem J 340:385–391
Sørensen TF, Cheng CC-H, Ramløv H (2006) Isolation and some characterisation of an antifreeze protein from the eelpout *Zoarces viviparus*. CryoLetters 27:387–399
Ustun NS, Turhan S (2015) Antifreeze proteins: characteristics, function, mechanism of action, sources and application to foods. J Food Process Preserv 39:3189–3197
Venketesh S, Dayananda C (2008) Properties, potentials, and prospects of antifreeze proteins. Crit Rev Biotechnol 28:57–82
Wang Y, Qiu L, Dai C, Wang J, Luo J, Zhang F, Ma J (2008) Expression of insect (*Microdera puntipennis* Dzungarica) antifreeze protein MpAFP149 confers the cold tolerance to transgenic tobacco. Plant Cell Rep 27:1349–1358
Warren CJ, Mueller GM, McKown RL (1992) Ice crystal growth suppression polypeptides and methods of preparation. US Patent 5,118,7921992, June 2
Wilson CL, El Ghaouth A, Chalutz E, Droby S, Steres C, Lu JY, Khan V, Arul J (1994) Potential of induced resistance to control postharvest diseases of fruits and vegetables. Plant Dis 78:837–843
Xu H-N, Huang W, Jia C, Kim Y, Liu H (2009a) Evaluation of water holding capacity and breadmaking properties for frozen dough containing ice structuring proteins from winter wheat. J Cereal Sci 49:250–253
Xu HN, Huang W, Wang Z, Rayas-Duarte P (2009b) Effect of ice-structuring proteins from winter wheat on thermophysical properties of dough during freezing. J Cereal Sci 50:410–413
Yeh CM, Kao BY, Peng HJ (2009) Production of a recombinant type 1 antifreeze protein analogue by L.lactis and its applications on frozen meat and frozen dough. J Agric Food Chem 57:6216–6223

Zhang C, Zhang H, Wang L (2007) Effect of carrot (*Daucus carota*) antifreeze proteins on the fermentation capacity of frozen dough. Food Res Int 40:763–779

Zhang C, Zhang H, Wang L, Gao H, Guo XN, Yao HY (2008) Effect of carrot (*Daucus carota*) antifreeze proteins on texture properties and frozen dough and volatile compounds of crumb. LWT Food Sci Technol 41:1029–1036

Zhang YJ, Zhang H, Wang L, Qian HF, Qi XG (2015) Extraction of oat (*Avena sativa* L.) antifreeze proteins and evaluation of their effects on frozen dough and steamed bread. Food Bioprocess Technol 8:2066–2075

Zhang Y, Zhang H, Ding X, Cheng L, Wang L, Qian H, Qi X, Song C (2016) Purification and identification of antifreeze protein from cold-acclimated oat (*Avena sativa* L.) and the cryoprotective activities in ice cream. Food Bioprocess Technol 9:1746–1755

Chapter 11
Cell, Tissue, and Organ Preservation with Insect-Derived Antifreeze Peptides

Kelvin G. M. Brockbank, John D. Duman, Zhen Chen, Elizabeth D. Greene, Henry M. Vu, and Lia H. Campbell

Abbreviations

AFGL	Antifreeze glycolipid
AFP	Antifreeze protein
ALT	Alanine aminotransferase
CAFP	Cucujus antifreeze protein
DAFP	Dendroides antifreeze protein
DMSO	Dimethylsulfoxide
G/T	0.5 M glycerol and 150 mM trehalose supplement
LDH	Lactate dehydrogenase

K. G. M. Brockbank (✉)
Tissue Testing Technologies LLC, North Charleston, SC, USA

Department of Bioengineering, Clemson University at Charleston, Charleston, SC, USA

J. D. Duman
University of Notre Dame, Notre Dame, IN, USA
e-mail: jduman@nd.edu

Z. Chen · E. D. Greene · L. H. Campbell
Tissue Testing Technologies LLC, North Charleston, SC, USA
e-mail: lcampbell@t3-tissuetestingtechnologies.com

H. M. Vu
W.M. Keck Center for Transgene Research, University of Notre Dame, Notre Dame, IN, USA
e-mail: hvu@nd.edu

H. Ramløv, D. S. Friis (eds.), *Antifreeze Proteins Volume 2*,
https://doi.org/10.1007/978-3-030-41948-6_11

11.1 Introduction

This chapter summarizes our experience using insect-derived antifreeze proteins (AFPs) for preservation of cells, tissues, and organs. Preservation is crucial as an enabling technology for transplantation medicine, regenerative medicine, tissue engineering, and research products that contain living cells. Living biological materials can be short-term preserved using hypothermic storage at 0–6 °C and this is common practice for complex vascularized allotransplant tissues, such as hands and faces, as well as internal organs, such as hearts and livers, for transplantation but is limited in most cases, except kidneys, to <10 h post-procurement from donors. Longer term preservation of cells and some relatively simple tissues can be obtained by either freezing or vitrification methods. Cryopreservation by freezing would provide an ideal solution to this need except that not all cells can be cryopreserved adequately employing existing, popular dimethylsulfoxide (DMSO)-based methods. However, some cell types are difficult to preserve, such as hepatocytes and cardiac myocytes, under most conditions. Others are difficult in adherent mode in vitro or in three-dimensional tissues but not when frozen as single cell suspensions.

New hypothermic storage and cryopreservation methods may be possible by mimicking the strategies employed by animals to survive subzero temperatures (Brockbank et al. 2011). These include insects that combine increases in cryoprotectant content (such as glycerol, proline, trehalose) with the onset of cold environmental conditions and production of antifreeze proteins (AFPs) and glycolipids (AFGLs) to survive temperatures as low as -80 °C. Furthermore, Vu and Duman (2017) recently reported the paradoxical phenomenon of increased heat tolerance of insects in the winter months. For biomedical and research applications, understanding the mechanisms that permit increased survivorship at both high and low temperatures may be beneficial for preservation of mammalian cells and tissues or even organs, 1 day, for transplantation purposes. With the drastic changes in temperature from 37 to <-135 °C and back to 37 °C common during cryopreservation, the cells, tissues, and organs undergo extreme temperature stress. If the mechanisms underlying the insects' winter phenomenon of increased heat tolerance can be understood, the compounds driving the insects' winter survivorship could be used to increase the yield and success in cryopreservation and mitigate damage resulting from temperature stress.

Recombinant copies of some insect-derived AFPs are available and they are much more effective in the modulation of ice formation in solution (thermal hysteresis, indicative of AFP activity) and promotion of survival of the organisms in which they are found than the better-known fish-derived AFPs (Davies 2014; Duman 2015). This chapter is organized in sections on our experiences using insect-derived AFPs for preservation of cells, tissues, and organs followed by a general discussion.

11.2 Experimental Design and Research Strategy

The overall objective of our experiments was to determine the impact of preservation treatments ± AFPs on living mammalian biological materials compared with untreated fresh materials. Details of the preparation of these AFPs, which included a His-tag, were previously presented (Wang and Duman 2005; Halwani et al. 2014). In short, recombinant AFPs were expressed in transformed *E. coli* BL21(DE3) pLysS cells grown in 1 L shaker flasks in early experiments or in a 7.5 L bio-fermenter in the later experiments. The crude protein extracted from the cells was purified with either a His-bind purification kit (Novagen) or a cobalt resin column (Clontech, Mountain View CA, USA). The purified protein was run on 10% SDS gels to check purity. The His-tag was not removed prior to testing the DAFPs.

Our experiments were performed at different levels of biological complexity using assays that are customary for either isolated cells, ex vivo blood vessels and livers. All aspects of animal research were approved by the Medical University of South Carolina Animal Care and Use Committee prior to initiation. Statistical differences were assessed by *t*-test and/or one way analysis of variance (ANOVA). p-values <0.05 were regarded as significant.

In experiments with either isolated cells or small pieces of tissue we employed a resazurin assay that is a nontoxic fluorometric indicator based on detection of metabolic activity (Halwani et al. 2014). The resazurin assay incorporates a water-soluble fluorometric viability oxidation–reduction indicator that detects metabolic activity by both fluorescing and changing color in response to chemical reduction. Fluorescence was measured at an excitation wavelength of 544 nm and an emission wavelength of 590 nm. This evaluation can be performed after a short period of recovery following rewarming to physiologic temperature for evaluation of relative cell viability. Alternatively, growth or survival curves can be generated to allow characterization of rewarmed isolated cells or cells in tissues. After each reading the cells or tissues are washed and placed back in tissue culture. Results shortly after rewarming (day 0) demonstrate cell viability. Decreases over 1–2 days of culture are indicative, suggestive, of cell death due to apoptosis and increases are due to cell recovery or proliferation. Isolated cells in control and experimental treatment groups were always seeded at the same number so the final outcome could be expressed as either relative fluorescent units (RFU) or in percent of control. Control and experimental tissues were dried at the end of each experiment and results expressed as RFU/mg tissue dry weight or in percent of untreated controls.

In the case of vascular tissues, veins, and arteries, we also included physiological assays to determine whether or not the vessels were still able to perform physiologically normal contraction and or relaxation (Song et al. 2000; Brockbank et al. 2007). Blood vessel rings, approximately 2 mm in length, were prepared for vascular physiology studies in a Radnoti organ bath system using constrictive and dilatory drugs. Isometric contractile and relaxation tensions were measured after addition of increasing concentrations of contractile drugs (such as norepinephrine), endothelial-dependent (calcium ionophore A21837) and -independent (sodium nitroprusside) smooth muscle relaxation drugs and expressed as grams tension/mg tissue dry weight.

Control and experimental liver function were assessed on an ex vivo perfusion circuit under simulated physiological conditions. Adult domestic Yorkshire cross farm pigs (25–30 kg) were used as heart beating donors. The animals were weighed and pre-anesthetized with a mixture of ketamine (22 mg/kg), acepromazine (1.1 mg/kg), and atropine (0.05 mg/kg) given IM. After establishing an ECG, the pigs were intubated, and placed on isoflurane anesthesia at 1.5–2%. The pigs were anticoagulated with heparin (400 IU/kg, IV). After a midline incision from the xiphoid process to the pubis, the dorsal aorta was cannulated below the liver, and clamped off above the liver. The livers were perfused with 4 L of Lactated Ringers, flushed with Belzer's machine perfusion solution, excised and placed in a plastic bag on ice for 2 h during transport and preparation for studies. Following liver removal, the donor's heparinized blood was collected for use in the perfusate during normothermic evaluation. The pigs were euthanized by exsanguination under anesthesia. Subzero liver perfusions were performed in an insulated chilled bath in which the temperature was regulated by recirculation from a temperature controlled bath. Hypothermic perfusions above zero and normothermic evaluations were performed in a liver perfusion prototype as previously described (Brockbank et al. 2012). Physiological data was collected during the normothermic perfusion and perfusate samples were collected every 30 min for evaluation of pCO_2, pO_2 electrolytes (Na^+, K^+, Ca^{++}, $NH4^{++}$), metabolites (lactate and glucose), nonspecific indicators of tissue damage (lactate dehydrogenase and alanine aminotransferase). Specific liver functions evaluated included bile production, Factor V, indocyanine green clearance, and total bile bilirubin. Cytokines were assessed using enzyme-linked immunosorbent assays and the manufacturer's instructions. Please see our prior publication for full details on procedures (Brockbank et al. 2012).

11.3 Cell Preservation

In our initial studies we demonstrated feasibility of improving mammalian cell viability using recombinant antifreeze compounds from *Dendroides canadensis* during cryopreservation by freezing and characterized the impact of these compounds on nucleation and recrystallization using cryomicroscopy (Halwani et al. 2014). A rat-derived smooth muscle cell line, A10, was used as a model because we usually obtained about 30% cell survival when the cells were cryopreserved in adherent mode with dimethylsulfoxide (DMSO) alone on a tissue culture plastic substrate. Therefore, improvements in viability as well as decreases could be detected in response to cryopreservation in the presence of experimental formulations containing AFPs.

AFPs, lower the non-colligative freezing point of water, inhibit ice nucleators, modify ice structure, inhibit recrystallization, and modify the fluid properties of solutions, thereby extending survival of organisms in subzero environments (Davies

2014; Duman 2015). The antifreeze molecules are diverse in structure, and have been identified in a variety of organisms including fish, insects, and other terrestrial arthropods, plants, fungi, and bacteria (DeVries 1971; Cheng and Devries 2002; Duman 2001; Duman and Olsen 1993). The first to be discovered and best characterized are the antifreeze glycoproteins and proteins found in Antarctic and northern fish species. The fish-derived AFPs are thought to adsorb preferentially to the prism face of ice or to internal planes that also result in inhibition of ice crystal growth perpendicular to the prism face (Davies 2014; Olive et al. 2016). The insect AFPs actively bind both the basal and prism faces of ice and are, by far, the most active, and consequently, may offer the best opportunity for success in applications (Huang et al. 2002; Nicodemus et al. 2006). AFPs such as those derived from certain beetle larvae, such as *Dendroides canadensis*, have much greater effects on ice formation than fish-derived AFPs (up to an order of magnitude greater thermal hysteresis). In addition, Duman and his colleagues showed that larvae of an Alaskan beetle, *Cucujus clavipes*, can survive −100 °C without freezing, and that they vitrify at temperatures of −60 to −70 °C (Sformo et al. 2010). *C. clavipes* larvae produce AFPs similar to those of *D. canadensis*, and it is likely that these AFPs also enhance the ability of the larvae to vitrify (unpublished), probably at least partially because of the inhibition of ice nucleators by the AFPs (Duman 2001), especially when enhanced by glycerol (Duman 2002).

The study of how organisms survive extreme temperatures has revealed that multiple phenomena occur in synchrony to promote survival (reviewed: Brockbank et al. 2011). Insects, in particular, have revealed several strategies, besides the production of antifreeze proteins, which help insects either avoid freezing or tolerate freezing until warmer temperatures are available. Considering all the insects in the world and how they successfully emerge from winter (even when subjected to −60 °C in northern Alaska), it is obvious that they have developed strategies for surviving adverse conditions that are very effective. An objective of cryobiologists World Wide has been learning from that which has been observed to be successful in nature and translating this information for the preservation of mammalian cells, tissues, and organs. For example, sugars have always been classified as cryoprotectants but they were never used extensively in preservation strategies because of their perceived limited benefits for the survival of preserved cells and tissues. However, as has been observed in the recent literature, there is renewed interest in sugars, more specifically disaccharides such as trehalose (a natural blood sugar in insects), and their ability to protect cells and tissues during preservation including preservation of platelets (Crowe et al. 2001, 2003). The use of sugars is only one of several strategies used by insects to survive adverse conditions. These may include the use of antifreeze proteins, polyols, and loss of water (cryoprotective dehydration as in *C. clavipes*) among others (Brockbank et al. 2011) providing an opportunity to develop preservation methods for a variety of cell and tissue types that are not readily preserved using existing methods.

Our first published observations using insect AFPs was that insect AFPs from the beetle *Dendroides canadensis* have significant thermal hysteresis and supercooling (Wang and Duman 2005, 2006), and are able to improve the viability of

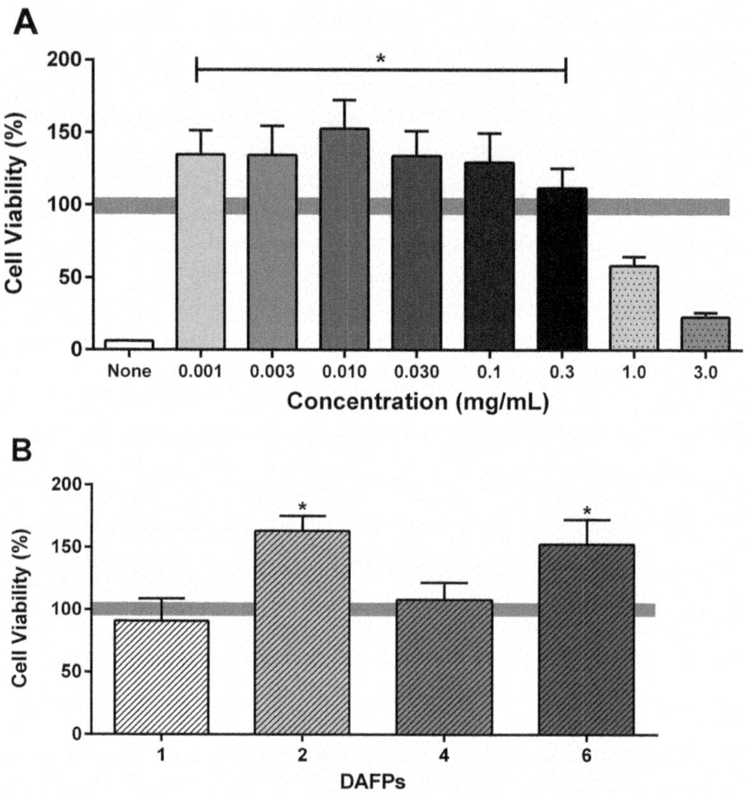

Fig. 11.1 A10 cell viability after cryopreservation with DAFPs combined with 1 M DMSO. (**a**) A10 cells were cryopreserved in the indicated concentrations of DAFP-6. (**b**) Peak cell viability observed in DAFP cryopreservation experiments. DAFP-1 and -2 at 1 μg/mL, and DAFP-4 and -6 at 10 μg/mL. Data is presented as the mean (±1 SEM) of 12 replicates in percent of 1 M DMSO controls, $*p < 0.05$. Reproduced with permission Halwani et al. (2014)

cryopreserved cells (Fig. 11.1). Four AFPs derived from *D. canadensis* (DAFP-1, -2, -4 & -6) were assessed using an adherent rodent smooth muscle cell line (A10). The cells were maintained in standard culture flasks and experiments were performed in 24 well plates. After overnight incubation with cell culture medium the plates were placed on ice and increasing concentrations of cryoprotectant added. Then the final cryoprotectant solution was added and left for 10 min on ice before being cooled at a controlled rate to −80 °C, and stored at ≤−135 °C overnight. The next day, the plates were thawed by placement in a −20 °C freezer for 30 min followed by rapid warming at 37 °C as previously described (Campbell et al. 2003). The cryoprotectant was then removed and the cells were allowed to recover for 1 h at 37 °C in cell culture medium before assessment of cell viability. The cells from plates were then cultured and assayed using a resazurin-based metabolic assay. Significantly improved smooth muscle cell viability was observed in cryopreservation experiments with low DAFP-6 and DAFP-2 concentrations in combination with DMSO.

No significant improvement in viability was observed with either DAFP-1 or DAFP-4. These findings support the potential of incorporating DAFPs in cryoprotectant formulations used to cryopreserve cells and tissues. Dilutions as low as 1 μg/mL were sufficient to trigger this effect. Low and effective DAFP concentrations are advantageous because they minimize concerns regarding cell cytotoxicity and manufacturing cost.

Curiously, solutions containing multiple DAFP types that demonstrated significant supercooling point depressing activity (~9 °C difference) along with inhibition of ice recrystallization during rewarming when compared to single DAFPs and/or DMSO-only solutions (Fig. 11.2, $p < 0.05$, Halwani et al. 2014) failed to demonstrate benefits for cell survival post-cryopreservation (Fig. 11.3). Each DAFP was evaluated alone or in combinations of two, three, or four DAFPs. All possible combinations were tested in 96 well plates using a high throughput screening strategy in at least four independent experiments. Concentrations as low as 1 μg/mL effectively increased cell viability over DMSO alone controls ($p < 0.05$), while concentrations above 300 μg/mL resulted in decreased cell viability after cryopreservation. Viability was dependent on the individual DAFP or combination of DAFPs used. A significant increase in cell viability was observed across several DAFP-6 concentrations (Fig. 11.1a). DAFP-1 and DAFP-4 had no benefit for viability, while DAFP-2, DAFP-6, and some combinations of two DAFPs were also effective at increasing cell viability (Fig. 11.3). Unlike the freezing point depression studies (Fig. 11.2, Halwani et al. 2014) combinations of three or more DAFPs did not produce better cell viability than single DAFPs type 2 or 6 (Fig. 11.3). In fact there was a decreased viability compared with DMSO-only treated controls. These results demonstrated the feasibility of employing some DAFPs, particularly DAFP-2 and DAFP-6, during cryopreservation of adherent A10 cells because statistically significant improvements in cell viability were observed.

There is much more to be learned about these insect AFPs, for example the other 30 or so DAFPs identified in *D. canadensis* and at least ten identified in *C. claviceps* have not been studied adequately. Some of the *D. canadensis* and *C. claviceps* AFPs have been cloned, sequenced, and expressed in bacteria. The composition, size, and stability of the complexes formed by combinations will likely vary with protein concentration, protein ratios, or other factors that impact insect survival in nature. New AFPs from a variety of sources are being identified continuously and most of them have not been studied yet for their potential impact on cell survival during cryopreservation. It is anticipated that many of them will have benefits for mammalian cell preservation.

11.4 Tissue Preservation

The use of hypothermia as the principal means to suppress metabolism in a reversible way is the foundation of most of the effective methods for complex tissue and organ storage prior to transplantation. Antifreeze proteins depress the

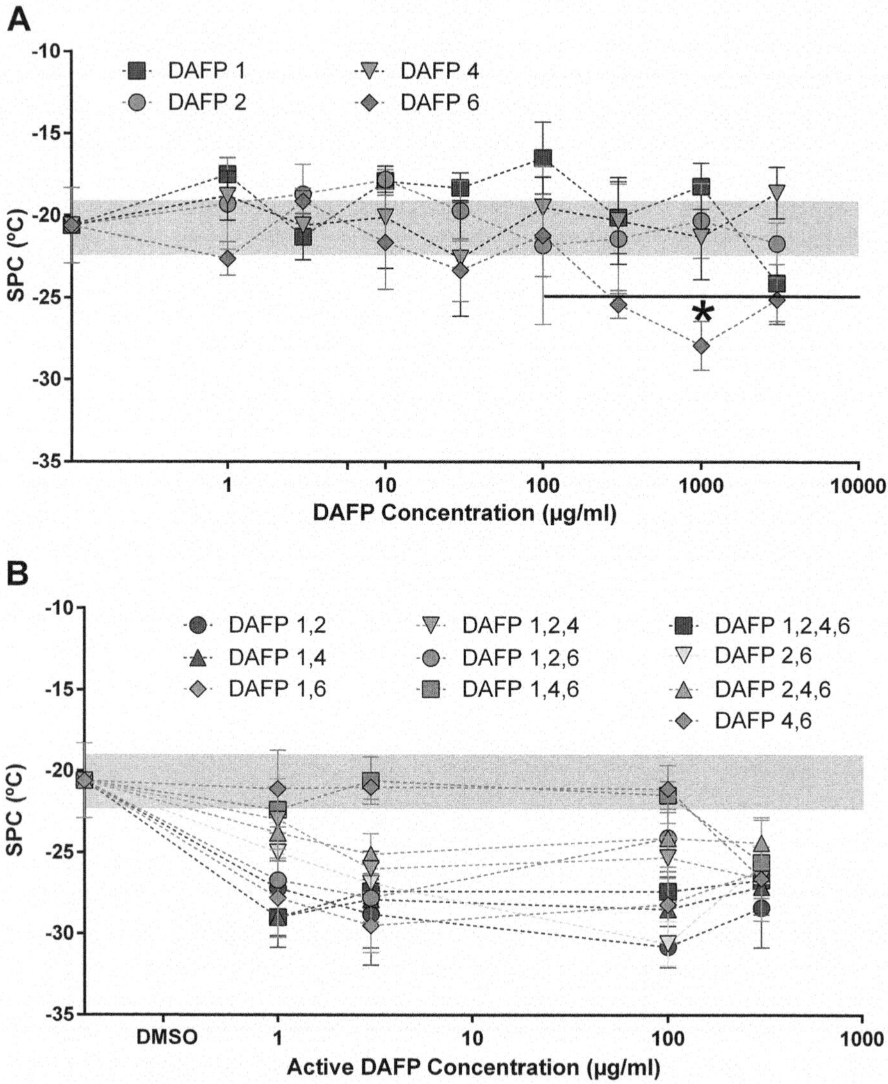

Fig. 11.2 Nucleation temperatures of solutions containing 1 M DMSO with the indicated DAFPs in single (**a**) and multiple combinations (**b**). Gray band represents one standard deviation for the DMSO control. Data are expressed as mean ± SE for at least three replicates. $*p < 0.05$ vs. DMSO control. Reproduced with permission Halwani et al. (2014)

nonequilibrium freezing point of water below the melting point, thereby producing a characteristic difference between the melting and hysteretic freezing points that is termed thermal hysteresis. Insect antifreeze proteins, including those from beetles, are the most active AFPs known. Another critical antifreeze effect of beetle AFPs is promotion of supercooling by inhibition of ice nucleation sites. The overall goal of

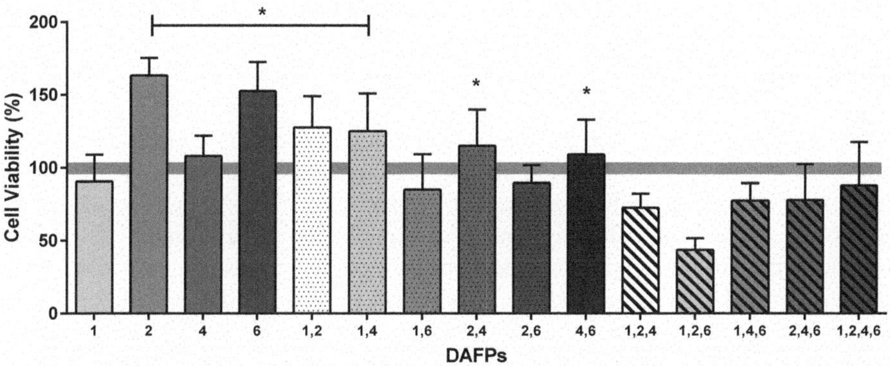

Fig. 11.3 Cell viability after cryopreservation. A10 cells were exposed to solutions with 1 M DMSO and DAFPs then cryopreserved. Upon thawing, metabolic activity was measured using resazurin dye. Cell viability was calculated based on the DMSO control (purple bar) and is the mean (±SEM) of 12 replicates. All combinations of DAFP-1, -2, -4, and -6 were used at their best concentration based upon earlier experiments, $*p < 0.05$ versus the DMSO control

our hypothermic tissue preservation project was to test antifreeze proteins from the beetle *Dendroides canadensis* for the ability to protect mammalian tissue by preventing ice formation while holding the tissues at −7 °C. Certain of the *Dendroides* antifreeze proteins are known from previous work to self-enhance their thermal hysteresis antifreeze activity such that the hysteretic freezing point of aqueous solutions of the DAFPs is as much as 10 °C below the melting point. In the case of *Dendroides* AFPs maximum thermal hysteresis and inhibition of ice nucleators requires the presence of enhancers, low molecular weight enhancers include glycerol and trehalose, while large molecular weight enhancers include other DAFPs and endogenous thaumatin-like protein (Li et al. 1998; Wang and Duman 2005, 2006).

The major anticipated benefit of mammalian tissue storage at −7 °C, in the presence of DAFPs, was that the tissues may experience much lower levels of endogenous metabolic activity, essentially zero, compared with ~4% under current hypothermic storage practices at +4 °C. Physiologists have long recognized that biochemical processes, such as metabolism, decline at low temperatures, dropping by at least half with each 10 °C drop, i.e., $Q_{10} > 2$ (where the temperature coefficient Q_{10} is the factor by which a rate increases with a 10 °C increase in temperature or vice versa). This generalization is supported by the Arrhenius' equation. For example, this decline has been found in a diversity of muscles from a vast array of animals, including vertebrates and invertebrates, endotherms and ectotherms (Deban and Lappin 2011). Our working hypothesis was that this reduced metabolic activity might translate to longer cell and tissue storage times provided that there were no unanticipated side effects of either DAFPs and/or −7 °C storage.

The recombinant DAFPs were produced in *E. coli* and validated for thermal hysteresis activity by Professor Duman's group at Notre Dame prior to shipping to

Charleston, SC, for further testing. Ice-free storage was observed for temperatures down to −7 °C, spontaneous nucleation and freezing was observed at lower temperatures.

Insect-derived antifreeze protein formulations were tested subzero at −7 °C and at +4 °C, to detect antifreeze protein concentration-dependent cytotoxicity effects, and compared with controls without antifreeze proteins in order to increase the duration of hypothermic storage. The test temperature, −7 °C, was selected on the basis of the greater non-colligative freezing point depression produced by the insect antifreeze proteins that provide much greater activity compared with antifreeze proteins derived from other sources. The vascular tissue models selected for this proposal, the rabbit jugular vein and carotid artery (Song et al. 2000; Taylor et al. 2004; Brockbank et al. 2007), were chosen because of our extensive research experience with these models and because the vasculature represents a crucial component of major organs, such as the liver, upon which organ survival depends. Furthermore, these models permit dissection of hypothermic storage effects upon the morphology and functional viability of both smooth muscle and endothelial cells because these studies can be performed with the small (mg) quantities of insect-derived antifreeze proteins that can be produced with our current protein expression methods. In our initial experiments vessel rings from rabbit jugular veins were stored in aqueous solutions consisting of recombinant DAFPs-1, -2, and -4 plus thaumatin-like protein (at a concentration of 2.7 mg/mL of each of the four proteins combined) at +4 and −7 °C for up to 6 days. Smooth muscle and endothelial physiology were studied employing a series of drugs that induced smooth muscle contraction and endothelial-dependent and -independent smooth muscle relaxation as described in the figure legends in an organ bath test system. Approximately 30% of samples stored at −7 °C without AFPs froze during storage and when tested these samples demonstrated no functions. The insect-derived antifreeze proteins inhibited ice nucleation at −7 °C, none froze at this temperature.

Both smooth muscle and endothelial functions were preserved in jugular veins and carotid artery rings stored at −7 °C for up to 6 days. Generally, the responses of blood vessel rings stored at −7 °C were more similar to fresh controls than those stored at +4 °C suggesting that the additional reduction in metabolism provided by the reduced temperature benefited the tissue, for example, the +4 °C vein tissue demonstrated lower maximum bradykinin-induced maximal contractions and less acetylcholine-induced endothelial mediated smooth muscle relaxation (Fig. 11.4). Bradykinin has a potent dose-dependent constrictive effect on the smooth muscle of many blood vessels while acetylcholine has the opposite relaxing, dilatory effect. The ability of such drugs to cause constriction or dilation is lost as cell viability and associated tissue function is lost during storage. In marked contrast to jugular veins, carotid arteries demonstrated only minor changes during up to 12 days of storage (data not shown) with all storage groups being very similar to fresh control values for contractile (norepinephrine and phenylephrine) and dilatory (endothelial mediated calcium ionophore) drugs. All artery storage groups demonstrated an increased response to sodium nitroprusside, that has a direct dilatory effect on smooth muscle, with some significant differences ($p < 0.05$) after 3 and 6 days of hypothermic

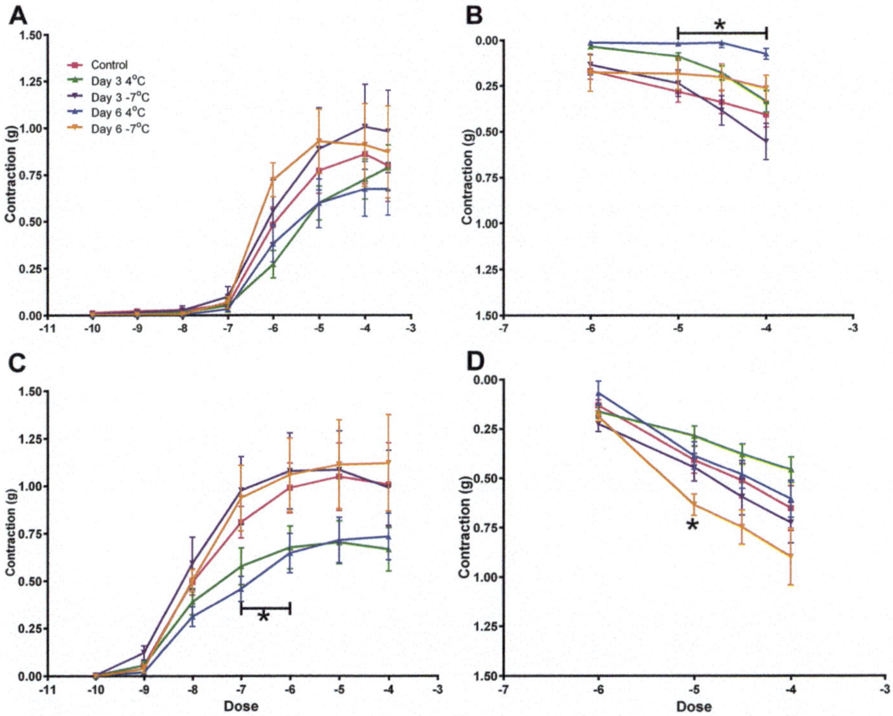

Fig. 11.4 Summary of jugular vein functions after storage in the presence of DAFPs-1, -2, and -4 plus thaumatin-like protein, (**a**) histamine and (**c**) bradykinin and relaxation, (**b**) acetylcholine and (**d**) sodium nitroprusside studies after storage in DAFP formulations at +4 and −7 °C. Drug doses indicated as 10^{-X} M. The histamine and bradykinin results for −7 °C were similar to fresh controls. The 4 °C histamine functions were also not significantly different compared with fresh controls. However, the +4 °C groups were significantly different in their contractile responses compared with both fresh and −7 °C groups at 10^{-7} and 10^{-6} M bradykinin. The endothelium-dependent acetycholine response from 10^{-4} to 10^{-5} M was significantly different after 6 days of storage at +4 °C for most drug concentration comparisons. In contrast the relaxation response of smooth muscle to sodium nitroprusside of 6 day −7 °C stored rings was significantly greater than fresh controls and +4 °C at the 10^{-5} M dose. The data is expressed as the mean ± 1 SEM of 6–12 replicates derived from three or more rabbits. All *p*-values mentioned were $p < 0.05$ using the Mann–Whitney, nonparametric *t*-test, asterisk indicates range in which most statistical significance was observed. Analysis of variance using the nonparametric Kruskal–Wallis test with Dunns post-test also resulted in statistical significance *p*-values ranging from $p < 0.001$ at 10^{-5} M to $p < 0.05$ at 10^{-4} M, for the 6 day acetycholine data from the +4 °C storage group

storage. There was a trend toward an increase in sensitivity and maximum tension generated in response to contractile drugs, norepinephrine and phenylephrine, for artery rings stored at +4 °C with DAFPs compared with both fresh controls and −7 °C stored tissues. The benefits of storage with DAFPs were most pronounced in jugular veins (Fig. 11.4). The results demonstrated that both smooth muscle and endothelial vascular tissue functions after 3–6 days at −7 °C and, with one exception, were similar to fresh tissues. The exception was a trend for the sixth day −7 °C group responses to sodium nitroprusside to be greater than the other test groups at 10^{-5}–10^{-4} M achieving significance at 10^{-5} M. There was also a trend for +4 °C contractile functions to decrease at both 3 and 6 days, statistically significant for certain doses of bradykinin, but not histamine. Day 6 acetycholine responses were very low for the +4 °C stored group ($p < 0.05$) at each of the three highest doses, demonstrating deterioration of endothelial cell-mediated functions. Overall the +4 °C stored rings appeared to be deteriorating faster than the −7 °C stored vein rings. This is probably an effect of reduced metabolism at −7 °C compared with +4 °C, although we cannot rule out an effect of the DAFPs from our present data. The histology of 6 day vessels stored at −10 and +4 °C with DAFPs appeared normal with good endothelial cell retention, no cell swelling, and no signs of nuclear fragmentation or necrosis (Fig. 11.5).

We also evaluated AFPs with 150 mM trehalose and 0.5 M glycerol because we observed protein aggregates in some AFP formulations. Trehalose is commonly used

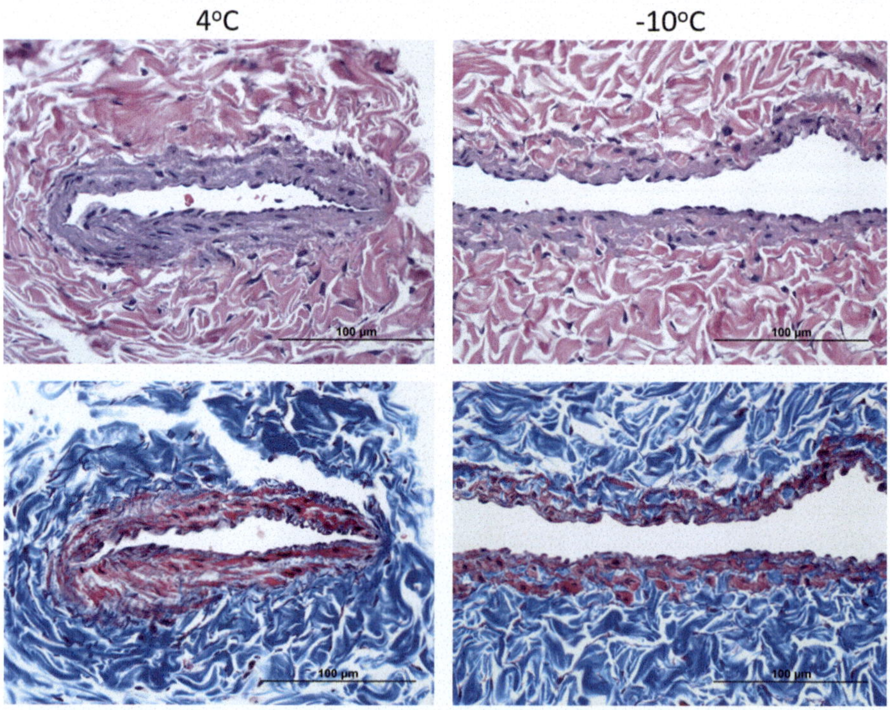

Fig. 11.5 Representative histology of DAFP stored blood vessels. Minimal changes were observed in either +4 °C (left) or −7 °C (right) vessels (jugular veins shown) with DAFPs. The examples are vein sections from samples fixed after 6 days of storage and stained with either Hematoxylin & Eosin (top) or Masson's Trichrome (bottom). Please note the integrity of the endothelial lining in the central lumens of these cross sections at 40× magnification and the absence of pycnotic nuclei, nuclear fragmentation, or necrosis. These sections are as good as can be expected for veins that are fixed in this manner [perfusion fixation is not possible because of the short length of the rings (~2 mm)]

to prevent protein aggregation in frozen and freeze dried protein formulations (Carpenter et al. 2002; Jain and Roy 2010). 150 mM trehalose was selected. Complex DAFP formulations and less complex AFP formulations based upon either DAFP-1,2 DAFP-10, CuAFP-3, or CuAFP-4 combined with 0.5 M glycerol and 150 mM trehalose (G/T) were tested by storing blood vessels subzero at −7 C for days. In the 1–3 day time frames vessel storage in various AFP/T/G formulations and concentrations consistently resulted in equal or better contractile functions than storage controls, but not statistically different from fresh untreated controls ($p < 0.05$, not shown). Longer storage studies up to 12 days were also performed with two formulations DAFP-1,2,4,6 + thaumatin-like protein +G/T, DAFP-1,2 (1.0 and 3.0 mg/mL) +G/T with excellent retention of contractile functions (data not shown). The simplest formulations that provided retention of vascular tissue functions were *Cujucus* AFP-3 (CuAFP-3) and *Dendroides* AFP-10 (DAFP-10) formulations in combination with trehalose and glycerol that demonstrated significantly higher maximum contractions than storage controls. Both formulations tended to be higher than fresh controls at some drug doses ($p < 0.05$, Fig. 11.6) suggesting an unanticipated induction of membrane receptor stabilization by these AFPs.

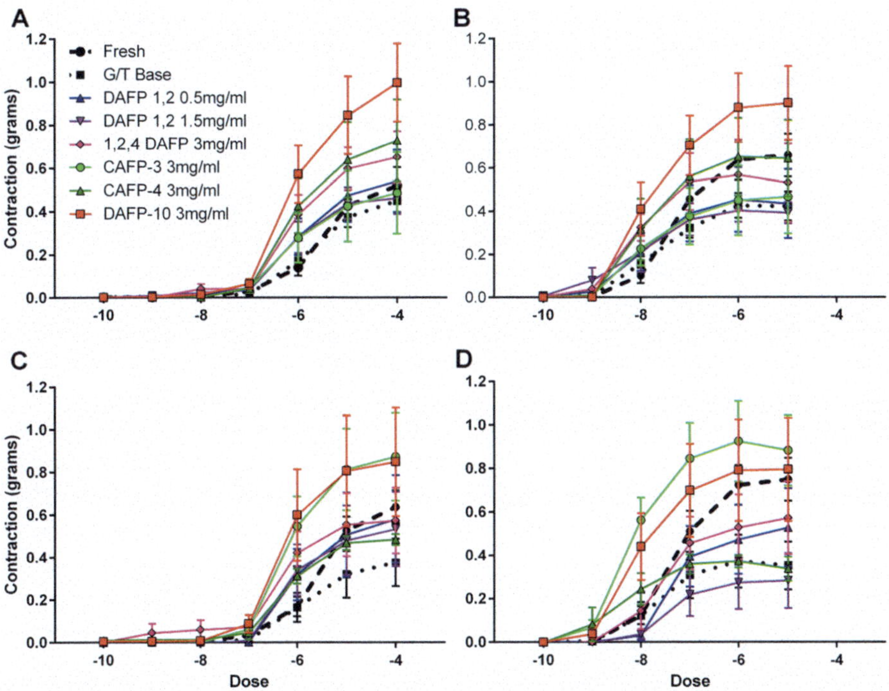

Fig. 11.6 Jugular vein storage with AFPs in combination with glycerol and trehalose were evaluated with (**a**) histamine and (**b**) bradykinin after 1 day and (**c**) histamine and (**d**) bradykinin after 2 days of subzero storage. CAFP-3 and DAFP-10 stored veins demonstrated significantly higher contractile responses than storage controls at lower effective drug doses (lower dotted black line). Drug doses indicated as 10^{-X} M

We focused on DAFP-10 for longer blood vessel storage studies. Veins stored with DAFP-10 demonstrated DAFP-10 concentration-dependent preservation of smooth muscle functions in vessel rings stored for 6 days at −7 °C, endothelial-mediated relaxations were observed in all storage groups but less than fresh controls ($p < 0.05$, not shown). The best preservation, similar to untreated fresh controls, was observed at 1.5 mg/mL (Fig. 11.7a). On day 12 of storage the contractions were reduced to ~50% of fresh controls ($p < 0.05$, Fig. 11.7d) and decreased levels of endothelial-mediated relaxation were still present (Fig. 11.7c, e). Direct sodium nitroprusside-mediated relaxation was retained at fresh control values (Fig. 11.7b). In contrast, there was no effect of DAFP-10 on artery physiology after 6 (Fig. 11.8a–c) or 12 days (Fig. 11.8d–f) of storage. However, the storage solution control group without DAFP-10 demonstrated similar physiological functions at both storage time points compared with fresh untreated controls and DAFP appeared to have a dose-dependent inhibition of function (Fig. 11.8c, d).

An alternative strategy for preserving tissues is vitrification, the solidification of a liquid without crystallization (Brockbank et al. 2012, 2014). Some insects such as the Alaskan *C. clavipes* larvae use a combination of cryoprotective compounds, extensive dehydration and AFPs to survive temperatures as low as −100 °C without freezing and instead vitrify (Sformo et al. 2010). Recent evidence demonstrated that vitrification of the residual solution after slow cooling and freezing to −20 °C is associated with the survival of cold acclimated *C. costata* larvae when subsequently cryopreserved in liquid nitrogen (Rozsypal et al. 2018). Vitreous cryopreservation is an alternative cryopreservation strategy that stabilizes the biological materials in a glass, thereby avoiding ice crystallization. However, there is the risk of cytotoxicity if the sample is not managed correctly during both addition and removal of the high concentrations of cryoprotectants required for vitrification (Brockbank et al. 2014, 2015). Cryopreservation formulations for mammalian tissue vitrification strategies usually consist of 55–83% volume cryoprotectant combinations unless the sample size is small. Cryoprotectants have been associated with risks of patient reactions and cytotoxic effects of residual post-processing concentrations. Banking of both natural and engineered tissues using either freezing or vitrification methods of cryopreservation will be achieved only if adequate methods of controlling ice growth are devised that are less dependent upon cooling and warming rates while reducing the viscosity of the formulations and concentrations of the potentially cytotoxic cryoprotectants employed. Research on ice-free vitrification of tissues with complex cryoprotectant formulations and insect-derived recombinant AFPs is in progress.

11.5 Organ Preservation

Static storage at subzero (−2.5 not −7 °C as originally planned due to nucleation in larger solution volumes) ±1.5 mg/L *Dendroides* AFP-10 in Unisol™ with 0.5 M glycerol and 150 mM trehalose (G/T) solution was compared with hypothermic storage using gold standard Belzer's for varying times. Evaluation was performed on

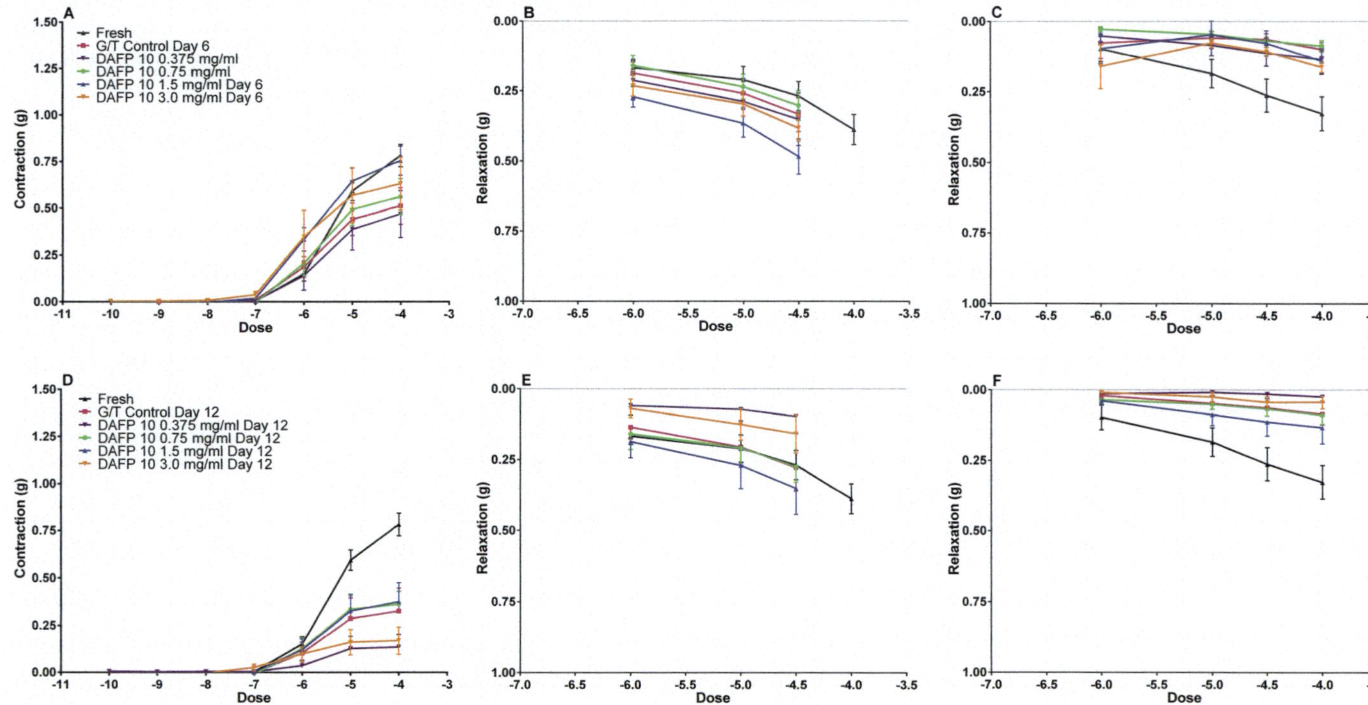

Fig. 11.7 Rabbit jugular vein function after storage in DAFP-10 with glycerol and trehalose for 6 days. (**a**) Histamine-induced contraction, (**b**) sodium nitroprusside-induced relaxation, and (**c**) acetycholine-induced endothelial mediated relaxation. Function after 12 days of storage (**d**) histamine-induced contraction, (**e**) sodium nitroprusside-induced relaxation and (**f**) acetycholine-induced endothelial mediated relaxation. Drug doses indicated as 10^{-X} M

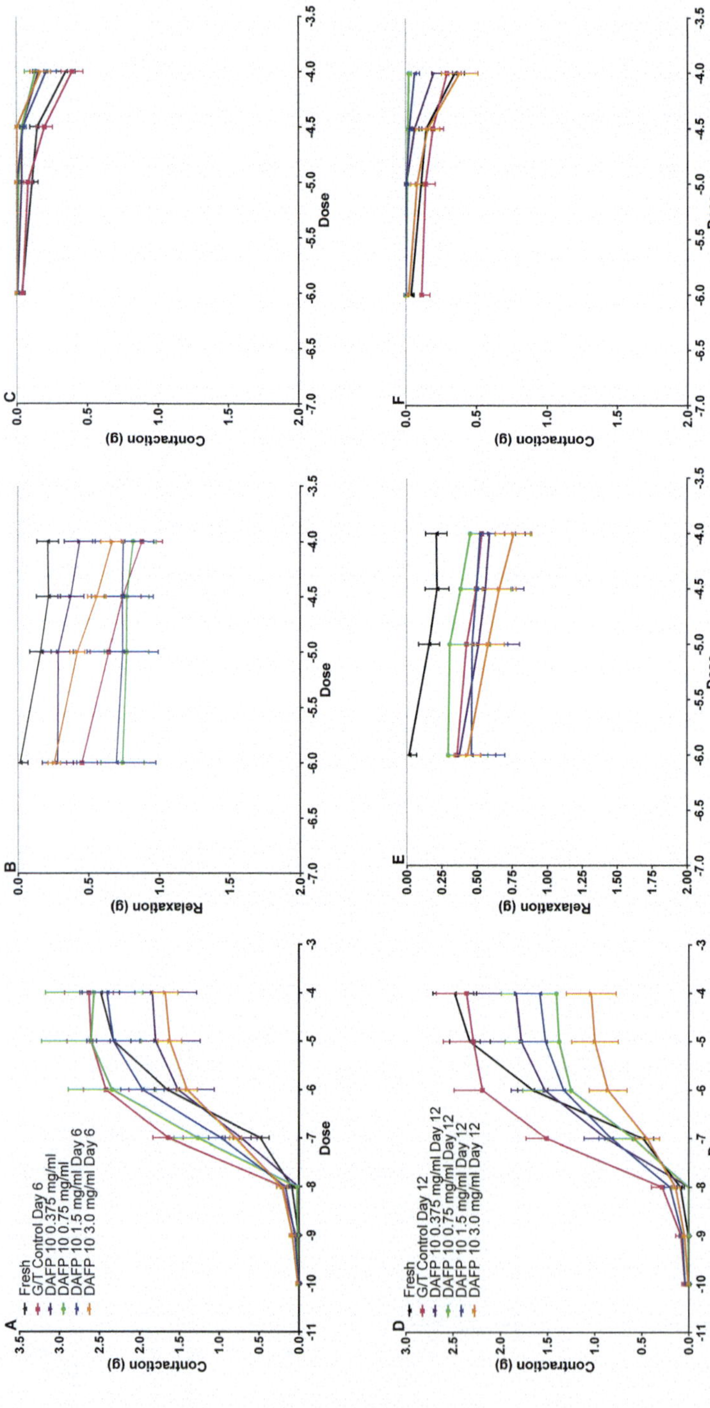

Fig. 11.8 Rabbit carotid artery function after storage in DAFP-10 with glycerol and trehalose for 6 days. **(a)** Histamine-induced contraction, **(b)** sodium nitroprusside-induced relaxation, and **(c)** acetycholine-induced endothelial mediated relaxation. Function after 12 days of storage **(d)** histamine-induced contraction, **(e)** sodium nitroprusside-induced relaxation, and **(f)** acetycholine-induced endothelial mediated relaxation. Drug doses indicated as 10^{-X} M

an ex vivo normothermic oxygenated blood-based perfusion circuit under physiological conditions. The methods have previously been described at length (Brockbank et al. 2012). The 2 h storage group in Belzer's at 4 °C was considered to be the positive control because livers treated this way perform well in vivo and the 24 h storage group in Belzer's was considered the negative control because these livers would be anticipated to perform very poorly in vivo. Perfusate samples were collected during normothermic organ reperfusion using a bench top apparatus and gas composition (pCO_2, pO_2), pH, osmolarity, electrolytes (Na^+, K^+, Ca^{++}, $NH4^{++}$) and metabolites (lactate, glucose, glutamine, glutamate) were tested immediately and every 30 min for 3 h. Perfusate samples and bile were stored for future assays. Hepatocellular damage was assessed by measuring two liver enzymes, lactate dehydrogenase (LDH) and alanine aminotransferase (ALT), both commonly used indicators of hepatocellular damage. Several liver functional assays and cytokine release assays were also performed. Fourteen static storage livers at 2, 6, or 24 h of storage were compared with 14 subzero stored livers without AFP at either 6 or 24 h of storage. Four livers were evaluated at 24 h of subzero storage with AFPs. We were limited to four livers by the quantity of AFP-10 that was available.

Thirty-two livers from 25 to 30 kg pigs were successfully used for experiments. Animal use was approved by the Institutional Animal Care and Use Committee of the Medical University of South Carolina. Some livers were obtained from animals sacrificed in other approved studies for procurement of other organs using the same protocol. Subzero temperatures were achieved by placing the livers in a custom-designed alcohol-cooled cold bath maintained at −2.5 °C. Thermocouples were placed throughout each liver. The livers ranged from 650 to 800 g in weight and took between 3 and 4 h to reach the target storage temperature of −2.5 °C (Fig. 11.9a). The mean hepatic artery resistance observed increased with storage time and subzero exposure (Fig. 11.9b). The presence of AFP subzero made no impact on hepatic artery resistance or flow rate. In contrast, the pulmonary vein mean resistance and flow were similar in the positive control and 24 h AFP subzero group (Fig. 11.9c). *Note*: this correlated with the benefits observed for veins but not arteries in earlier experiments. Most interesting were the liver enzyme assay results that demonstrated trends but not statistically significant differences between other storage groups and the AFP group for both LDH and ALT (Fig. 11.10). The AFP group also had excellent albumin synthesis (Fig. 11.10), a measure of normal liver function, compared with other storage groups that almost achieved statistical significance using a paired *t*-test ($p < 0.02$) versus other 24 h storage groups. The cytokines, interleukin-8, tumor necrosis factor-α, and transforming growth factor-β release decreased in all storage groups at 24 h (Fig. 11.10). The mean β-galactosidase level observed, a measure of monocyte/macrophage activation, was lower than any other storage group except the 2 h positive control storage group (Fig. 11.10). Increased cytokines and/or β-galactosidase would have been considered a negative outcome. These observations indicate that liver macrophages, Kupffer cells, were not activated by either cold storage method.

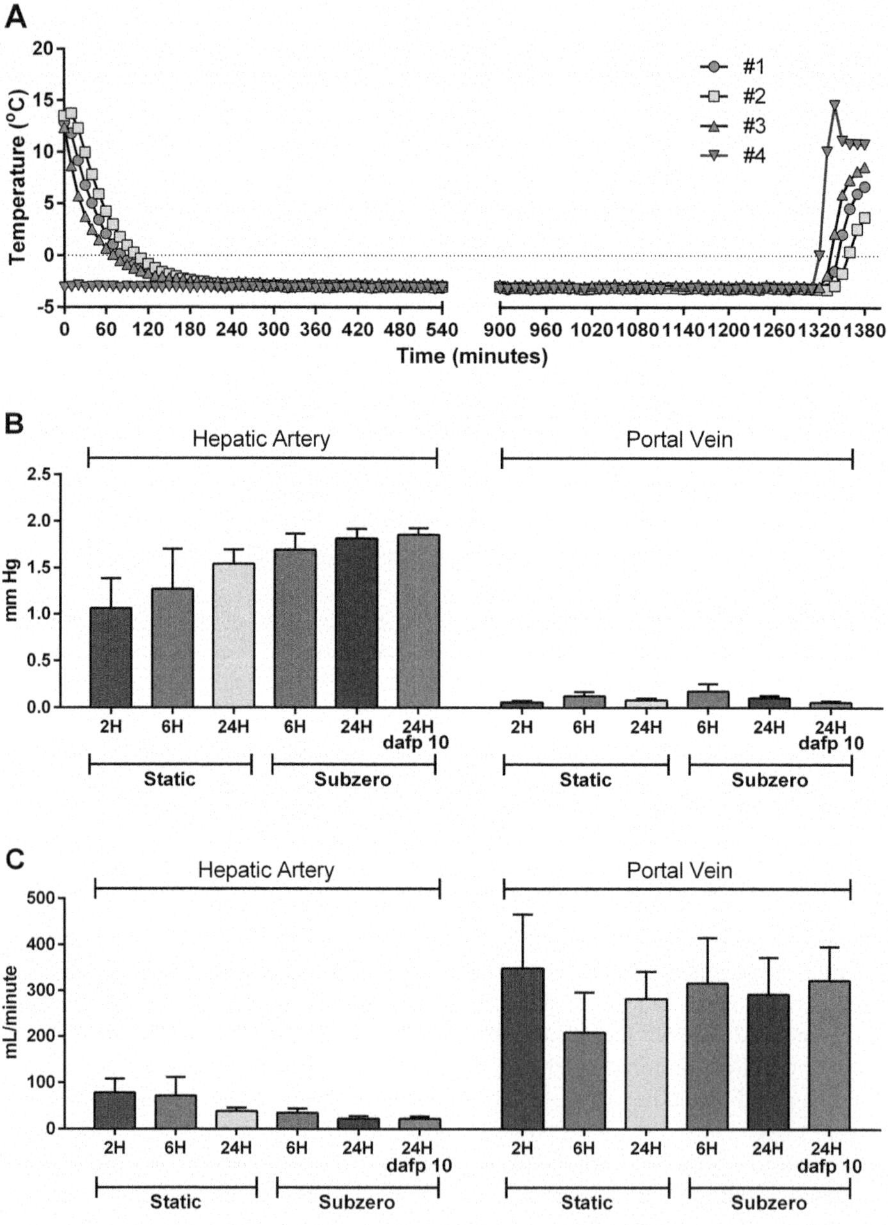

Fig. 11.9 Liver study results. (**a**) Representative liver cooling and storage temperatures observed during an experiment. Thermocouples were placed at four locations in each liver during cooling and storage in an alcohol bath. (**b**) Mean resistance during posttreatment normothermic perfusion post-hypothermic storage, hepatic artery data on the left and portal vein on the right. (**c**) Mean flow rate during posttreatment normothermic perfusion in the hepatic artery on the left and portal vein on the right

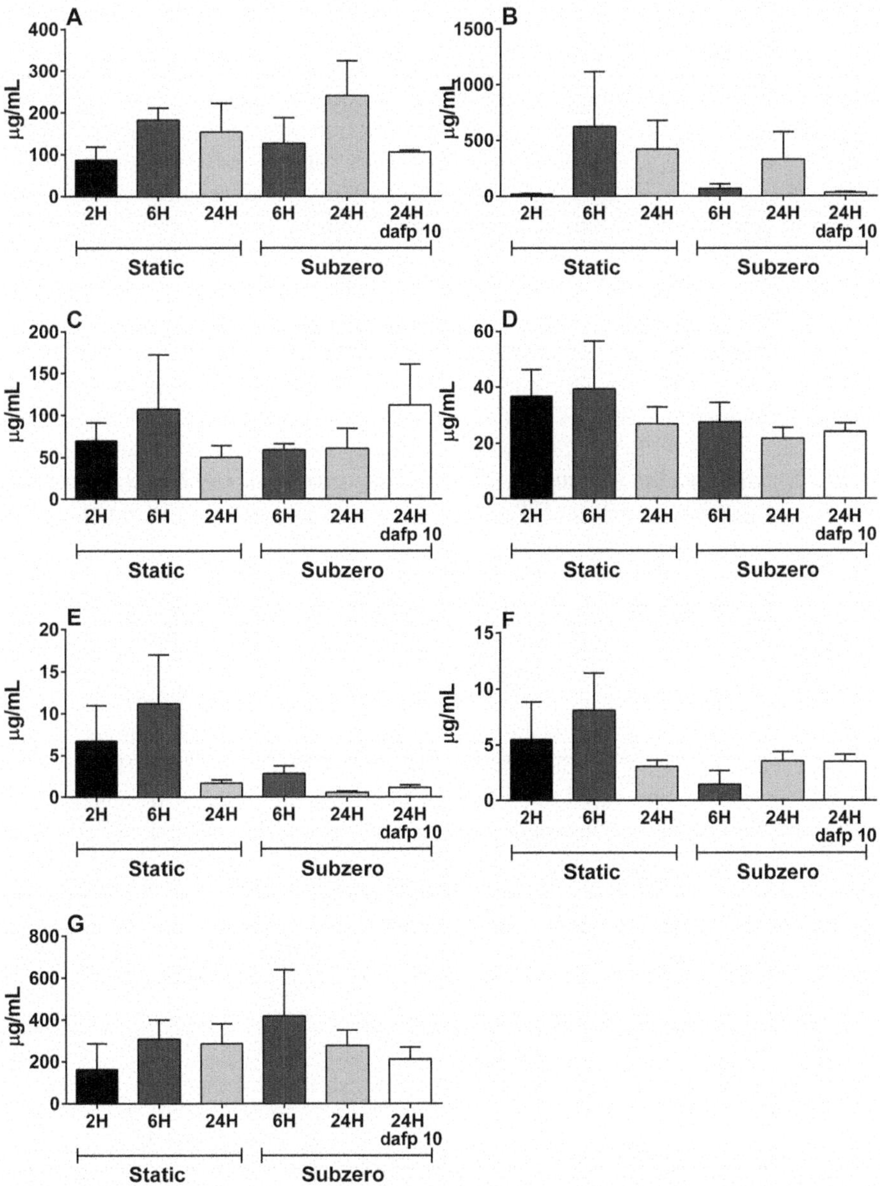

Fig. 11.10 Overview of liver storage results. The results shown in these figures were obtained from the last perfusate sample collected after 3 h of normothermic perfusion after either 4 °C (static) or −2.5 °C (subzero) storage experiments. (**a**) Lactate dehydrogenase, (**b**) alanine aminotransferase, (**c**) albumin, (**d**) interlukin-8, (**e**) tumor necrosis factor-α, (**f**) transforming growth factor-β, and (**g**) β-galactosidase. No statistically significant differences were observed due to the large standard errors observed

11.6 Discussion

The hypothesis that naturally occurring AFPs could be used to limit problems associated with ice formation during cryopreservation was previously proposed by Knight et al. (1984). The utilization of fish AFPs for cryopreserving mammalian cells has since been investigated with varied positive and negative results with respect to cell viability in different cell types (Rubinsky et al. 1991, 1992a, 1994; Carpenter and Hansen 1992; Pegg 1992; Hansen et al. 1993; Arav et al. 1993; Wang et al. 1994; Mugnano et al. 1995; Chao et al. 1996; Younis et al. 1998; Langeaux et al. 1997). Positive studies include cryopreservation of oocytes and two cell embryos of mice and pigs (Rubinsky et al. 1991, 1992b) bovine and ovine embryos at morula/blastocyst stage (Arav et al. 1993), and cryoprotection of chimpanzee sperm (Younis et al. 1998). Conversely, several experiments cast doubts on the cryoprotective effect of AFPs, as no benefit was demonstrated by addition of AFPs during cryopreservation of equine embryos (Langeaux et al. 1997) and human red blood cells in glycerol (Pegg 1992). Carpenter and Hansen (1992) reported that higher AFP type 5 concentrations almost completely inhibited ice recrystallization in extracellular spaces, but destructive ice formed in association with the cells. Another study (Hansen et al. 1993) showed that the addition of type I fish AFP to DMSO for myelogenous leukemia cell preservation significantly decreased cell recovery at all concentrations. In another example, both positive and negative outcomes were reported, cryopreservation of red blood cells using winter flounder type I AFP and hydroxyethyl starch improved viability outcomes at low AFP concentrations but not at high concentrations (Carpenter and Hansen 1992; Chao et al. 1996). However, in the meantime modeling based upon fish antifreeze glycoprotein has resulted in the development of a library containing different classes of molecules that are effective inhibitors of ice recrystallization (Capicciotti et al. 2012; Balcerzak et al. 2013), some of which function as cryoprotectants for the freezing of human red blood cells and human hematopoietic stems cells, and also protect against transient warming injuries (Capicciotti et al. 2015). These synthetic compounds represent novel strategies to potentially improve cell viability during cell, tissue, and organ preservation.

In contrast with insect AFPs, fish AFPs and antifreeze glycoproteins (AFGPs) are known to express a relatively low thermal hysteresis activity (1–2 °C). We have shown that some individual insect-derived AFPs can improve freezing of mammalian cells, DAFP-2 and 6, while others, DAFP-2 and 4, either had little or no impact (Figs. 11.1 and 11.3). Even more interesting is the observation that increasing the complexity of DAFP formulations actually led to loss of benefits associated with DAFPs 2 and 6 and in the case of DAFPs 1, 2, and 6 combined appeared to promote loss of cell viability (Fig. 11.3). Obviously, the correlation of viability enhancement with formulation simplicity has no correlation with formulation ability to induce super cooling (Fig. 11.2). More complex formulations generally promote supercooling more effectively than single DAFP formulations (Fig. 11.2) while single DAFP formulations were the best for preservation of cell viability

(Fig. 11.3). What we have shown here is just the tip of the iceberg for insect AFPs; there are many more AFPs available, plus those not yet discovered, and most have never been evaluated for promotion of cell viability during cryopreservation. We obtained several more from *Dendroides* and *Cajucus* AFPs since performing the cryopreservation work shown here but have not had the opportunity to cryopreserve cells using them. There are also many cryopreservation variables that could be tested instead of the fixed screening method that we have employed including the incorporation of other cryoprotectants such as glycerol and trehalose that are commonly employed by insects as part of Nature's cold survival strategies.

The storage and transport of engineered tissues and organs are a vital and required part of the tissue engineering supply chain. Tissue storage has already become a major barrier for tissue engineering companies working to manufacture skin graft substitutes, blood vessels, and other engineered products (Alvarez 2015). If measures are not taken to develop shelf-lives for engineered tissues and organs, storage constraints are likely to plague regenerative medicine just as the organ shortage limits transplantation today. End-stage organ disease accounts for over 700,000 deaths per year in the United States, dwarfing the number of patients added to transplant waitlists (Report-1 2015), and according to some estimates over 30–35% of all US deaths could be prevented or delayed by an organ transplant and broader tissue engineering (Fahy et al. 2006). In the United States, organ impairment leads to more deaths than cancer, and chronic organ impairment makes up a substantial fraction of these deaths (Report-2 2015). The World Health Organization estimates that organ transplantation meets less than 10% of the global demand (Report-3 2012). The huge need for organ banking has recently been reviewed (Giwa et al. 2017). The development of effective cryopreservation methods for large complex tissues and organs would enable upon demand availability of immunologically matched banked donor-derived tissues and organs.

Positive impacts of fish AFPs during hypothermic nonfrozen storage have been reported, including survival of mammalian cells at 4 °C for 1–3 days (Rubinsky et al. 1991; Lee et al. 1992) and subzero storage rat organs at −1 to −4 °C (Soltys et al. 2001; Amir et al. 2003, 2004). The performance in the hypothermic/high-subzero temperature range relative to cryogenic temperatures observed for fish AFPs is consistent with the function of fish AFPs within cold seawater. We thought it well worth evaluation of insect AFPs for possible benefits during hypothermic storage of tissue and organs at 4 °C or subzero temperatures. Our blood vessel studies demonstrated very different outcomes for veins versus arteries. Better functions were demonstrated for veins compared with controls using both DAFP-10 and CAFP-3 in combination with trehalose and glycerol compared with AFP-free controls at −7 °C. Preservation appeared to be better at −7 °C than at +4 °C. In marked contrast, there was no benefit for arteries and there appeared to be a dose-dependent inhibition of arterial functions (both contraction and relaxation). We then went on to investigate the potential impact of DAFP-10 with trehalose and glycerol on subzero porcine liver storage. The portal vein mean resistance and flow were similar in the positive control and 24 h AFP subzero group (Figs. 11.6 and 11.7) while there was no impact on the hepatic artery (Fig. 11.6). These results, like the blood vessel studies, suggest a positive impact on veins but not arteries. The posttreatment assays employed on samples obtained during ex vivo perfusion under physiological conditions were generally supportive of AFP utilization during subzero storage; however, the livers would need to be transplanted to truly determine effectiveness after varying periods of subzero storage with DAFP 10, trehalose and glycerol. Based upon observations with fish AFPs and insect AFPs we believe it likely that AFPs may have benefits for cell, tissue, and organ preservation at hypothermic, nonfrozen temperatures, but both the published literature and our results are inconclusive, merely suggestive of potential benefits. However, we should note that there have been several reports that simply storing organs or cells at colder temperatures than +4 °C, without freezing, results in better preservation (Hertl et al. 1994; Sakaguchi et al. 1996; Yoshida et al. 1999; Matsuda et al. 1999; Vajdova et al. 2000). However, just as with cryopreservation, there are many AFPs and other experimental variables that need to be studied to find clinically relevant formulations and protocols.

11.7 Conclusions

Our studies demonstrate the complexity of research on utilization of AFPs for preservation of cells, complex tissues, and organ storage. However, the ability to replace organs and tissues on demand from organ banks could save or improve millions of lives each year globally and create public health benefits on par with curing cancer (reviewed, Giwa et al. 2017). Organ transplantation is one of the most impressive medical achievements of the past century. In about 25 years it has added over 2 million life-years to patients in the United States (Rana et al. 2015). However, access to transplantation and its efficacy are still constrained with one in five patients dying while waiting for an organ in the United States and most patients dying without ever being considered for an organ in most of the world (Giwa et al. 2017). In addition to transplantation organ preservation impacts cancer treatment and posttreatment fertility, development of stockpiles of stem cells and tissues for emergency preparedness, limb recovery for transplantation, trauma care, tissue engineering and regenerative medicine. Coordinated federal funding and research initiatives are needed to solve for organ banking and naturally occurring AFPs or synthetic analogues may be an important part of the solution(s).

Acknowledgments We would like to thank Dr. Dina O. Halwani and Ms. Lindsay K. Freeman for their technical support on parts of the unpublished research described in the chapter. Support for this work was obtained from US Public Health Grants 5R44DK081233, R43GM088900 and SCLaunch Matching Grant Agreement #2010-129. The views, opinions, and findings contained in this report are those of the authors and should not be construed as an official National Institutes of Health position, policy, or decision unless so designated by other documentation.

References

Alvarez L (2015) Organ and tissue banking compendium. http://science.dodlive.mil/files/2015/01/Organ-and-Tissue-Banking-Compendium-2015-Jan.pdf. Accessed 3 July

Amir G, Rubinsky B, Kassif Y, Horowitz L, Smolinsky AK, Lavee J (2003) Preservation of myocyte structure and mitochondrial integrity in subzero cryopreservation of mammalian hearts for transplantation using antifreeze proteins—an electron microscopy study. Eur J Cardiothorac Surg 24:292–297

Amir G, Horowitz L, Rubinsky B, Yousif BS, Lavee J, Smolinskyd AK (2004) Subzero nonfreezing cryopreservation of rat hearts using antifreeze protein I and antifreeze protein III. Cryobiology 48:273–282

Arav A, Ramsbottom G, Baguis A, Rubinsky B, Roche JF, Boland MP (1993) Vitrification of bovine and ovine embryos with the MDS technique and antifreeze proteins. Cryobiology 30:621–622

Balcerzak AK, Febbraro M, Ben RN (2013) The importance of hydrophobic moieties in ice recrystallization inhibitors. RSC Adv 3:3232–3236

Brockbank KGM, Song YC, Greene ED, Taylor MJ (2007) Quantitative analyses of vitrified autologous venous arterial bypass graft explants. Cell Preserv Technol 5:68–76

Brockbank KGM, Campbell LH, Greene ED, Brockbank MCG, Duman JG (2011) Lessons learned from nature for preservation of mammalian cells, tissues, and organs. In Vitro Cell Dev Biol Animal 47:210–217

Brockbank KGM, Lee CY, Greene ED, Chen Z, Freeman LK, Campbell LH (2012) *Ex vivo* evaluation of porcine livers post-hypothermic machine perfusion at 4–6°C and 12–14°C. J Regen Med Tissue Eng. www.hoajonline.com/jrmte/1/1/2

Brockbank KGM, Chen Z, Greene ED, Campbell LH (2014) Ice-free cryopreservation by vitrification. Mini review. MOJ Cell Sci Rep 1:27–31. http://medcraveonline.com/MOJCSR/MOJCSR-01-00007.php

Brockbank KGM, Chen Z, Greene ED, Campbell LH (2015) Vitrification of heart valve tissues, Chapter 20. In: Wolkers WF, Oldenhof H (eds) Methods in cryopreservation and freeze-drying. Methods in molecular biology. Springer, Cham. http://www.springer.com/chemistry/biotechnology/book/978-1-4939-2192-8

Campbell LH, Taylor MJ, Brockbank KGM (2003) U.S. Patent #6,596,531. Two stage method for thawing cryopreserved cells

Capicciotti CJLM, Perras FA, Bryce DL, Paulin H, Harden J, Yiu Y, Ben RN (2012) Potent inhibition of ice recrystallization by low molecular weight carbohydrate-based surfactants and hydrogelators. Chem Sci 3:1408–1416

Capicciotti CJ, Kurach JD, Turner TR, Mancini RL, Acker JP, Ben RN (2015) Small molecule ice recrystallization inhibitors enable freezing of human red blood cells with reduced glycerol concentrations. Nat Sci Rep 5:1–10

Carpenter JF, Hansen TN (1992) Antifreeze protein modulates cell survival during cryopreservation: mediation through influence on ice crystal growth. Proc Natl Acad Sci 89:8953–8957

Carpenter JF, Chang BS, Garzon-Rodriguez W, Randolph TW (2002) Rational design of stable lyophilized protein formulations: theory and practice. Pharm Biotechnol 13:109–133

Chao HM, Davies PL, Carpenter JF (1996) Effect of antifreeze proteins on red blood cell survival during cryopreservation. J Exp Biol 199:2071–2076

Cheng CCM, DeVries AL (2002) Origins and evolution of fish antifreeze proteins. In: Ewart KV, Hew CL (eds) Fish antifreeze proteins. World Scientific, New Jersey, pp 83–108

Crowe JH, Crowe LM, Oliver AE, Tsvetkova N, Wolkers W, Tablin F (2001) The trehalose myth revisited: introduction to a symposium on stabilization of cells in the dry state. Cryobiology 43:89–105

Crowe JH, Tablin F, Wolkers WF, Gousset K, Tsvetkova NM, Ricker J (2003) Stabilization of membranes in human platelets freeze-dried with trehalose. Chem Phys Lipids 122:41–52

Davies PL (2014) Ice-binding proteins: a remarkable diversity of structures for stopping and starting ice growth. Trends Biochem Sci 39:548–555

Deban SM, Lappin AK (2011) Thermal effects on the dynamics and motor control of ballistic prey capture in toads: maintaining high performance at low temperature. J Exp Biol 214:1333–1346

DeVries AL (1971) Glycoproteins as biological antifreeze agents in Antarctic fishes. Science 172:1152–1155

Duman JG (2001) Antifreeze and ice nucleator proteins in terrestrial arthropods. Annu Rev Physiol 63:327–357

Duman JG (2002) The inhibition of ice nucleators by insect antifreeze proteins is enhanced by glycerol and citrate. J Com Physiol B 172:163–168

Duman JG (2015) Animal ice-binding (antifreeze) proteins and glycolipids: an overview with emphasis on physiological function. J Exp Biol 218:1846–1855

Duman JG, Olsen TM (1993) Thermal hysteresis activity in bacteria, fungi and primitive plants. Cryobiology 30:322–328

Fahy GM, Wowk B, Wu J (2006) Cryopreservation of complex systems: the missing link in the regenerative medicine supply chain. Rejuvenation Res 9:279–291

Giwa S, Lewis JK, Alvarez L, Langer R, Roth AE, Church GM, Markmann JF, Sachs DH, Chandraker A, Wertheim JA, Rothblatt M, Boyden ES, Eidbo E, Lee WPA, Pomahac B, Brandacher G, Weinstock DM, Elliott G, Nelson D, Acker JP, Uygun K, Schmalz B, Weegman BP, Tocchio A, Fahy GM, Storey KB, Rubinsky B, Bischof J, Elliott JAW, Woodruff TK, Morris GJ, Demirci U, Brockbank KGM, Woods EJ, Ben RN, Baust JG, Gao D, Fuller B, Rabin Y, Kravitz DC, Taylor MJ, Toner M (2017) The promise of organ and tissue preservation to transform medicine. Nat Biotechnol 35:530–542

Halwani DO, Brockbank KGM, Duman JG, Campbell LH (2014) Recombinant Dendroides canadensis antifreeze proteins as potential ingredients in cryopreservation solutions. Cryobiology 68:411–418

Hansen TN, Smith KM, Brockbank KGM (1993) Type I antifreeze protein attenuates cell recoveries following cryopreservation. Transplant Proc 24:3186–3188

Hertl M, Chartrand PB, West DD, Harvey PR, Strasberg SM (1994) The effects of hepatic preservation at 0 degrees C compared to 5 degrees C: influence of antiproteases and periodic flushing. Cryobiology 31:434–440

Huang T, Nicodemus J, Zarka DG, Thomashow MF, Duman JG (2002) Expression of an insect (Dendroides canadensis) antifreeze protein in Arabidopsis thaliana results in a decrease in plant freezing temperature. Plant Mol Biol 50:333–344

Jain NK, Roy I (2010) Trehalose and protein stability. Curr Protoc Protein Sci. Chapter 4:Unit 4.9. https://doi.org/10.1002/0471140864.ps0409s59

Knight CA, DeVries AL, Oolmann LD (1984) Fish antifreeze proteins and the freezing and recrystallization of ice. Nature 308:295–296

Langeaux D, Huhtinen M, Koskinen E, Palmer E (1997) Effect of antifreeze protein on the cooling and freezing of equine embryos as measured by DAPI-staining. Equine Vet J 25:85–87

Lee CY, Rubinsky B, Fletcher GL (1992) Hypothermic preservation of whole mammalian organs with antifreeze proteins. Cryo-Letters 13:59–66

Li N, Andorfer CA, Duman JG (1998) Enhancement of insect antifreeze protein activity by low molecular weight solutes. J Exp Biol 201:2243–2251

Matsuda H, Yagi T, Matsuoka J, Yamamura H, Tanaka N (1999) Subzero nonfreezing storage of isolated rat hepatocytes in University of Wisconsin solution. Transplantation 67:186–191

Mugnano JA, Wang T, Layne JR Jr, De Vries AL, Lee RE (1995) Antifreeze glycoproteins promote lethal intracellular freezing of rat cardiomyocytes at high subzero temperatures. Cryobiology 32:556–557

Nicodemus J, OfTousa JE, Duman JG (2006) Expression of a beetle, Dendroides canadensis, antifreeze protein in Drosophila melanogaster. J Insect Physiol 52:888–896

Olive LC, Meister K, DeVries AL, Duman JG, Guo S, Bakker HJ, Voets IK (2016) Blocking rapid ice crystal growth through non-basal plane adsorption of antifreeze proteins. Proc Natl Acad Sci USA 113:3740–3745

Pegg DE (1992) Antifreeze proteins. Cryobiology 29:774–782
Rana A, Gruessner A, Agopian VG, Khalpey Z, Riaz IB, Kaplan B, Halazun KJ, Busuttil RW, Gruessner RW (2015) Survival benefit of solid-organ transplant in the United States. JAMA Surg 150:252–259
Report-1 (2015) Centers for disease control. http://www.cdc.gov/nchs/data/nvsr/nvsr62/nvsr62_06.pdf. Accessed 12 Jun 2015
Report-2 (2015) The impact of ex-vivo organ perfusion. Perfusix.com. http://www.perfusix.com/impact-of-ex-vivo.html. Accessed 3 Jul 2015
Report-3 (2012) Keeping kidneys. Bull World Health Organ 90(10):713–792
Rozsypal J, Moos M, Simek P, Kostal V (2018) Thermal analysis of ice and glass transitions in insects that do and do not survive freezing. J Exp Biol 221:jeb170464
Rubinsky B, Arav A, Fletcher GL (1991) Hypothermic protection—a fundamental property of "antifreeze" proteins. Biochem Biophys Res Commun 180:566–571
Rubinsky B, Arav A, De Vries AL (1992a) The cryoprotective effect of antifreeze glycopeptides from Antartic fishes. Cryobiology 29:69–79
Rubinsky B, Mattioli M, Arav A, Barboni B, Fletcher GL (1992b) Inhibition of Ca++ and K+ currents by "antifreeze" proteins. Am J Phys 262:R542–R545
Rubinsky B, Arav A, Hong JS, Lee CY (1994) Freezing of mammalian livers with glycerol and antifreeze proteins. Biochem Biophys Res Commun 200:732–741
Sakaguchi H, Kitamura S, Kawachi K, Kobayashi S, Yoshida Y, Niwaya K, Gojo S (1996) Preservation of myocardial function and metabolism at subzero nonfreezing temperature storage of the heart. J Heart Lung Transplant 15:1101–1107
Sformo T, Walters K, Jeannet K, Wowk B, Fahy GM, Barnes BM, Duman JG (2010) Deep supercooling, vitrification and limited survival to -100°C in the Alaskan beetle Cucujus clavipes puniceus larvae. J Exp Biol 213:502–509
Soltys KA, Batta AK, Koneru B (2001) Successful nonfreezing, subzero preservation of rat liver with 2,3-butanediol and type I antifreeze protein. J Surg Res 96:30–34
Song YC, Khirabadi BS, Lightfoot F, Brockbank KGM, Taylor MJ (2000) Vitreous cryopreservation maintains the function of vascular grafts. Nat Biotechnol 18(296–299):2000
Taylor MJ, Song YC, Chen ZZ, Lee F, Brockbank KGM (2004) Interactive determinants for optimized stabilization of autologous vascular grafts during surgery. Cell Preserv Technol 2:198–208
Vajdova K, Smrekova R, Mislanova C, Kukan M, Lutterova M (2000) Cold-preservation-induced sensitivity of rat hepatocyte function to rewarming injury and its prevention by short-term reperfusion. Hepatology 32:289–296
Vu HM, Duman JG (2017) Upper lethal temperatures in three cold-tolerant insects are higher in winter than in summer. J Exp Biol 220:2726–2732
Wang L, Duman JG (2005) Antifreeze proteins of the beetle Dendroides canadensis enhance one another's activities. Biochemistry 44:10305–10312
Wang L, Duman JG (2006) A thaumatin-like protein from larvae of the beetle Dendroides canadensis enhances the activity of antifreeze proteins. Biochemistry 45:1278–1284
Wang T, Zhu Q, Yang X, Layne JR Jr, DeVries AL (1994) Antifreeze glycoproteins from the antarctic notothenoid fishes fail to protect the rat cardiac explant during hypothermic and freezing preservation. Cryobiology 31:185–192
Yoshida K, Matsui Y, Wei T, Kaibori M, Kwon AH, Yamane A, Kamiyama Y (1999) A novel conception for liver preservation at a temperature just above freezing point. J Surg Res 81:216–223
Younis AI, Rooks B, Khan S, Gould KB (1998) The effect of antifreeze peptide III and insulin transferrin selenium (ITS) on cryopreservation of chimpanzee spermatozoa. J Androl 19:207–214

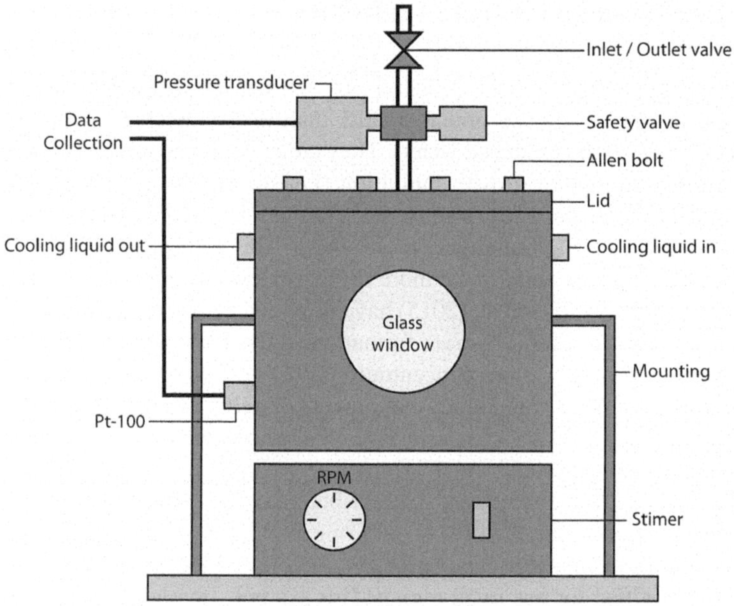

Fig. 12.5 Schematic of an experimental setup for testing hydrate formation. The setup is stirred reactor with a sapphire window. Hydrate formation can be observed visually and can also be detected as a sudden drop in pressure. Vigorous mixing ensures that hydrates are rapidly formed (Jensen et al. 2011)

differential scanning microcalorimeter (DSC). While the last two are quite specialized pieces of equipment, the stirred tank (with variations) is a standard tool for studying many chemical engineering phenomena. The detection of hydrate formation can be done in a number of ways. The simplest is the visual observation of hydrate crystal formation, which can be supported by more sophisticated imaging techniques such as magnetic resonance imaging (Zeng et al. 2006). Typically hydrate detection is based on a sudden drop in pressure, such as is observed in Fig. 12.8 top and bottom. This sudden pressure decrease upon formation of hydrates occurs because hydrate-forming components such as methane are much more densely packed in the hydrate phase than in a gas phase at similar temperature and pressure. A related method is the measurement of gas consumed to form hydrate at constant pressure. For example, Jensen et al. (2011) fitted the cell shown in Fig. 12.5 with a pressure controller and flowmeter to measure gas consumed in hydrate formation, the principle being that when hydrates are formed initially, the pressure drops slightly. This drop in pressure causes the controller to allow more gas into the cell to maintain the pressure as hydrates form. This gas flow is then measured and the amount of hydrate formed can then readily be calculated. The same pressure principle is used in the rocking cell of Figs. 12.6 and 12.8 shows examples of the kinds of results obtained with this setup. In the top part of the figure, a natural gas and water mixture is pressurized to 80 bar and the system is cooled at a constant rate.

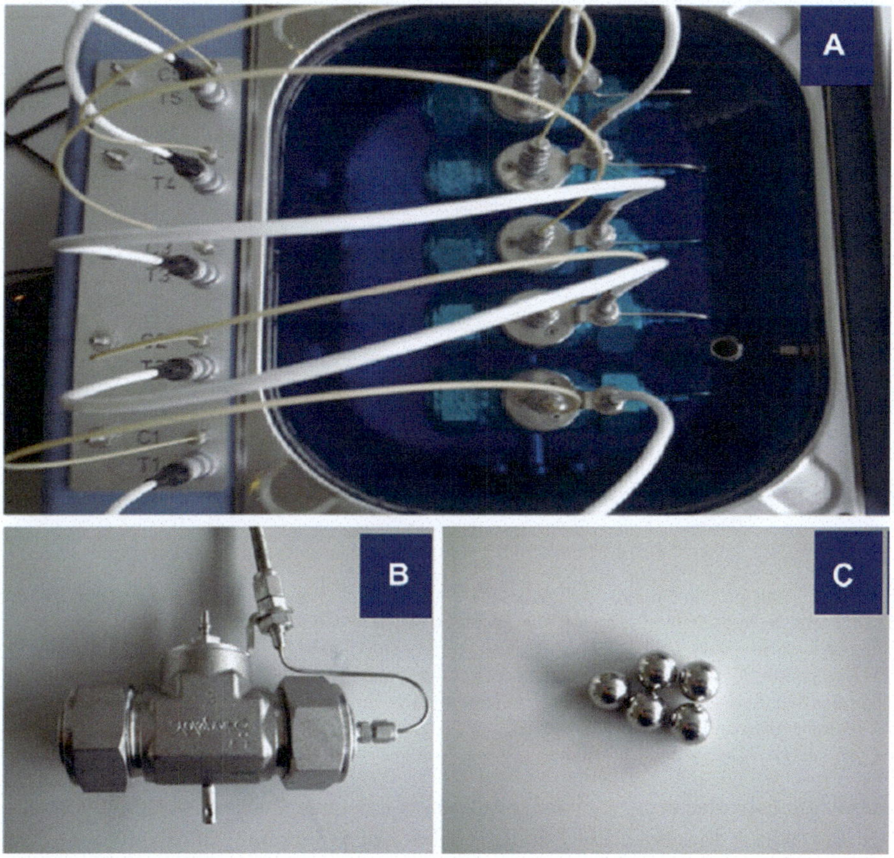

Fig. 12.6 Rocking cell (PSL Systemtechnik). (**a**) The rocking cell setup consisting of the cooling bath with the five cells and the high-pressure panel. (**b**) One of the test cells (Daraboina et al. 2013a). The five chambers each contain a steel ball bearing (**c**) which moves back and forth while the cells rock in a bath at a controlled temperature. Hydrate formation is seen as a sudden drop in pressure

While under agitation the cells are rocked back and forth and a ball bearing placed in the cell provides mixing. The lower part of the figure shows a constant temperature run where the inhibitor Luvicap was added to the liquid. The high-pressure micro-calorimeter in Fig. 12.7 reveals the formation and decomposition of hydrates as a heat signal, since hydrate formation is an exothermic process. An example of a typical experiment is shown in Fig. 12.9. Ice or hydrate formation is identified by the exothermic peaks (A) during the cooling cycle of a temperature ramping experiment. Subsequently, endothermic peaks (B) are observed as first the ice and then the hydrate melts. The formation process is stochastic—freezing/hydrate formation occurs as a number of events at different times. The amount of ice and hydrate formed is only unequivocally revealed upon melting—the larger ice-melting peak at 0 °C and the smaller hydrate melting peak at around 13 °C. This difficulty in

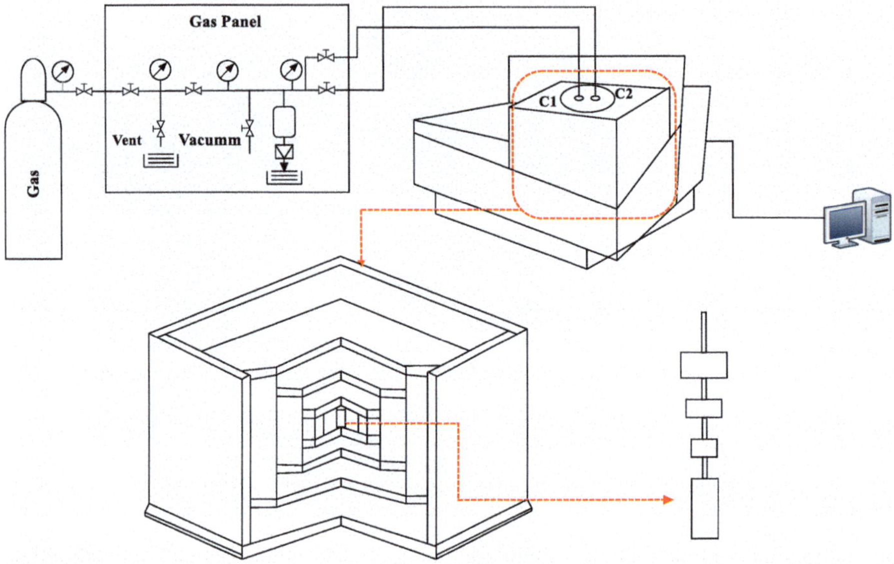

Fig. 12.7 High-pressure microcalorimeter (Setaram instrumentation). The calorimeter measures phase changes (hydrate/ice formation and decomposition) as a heat signal. It has the advantage of being able to use very small samples (micrograms). This particular unit can also operate up to 150 bar so conditions match closely the production of natural gas. The drawback to this type of method is that there is no stirring, so even without hydrate inhibitor present it can be difficult to form gas hydrates (Mu and von Solms 2018)

unambiguously observing hydrate formation is a definite drawback of the microcalorimeter, which does not allow for stirring. Lack of agitation means that hydrates are not readily formed, even for systems without inhibitor, making an assessment of the efficacy of an inhibitor difficult. This disadvantage was mitigated to some extent in Ohno et al. (2010) and Daraboina et al. (2015a) where the liquid sample was placed in silica gel inside tiny capillary tubes, providing a greater liquid surface area, which promoted hydrate formation. Despite this drawback the high-pressure microcalorimeter is an important tool where very small samples (in this case 1 μL) are required, such as for antifreeze proteins where only a limited amount of material may be available for testing. A common characteristic for all three setups shown in Figs. 12.5, 12.6 and 12.7 is that they can be operated at elevated pressures—in the cases presented here up to 200 bar. In this way the experiments more nearly match conditions under which oil and gas are produced. Tests with actual crude oil have also been done (Daraboina et al. 2015b, c) for synthetic inhibitors but not yet for AFPs. Experiments can also be done at atmospheric pressure (for example, in a beaker) using a hydrate promoter such as tetrahydrofuran (THF). THF solutions form hydrates near atmospheric pressure at temperatures above the freezing point of water, so this is a convenient system to study (Zeng et al. 2003), although ideally one would like to match the relevant industrial conditions of interest (temperature and pressure) as closely as possible. The detection methods for hydrate formation have

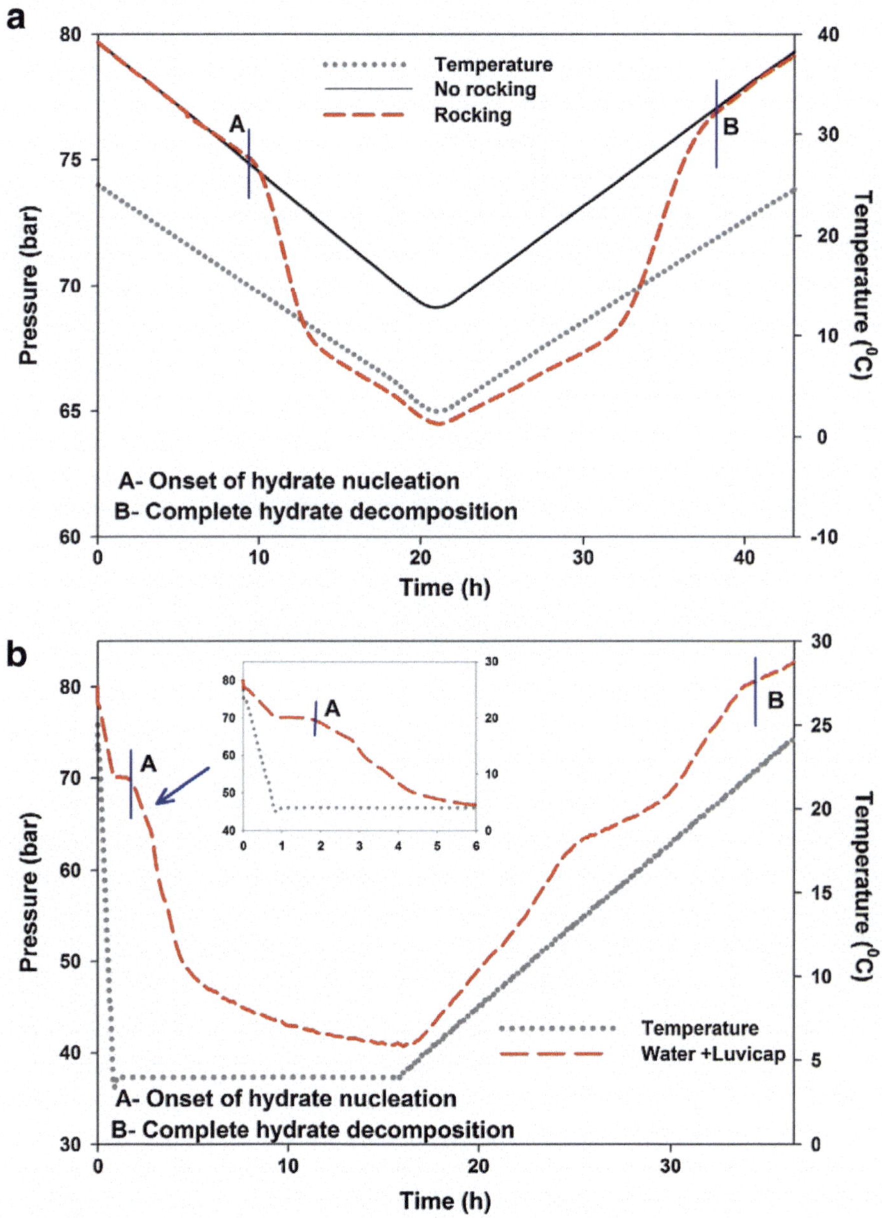

Fig. 12.8 Types of hydrate tests that can be performed. A temperature ramping (constant cooling and heating rate) with water and natural gas only (no inhibitor) (top) and a constant temperature experiment with water, natural gas, and the commercial inhibitor Luvicap (bottom) (Daraboina et al. 2013a)

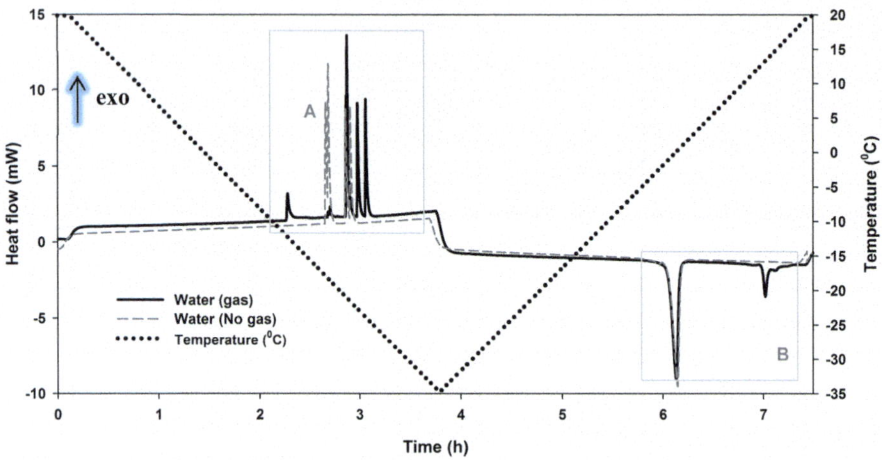

Fig. 12.9 A typical heat flow and temperature ramp during experiments with water (gray dashed line) and methane and water (solid black line) at 100 bar. Hydrate and ice formation are exothermic processes; ice/hydrate formation is identified by the exothermic peaks (A) during the cooling cycle of a ramping experiment. Subsequently, endothermic peaks (B) are observed as first the ice and then the hydrate melts. The formation process is stochastic—freezing/hydrate formation occurs as a number of events at different times. The amount of ice and hydrate formed is only unequivocally revealed upon melting—the larger ice-melting peak at 0 °C and the smaller hydrate melting peak at around 13 °C

been described qualitatively above. Naturally it is desirable to be able to assess and compare inhibitors quantitatively. Two types of hydrate inhibition experiments are usually performed—temperature ramping or constant temperature. Figure 12.8 illustrates both methods (both experiments carried out in a rocking cell): the top part of the figure shows a temperature ramping (also called a constant cooling rate) experiment. As the temperature decreases hydrates form, at point A. The temperature of hydrate formation is noted as the onset temperature for hydrate formation (in this case around 13 °C). For these kinds of experiments the onset temperature is then the metric for comparison of inhibitors. The lower part of the figure shows a constant temperature experiment (also known as a constant subcooling experiment since the temperature is fixed at a constant value below the equilibrium temperature for hydrate formation). In the figure the temperature is rapidly reduced to 4 °C and hydrate nucleation occurs after about an hour. In this kind of experiment the metric for inhibitor efficacy is therefore nucleation time. As a rule temperature ramping experiments allow more rapid screening of candidate inhibitors, although one should be aware that constant temperature and time to nucleation are the parameters that more closely match what is happening in a pipeline (or what the operator is trying to avoid happening).

12.5 Antifreeze Proteins as Gas Hydrate Inhibitors

The initial research on antifreeze proteins as hydrate inhibitors (and indeed the history of low-dosage hydrate inhibitors in general) is compellingly described by Kelland (2006). According to Kelland (2006), the company Shell pursued a research program into low-dosage hydrate inhibitors inspired by the function and mechanisms of antifreeze proteins (for example, the work of Yeh and Feeney 1996) which led to the discovery of the polymer PVP (Fig. 12.4) independently of the Sloan group (Sloan 1995a, b; Sloan et al. 1997). While PVP and similar compounds such as PVCap are effective low-dosage inhibitors, these synthetic polymers are not biodegradable to an extent which is acceptable for their use, for example, they are prohibited in the North Sea. For this reason, more environmentally friendly compounds are being sought. Antifreeze proteins are examples of such compounds—either for direct application or in order to discover their structures and mechanisms which could inspire the discovery of effective, environmentally friendly, synthetic compounds. We know for example that the action of antifreeze proteins is different from PVP. For example, it was found that while synthetic kinetic inhibitors inhibit the formation of hydrate initially, once the hydrate does form, it is in fact stabilized by the inhibitor, meaning that the hydrate remains stable above its equilibrium melting point for some time (Sharifi et al. 2016). On the other hand, hydrates formed with antifreeze proteins present are readily melted as soon as their equilibrium dissociation temperature is reached (Sharifi et al. 2016; Daraboina et al. 2015a). Such behavior may make antifreeze proteins a more attractive option for hydrate remediation (removal of a hydrate plug once formed). The state-of-the-art as of 2015 is described and organized in the review article by Walker et al (2015). Table 12.1 summarizes the various antifreeze proteins studied for gas hydrate inhibition.

Zeng et al. (2003) performed the first study of the effects of antifreeze proteins on hydrate inhibition. They considered the systems water–THF and water–propane and found that proteins from winter flounder and spruce budworm inhibited hydrate formation (longer induction times) and changed the morphology of the hydrate crystals formed, compared with a protein without antifreeze activity. After a gap of 3 years where no new literature appeared on the topic, Zeng et al. (2006) extended their 2003 study to methane hydrate inhibition with winter flounder AFP. They introduced magnetic resonance imaging as a tool to study growth in multiple droplets. Importantly, they also reported the need for several samples/experiments in order to obtain statistically meaningful results. This stochasticity of hydrate formation is a continual challenge to research on hydrate formation and inhibition, partially resolved by setups such as the rocking cell (Fig. 12.6) where five experiments can be performed simultaneously. Ke and Kelland (2016) report results performed on a similar rig with 20 sapphire cells.

Al-Adel et al. (2008) performed experiments with winter flounder AFP on methane hydrate inhibition and found that the AFP performed better than a PVP/PVCap copolymer as an inhibitor, but that the AFP promoted hydrate growth in a certain range of temperature and pressure. Bruusgaard et al. (2009) found that

Table 12.1 Summary of gas hydrates experiments using antifreeze proteins

Antifreeze protein	Specie	Hydrate former	Summary	References
Type I, winter flounder	*Pleuronectes americanus*	Tetrahydrofuran (THF); propane	Morphology change, induction time increase	Zeng et al. (2003)
Spruce budworm	*Choristoneura fumiferana*			
Type I, winter flounder		Methane; propane	Nucleation and growth inhibition; magnetic resonance imaging	Zeng et al. (2006)
Type I, winter flounder		Methane	Nucleation inhibition, growth promotion	Al-Adel et al. (2008)
Type I		Methane	Morphology changes in drops	Bruusgaard et al. (2009)
Perennial rye grass	*Lolium perenne*	Methane–ethane–propane mixture	High-pressure differential scanning calorimetry (HP-DSC)	Ohno et al. (2010)
Mealworm	*Tenebrio molitor*			
Ocean pout (type III)	*Macrozoarces americanus*		Gas uptake in a pressurized vessel	
Perennial rye grass	*Lolium perenne*	Methane–ethane–propane mixture	Single crystal studies	Gordienko et al. (2010)
Ocean pout (type III)		THF	Gas uptake in a pressurized vessel	
Type III HPLC12 (Ocean pout)		Methane	Pressurized stirred vessel	Jensen et al. (2010a, b)
Mealworm	*Tenebrio molitor*	Natural gas		Jensen et al. (2011)
Type I AFP		Methane–ethane–propane mixture	Pressurized stirred vessel HP-DSC	Daraboina et al. (2011a, b, c, 2013c)
Type III AFP			High pressure powder X-ray diffraction, NMR, Raman	Ohno et al. (2012)
Type I AFP		Natural gas in saline solution, heptane	HP-DSC	Sharifi et al. (2014a, b, c, d)
Type III AFP			Pressurized stirred vessels	
Bark beetle	*Rhagium mordax*	Methane	Rocking cell RC-5	Perfeldt et al. (2014)
			HP-DSC	Daraboina et al. (2015a)

(continued)

Table 12.1 (continued)

Antifreeze protein	Specie	Hydrate former	Summary	References
Type III AFP		Carbon dioxide	Hydrate slurry loop	Zhou and Ferreira (2017)
"Maxi" fish AFP			Stirred tank morphology	Udegbunam et al. (2017)
Mealworm	*Tenebrio molitor*			
Perennial rye grass	*Lolium perenne*			

the morphology and translucency of hydrates were affected by kinetic inhibitors including type I AFPs.

Ohno et al. (2010) performed a systematic study including four different AFPs or AFP-GFPs (AFPs with a green fluorescent protein tag) based on perennial rye grass, mealworm, and ocean pout (type III). Experiments were performed both on a high-pressure DSC to observe nucleation temperatures as well as on a high-pressure vessel. A synthetic natural gas composed of methane, ethane, and propane was used. Both setups confirmed the efficacy of the proteins compared with PVP. Gordienko et al. (2010) confirmed this conclusion and extended the study of Ohno et al. (2010) to single-crystal experiments on THF.

Jensen et al. (2010a, b, 2011) used proteins from ocean pout (type III) and from mealworm to confirm inhibition and growth inhibition in structure I hydrates (methane) and structure II hydrates (natural gas), also in the presence of heptane as a model for light crude oil. A pressurized stirred tank was used in these experiments (Fig. 12.5).

In a series of papers Daraboina et al. (2011a, b, c, 2013c) and Ohno et al. (2012) studied the effect of type I and type III AFPs on inhibition of gas hydrate formation from a synthetic natural gas (methane, ethane, propane) mixture and compared results with synthetic inhibitors. In these studies, stirred tank and high-pressure microcalorimeter (HP-DSC) experiments were performed. They also probed structural and compositional differences between hydrates formed from synthetic inhibitors and from the AFP using X-ray diffraction, NMR, Raman spectroscopy, and gas chromatography. They observed that hydrates formed from the AFPs melted more readily than the synthetic inhibitors, possibly due to a lower cage occupancy for the AFP compared with the synthetic inhibitors. Cage occupancy refers to the fraction of gas molecules occupying hydrate cages compared to "full" occupancy. Referring to Fig. 12.1, one could imagine that every cage contains a gas molecule. In fact it is not necessary for every cage to be filled in order to stabilize the hydrate structure. Jensen et al. (2010a, b) report cage occupancies as low as 86% based on molecular simulations. NMR spectroscopy of hydrates formed in the presence of AFP confirmed the presence of both sI and sI hydrates initially although there was a conversion with time to the more stable sII.

In an attempt to more closely match conditions in real oil and gas processing Sharifi et al. (2014a, b, c, d) conducted several inhibition studies with saline solution and in the presence of heptane as a model for a light crude. As expected, the addition of heptane increased induction time and reduced growth (the heptane selectively dissolves heavier hydrocarbons in the gas, rendering the gas more methane-rich and thus increasing the pressure for hydrate formation at a fixed temperature—see Fig. 12.2 for an example of this). An important conclusion of this work is that the effect of salts may reduce the efficacy of some AFPs.

Perfeldt et al. (2014) and Daraboina et al. (2015a) investigated the inhibition effect of the hyperactive insect AFP, RmAFP1 from the bark beetle *Rhagium mordax*. The AFP was found to be a nucleation and growth inhibitor comparable to PVP. Perfeldt et al. (2015) found that hydrophobic coatings on the walls of a stirred vessel inhibited hydrate growth in the presence of 0.02 wt% of type I AFP. For an uncoated vessel 0.2 wt% AFP was needed to achieve the same growth inhibition (Fig. 12.10).

Zhou and Ferreira (2017) found that the addition of a type III AFP reduced the formation rate of CO_2 hydrate in a hydrate slurry loop setup. Very recently Udegbunam et al. (2017) reported a so-called three-in-one setup for studying hydrate inhibition with antifreeze proteins. This setup can reportedly study phase equilibrium, kinetics, and morphology of hydrates simultaneously.

Also recently *Rhagium mordax* protein was produced using a common fruit fly-based insect cell line (*Drosophila melanogaster* Schneider S2) by Expres^{2}ion Biotechnologies (Hørsholm, Denmark). Initial testing on a microcalorimeter with water and natural gas suggested that the protein produced in this way is as active as the native protein (Fig. 12.11).

Some authors (Sa et al. 2013, 2015; Perfeldt et al. 2014; Bhattacharjee et al. 2016) have studied the effects of natural amino acids on hydrate formation, possibly with a view to developing a new class of synthetic inhibitors. However, the inhibiting effect of amino acids seems to be small and certain amino acids may even promote hydrate

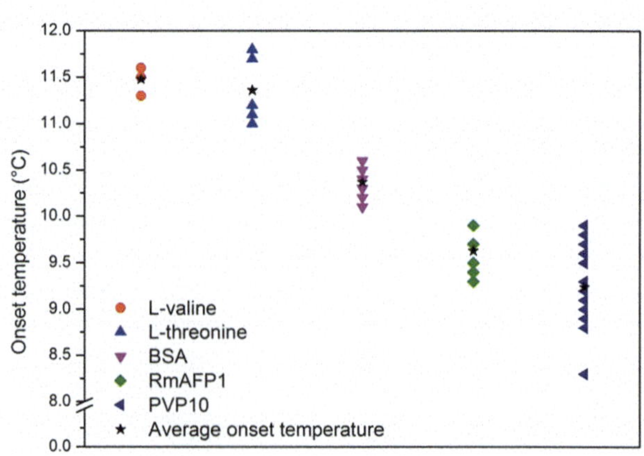

Fig. 12.10 Onset hydrate nucleation temperatures at a concentration of 2770 ppm for all inhibitors. Methane gas and water were pressurized to 95 bar in a rocking cell apparatus. The *x*-axis merely represents a ranking of the various substances tested for their inhibition effect. The data in the figure are from Perfeldt et al. (2014)

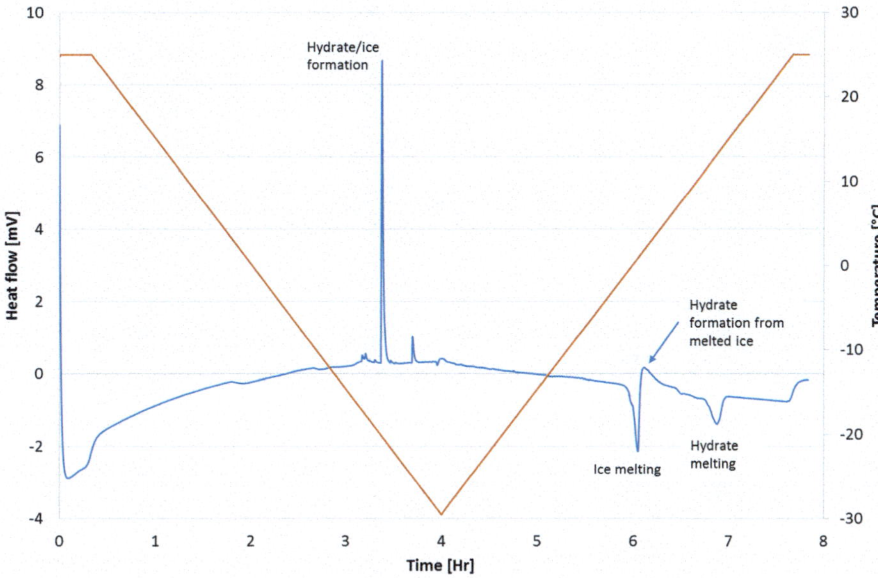

Fig. 12.11 Constant cooling rate test on a high pressure microcalorimeter. The ratio of ice to hydrate formed is higher in the case of RmAFP than for PVP and much higher than for water. This indicates that the AFP is a better hydrate inhibitor since the hydrates form at higher temperatures (12 °C) than ice, so hydrate formation is limited as the system cools

formation (Bhattacharjee et al. 2016). It is likely that simple amino acids are too simplistic a model to explain the inhibiting effects of AFPs.

In addition to the experiments discussed above, molecular simulation studies have also been performed in order to explain or discover supposed mechanisms of hydrate inhibition (Bhattacharjee et al. 2016; Bagherzadeh et al. 2015; Sun et al. 2015; Zhao et al. 2014; Zhang et al. 2012; Nada 2009). Bagherzadeh et al. (2015) hypothesized (based on molecular dynamics simulation) that winter flounder AFP adsorbs onto the methane hydrate surface via cooperative binding of a set of hydrophobic methyl pendant groups to the empty half-cages at the hydrate/water interface. This so-called binding set is composed of the methyl side chain of threonine and two alanine residues in the sequence of the protein (Fig. 12.12). Sun et al. (2015) used molecular modeling to support the hypothesis that AFPs, like KHIs, act by absorbing to a growing hydrate surface (absorbance inhibition).

12.6 Conclusions

Initial testing on idealized systems (pure water, single gases, synthetic natural gases) has shown that antifreeze proteins have potential as gas hydrate inhibitors. Research directions which could help take the next step toward possible use of antifreeze proteins in oil and gas applications or commercialization of the technology include:

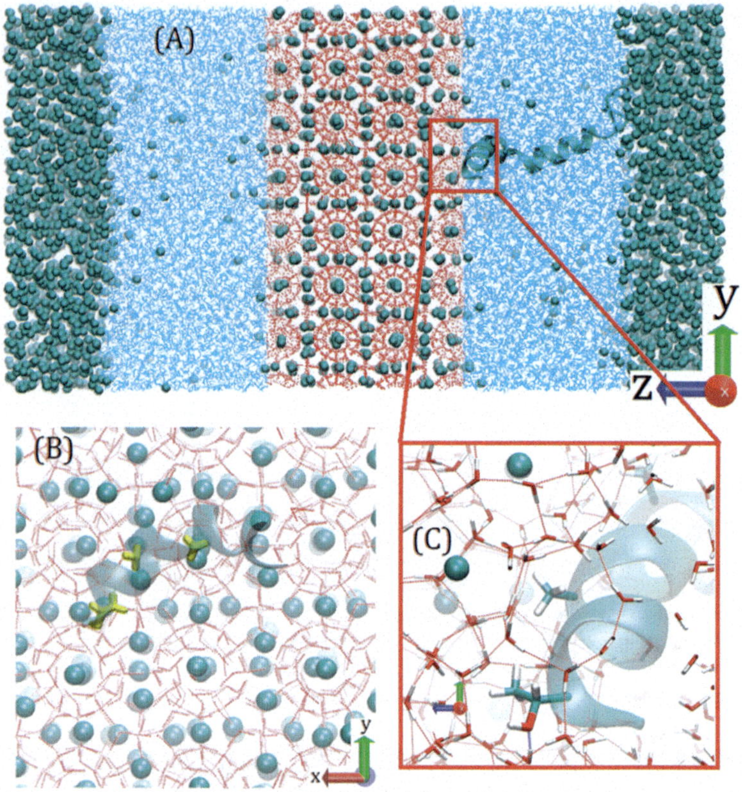

Fig. 12.12 Simulation of hydrate inhibition by antifreeze proteins (**a**). Binding residues are shown in yellow in panel (**b**) and as thick bars in panel (**c**). Liquid water is not shown in panel (**b**) for clarity. The entrapment of binding residues in the empty half-cages is evident in panel (**c**). Hydrogen bonds are shown as red dashed lines. Figure from Bagherzadeh et al. (2015) Other authors who have considered various aspects of hydrate inhibition with antifreeze proteins include: "Booker et al. (2011), Daraboina et al. (2013b), Del Villano et al. (2008), Frostman et al. (2003), Goodwin et al. (2015), Gulbrandsen and Svartaas (2017), Ivall et al. (2015), Koh (2002), Kumar et al. (2015), Lauersen et al. (2013), Lee and Englezos (2005, 2006), Lee et al. (2007, 2016), Marshall et al. (1964), McLeod and Campbell (1961), Ng and Robinson (1983, 1984), Ota et al. (2009), Ripmeester and Alavi (2016), Roberts et al. (1941), Saikia and Mahto (2016), Sharifi and Englezos (2015), Stoporev et al. (2016), Sun et al. (2012), Sundramoorthy et al. (2015), Talaghat et al. (2009), Tonelli et al. (2015), Townson et al. (2012) and Wilson et al. (2014)"

- Testing of antifreeze proteins on realistic reservoir fluids, such as actual live crude oils (with and without production chemicals such as corrosion inhibitors and so on)
- Production of antifreeze proteins on larger scale at lower cost
- Deeper understanding of the mechanism of antifreeze proteins for hydrate inhibition in order to inspire the production of synthetic (but biodegradable) polymers

As a best case scenario one could imagine inexpensive, large-scale production of antifreeze proteins resulting in robust formulations applied as environmentally friendly gas hydrate inhibitors. Ideally produced water would further be treated to recover the proteins, further bringing down the cost of their use.

References

Al-Adel S, Dick JAG, El-Ghafari R, Servio P (2008) The effect of biological and polymeric inhibitors on methane gas hydrate growth kinetics. Fluid Phase Equilib 267:92–98

Anderson FE, Prausnitz JM (1986) Inhibition of gas hydrates by methanol. AIChE J 32:1321–1333

Bagherzadeh SA, Alavi S, Ripmeester JA, Englezos P (2015) Why ice-binding type I antifreeze protein acts as a gas hydrate crystal inhibitor. Phys Chem Chem Phys 17:9984–9990

Bhattacharjee G, Choudhary N, Kumar A, Chakrabarty S, Kumar R (2016) Effect of the amino acid L-histidine on methane hydrate growth kinetics. J Nat Gas Sci Eng 35:1453–1462

Bishnoi PR, Natarajan V (1996) Formation and decomposition of gas hydrates. Fluid Phase Equilib 117:168–177

Booker RD, Koh CA, Sloan ED, Sum AK, Shalaev E, Singh SK (2011) Xenon hydrate dissociation measurements with model protein systems. J Phys Chem B 115:10270–10276

Bruusgaard H, Lessard LD, Servio P (2009) Morphology study of structure I methane hydrate formation and decomposition of water droplets in the presence of biological and polymeric kinetic inhibitors. Cryst Growth Des 9:3014–3023

Daraboina N, Ripmeester J, Walker VK, Englezos P (2011a) Natural gas hydrate formation and decomposition in the presence of kinetic inhibitors. 1. High pressure calorimetry. Energy Fuels 25:4392–4397

Daraboina N, Linga P, Ripmeester J, Walker VK, Englezos P (2011b) Natural gas hydrate formation and decomposition in the presence of kinetic inhibitors. 2. Stirred reactor experiment. Energy Fuels 25:4384–4391

Daraboina N, Ripmeester J, Walker VK, Englezos P (2011c) Natural gas hydrate formation and decomposition in the presence of kinetic inhibitors. 3. Structural and compositional changes. Energy Fuels 25:4398–4404

Daraboina N, Malmos C, von Solms N (2013a) Synergistic kinetic inhibition of natural gas hydrate formation. Fuel 108:749–757

Daraboina N, Malmos C, von Solms N (2013b) Investigation of kinetic hydrate inhibition using a high pressure micro differential scanning calorimeter. Energy Fuel 27:5779–5786

Daraboina N, Moudrakovski I, Ripmeester JA, Walker VK, Englezos P (2013c) Assessing the performance of commercial and biological gas hydrate inhibitors using nuclear magnetic resonance microscopy and a stirred autoclave. Fuel 105:630–635

Daraboina N, Perfeldt CM, von Solms N (2015a) Testing antifreeze protein from the longhorn beetle *Rhagium mordax* as a kinetic gas hydrate inhibitor using a high-pressure micro differential scanning calorimeter. Can J Chem 93:1025–1030

Daraboina N, Pachitsas S, von Solms N (2015b) Experimental validation of kinetic inhibitor strength on natural gas hydrate nucleation. Fuel 139:554–560

Daraboina N, Pachitsas S, von Solms N (2015c) Natural gas hydrate formation and inhibition in gas/crude oil/aqueous systems. Fuel 148:186–190

Del Villano L, Kommedal R, Kelland MA (2008) Class of kinetic hydrate inhibitors with good biodegradability. Energy Fuel 22:3143–3149

Frostman L, Thieu V, Crosby D, Downs H (2003) Low-dosage hydrate inhibitors (LDHIs): reducing costs in existing systems and designing for the future. International symposium on oilfield chemistry 2003, Houston, TX, SPE 80269

Goodwin MJ, Musa OM, Steed JW (2015) Problems associated with sour gas in the oilfield industry and their solutions. Energy Fuel 29:4667–4682
Gordienko R, Ohno H, Singh VK, Jia ZC, Ripmeester JA, Walker VK (2010) Towards a green hydrate inhibitor: imaging antifreeze proteins on clathrates. PLoS One 5:e8953
Gulbrandsen AC, Svartaas TM (2017) Influence on hydrate dissociation for methane hydrates formed in the presence of polyvinylcaprolactam versus polyvinylcaprolactam plus butyl glycol ether. Energy Fuel 31:6352–6357
Ivall J, Pasieka J, Posteraro D, Servio P (2015) Profiling the concentration of the kinetic inhibitor polyvinylpyrrolidone throughout the methane hydrate formation process. Energy Fuel 29:2329–2335
Jensen L, Thomsen K, von Solms N, Wierzchowski S, Walsh MR, Koh CA, Sloan ED, Wu DT, Sum AK (2010a) Calculation of liquid water-hydrate-methane vapor phase equilibria from molecular simulations. J Phys Chem B 114:5775–5782
Jensen L, Ramlov H, Thomsen K, von Solms N (2010b) Inhibition of methane hydrate formation by ice-structuring proteins. Ind Eng Chem Res 49:1486–1492
Jensen L, Thomsen K, von Solms N (2011) Inhibition of Structure I and II Gas Hydrates using Synthetic and Biological Kinetic Inhibitors. Energy Fuel 25:17–23
Kamal MS, Hussein IA, Sultan SA, von Solms N (2016) Application of various water soluble polymers in gas hydrate inhibition. Renew Sust Energ Rev 60:206–225
Ke W, Kelland MA (2016) Kinetic hydrate inhibitor studies for gas hydrate systems: a review of experimental equipment and test methods. Energy Fuel 30:10015–10028
Kelland MA (2006) History of the development of low dosage hydrate inhibitors. Energy Fuel 20:825–847
Koh CA (2002) Towards a fundamental understanding of natural gas hydrates. Chem Soc Rev 31:157–167
Kumar A, Sakpal T, Kumar R (2015) Influence of low-dosage hydrate inhibitors on methane clathrate hydrate formation and dissociation kinetics. Energy Technol 3:717–725
Lauersen KJ, Vanderveer TL, Berger H, Kaluza I, Mussgnug JH, Walker VK, Kruse O (2013) Ice recrystallization inhibition mediated by a nuclear-expressed and -secreted recombinant ice-binding protein in the microalga *Chlamydomonas reinhardtii*. Appl Microbiol Biotechnol 97:9763–9772
Lee JD, Englezos P (2005) Enhancement of the performance of gas hydrate kinetic inhibitors with polyethylene oxide. Chem Eng Sci 60:5323–5330
Lee JD, Englezos P (2006) Unusual kinetic inhibitor effects on gas hydrate formation. Chem Eng Sci 61:1368–1376
Lee JD, Wu HJ, Englezos P (2007) Cationic starches as gas hydrate kinetic inhibitors. Chem Eng Sci 62:6548–6555
Lee W, Shin JY, Kim KS, Kang SP (2016) Synergetic effect of ionic liquids on the kinetic inhibition performance of poly(N-vinylcaprolactam) for natural gas hydrate formation. Energy Fuel 30:9162–9169
Long J, Lederhos J, Sum A, Christiansen RL, Sloan ED (1994) Kinetic inhibitors of natural gas hydrates. In: Proceedings of the 73rd Annual Gas Processors Association (GPA) Convention, New Orleans, LA, March 7–9
Marshall DR, Saito S, Kobayashi R (1964) Hydrates at high pressures. 1: methane-water, argon-water, and nitrogen-water systems. AIChE J 10:202
McLeod HO, Campbell JM (1961) Natural gas hydrates at pressures to 10,000 psia. Trans Soc Petrol Eng AIME 222:590–594
Mu L, von Solms N (2018) Hydrate thermal dissociation behavior and dissociation enthalpies in methane-carbon dioxide swapping process. J Chem Thermodyn 117:33–42
Nada H (2009) Anisotropy in growth kinetics of tetrahydrofuran clathrate hydrate: a molecular dynamics study. J Phys Chem B 113:4790–4798
Ng HJ, Robinson DB (1983) Equilibrium phase composition and hydrating conditions in systems containing methanol, light hydrocarbons, carbon dioxide, and hydrogen sulfide. Research

report. RR-74, Joint research report for GPA and Canadian Gas Processors Association, Tulsa, OK
Ng HJ, Robinson DB (1984) The Influence of methanol on hydrate formation at low temperatures. Research report RR-74, Gas Processors Association, Tulsa, OK
Ohno H, Susilo R, Gordienko R, Ripmeester J, Walker VK (2010) Interaction of antifreeze proteins with hydrocarbon hydrates. Chemistry 16:10409–10417
Ohno H, Moudrakovski I, Gordienko R, Ripmeester J, Walker VK (2012) Structures of hydrocarbon hydrates during formation with and without inhibitors. J Phys Chem A 116:1337–1343
Ota M, Qi YX, Murakami K, Ferdows M (2009) Effects of additive solutions on gas hydrate formation. Proc Int Conf Power Eng 1:305–310
Perfeldt CM, Chua PC, Daraboina N, Friis D, Kristiansen E, Ramløv H, Woodley JM, Kelland MA, von Solms N (2014) Inhibition of gas hydrate nucleation and growth: efficacy of an antifreeze protein from the longhorn beetle *Rhagium mordax*. Energy Fuel 28:3666–3672
Perfeldt CM, Sharifi H, von Solms N, Englezos P (2015) Oil and gas pipelines with hydrophobic surfaces better equipped to deal with gas hydrate flow assurance. J Nat Gas Sci Eng 27:852–861
Perrin A, Musa OM, Steed JW (2013) The chemistry of low dosage clathrate hydrate inhibitors. Chem Soc Rev 42:1996–2015
Ripmeester JA, Alavi S (2016) Some current challenges in clathrate hydrate science: nucleation, decomposition and the memory effect. Curr Opin Solid State Mater Sci 20:344–351
Roberts OL, Brownscombe ER, Howe LS (1941) Phase diagrams of methane and ethane hydrates. Petr Eng 12:56
Sa JH, Kwak GH, Lee BR, Park DH, Han K, Lee KH (2013) Hydrophobic amino acids as a new class of kinetic inhibitors for gas hydrate formation. Sci Rep 3:2428
Sa JH, Kwak GH, Han K, Ahn D, Lee KH (2015) Gas hydrate inhibition by perturbation of liquid water structure. Sci Rep 5:11526
Saikia T, Mahto V (2016) Experimental investigations of clathrate hydrate inhibition in water based drilling fluid using green inhibitor. J Pet Sci Eng 147:647–653
Sharifi H, Englezos P (2015) Accelerated hydrate crystal growth in the presence of low dosage additives known as kinetic hydrate inhibitors. J Chem Eng Data 60:336–342
Sharifi H, Hatzikiriakos SG, Englezos P (2014a) Rheological evaluation of kinetic hydrate inhibitors in NaCl/n-heptane solutions. AIChE J 60:2654–2659
Sharifi H, Walker VK, Ripmeester J, Englezos P (2014b) Inhibition activity of antifreeze proteins with natural gas hydrates in saline and the light crude oil mimic, heptane. Energy Fuels 28:3712–3717
Sharifi H, Ripmeester J, Walker VK, Englezos P (2014c) Kinetic inhibition of natural gas hydrates in saline solutions and heptane. Fuel 117:109–117
Sharifi H, Walker VK, Ripmeester J, Englezos P (2014d) Insights into the behavior of biological clathrate hydrate inhibitors in aqueous saline solutions. Cryst Growth Des 14:2923–2930
Sharifi H, Ripmeester J, Englezos P (2016) Recalcitrance of gas hydrate crystals formed in the presence of kinetic hydrate inhibitors. J Nat Gas Sci Eng 35:1573–1578
Sloan ED (1995a) Method for controlling clathrate hydrates in fluid systems. U.S. Patent 5420370
Sloan ED (1995b) Method for controlling clathrate hydrates in fluid systems. U.S. Patent 5432292
Sloan ED Jr (2003) Fundamental principles and applications of natural gas hydrates. Nature 426:353–359
Sloan ED, Koh C (2007) Clathrate hydrates of natural gases, 3rd edn. CRC Press, Hardcover
Sloan ED, Christiansen RL, Lederhos J, Panchalingam V, Du Y, Sum AKW, Ping J (1997) Additives and method for controlling clathrate hydrates in fluid systems. U.S. Patent 5639925
Stoporev AS, Manakov AY, Kosyakov VI, Shestakov VA, Altunina LK, Strelets LA (2016) Nucleation of methane hydrate in water-in-oil emulsions: role of the phase boundary. Energy Fuel 30:3735–3741
Sun MW, Wang Y, Firoozabadi A (2012) Effectiveness of alcohol cosurfactants in hydrate antiagglomeration. Energy Fuel 26:5626–5632

Sun TJ, Davies PL, Walker VK (2015) Structural basis for the inhibition of gas hydrates by alpha-helical antifreeze proteins. Biophys J 109:1698–1705
Sundramoorthy JD, Hammonds P, Sabil KM, Foo KS, Lal B (2015) Macroscopic observations of catastrophic gas hydrate growth during pipeline operating conditions with or without a kinetic hydrate inhibitor. Cryst Growth Des 15:5919–5929
Talaghat MR, Esmaeilzadeh F, Fathikalajahi J (2009) Prediction of gas consumption during hydrate formation with or without the presence of inhibitors in a batch system using the Esmaeilzadeh-Roshanfekr equation of state. Chem Biochem Eng Q 23:295–308
Tonelli D, Capicciotti CJ, Doshi M, Ben RN (2015) Inhibiting gas hydrate formation using small molecule ice recrystallization inhibitors. RSC Adv 28:21728–21732
Townson I, Walker VK, Ripmeester JA, Englezos P (2012) Bacterial inhibition of methane clathrate hydrates formed in a stirred autoclave. Energy Fuel 26:7170–7175
Udegbunam LU, DuQuesnay JR, Osorio L, Walker VK, Beltran JG (2017) Phase equilibria, kinetics and morphology of methane hydrate inhibited by antifreeze proteins: application of a novel 3-in-1 method. J Chem Thermodyn. https://doi.org/10.1016/j.jct.2017.08.015
Walker VK, Zeng H, Ohno H, Daraboina N, Sharifi H, Bagherzadeh SA, Alavi S, Englezos P (2015) Antifreeze proteins as gas hydrate inhibitors. Can J Chem 93:839–849
Wilson SL, Voordouw G, Walker VK (2014) Towards the selection of a produced water enrichment for biological gas hydrate inhibitors. Environ Sci Pollut Res Int 21:10254–10261
Yeh Y, Feeney R (1996) Antifreeze proteins: structures and mechanisms of function. Chem Rev 96:601–618
Zeng H, Wilson LD, Walker VK, Ripmeester JA (2003) The inhibition of tetrahydrofuran clathrate-hydrate formation with antifreeze protein. Can J Phys 81:17–24
Zeng H, Moudrakovski IL, Ripmeester JA, Walker VK (2006) Effect of antifreeze protein on nucleation, growth and memory of gas hydrates. AIChE J 52:3304–3309
Zhang JF, Di Lorenzo M, Pan ZJ (2012) Effect of surface energy on carbon dioxide hydrate formation. J Phys Chem B 116:7296–7301
Zhao WH, Wang L, Bai J, Yuan LF, Yang JL, Zeng XC (2014) Highly confined water: two-dimensional ice, amorphous ice, and clathrate hydrates. Acc Chem Res 47:2505–2513
Zhou HX, Ferreira CI (2017) Effect of type-III Anti-Freeze Proteins (AFPs) on CO_2 hydrate formation rate. Chem Eng Sci 167:42–53

Chapter 13
Antifreeze Protein-Covered Surfaces

Woongsic Jung, Young-Pil Kim, and EonSeon Jin

13.1 Introduction

Antifreeze proteins (AFPs), which lower freezing temperature below the melting point, have been studied from various organisms including psychrophilic microorganisms living in polar regions (Raymond et al. 1994; Gwak et al. 2010; Lee et al. 2010; Jung et al. 2014, 2016). Due to the ability of AFPs to inhibit the growth of ice and frost, they have been known as fascinating substances for medical, biological, and industrial uses such as food and cryopreservation of biological resources (Lee et al. 1992; Warren et al. 1993; Payne and Young 1995; Koushafar et al. 1997; Venketesh and Dayananda 2008; Bakhach 2009; Rahman et al. 2010). AFPs from ocean pout (*Zoarces americanus*) living in cold water are used to cover minute ice crystals and prevent their recrystallization during storage and delivery of ice cream (QYR Chemical and Material Research Center 2017). AFPs are also used in a biomedical field called cryonics, which aims to freeze body tissues for long-term preservation (QYR-Chemical and Material Research Center 2017).

There has been a recent increase in the interest in undesirable ice formation causing severe problems such as construction damage, safety risks, and loss of energy at industrial facilities and transportation. Ice accumulation on the wings and other parts of aircrafts leads to economical and safety problems. Aircraft icing occurs during flight in clouds at or below freezing temperatures as supercooled water droplets freeze on unprotected parts (Gent et al. 2000). To solve this problem, aircraft icing was first analyzed at the NASA Lewis Research Center in the United

W. Jung
Repigen Co. Ltd., Daejeon, South Korea
e-mail: wschung@repigen.com

Y.-P. Kim · E. Jin (✉)
Hanyang University, Seoul, South Korea
e-mail: ypilkim@hanyang.ac.kr; esjin@hanyang.ac.kr

H. Ramløv, D. S. Friis (eds.), *Antifreeze Proteins Volume 2*,
https://doi.org/10.1007/978-3-030-41948-6_13

States and the Royal Aerospace Establishment and the Defense Evaluation and Research Agency in the United Kingdom (Gent et al. 2000). The number of agencies performing aircraft icing analysis increased significantly since the early 1990s. The icing of power transmission lines causes outages and results in repair costs of millions of dollars. Most of the methods to prevent icing suggested in research papers and patents rely on melting ice by high-frequency excitation (Gent et al. 2000; Sullivan et al. 2003). At high frequencies of the transmission lines, ice tends to be a wasting dielectric, resulting in heating transmission lines in ice. In addition, electric current in a thin layer on the surface of the lines known as the skin effect causes resistive losses and results in heating. A combination of dielectric heating and skin-effect heating could be used to accomplish the uniform heating of transmission lines and has been proposed as a way to solve the problem of power transmission line icing (Gent et al. 2000; Sullivan et al. 2003; Kreder et al. 2016). Wind turbines are environmental-friendly power plants that are usually located at high altitudes and cold areas because wind speed generally increases by 0.1 m/s per 100 m of altitude above the first 1000 m. Due to their locations in windy and cold environments, wind farms can experience icing (Parent and Ilinca 2011). Wind turbines can be damaged by icing from precipitation and cold water in the atmosphere and clouds. Ice accretion (accumulation) on wind turbines results in measurement errors (Laakso et al. 2003; Fortin et al. 2005), power losses due to a changed air flow across the wings (Marjaniemi and Peltola 1998), overproduction of the parts of wind turbines for maintaining the facilities (Jasinski et al. 1997), and mechanical and electrical failures (Laakso et al. 2003; Laakso and Peltola 2005). To alleviate ice accretion effects, anti-icing and deicing systems (ADIS) were adopted (Parent and Ilinca 2011). Anti-icing strategies prevent ice accumulation on surfaces, whereas deicing strategies eliminate accumulated ice layers (Parent and Ilinca 2011). Passive ADIS use the physical characteristics of blade surfaces of turbines. Passive anti-icing systems are composed of special coating, black paints, and the use of chemicals; these systems adopt flexible blades and active pitching methods. Active ADIS require external systems to provide thermal and chemical energy (Dalili et al. 2009). These systems rely on thermal heating, formation of air layers, or generation of microwaves. Heating resistance, warm air and radiators, and electro-impulsion/expulsion methods are used in active deicing strategies (Laakso and Peltola 2005; Mayer et al. 2007; Dalili et al. 2009).

Many studies are being reported that aim to reduce damage to industrial facilities and buildings by developing more effective and environment-friendly substances to solve the problem of icing (Laakso et al. 2003; Dalili et al. 2009; Parent and Ilinca 2011). Here we review recent developments and advances in anti-icing surfaces.

13.2 Coating Surfaces to Prevent Icing

Ice accumulation on the surfaces of industrial materials causes a number of problems of safety and efficacy and increases operation costs (Alizadeh et al. 2012). Recent active ice control methods are based on melting of ice layers that have already formed. However, these protocols usually consume considerable amounts of electric energy. Recently, passive icephobic coatings, which strongly inhibit ice adhesion, have been actively promoted (Ehre et al. 2010; Mishchenko et al. 2010; Jung et al. 2012; Bi et al. 2016; Golovin et al. 2016; He et al. 2016). Passive ice control methods attract attention as a better anti-icing strategy, which relies on inhibition of ice accumulation. Water-repellent characteristics of superhydrophobic surfaces (SHSs) can be used for this purpose because the time needed for withdrawal of water droplets from these surfaces is shorter than that of ice nucleation (Fig. 13.1; Mishchenko et al. 2010). Similar to former passive ice control methods, visual icing analysis was performed with supercooled water droplets and icing was modeled under the droplet effect in order to evaluate the anti-icing characteristics of SHSs from the water droplets (Cao et al. 2009).

Ice nucleation triggers the transition from water to ice (Ehre et al. 2010; Lupi et al. 2014; Bi et al. 2016; He et al. 2016). To avoid heterologous ice nucleation, many passive anti-icing methods control the conditions such as the conformational match of the lattice between the ice and the materials, wettability, and surface roughness (Jung et al. 2012; Li et al. 2012; Zielke et al. 2015; Kreder et al. 2016). A synthetic AFP possessing polyampholyte characteristics, which have both negatively and positively

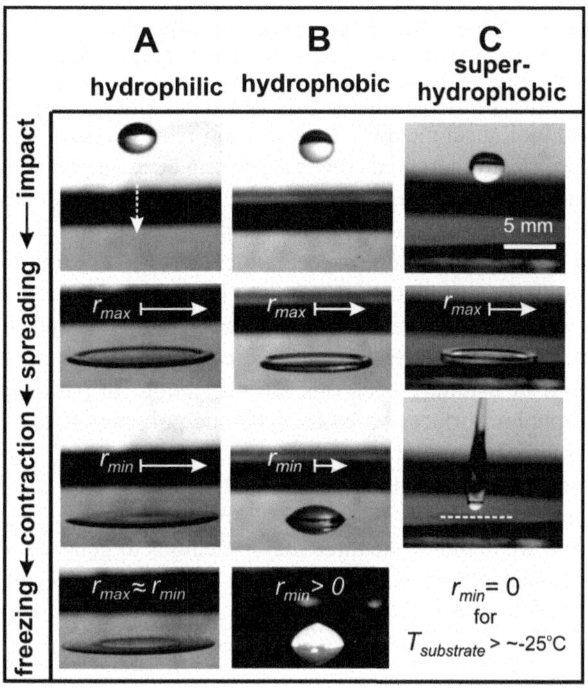

Fig. 13.1 Sequential images of the dynamic behavior of approximately 15 μl droplets impacting cooled ($T_{substrate} < 0$ °C) horizontal surfaces from a 10 cm height. Images from top to bottom depict droplet impact, maximum spreading (r_{max}), maximum retraction (r_{min}), and freezing. Pinning and ice formation were observed on hydrophilic (**a**) and smooth hydrophobic (**b**) surfaces, while full retraction was observed on a superhydrophobic surface (**c**) (Mishchenko et al. 2010)

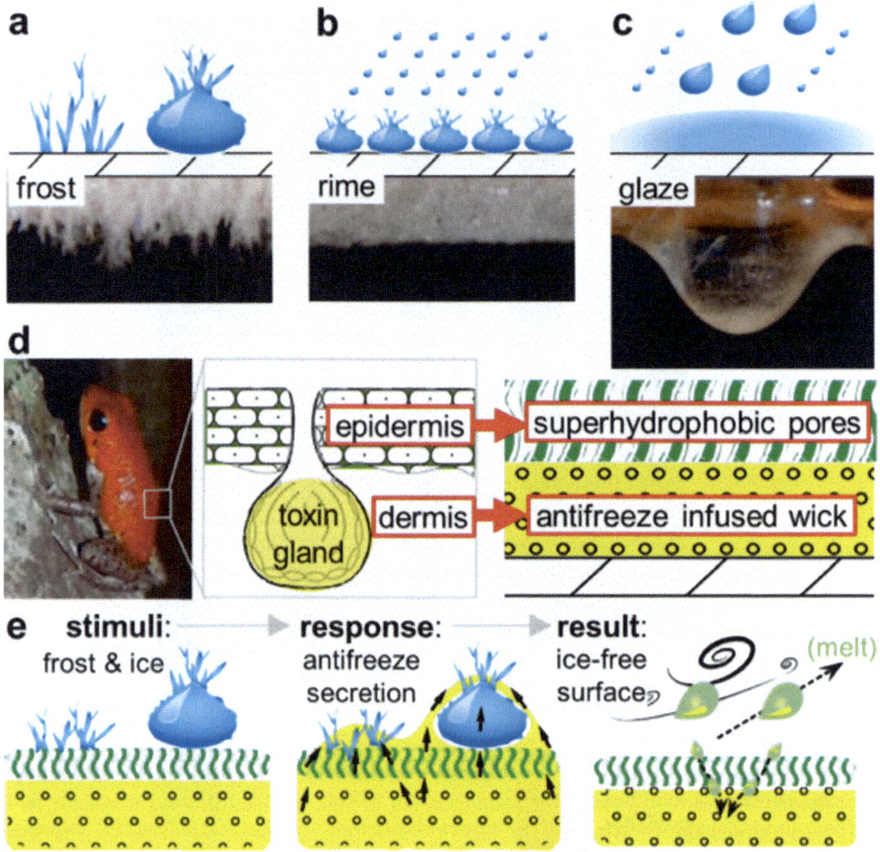

Fig. 13.6 Schemes and representative images of ice types formed in different atmospheric conditions: (**a**) frost formed by desublimation or condensation followed by freezing, (**b**) rime formed in freezing fog, and (**c**) glaze formed in freezing rain. (**d**) Scheme of stimuli-responsive antifreeze-secreting anti-icing coating (right) inspired by the function of the poison dart frog skin (left), which has a bilayer architecture: the porous superhydrophobic epidermis separates the antifreeze-infused dermis from the environment. (**e**) Scheme of antifreeze liquid secretion through the pores of the dart frog skin in response to contact with ice or frost forming on the surface, which results in ice or frost melting. Subsequently, the melt is removed via air motion or is partially wicked back into the dermis. Adapted from Sun et al. (2015)

epidermis (which functions as a barrier to the environment) and the wick-like underlying dermis infused with antifreeze liquid, the surfaces showed better anti-icing performance in condensation frosting, simulated freezing fog, and freezing rain experiments than did SHSs and lubricant-impregnated surfaces (Fig. 13.6).

As summarized by the Aizenberg group (Kreder et al. 2016), icephobic surfaces can be categorized according to ice forms and passive prevention strategies (Fig. 13.7). These include dry and lubricated surfaces, spanning a range of chemical functionalities and length scales.

Topology of surface	Type of surface	Properties
Dry/Smooth	Self-assembled monolayer 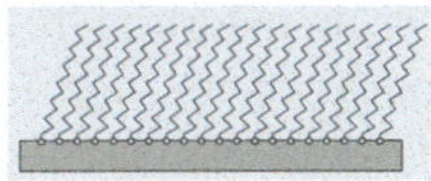	• Environmentally tolerant • Limited surface compatibility • Lower performance than state of the art
	Bulk coatings 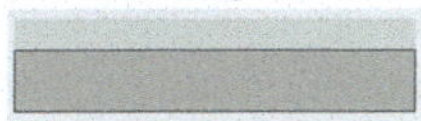	• Environmentally tolerant • Versatile and durable • Lower performance than state of the art
Dry/Textured	Microstructured 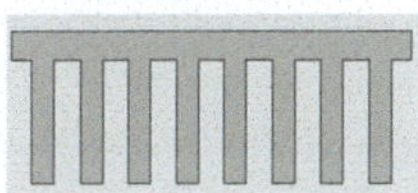	• Rapid shedding of droplets prevents ice nucleation • Poor pressure and humidity tolerance • Poor durability
	Nanostructured 	• Improved pressure stability • Improved humidity tolerance (jumping droplet effect) • Poor durability
Wet/Smooth	Microstructured 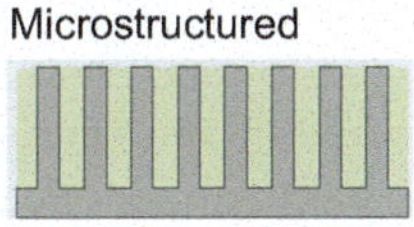	• Low ice adhesion and droplet contact a ngle hysteresis • High humidity and pressure tolerance • Poor resistance to lubricant depletion
	Nanostructured 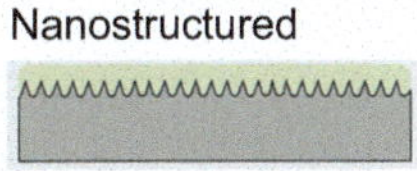	• Improved lubricant retention • Poor mechanical robustness
	Infused polymer 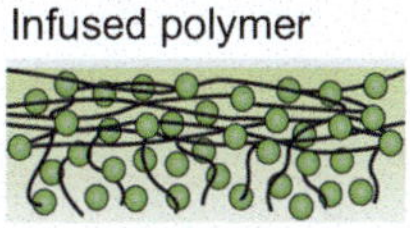	• Increased lubricant content • Kinetics of lubricant depletion and replenisment unknown
	Hydrated	• Low ice adhesion without need for lubricant replenishment

Fig. 13.7 Topology and structure of surfaces used to achieve icephobicity. Adapted from Kreder et al. (2016)

13.4 Effective Anti-icing Antifreeze Proteins from an Antarctic Diatom

Al is an important and versatile material in such industrial fields as building construction, electronic goods, freezers, and transportation. Ice formation on Al surfaces causes severe problems in their applications in air conditioners, transportation (especially aircraft), and power plants due to the reduction in the cold resistance of facilities resulting in high energy waste. Lubricant impregnation was proposed to alleviate ice formation of solid surfaces (Rykaczewski et al. 2013). Rykaczewski and colleagues have demonstrated that perfluorinated oils used as lubricants showed positive condensation dynamics by cloaking water droplets (Anand et al. 2012; Rykaczewski et al. 2013). These authors insisted that this cloaking mechanism might play a key role in the repression of ice growth on lubricant-impregnated surfaces (Rykaczewski et al. 2013).

Many surface-coating methods as the part of passive ADIS (anti-icing and deicing system) that use thermal, chemical, and mechanical principles have been used to confer the anti-icing property, but most of these techniques depend on complex processes requiring expensive instruments and labor-intensive steps (Gwak et al. 2015). Several chemical cryoprotectants such as glycerol, propylene glycols, and DMSO, are still applied to preserve biological materials even though these substances exhibit various levels of toxicity (Best 2015). AFPs originating from psychrophilic organisms have attracted interest as exciting substances that can be used in a variety of industrial, medical, and biological fields due to their ability to inhibit ice formation in the non-colligative characteristics (Raymond and DeVries 1977; Lee et al. 1992; Warren et al. 1993; Payne and Young 1995; Koushafar et al. 1997; Storey et al. 1998; Uchida et al. 2007; Venketesh and Dayananda 2008). In this section, we will introduce practical anti-icing Al surfaces coated with AFPs originating from an Antarctic diatom and a potent protocol for AFP immobilization on Al surfaces.

The AFP from the Antarctic diatom *C. neogracile* (*Cn*-AFP) was first reported by Gwak et al. (2010). The TH activity of 0.8 °C of wild-type recombinant *Cn*-AFP was enhanced to that of 1.8 °C by a site-directed mutation that increased the flatness of the ice-binding surface (Gwak et al. 2014). The G124Y mutant (*Cn*-AFP_{G124Y}) showed improved TH activity, which was formerly described, due to the outward direction of the functional group of Y124 and formation of a hydrogen bond with D122. This conformation was predicted to expand the ice-binding surface and increase its flatness to bind to the ice surface via the tyrosyl functional group (Gwak et al. 2014). Gwak et al. (2015) proposed a simple and useful way to fabricate anti-icing Al surfaces coated with *Cn*-AFP_{G124Y} by adding an Al-binding peptide (ABP) sequence at the N-terminus of *Cn*-AFP_{G124Y}. The ABP sequence (VPSSGPQDTRTT) was initially reported by Zuo et al. (2005), who used phage display to discover novel peptides capable of binding to aluminum and mild steel to protect them from deterioration. To check stability of peptide-coated materials, the authors attached bacterial biofilms to metal surfaces and the bacteria were allowed to express metal-binding peptides (Zuo et al. 2005). Among 12 candidate peptides, VPSSGPQDTRTT was chosen as the best peptide showing fourfold intensity to other peptides of interaction with metal surfaces (Zuo et al. 2005). The immobilization of ABP-*Cn*-AFP$_{G124Y}$ was checked colorimetrically by using the interaction between the hexa-histidine tag at the N-terminus of AFP-*Cn*-AFP$_{G124Y}$ and Ni(II)/horseradish peroxidase (HRP); the interaction was evaluated as blue coloration due to HRP-catalyzed oxidation of its substrate (tetramethylbenzidine, TMB) (Fig. 13.8; Gwak et al. 2015).

To overcome protein denaturation and resultant reduction in antifreeze activity, ABP-*Cn*-AFP$_{G124Y}$ were developed to improve protein stability by using uncomplicated sugar-coating steps. Under dehydration, protein conformation can be efficiently sustained by keeping the water molecules by adoption of trehalose, a nonreducing sugar consisting of two glucose molecules (Scotter et al. 2006). Al surfaces coated with ABP-*Cn*-AFP$_{G124Y}$ fusion proteins and immersed in trehalose solution showed continuous immobilization of the fusion proteins and stable antifreeze activity up to 12 days at room temperature, as evidenced by the maintenance

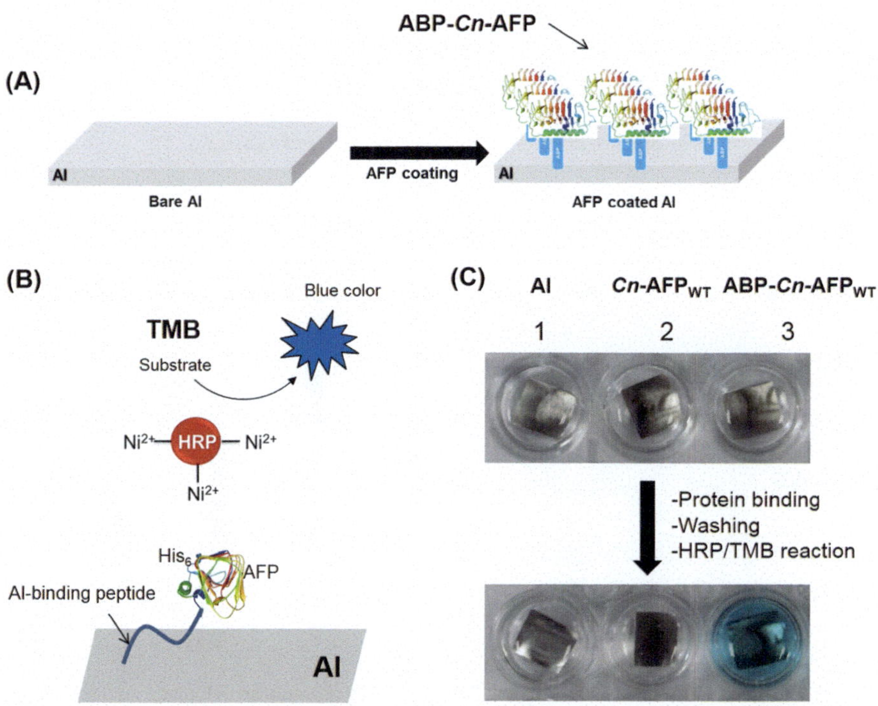

Fig. 13.8 Coating of an Al substrate with *Cn*-AFP. (**a**) The overall design. Blue bars indicate the Al-binding peptide. (**b**) Tetramethylbenzidine (TMB) assay. In the presence of HRP, TMB and peroxide present in the substrate solution react to produce a blue product. The color intensity is proportional to HRP activity. (**c**) Visualization of ABP-*Cn*-AFP binding to an Al substrate by the TMB assay. In the presence of ABP-*Cn*-AFP-coated Al, the solution color changed to blue, which indicated ABP-*Cn*-AFP binding to Al (Gwak et al. 2015)

of blue color in TMB assay, whereas those not treated with trehalose exhibited a fast color change from blue to clear, which indicated a quick reduction of antifreeze activity and prompt protein denaturation (Fig. 13.9; Gwak et al. 2015). These results demonstrate that trehalose coating is an efficient method to preserve the antifreeze activity of ABP-*Cn*-AFP_{G124Y} fusion proteins. The supercooling temperature of ABP-*Cn*-AFP_{G124Y} fusion proteins coated with trehalose (trehalose-ABP-*Cn*-AFP_{G124Y}) and those without trehalose was similar (−7.92 °C and −8.43 °C, respectively) (Fig. 13.10; Gwak et al. 2015), indicating that coating with trehalose does not affect the antifreeze and anti-icing activity of ABP-*Cn*-AFP_{G124Y}.

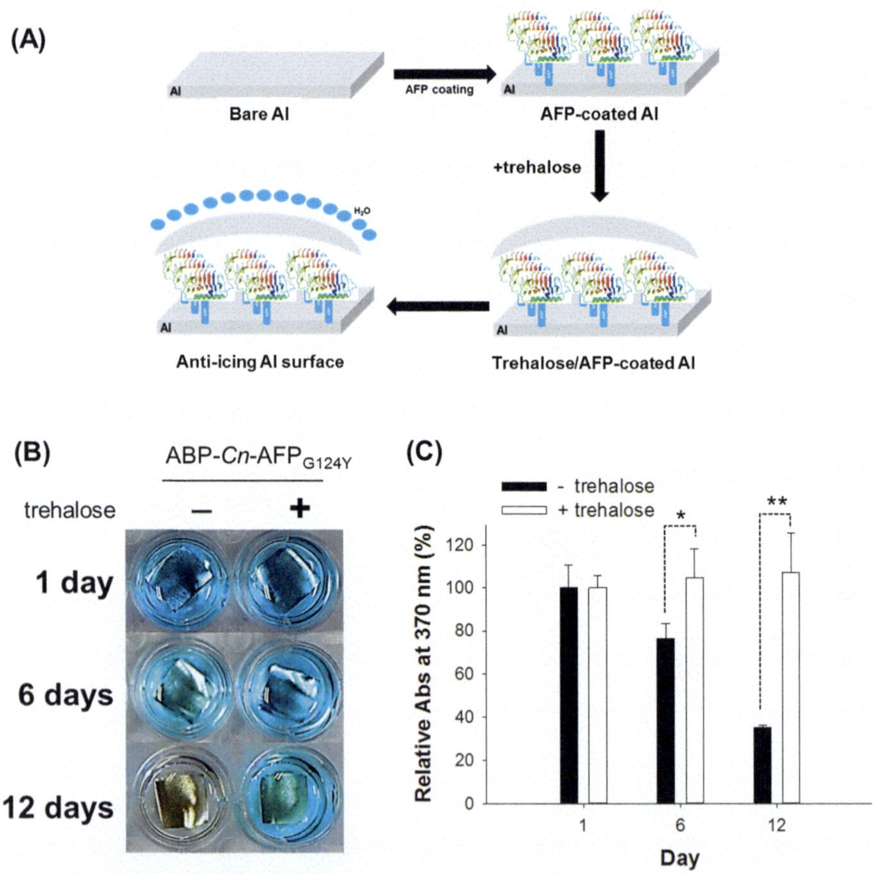

Fig. 13.9 Secondary trehalose coating of ABP-*Cn*-AFP_{G124Y} on an Al plate. (**a**) Schematic diagram of the trehalose coating procedure. (**b**) TMB assay. Six days after coating, the protein was denatured in the absence of trehalose coating. Secondary trehalose coating preserved protein structure. (**c**) Spectrophotometry of the TMB solutions. Without trehalose coating, protein on the Al plate was denatured after 6 days. In the presence of trehalose coating, the protein was still not denatured 12 days after coating. Asterisks denote statistical significance of the differences ($*p < 0.01$ and $**p < 0.001$, paired *t*-test, $n = 4$). Error bars indicate standard deviation (Gwak et al. 2015)

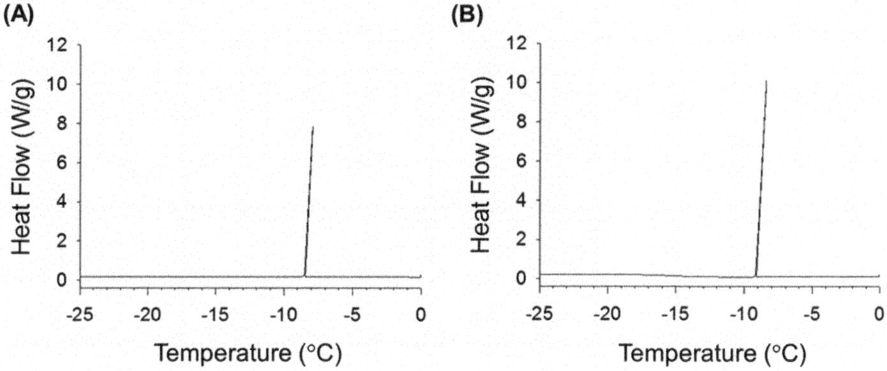

Fig. 13.10 Supercooling points of ABP-*Cn*-AFPG$_{124Y}$ on Al with and without trehalose coating. The supercooling point was −7.92 °C with trehalose coating (**a**) and −8.43 °C without trehalose coating (**b**) (Gwak et al. 2015)

Remarkably, Al surfaces coated with trehalose-ABP-*Cn*-AFP$_{G124Y}$ inhibited ice formation, whereas ice and frost formation were observed on bare Al and hydrophilic Al surfaces coated with ZrO_2 (Fig. 13.11; Gwak et al. 2015). This experiment strongly suggests that direct immobilization of a biological AFP fused with ABP and coating with trehalose efficiently prevent ice formation on Al surfaces (Gwak et al. 2015).

There are many reports on different peptides that bind to metals such as silver, gold, and mild steel and can be used for protein immobilization on corresponding metal surfaces (Brown 1997; Naik et al. 2002; Zuo et al. 2005).

Metal-binding peptides fused with AFPs have been developed for use in numerous industrial fields (Brown 1997; Naik et al. 2002; Zuo et al. 2005). Gwak et al. (2015) presented a useful two-step method to coat Al surfaces with AFPs for industrial applications, which has not been available to date. The AFP from an Antarctic marine diatom not only can be produced at high concentrations in a prokaryotic expression system (Gwak et al. 2010), but it also has enormous ability to lower the freezing temperature below the melting temperature and to inhibit ice formation at the concentration of 5 mg/ml (Gwak et al. 2010, 2014). Trehalose-ABP-*Cn*-AFP$_{G124Y}$ coating definitely hinders ice formation on Al surfaces (Gwak et al. 2015).

The results of AFP coating on Al fortified with trehalose can offer notable benefits and advantages for the industrial application of AFPs. Especially recombinant ABP-*Cn*-AFP proteins can be feasible for production on an industrial scale; this approach is an environmental-friendly system because the production of recombinant proteins is bio-based. Furthermore, Al surface coating with AFPs is accomplished speedily via a one-step dipping method without difficult surface modification. ABP-fused AFPs are capable of maintaining the appropriate orientation of the AFP on a surface, which allows for application of their full anti-icing properties. Above all, AFPs clearly hampered ice formation on the Al surface when compared to bare Al or traditional hydrophilic Al coatings (Gwak et al. 2015). However, for commercial application of surface coating with AFP, optimization of large-scale process to obtain recombinant AFP protein should be required.

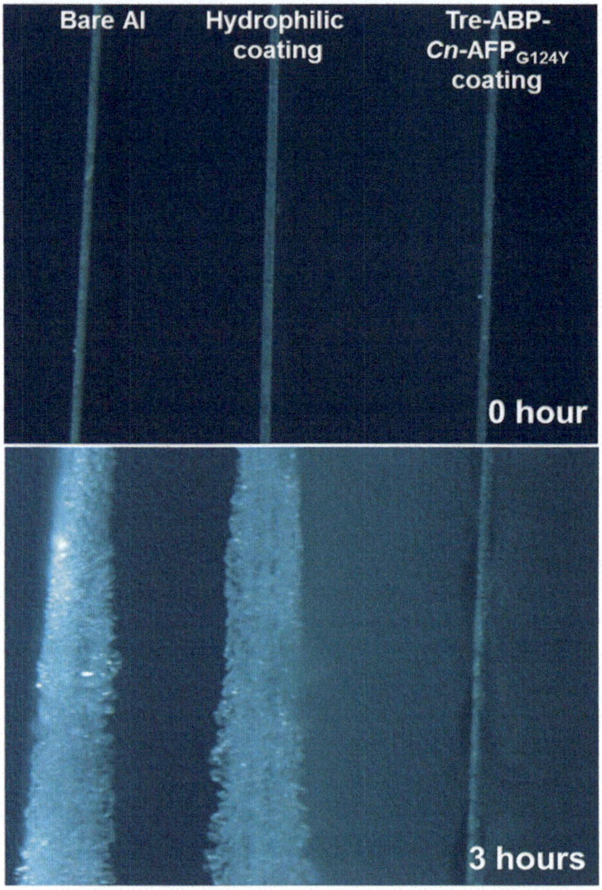

Fig. 13.11 The anti-icing effect of Tre-ABP-*Cn*-AFPG124Y on Al as the passive ADIS. ABP-*Cn*-AFPG124Y coating fortified by trehalose inhibited ice crystal growth and impeded frost formation, while large amounts of ice and frost accumulated on bare Al and on Al with a hydrophilic ZrO2 coating (Gwak et al. 2015)

13.5 Conclusions

Creating anti-icing surfaces is increasingly demanded due to their significant roles in economic and energy strategies in both scientific and industrial aspects. Bio-inspired structures, materials, and their combinations are of significant interest for generating anti-icing surfaces, which can overcome limitations of traditional anti-icing strategies. With recently developed methods to monitor IRI and TH activity (Park et al. 2013; Mitchell et al. 2015), continuing investigations into anti-icing surfaces and the simplicity and durability of these surfaces will make their use feasible in many industrial applications such as refrigerator, aircraft, and cold storage industries where protection against ice and frost formation is required.

Acknowledgments This work was funded by Korea Polar Research Institute (KOPRI, PE17180). We thank the copyright holders to use their material.

References

Alizadeh A, Yamada M, Li R, Shang W, Otta S, Zhong S, Ge L, Dhinojwala A, Conway KR, Bahadur V, Vinciquerra AJ, Stephens B, Blohm ML (2012) Dynamics of ice nucleation on water repellent surfaces. Langmuir 28:3180–3186

Anand S, Paxson AT, Dhiman R, Smith JD, Varanasi K, Kajiwara S (2012) Enhanced condensation on lubricant-impregnated nanotextured surfaces. ACS Nano 6:10122–10129

Bakhach J (2009) The cryopreservation of composite tissues: principles and recent advancement on cryopreservation of different type of tissues. Organogenesis 5:119–126

Balcerzak AK, Capicciotti CJ, Briard JG, Ben RN (2014) Designing ice recrystallization inhibitors: from antifreeze (glyco)proteins to small molecules. RSC Adv 4:42682–42696

Best BP (2015) Cryoprotectant toxicity: facts, issues, and questions. Rejuvenation Res 18:422–436

Bi Y, Cabriolu R, Li T (2016) Heterogeneous ice nucleation controlled by the coupling of surface crystallinity and surface hydrophilicity. J Phys Chem C 120:1507–1514

Brown S (1997) Metal-recognition by repeating polypeptides. Nat Biotechnol 15:269–272

Cao L, Jones AK, Sikka VK, Wu J, Gao D (2009) Anti-icing superhydrophobic coatings. Langmuir 25:12444–12448

Czechura P, Tam RY, Dimitrijevic E, Murphy AV, Ben RN (2008) The importance of hydration for inhibiting ice recrystallization with C-linked antifreeze glycoproteins. J Am Chem Soc 130:2928–2929

Dalili N, Edrisy A, Carriveau R (2009) A review of surface engineering issues critical to wind turbine performance. Renew Sust Energ Rev 13:428–438

Deller RC, Vatish M, Mitchell DA, Gibson MI (2014) Synthetic polymers enable non-vitreous cellular cryopreservation by reducing ice crystal growth during thawing. Nat Commun 5:3244

DeMerlis CC, Schoneker DR (2003) Review of the oral toxicity of polyvinyl alcohol (PVA). Food Chem Toxicol 41:319–326

Duman JG (2015) Animal ice-binding (antifreeze) proteins and glycolipids: an overview with emphasis on physiological function. J Exp Biol 218:1846–1855

Ehre D, Lavert E, Lahav M, Lubomirsky I (2010) Water freezes differently on positively and negatively charged surfaces of pyroelectric materials. Science 327:672–675

Eniade A, Purushotham M, Ben RN, Wang JB, Horwath K (2003) A serendipitous discovery of antifreeze protein-specific activity in C-linked antifreeze glycoprotein analogs. Cell Biochem Biophys 38:115–124

Esser-Kahn AP, Trang V, Francis MB (2010) Incorporation of antifreeze proteins into polymer coatings using site-selective bioconjugation. J Am Chem Soc 132:13264–13269

Farhadi S, Farzaneh M, Kulinich SA (2011) Anti-icing performance of superhydrophobic surfaces. Appl Surf Sci 257:6264–6269

Fortin G, Perron J, Ilinca A (2005) Behaviour and modeling of up anemometers under icing conditions. IWAIS Xi, Montreal, p 6

Fu QT, Wu XH, Kumar D, Ho JWC, Kanhere PD, Srikanth N, Liu E, Wilson P, Chen Z (2014) Development of sol-gel icephobic coatings: effect of surface roughness and surface energy. ACS Appl Mater Interfaces 6:20685–20692

Ge L, Ding GF, Wang H, Yao JY, Cheng P, Wang Y (2013) Anti-icing property of superhydrophobic octadecyltrichlorosilane film and its ice adhesion strength. J Nanomater 2013:1–5

Gent RW, Dart NP, Cansdale JT (2000) Aircraft icing. Philos Trans R Soc Lond A 358:2873–2911

Gibson MI (2010) slowing the growth of ice wth synthetic macromolecules: beyond antifreeze (glyco) proteins. Polym Chem 1:1141–1152

Gibson MI, Barker CA, spain SG, Albertin L, Cameron NR (2009) Inhibition of ice crystal growth by synthetic glycopolymers: implications for the rational design of antifreeze glycoprotein mimics. Biomacromolecules 10(2):328–333

Golovin K, Kobaku SPR, Lee DH, DiLoreto ET, Mabry JM, Tuteja A (2016) Designing durable icephobic surfaces. Sci Adv 2:1501496

Guo P, Zheng YM, Wen MX, Song C, Lin YC, Jiang L (2012) Icephobic/anti-icing properties of micro/nanostructured surfaces. Adv Mater 24:2642–2648

Gwak IG, Jung W, Kim HJ, Kang SH, Jin ES (2010) Antifreeze protein in Antarctic marine diatom, *Chaetoceros neogracile*. Mar Biotechnol 12:630–639

Gwak Y, Jung W, Lee Y, Kim JS, Kim CG, Ju JH, Song C, Hyun JK, Jin ES (2014) An intracellular antifreeze protein from an Antarctic microalga that responds to various environmental stresses. FASEB J 28:4924–4935

Gwak Y, Park J, Kim M, Kim HS, Kwon MJ, Oh SJ, Kim YP, Ji ES (2015) Creating anti-icing surfaces via the direct immobilization of antifreeze proteins on aluminum. Sci Rep 5:12019

Hayward JA, Haymet ADJ (2001) The ice/water nterface: molecular dynamics simulations of the basal, prism, {20(2)over-bar1}, and {2(11)over-bar0} interfaces of ice Ih. J Chem Phys 114:3713–3726

He Z, Xie WJ, Liu Z, Liu G, Wang Z, Gao YQ, Wang J (2016) Tuning ice nucleation with counterions on polyelectrolyte brush surfaces. Sci Adv 2:1600345

Huang ML, Ehre D, Jiang Q, Hu CH, Kirshenbaum K, Ward MD (2012) Biomimetic peptoid oligomers as dual-action antifreeze agents. Proc Natl Acad Sci USA 109:19922–19927

Jasinski WJ, Noe SC, Selig MS, Bragg MB (1997) Wind turbine performance under icing conditions. Aerospace Sciences Meeting & Exhibit. AIAA, Reno, p 8

Jung S, Tiwari MK, Doan NV, Poulikakos D (2012) Mechanism of supercooled droplet freezing on surfaces. Nat Commun 3:615

Jung W, Gwak Y, Davies PL, Kim HJ, Jin ES (2014) Isolation and characterization of antifreeze proteins from the Antarctic marine microalga *Pyramimonas gelidicola*. Mar Biotechnol 16:502–512

Jung W, Campbell RL, Gwak Y, Kim JI, Davies PL, Jin ES (2016) New cysteine-rich ice-binding proteini secreted from Antarctic microalgal, *Chloromonas* sp. PLoS One 11:0154056

Karim OA, Haymet ADJ (1988) The ice water interface—a molecular-dynamcs simulation study. J Chem Phys 89:6889–6896

Koushafar H, Pham L, Lee C, Rubinsky B (1997) Chemical adjuvant cryosurgery with antifreeze proteins. J Surg Oncol 66:114–121

Kreder MJ, Alvarenga J, Kim P, Aizenberg J (2016) Design of anti-icing surfaces: smooth, textured or slippery? Nat Rev Mater 1:15003

Kulinich SA, Farzaneh M (2009) How wetting hysteresis influences ice adhesion strength on superhydrophobic surfaces. Langmuir 25:8854–8856

Laakso T, Peltola E (2005) Review on blade heating technology and future prospects. BOREAS VI, FMI, Saariselka, p 12

Laakso T, Baring-Gould I, Durstewitz M, Horbaty R, Lacroix A, Peltola E, Ronsten G, Tallhaug L, Wallenius T (2003) State-of-the-art of wind energy in cold climates. IEA Wind Annex XIX, p 53

Lee C, Rubinsky B, Fletcher G (1992) Hypothermic preservation of whole mammalian organs with antifreeze proteins. Cryo-Letters 13:59–66

Lee JK, Park KS, Park S, Park H, Song YH, Kang SH, Kim HJ (2010) An extracellular ice-binding glycoprotein from an Arctic psychrophilic yeast. Cryobiology 60:222–228

Li K, Xu S, Shi W, He M, Li H, Li S, Zhou X, Wang J, Song Y (2012) Investigating the effects of solid surfaces on ice nucleation. Langmuir 28:10749–10754

Liu SH, Ben RN (2005) C-linked galactosyl serine AFGP analogues as potent recrystallization inhibitors. Org Lett 7:2385–2388

Liu K, Wang C, Ma J, Shi G, Yao X, Fang H, Song Y, Wang J (2016) Janus effect of antifreeze proteins on ice nucleation. Proc Natl Acad Sci USA 113:14739–14744

Lupi L, Hudait A, Molinero V (2014) Heterogeneous nucleation of ice on carbon surfaces. J Am Chem Soc 136:3156–3164

Lv J, Song Y, Jiang L, Wang J (2014) Bio-inspired strategies for anti-icing. ACS Nano 8:3152–3169

Marjaniemi M, Peltola E (1998) Blade heating element design and practical experiences. BOREAS IV, FMI, Hetta, pp 197–209

Matsumura K, Hyon SH (2009) Polyampholytes as low toxic efficient cryoprotective agents with antifreeze protein properties. Biomaterials 30:4842–4849

Mayer C, Ilinca A, Fortin G, Perron J (2007) Wind tunnel study of electro-thermal deicing of wind turbine blades. Int J Offshore Polar Eng 17:182–188

Meryman HT (2007) Cryopreservation of living cells: principles and practice. Transfusion 47:935–945

Mishchenko L, Hatton B, Bahadur V, Taylor JA, Krupenkin T, Aizenberg J (2010) Design of ice-free nanostructured surfaces based on repulsion of impacting water droplets. ACS Nano 4:7699–7707

Mitchell DE, Lilliman M, Spain SG, Gibson MI (2014) Quantitative study on the antifreeze protein mimetic ice growth inhibition properties of poly(ampholytes) derived from vinyl-based polymers. Biomater Sci 2:1787–1795

Mitchell DE, Cameron NR, Gibson MI (2015) Rational, yet simple, design and synthesis of an antifreeze-protein inspired polymer for cellular cryopreservation. Chem Commun 51:12977–12980

Naik RR, Stringer SJ, Agarwal G, Jones SE, Stone MO (2002) Biomimetic synthesis and patterning of silver nanoparticles. Nat Mater 1:169–172

Norgren AS, Budke C, Majer Z, Heggemann C, Koop T, Sewald N (2009) On-resin click-glycoconjugation of peptoids. Synthesis-Stuttgart 3:488–494

Parent O, Ilinca A (2011) Anti-icing and de-icing techniques for wind turbines: critical review. Cold Reg Sci Technol 65:88–96

Park JI, Lee JH, Gwak Y, Kim HJ, Jin E, Kim YP (2013) Frozen assembly of gold nanoparticles for rapid analysis of antifreeze protein activity. Biosens Bioelectron 41:752–757

Payne SR, Young OA (1995) Effects of pre-slaughter administration of antifreeze proteins on frozen meat quality. Meat Sci 41:147–155

QYR Chemical & Material Research Center (2017) Global antifreeze proteins (AFP) market 2016 industry trend and forecast 2022

Rahman AS, Parvinjah S, Hanna MA, Helguera PR, Busciglio J (2010) Cryopreservation of cortical tissue blocks for the generation of highly enriched neuronal cultures. J Vis Exp 45:2384

Raymond JA, DeVries AL (1977) Adsorption inhibition as a mechanism of freezing resistance in polar fishes. Proc Natl Acad Sci USA 74:2589–2593

Raymond JA, Sullivan CW, DeVries AL (1994) Release of an ice-active substance by Antarctic sea ice diatoms. Polar Biol 14:71–75

Rykaczewski K, Anand S, Subramanyam SB, Varanasi KK (2013) Mechanism of frost formation on lubricant-impregnated surfaces. Langmuir 29:5230–5238

Saito H, Takai K, Yamauchi G (1997) Water- and ice-repellent coatings. Surf Coat Int 80:168–171

Scotter AJ, Marshall CB, Graham LA, Gilbert JA, Garnham CP, Davies PL (2006) The basis for hyperactivity of antifreeze proteins. Cryobiology 53:229–239

Shih IL, Van YT, Sau YY (2003) Antifreeze activities of poly (gamma-glutamic acid) produced by *Bacillus licheniformis*. Biotechnol Lett 25:1709–1712

Storey BT, Noiles EE, Thompson KA (1998) Comparison of glycerol, other polyols, trehalose, and raffinose to provide a defined cryoprotectant medium for mouse sperm cryopreservation. Cryobiology 37:46–58

Subramanyam SB, Rykaczewski K, Varanasi KK (2013) Ice adhesion on lubricant-impregnated textured surfaces. Langmuir 29:13414–13418

Sullivan CR, Petrenko VF, McCurdy JD, Kozliouk V (2003) Breaking the ice [Transmission line icing]. IEEE Ind Appl Mag 9:49–54

Sun TL, Qing GY (2011) Biomimetic smart interface materials for biological applications. Adv Mater 23:H57–H77

Sun XD, Damle VG, Liu SLZ, Rykaczewski K (2015) Bioinspired stimuli-responsive and antifreeze-secreting anti-icing coatings. Adv Mater Interf 2:1400479

Uchida T, Nagayama M, Shibayama T, Gohara K (2007) Morphological investigations of disaccharide molecules for growth inhibition of ice crystals. J Cryst Growth 299:125–135
Venketesh S, Dayananda C (2008) Properties, potentials, and prospects of antifreeze proteins. Crit Rev Biotechnol 28:57–82
Vorontsov DA, Sazaki G, Hyon SH, Matsumura K, Furukawa Y (2014) Antifreeze effect of carboxylated epsilon-poly-L-lysine on the growth kinetics of ice crystals. J Phys Chem B 118:10240–10249
Walters KR, Serianni AS, Voituron Y, Sformo T, Barnes BM, Duman JG (2011) A thermal hysteresis-producing xylomannan glycolipid antifreeze associated with cold tolerance is found in diverse taxa. J Comp Physiol B 181:631–640
Wang N, Xiong DS, Pan S, Wang K, Shi Y, Deng YL (2017) Robust superhydrophobic coating and the anti-icing properties of its lubricants-infused-composite surface under condensing condition. New J Chem 41:1846–1853
Warren GJ, Hague CM, Corotto LV, Mueller GM (1993) Properties of engineered antifreeze peptides. FEBS Lett 321:116–120
Wier KA, McCarthy TJ (2006) Condensation on ultrahydrophobic surfaces and its effect on droplet mobility: ultrahydrophobic surfaces are not always water repellant. Langmuir 22:2433–2436
Yang DSC, Hon WC, Bubanko S, Xue Y, Seetharaman J, Hew CL, Sicheri F (1998) Identification of the ice-binding surface on a type III antifreeze protein with a "flatness function" algorithm. Biophys J 74:2142–2151
Zhang YF, Yu XQ, Zhou QH, Li KN (2010) Fabrication and anti-icing performance of a superhydrophobic copper surface with low adhesion. Acta Phys Chim Sin 2010:1457–1462
Zheng SL, Li C, Fu QT, Xiang TF, Hu W, Wang J, Ding S, Liu P, Chen C (2016) Fabrication of a micro-nanostructured superhydrophobic aluminum surface with excellent corrosion resistance and anti-icing performance. RSC Adv 6:79389–79400
Zielke SA, Bertram AK, Patey GN (2015) A molecular mechanism of ice nucleation on model AgI surfaces. J Phys Chem B 119:9049–9055
Zuo R, Ornek D, Wood TK (2005) Aluminum-and mild steel-binding peptides from phage display. Appl Microbiol Biotechnol 68:505–509

Chapter 14
Mutational Studies on Antifreeze Proteins

Dennis Steven Friis and Hans Ramløv

14.1 Introduction

Genetic modification is an important tool when investigating proteins. This has also been the case for antifreeze proteins (AFPs), where especially site-directed mutagenesis (SDM) has been applied. The concept of this method is further described below. During the 1990s the mutagenesis studies on AFPs took their entry, where especially the winter flounder, *Pseudopleuronectes americanus*, type I AFP HLPC-6 isoform was investigated. Many studies focused on the mechanism by which the AFPs bind to ice, the forces involved and the chemical properties of the amino acids responsible for the ice interaction. During the discovery of new AFPs, SDM studies have also been focused on the putative ice-binding domains (IBDs) to confirm or reject their role in ice interaction. Another strategy of SDM has been to improve the activity of AFPs by increasing the ice-binding capability of the IBDs, either by increasing the size of the IBD or making the amino acids in the IBD more uniform. A different type of protein modification is adding tags or fusion proteins to AFPs. This method has been used mainly to assist the purification possibilities or by fusing a fluorescent protein to the AFPs and make fluorescent ice purification assays (FIPA) and visualising the AFPs on the ice, described in Chap. 9.

The possibility of genetic modification has thus given a variety of options to investigate proteins, which has also been applied on AFPs. Before going in detail with the conducted experiments, a short overview of genetic manipulation is presented.

D. S. Friis (✉)
Copenhagen, Denmark

H. Ramløv
Department of Natural Sciences, Roskilde University, Roskilde, Denmark
e-mail: hr@ruc.dk

H. Ramløv, D. S. Friis (eds.), *Antifreeze Proteins Volume 2*,
https://doi.org/10.1007/978-3-030-41948-6_14

14.2 Genetic Manipulation

When the DNA or amino acid sequence coding for a mature AFP is obtained, genetic manipulation can begin. The host for the genetic manipulation is most often bacteria, e.g. *Escherichia coli*, where plasmids make it easy to introduce mutations as exemplified here; however, there are many other approaches. Today, many bacterial strains and plasmids prepared for inserting, expressing and purifying genes and their products are commercially available. The obtained DNA sequence (excluding exons), or in general any DNA sequence that will be translated to the obtained amino acid sequence, can be inserted into a cloning plasmid using standard molecular biology tools, from where this wild type can be expressed and purified.

After obtaining the protein structure, e.g. by crystallisation or computer modelling, the placement of the amino acids are visualised and hypotheses of their roles can be made. Then rational changes can be applied to the protein by genetic manipulation, or SDM, by designing primers introducing the desired changes. Today, gene synthesis is affordable; so instead of primers, the mutated gene can be readily ordered, and tailored for your plasmid. The changes to the protein can be anything from single amino acid substitutions to deletion or addition of protein domains. Single point substitutions are the most common changes and are, e.g. written A12T when the alanine being on the 12th position in the protein is substituted with threonine. Some software can simulate the changes and estimate the effects, i.e. if changes are thermodynamically sound (Jorgensen and Tirado-Rives 1988). Otherwise, trial and error is also a popular strategy to test if the mutants fold into functional proteins. Often the structure of a mutant is compared to the wild type, to make sure the overall structure is maintained and that the protein has folded correctly (Friis et al. 2014). A frequently used method for this is circular dichroism spectroscopy, in which spectrum gives a unique representation of the secondary structure of the protein. Thus, changes in, e.g. α-helix or β-sheet composition will change the spectrum (Provencher and Glöckner 1981).

An advantage of transferring a gene to, e.g. *E. coli*, besides making genetic manipulation, is that the systems are usually also tuned to express recombinant proteins in high quantities. Furthermore, the plasmids also introduce tags to the proteins, making the purification process much smoother, usually just consisting of a short series of histidines. These tags can be removed enzymatically post purification, though sometimes leaving a few residues if the cleavage site was not, or could not be, placed adjacent to the native sequence. Often the proteins are still active with the tag on (especially the small histidine tags), making some researchers inclined to keeping the tags on the protein during experiments, though taking the risk of it affecting the result. This, however, is often accommodated by having the same tags on both the wild type, mutants and controls, thus levelling out any potential influence of the tag.

The above strategies are commonly used to produce, isolate and investigate (altered) proteins in vitro. With regards to AFPs, researchers have also been interested in transferring AFPs to cold-susceptible organisms to make the organisms freeze tolerant, or at least more cold-hardy. Put simply, one of the processes of making transgenic multicellular organisms can be split into the following steps; Embryo cells of the target organisms must be obtained. The gene of interest (AFP) is then delivered to the embryo cells by the use of a vector. A successful delivery will entail that the gene is incorporated into the DNA in a non-lethal place (i.e. not disrupting essential genes) and after a promoter sequence to ensure expression, which can be regulated depending on the promoter. The gene will then be copied together with the rest of the genome, and be part of all the cells during the embryonic development. The grown organism will then have the gene of interest in all its cells, including germ cells, meaning that the trait will be passed on to offspring. There are many ways of construction these transgenic organisms as well as methods to ensure a correct gene delivery and even ways to control the expression, but it is not the scope of this book. The AFP-related experiments involving transgenic organisms are described in Sect. 14.7.

The ways and goals of changing the proteins by genetic manipulation are many, and in regard to AFPs, many different experiments have been carried out. The following sections describe the nature and results of these experiments.

14.3 Tags and Fusion Proteins

Before digging into the mutation studies on the antifreeze protein itself, this section describes how protein engineering can be exploited by addition of tags or whole proteins to the AFPs.

When the gene sequence of the protein of interest is obtained, it is fairly simple to add any C- or N-terminal amino acids to it, when expressing it in microorganisms in a plasmid-based system. By standard cloning techniques, DNA sequences can be inserted in the plasmid to be expressed in the same reading frame as the protein of interest, thus leading to an expression of a protein with additional amino acids. These can vary from one amino acid to complete proteins, in which the protein complex is called a fusion protein. In this case a linker of 6–10 small flexible amino acids is inserted in between the two fused proteins to reduce the risk of the two separate proteins of influencing the folding of each other. Hereby two separate, but linked and co-expressed functional proteins, are obtained.

14.3.1 Purification and Yield

There can be different objectives when adding additional amino acids or proteins to the AFPs (or proteins in general). Most often this is done to improve the possibility of purifying the AFPs. The preferred strategy here is by adding 6–8 histidines (referred to as a His-tag), to either end of the protein. This tag enables an efficient purification step by using immobilised metal ion affinity chromatography, where the

14.4 Localising Ice-Binding Domains

When an AFP gene has successfully been transferred to an organism for high protein expression, and a suitable purification method has been developed, one of the first mutation studies performed is often focusing on localising the IBD. Usually this is investigated by substituting an amino acid with a small side chain pointing outward in the putative IBD with an amino acid with a larger side chain. This results in a disrupted surface of the IBD, leading to an impaired ice interaction and thus a lowered thermal hysteresis. If no change in activity is observed, the substituted amino acid is most likely not part of the IBD.

Before making any of these substitutions, the structure of the protein or at least a model of a presumed structure must be obtained. As the protein has been transferred to an organism for high level expression, it is often possible to obtain large and pure amounts to obtain a crystal structure of the AFP. Another approach is to search for other AFPs (or proteins in general) with high sequence homology and for which the structure is already obtained. It is then possible to model or sculpture the structure of the novel AFP using the structure of the homolog as a template.

Once the structure is obtained, the IBD prospects can be investigated. The IBD is usually a flat area on the protein surface, which is hydrophobic relative to the rest of the protein surface. For the most active insect AFPs the IBD consists of solvent exposed threonine arrays, spaced very regularly. For fish AFPs the IBD is less ordered and recognisable. In the selected IBD candidates one or two amino acids with the side chain pointing out towards the solvent is then chosen for substitution. Usually these are threonines as these are associated with ice interaction and they are often substituted with the large aromatic amino acid tyrosine.

If the chosen amino acid was in fact part of the IBD, the substitution will disrupt the flatness if the IBD, causing a steric hindrance between ice and the ice-binding amino acids in the IBD (Graether et al. 2000). Depending on how pivotal the amino acid was in regard to ice interaction, the activity can decrease slightly or even totally diminish. Several studies can therefore be made, targeting different amino acids in the IBD to locate the most essential sites or amino acids of the IBD.

Before any conclusions can be made, it is important to ensure that the substitutions have not made any structural dislocations of the protein, which would be the cause for the impaired activity rather than a steric hindrance in the IBD. The effects of a mutation can be difficult to predict and could lead to impaired protein folding, disruption of salt bridges etc. Even small structural changes in the protein could be detrimental for the antifreeze activity, as this depends on a close match between the protein and ice/water molecules. Hence, substitutions outside the IBD could impair the antifreeze activity because of structural dislocation, thus leading to a false presumption that it is located in the IBD. Therefore, the studies are usually supported by structural studies such as circular dichroism spectroscopy to conclude that none, or very little, structural change is caused by the substitution(s).

In the sections below are examples of investigations of the IBD for some of the different AFP types.

Antifreeze glycoproteins are a special case as they consist of both a peptide backbone and sugar moieties, described in detail in Chap. 5 of Volume 1. They consist of a repeated tripeptide, AAT, with occasional arginine for threonine or proline for alanine substitutions. Though variations in the peptide sequence occur, substitution of amino acids with, e.g. serine or glycine impairs the antifreeze activity or the ice recrystallisation inhibition. However, the actual activity of AFGPs is likely to be ascribed to the sugar moieties, as changes in these has just as detrimental effect on activity. Much more studies have focused on the sugar moieties and analogues of these than mutation studies. A thorough overview of the studies on AFGPs can be found in the review by Bang et al. (2013).

Antifreeze protein type I is a group of rod-shaped helical proteins, which do not have an apparent flat side that could serve as the IBD. However, they have threonines aligned on the relatively hydrophobic side of the protein which long have been thought to be involved in the interaction with ice (Devries and Lin 1977). This has been supported by SDM on the SS-8 AFP isoform from shorthorn sculpin, *Myoxocephalus scorpius*. Here alanine to lysine substitution showed no or little effect when the substitutions were located on the hydrophilic side, while substitutions on the hydrophobic side diminished the activity (Baardsnes et al. 2001). Similarly, on the HPLC-6 AFP isoform from *P. americanus*, alanine to leucine substitutions close to the threonines showed a loss of activity (Baardsnes et al. 1999; Wen and Laursen 1993b). However, other alanine to leucine substitutions close to threonine showed only little effect on activity, indicating that specific alanines has a role in the ice binding as well (Baardsnes et al. 1999). All mutational studies still point to the hydrophobic side of the protein to be the IBD, and this has also been a widely investigated subject for the HPLC-6 isoform, which is further described later in this chapter.

Antifreeze protein type II is a group of proteins believed to have evolved from C-type lectins and has carbohydrate-recognition domains. In this group we find AFPs from smelt, *Osmerus mordax*, herring, *Clupea harengus*, and sea raven, *Hemitripterus americanus* (Ewart and Fletcher 1993; Ewart et al. 1992; Slaughter et al. 1981). Though they all have the same tertiary structure, the first two are much more alike with 83% homology on amino acid level, while the latter is only 40% identical to the other two (Loewen et al. 1998). Furthermore, the antifreeze activities of *O. mordax* and *C. harengus* AFPs are Ca^{2+} dependent, while the *H. americanus* AFP is not. This fact has also been the focal point when investigating the ice-binding domain.

In the Ca^{2+}-dependent *C. harengus* AFP, the Ca^{2+}-binding site and carbohydrate-binding site coincide. In an investigation of the IBD, the native carbohydrate-binding motif, QPD amino acids (galactose type), was substituted with another carbohydrate-binding motif, EPN (mannose type), by mutagenesis. By performing this kind of substitution, the integrity of the protein could still be demonstrated as it exhibited Ca^{2+} binding and had maintained its folding and stability. However, the protein had lost it antifreeze activity, and hardly even affected ice crystal morphology (Ewart et al. 1998). Another study supporting that the *C. harengus* AFP IBD is in fact the Ca^{2+}-binding site was performed by SDM of five residues believed to interact with Ca^{2+}. Here, especially N99 and D114 were shown to be important for antifreeze activity (Li et al. 2004).

While the Ca^{2+}-dependent type II AFPs seems to bind ice via the Ca^{2+}-binding site, the Ca^{2+}-independent type II AFPs does not. A study on *H. americanus* AFP shows that various mutations in the Ca^{2+}-binding site did not decrease antifreeze activity much, and the most extensive mutations only decreased the activity by 25%. However, a single point mutation, S120H, caused a 35% decrease in activity. This amino acid is not located in the *C. harengus* ice- or Ca^{2+}-binding site but instead in the region corresponding to the proposed $CaCO_3$ binding site of another homologous protein, the pancreatic stone protein (Loewen et al. 1998). A different Ca^{2+}-independent type II AFP from the longsnout poacher, *Brachyopsis rostratus*, having 64% amino acid homology with the *H. americanus* AFP, neither has its IBD located in the Ca^{2+}-binding site, based on mutational studies as well (Nishimiya et al. 2008).

Antifreeze protein type III has six amino acids in its putative IBD, located on two anti-parallel β-sheets—N14, A16, T18, S42, Q44 and N46. In a study on five of these (not including A16), especially the T18N and N14S and to some extent the Q44T mutants impaired the antifreeze activity, without altering the protein structure which was confirmed by NMR spectroscopy (Chao et al. 1994). This thus confirmed that the β-sheets, in particular the N14, A16, T18 β-sheet, play a role in regard to ice interaction. The study also showed that the effects of the impairing mutations are cumulative, as the N14S/Q44T double substitution hardly evoked any antifreeze activity, and the N14S/T18N/Q44T triple mutant had no activity at all (Chao et al. 1994). In subsequent studies, the A16 amino acid was substituted with seven different amino acids with different impact on antifreeze activity; M, V, L, T, R, Y and H ordered with M showing the least impact and H the largest (DeLuca et al. 1996).

Besides affecting the activity, the substitutions also caused changes in ice crystal morphology, where the least active mutants gave rise to the longest hexagonal bipyramidal ice growth in the hysteresis gap (DeLuca et al. 1996). A similar study on the A16 amino acid, where the protein structures are characterised by X-ray crystallography, shows that the A16H and A16Y substitutions might entail a shielding of Q44, contributing to a very low antifreeze activity. The study also shows that due to a tight packing of the residues in the IBD, substitution to bulkier amino acids is causing positional changes of adjacent ice-binding amino acids, further attributing to the reduced antifreeze activity (DeLuca et al. 1998b). Besides demonstrating the amino acids comprising the IBD, the studies also show that the antifreeze activity can be modified. With some steric hindrance implemented, the other ice-binding residues can still accommodate ice interaction, though at a lower affinity, or maybe affinity for a different ice plane, as some hysteresis is still observed together with a change in ice crystal morphology.

So far, no mutational studies have been conducted on the presumed *antifreeze protein type IV*.

Insect antifreeze proteins have very regular spaced threonines termed TxT motifs in their flat and ordered IBD. There is therefore usually not much doubt of the location of the IBD if the protein structure is obtained. The characteristic IBD structure is also an advantage when only the amino acid sequence is known, and the putative structure has to be modelled. Still, SDM is used to insert obstructive amino acids in different solvent exposed sites of the protein, to observe the impact on the antifreeze activity, both in order to verify the location of the IDM or evaluate the importance of the different amino acids in regard to antifreeze activity.

The IBD has been located by this method for AFPs from *T. molitor* (Marshall et al. 2002), the desert beetle *Microdera punctipennis* (Jiang et al. 2011) and *C. fumiferana* (Graether et al. 2000). The latter is a great example of a study on localising the IBD and the amino acids around it, and is described below.

The AFP from *C. fumiferana* is a 90 amino acids long β-helix consisting of five left-handed β-loops and one right-handed loop at the C-terminal. Each loop consists of 15 amino acids (17 for the right-handed loop), creating a triangular cylindrical structure with the characteristic TxT motif aligned on one face of the protein and two less structured faces. A study by Graether et al. involves threonine to leucine substitutions for four threonines in the four first TxT motifs (T7, T21, T38 and T51), as well as two threonines adjacent to the TxT motif (T48 and T66), and one on the opposite side of the TxT motifs (T86). The substitutions in the TxT motifs caused a decrease in antifreeze activity by 80–90%, while the substitutions adjacent to the TxT motifs caused a decrease in antifreeze activity of 30% (T48L) or 35% (T66L). The T86L mutant with the substitution located on the opposite site of the TxT motifs showed no change in activity (Graether et al. 2000).

Though tyrosine is most often used in these kinds of perturbation studies due to its size, the diminishing effect on antifreeze activity by leucine is also evident. However, the antifreeze activity is never completely diminished by single amino acid substitutions in the IBD, not just in this study but in general, indicating that the surrounding threonines can still mediate an interaction with ice, causing the protein to adhere and evoke antifreeze activity. In a study on *T. molitor* AFP, one of the threonines in the IDB was substituted with leucine, lysine or tyrosine. Here the leucine caused a 75% reduced antifreeze activity, while the other two amino acids caused a reduction of around 90%, indicating that the loss in activity is related to the size of the introduced amino acid (Marshall et al. 2002). The lowered activity could be an expression of whether the remaining threonines co-interact with ice long or strong enough or how well they can form a structured hydration shell to interact with the ice. The fact that these mutants still show antifreeze activity also shows that intrusive amino acids in the IBD can be buried into the ice while the rest of the flat side covers the ice surface.

The fact that the substitutions adjacent to the TxT motifs of the *C. fumiferana* AFP also had an impact on activity (Graether et al. 2000) indicates that they might have caused a small dislocation of the ice-binding motif, causing a small mismatch with the ice lattice. Another cause could be that these amino acids also play a role in the ice-binding mechanism. This could be in regard to formation or shape of the hydration shell, thus assisting to the ice binding, or somehow prevent the AFPs of being overgrown by ice once adhered to an ice surface.

Antifreeze proteins from other organisms have also been subjected to SDM in their IBD. Some of these studies are presented in this section. The AFP from the snow mould fungus *Typhula ishikariensis* has a right-handed triangular structure as *T. molitor*, with an additional α-helix lying parallel to one of the sides along the protein. The protein surface structure shows very poor regularity of the solvent exposed amino acids, as opposed to most other AFPs, as no typical TxT motifs are present. In this case it was therefore difficult to predict which of the three sides of the protein surface was the IDB. Amino acids located in the centre of each side of the protein were substituted with tyrosine to probe the IBD by steric mutations. The mutated protein structures were verified by circular dichroism spectroscopy. While two mutations showed a loss of antifreeze activity of around 15% and one a loss of around 60%, it was concluded that the latter mutation was located in the IBD (Kondo et al. 2012). Similarly, the AFP from the perennial ryegrass *Lolium perenne* folds into a β-roll with two flat β-sheets on opposite sides of the protein. By substitution with tyrosine on several sites on each side, and in the edge of the parallel β-strands, the antifreeze activity was clearly more impaired by substitutions on one specific side, thus concluding it to be the IBD. The structures were verified by circular dichroism spectroscopy (Middleton et al. 2009). The structures of the two flat β-sheet sides show no clear difference and even have almost identical consensus sequences (xxNxVxG and xxNxVx-G). It is therefore peculiar that only one side, and not both, is responsible for evoking antifreeze activity. As recrystallisation inhibition (described in Chap. 7) is an important function of antifreeze proteins in plants, the inactive side (in regard to antifreeze activity) might have a role in recrystallisation instead.

While it is a common strategy to insert tyrosine amino acids in the presumed IBD of various AFPs to cause a steric hindrance, the putative IBD of midge AFP actually consists of aligned tyrosines. In a study on localising the IBD, one of the aligned tyrosines was substituted with arginine or aspartic acid, both causing a drop in antifreeze activity from 0.75 °C to around 0.05 °C. Substitution of a tyrosine located on the opposite protein side of the aligned tyrosines had no effect on the activity, thus indicating that the IBD of the AFP is made up by the aligned tyrosines (Basu et al. 2016).

The method of introducing steric hindrance by SDM to localise the IBD seems very sound and reproducible. Substitutions inside the IBD shows detrimental effect on the antifreeze activity, which is often decreased by up to 90% by a single point mutation, and completely diminished by combination of two or three substitutions. Control mutations focused outside the IBD seems in general to have lesser effect the further it is from the IBD, yet usually still affecting the activity slightly which could be due to small dislocations in the protein. Some mutations might lead to a misfolded protein and make the AFP inactive, even if the mutations were located outside the IBD. This could be interpreted as a false positive (believing the mutation was in fact made in the IBD), if the investigation is not assisted with a structural investigation (e. g. circular dichroism spectroscopy) of the protein. In conclusion, when investigating localising the IBD, site-directed mutagenesis makes a great tool when supplemented with structural analysis of the proteins.

14.5 Exploring the Ice-Binding Mechanism

The applications of SDM are vast, and the method has also been used to investigate the unique ability of the AFPs—the binding to ice surfaces. At first, SDM was a great tool to pinpoint the amino acids that had a role in ice interaction, which was later followed by studies investigating the effect of substituting these amino acids with ones having slightly different side chains and observing the effect on antifreeze activity, ice morphology or even the ice recrystallisation inhibition efficiency. Especially one AFP, the *P. americanus* AFP type I isoform HPLC-6, has been investigated thoroughly in this manner. The investigations and results on this protein are presented in the following section.

14.5.1 Antifreeze Protein Type I, HPLC-6

The HPLC-6 AFP is a small α-helical alanine-rich protein consisting of 37 amino acids. Each turn in the helical structure consists of 3–4 amino acids, and the protein consists of ten complete turns. In this helical structure, four threonines are aligned on one side of the protein, with 11 amino acid interval. For an alternative way of illustrating the structure, Kim et al. have made a wheel presentation of the structure (Kim et al. 2013). These threonines are believed to be the key amino acids in the facilitation of the ice binding, and thus also the focal point of various mutagenic studies.

The general approach to the mutagenic studies has been to substitute some or all of the four threonines with amino acids which resemble the side chain of threonine. The substituting amino acids have usually consisted of valine, serine or alanine, but also glycine, isoleucine, allo-threonine and 2-amino butyric acid (B) has been investigated. The molecular structure of the side chains of these amino acids are presented in Fig. 14.1.

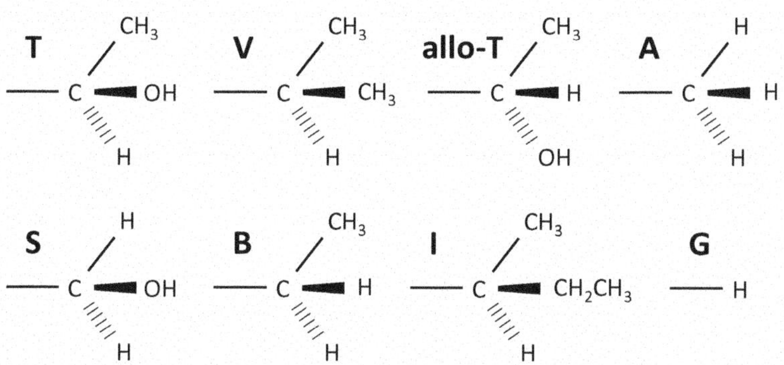

Fig. 14.1 Amino acid side chain structures. The amino acid substitutions performed on the wild-type HPLC-6 AFP includes: T (threonine, wild type), V (valine), allo-T (allo-threonine), A (alanine), S (serine), B (2-amino butyric acid), I (isoleucine) and G (glycine)

The wild-type threonines have a methyl group, a hydroxyl group and a hydrogen (CH_3, OH, H) as constituents of their side chain. The different substituent amino acids differ from this structure in the following ways: Valine has a methyl group instead of the hydroxyl group (CH_3, CH_3, H). Serine has a hydrogen instead of a methyl group (H, OH, H). Alanine has hydrogen instead of both hydroxyl and methyl groups (H, H, H). Glycine has no carbon in its side chain and thus has no groups other than a hydrogen on the peptide backbone (–). Isoleucine has an ethyl group instead of hydroxyl group (CH_3, CH_2CH_3, H). Allo-threonine is a stereoisomer of threonine, thus consisting of the same groups as threonine but in an invert arrangement (CH_3, H, OH). 2-amino butyric acid has a hydrogen instead of a hydroxyl group (CH_3, H, H).

Before presenting the results of the threonine substitutions, another modification made by SDM needs to be addressed. This entails the introduction of two additional salt bridges to the protein, in addition to the salt bridge between K18 and E22 in the wild-type protein. These two double substitutions are often made in combination with the threonine substitutions. The introduction of the two salt bridges (A7K-A11E and A29K-A33E) was originally performed by Chakrabartty and Hew to investigate the relationship between antifreeze activity and α-helicity of the protein (Chakrabartty and Hew 1991). In the literature, this mutant is listed with the suffix 2KE. While they observed an increased α-helicity of the mutant, the antifreeze activity was the same as the wild type. However, they did observe an effect on ice crystal growth rates at much lower concentrations by the 2KE mutant than the wild type, indicating a complex relation between proteins' antifreeze activity and the α-helicity (Chakrabartty and Hew 1991). Furthermore, the mutant showed increased solubility compared to the wild type, and increased the yield of the protein during expression and purification (Haymet et al. 1998), and the 2KE substitutions are therefore often used in combination with other mutations.

As mentioned, several SDM studies have been performed on the HPLC-6 AFPs threonines in order to investigate their effect and attempting to deduce the mechanism in the ice-binding ability. In Table 14.1, the substitutions and their effect on antifreeze activity of the majority of the existing studies are presented. These studies in the late 1990s have all focused on which molecular group of the threonine side chain was responsible for the ice interaction, and which forces that drives it. The theory of ice interaction preceding these studies stated that the hydroxyl group of the threonine was responsible for the interaction with ice via formation of hydrogen bonds (Wen and Laursen 1993a; Sicheri and Yang 1995; Knight et al. 1991); however, threonine side chains may orient themselves such that the methyl group are placed on the surface of the protein, making a hydrophobic surface of the protein (Haymet et al. 1998). To investigate whether the methyl group had any effect on ice interaction, researchers made different combinations of substitution of the threonines with serines (Zhang and Laursen 1998; Chao et al. 1997; Haymet et al. 1999). Serine differs from threonine by lacking the methyl group, but still has a hydroxyl group (Fig. 14.1). These substitutions led to great loss of activity, from 10 to 25% with single substitutions to complete loss with three or all four threonines substituted with serine (Table 14.1). The substitutions did not lead to any change in structure, as the

Table 14.1 Relative hysteresis of HPLC-6 mutants

Mutant	TH % 1 mM	TH % 2 mM	References
HPLC-6 (TTTT)	100	100	
D-HPLC-6[a]	100	100	Wen and Laursen (1993a)
TTTT (Allo)	0	29	Zhang and Laursen (1998)
TT*T*T	82	93	Zhang and Laursen (1998)
TTTT2KE	100	100	Chakrabartty and Hew (1991)
	100	100	Haymet et al. (1999)
BBBB2KE	5	5	Haymet et al. (2001)
AAAA2KE	20	27	Haymet et al. (1999)
IIII2KE	0	0	Haymet et al. (2001)
GGGG2KE	0	0	Haymet et al. (1999)
	0	0	Haymet et al. (1998)
SSSS2KE	0	0	Haymet et al. (1998)
	0	0	Haymet et al. (1999)
VVVV2KE	0	31	Zhang and Laursen (1998)
	90	65	Haymet et al. (2001)
TVVT	88	87	Chao et al. (1997)
VTVT	48	58	Ebbinghaus et al. (2012)
TSTT	81	84	Zhang and Laursen (1998)
TTST	72	76	Zhang and Laursen (1998)
TTTS	83	89	Zhang and Laursen (1998)
STTS	73	68	Zhang and Laursen (1998)
TSST	7	11	Zhang and Laursen (1998)
	0	0	Haymet et al. (1999)
	0	0	Chao et al. (1997)
STSS	0	0	Zhang and Laursen (1998)
SSSS	0	0	Zhang and Laursen (1998)
	0	0	Haymet et al. (1999)

Various mutations have been performed on the HPLC-6 wild type (TTTT), with diverse effect. The hysteresis of the mutants is given as a percentage of the wild type at 1 or 2 mM protein concentration
[a]This peptide is synthesised using D-enantiomer amino acids only

proteins maintained their helical structure (Haymet et al. 1998, 1999; Chao et al. 1997). This observation indicates both that the methyl group has a role in ice binding, and that hydrogen bond mediated by the hydroxyl group does not seem to be the dominant force to drive the ice binding. Substitutions of threonine to valine further investigated the role of the methyl and hydroxyl groups. This substitution entails replacement of the hydroxyl group with a second methyl group (Fig. 14.1). Contrary to serine substitutions, valine substitutions did not eliminate antifreeze activity even when all four threonines were substituted; however, the activity was decreased 30–50% (Haymet et al. 2001; Zhang and Laursen 1998). Substitution of the two central threonines with valine caused a loss of activity of only 10–15% (Zhang and Laursen 1998), while the same substitutions with serines caused a loss of

90–100% (Haymet et al. 1999; Zhang and Laursen 1998; Chao et al. 1997). This indicates that hydrogen bond of threonines is not necessary for antifreeze activity (Haymet et al. 1998), and that the methyl group plays a larger role in ice binding likely through hydrophobic or van der Waal's interaction (Zhang and Laursen 1998). Interestingly, mutants with all threonines substituted with alanine show activity of around 25% of the wild type (Haymet et al. 1999), but when a methyl group is added to them, i.e. the 2-amino butyric acid substitutions, the activity is almost completely lost (Haymet et al. 2001), and the addition of a second methyl group, i.e. valine substitution, some activity is restored. However, the 2-amino butyric acid mutant does not show the same degree of helicity as the wild type or other mutants, which per se could contribute to the lower antifreeze activity (Haymet et al. 2001). Mutants with glycine or isoleucine substitutions showed no activity (Haymet et al. 1998, 2001), likely due to the lack of functional groups to mediate any interaction with ice or because of a too large side chain causing steric hindrances, respectively (Fig. 14.1).

Though the experiments do not show a clear pattern in the molecular mechanism of the ice binding, Zhang and Laursen propose that there are two types of motifs on the ice surface that the protein can bind to—a dominant, hydrophobic one (where the methyl group interact) and a hydrogen bonding one (where the hydroxyl group interact). Only threonine can take advantage of both. Despite there is not much difference in energy between a hydrogen bond to water or to ice, the hydrogen bonds to ice could be favourable once the protein is docked to the ice due to proximity effects (Zhang and Laursen 1998).

Some mutational studies on the HPLC-6 AFP also look beyond the four notorious threonines. Wen and Laursen have performed a series of studies, looking into the role of neutral polar amino acids (Wen and Laursen 1992), charged amino acids (Wen and Laursen 1993c), and the effect of adding amino acids with bulky groups to the protein (Wen and Laursen 1993b). Amongst several observations, they find that the loss of activity is related to a loss of helicity, that the salt bridge (K18-E22) is not absolutely essential for the protein to evoke antifreeze activity and that the asparagine and aspartic acid amino acids are essential for optimal activity. Another study has shown that some substitution with leucine increased the activity of the AFP, however at the expense of solubility, indicating a delicate balance between these two parameters (Loewen et al. 1999). Finally, Ebbinghaus et al. have studied a series of mutants, and found that their activity can be related to their ability to form long range hydration shells, indicating this plays an important role for ice binding (Ebbinghaus et al. 2012).

As the threonines are common in the ice-binding domain of most AFP, the observations, discoveries, and theories from the mutational studies on HPLC-6 can in general be extended to other AFPs as well. However, there is more to the interaction between AFP and ice than can be deduced by the mutational studies, but they provided helpful insight to some of the possible mechanisms. The interaction between AFP and ice are described in detail in Chap. 4.

14.6 Improving Antifreeze Activity

After having deduced the important amino acids and their role in the interaction with ice, investigations have turned the focus to how the antifreeze activity can be improved. In general, two different approaches or strategies can be used to improve protein function by mutagenesis; Directed evolution is non-specific changes to the protein, creating large libraries, which is screened for mutants with enhanced function, i.e. antifreeze activity for AFPs. This strategy has not been used in regard to AFPs, probably due to the lack of appropriate screening methods for large libraries. The second strategy is called rational design, where proteins are changed based on insight into the structural and functional elements of the protein, and thereby hypothesising an expected effect by a given modification. The tool for the latter strategy is SDM, and the outcomes of the experiments of rational design on AFPs are presented below. The concepts of the experiments can be divided in three groups: polymerisation of AFPs, increasing the number of coils and fine-tuning of the IBD.

14.6.1 Polymerisation of Ice-Binding Domains

The mantra of this approach is basically 'why settle for one ice-binding domain when you can have two' (or three or four…). Thus, researchers have polymerised, or made tandem repeats, of an AFP giving each protein a larger ice-binding surface and thus more sites to interact with and bind to the ice, in order to increase the coverage of AFPs on the ice surface, again leading to a higher antifreeze activity. The idea of applying this approach on AFPs probably sprung from the observation that the type III AFP isoform RD3 from the Antarctic eel pout, *Lychodichthys dearborni*, essentially consists of two AFPs joined in tandem by a short linker. This protein showed the same activity as similar monomers on a weight basis, but thus showed twice the activity on a molar basis (Wang et al. 1995).

Inspired by this, Baardsnes et al. made a dimer of another type III AFP, the HPCL-12 isoform from ocean pout *M. americanus*, using the same linking sequence of the RD3 AFP. They too observed twice as high activity on the molar basis compared to the monomer AFP. In addition, when knocking out one of the IBDs they still observed an activity 1.2 times larger than the monomer. The same effect was observed when the dimer was constructed so that one AFP IBD turns the opposite direction, thus disabling both AFPs to interact with the ice simultaneously (Baardsnes et al. 2003). The mere enlargement of the protein thus caused the protein to evoke higher antifreeze activity (per protein), which could be caused by more extensive coverage of the ice surface, making less room for convex ice growth between the adhered proteins. Another or assisting cause could be that the larger proteins have a lower solubility thus driving more proteins to the ice–water interface, essentially causing more proteins to adhere. This effect is discussed in depth in

Chap. 6. Likewise, DeLuca et al. also observed increased activity of a 7 kDa type III AFP when fusing it to non-AFPs (thioredoxin, 12 kDa and maltose-binding protein, 42 kDa), supporting the evidence that the mere size of the AFP affect the antifreeze activity (DeLuca et al. 1998a).

Nishimiya et al. took the experiment a step further, creating also trimers and tetramers of the RD3 AFPs, also combining the composition of the two original (wild type) domains. As in the other experiment, they observed a positive correlation between the number of IBDs and thermal hysteresis. Though the tetramer evoked the highest antifreeze activity on a molar basis, the most active AFP constructs on a per-domain basis was the trimers, thus indicating some saturation point of the effect. Interestingly, it was also observed that the morphology of the formed ice crystals were different for each multimer, indicating different ice-binding planes amongst the AFPs (Nishimiya et al. 2003).

Taking the experiment to the last step, Stevens et al. used a cross-linker, with 16 potential active sites, to create molecules wielding 6–11 RD3 monomeric AFPs. The heterogeneous sample of the constructs showed a greater than fourfold increase in antifreeze activity, over that of the monomeric type III AFP. The dendritic protein complex also showed increased ice recrystallisation inhibition, likely due to an ability to bind multiple adjoining ice surfaces because of its structure (Stevens et al. 2015).

14.6.2 Increasing the Number of Coils

Insect AFPs have a β-coil structure, with a flat ice-binding side, made up of stacked β-sheets. The number of coils, and thus the size of the IBD, varies amongst the AFPs, not only between species, but also within isoforms found in one specie. For example, in *C. fumiferana*, the Cf-501 isoform consists of two coils more than the CfAFP-337 isoform, and it evokes more than twice the antifreeze activity on a weight basis than its smaller counterpart. By removing two coils, thus making the CfAFP-501 isoform the size of the CfAFP-337 isoform, the activity was decreased to a level similar to the smaller isoform. This demonstrates that the difference in activity of the two native isoforms is not due to small differences in amino acid sequence, but solely related to the size of the AFP and its IBD (Leinala et al. 2002).

Because of the repetitive structure of the β-coiled insect AFPs, and in the case of *T. molitor* also the very repetitive amino acid sequence in each coil, it has been possible to construct stable mutants with an increased (or decreased) number of coils. Marshall et al. has performed such study, creating AFPs with 6–11 coils from the seven-coil *T. molitor* wild-type AFP. The study shows that the mutant with only six coils is less active than the wild type, and that the mutants with seven or more coils were more active than the wild type, both on a weight and molar basis. However, the nine-coiled mutant was the most active, and more coils (10–11) led to only slightly improved activities over the wild type. The cause of this is believed to be a slight mismatch between the protein surface and ice matrix, which cumulates for each coil added (Marshall et al. 2004). These observations are in close agreement with a simulation study performed without any prior knowledge of the experimental results (Liu et al. 2005).

A similar study has been performed on the blackspotted pliers support beetle, *Rhagium mordax,* AFP. This protein does not have the same repetitive and rigid structure as the AFP of *T. molitor*, and it was only possible to obtain a correctly folded and active mutant consisting of just one extra coil. This enlarged mutant also showed the same tendency of improved activity compared to the wild type; however, at high concentrations the effect gets less significant, and at concentration above 270 mM, the wild type is predicted to evoke the highest antifreeze activity. Thus, at low concentrations the enlarged mutant is more active than the wild type, but at high concentrations the wild type supersedes the mutant and the wild type also reaches higher activity in total (Friis et al. 2014).

14.6.3 Fine-Tuning of the Ice-Binding Domain

The ice-binding motifs of insect AFPs consist of TxT (or TxTxTxT) motifs neatly aligned on a flat surface of the AFPs. However, all reported sequences have a few imperfections in the TxT motifs, where a random amino is found in a position expected to be occupied by threonine, as shown in Table 14.2. These deviations do not have a clear purpose, and seem only to obstruct the AFPs in making a perfect match with the ice surface. However, the investigations of this matter are few, making it difficult to deduce their function.

The *R. mordax* AFP has six TxTxTxT motifs in its IBD which includes three deviations where other amino acids are occupying a threonine's position (See Table 14.2). Two of these have been substituted with threonine both separately and combined (H21T, K23T and H21T/K23T), to make the IBD more uniform. All three mutants showed increased activity compared to the wild type and the effect seems cumulative as the double substitution was the most effective. However, the beneficial effect was only exerted at low and moderate concentrations, as the wild-type AFP was predicted to become more active at high concentrations, as also observed for the mutant with an extra coil. It is speculated that this tendency is due to a change in the binding pattern induced by the mutations, caused by an increased or decreased affinity for specific ice planes (Friis et al. 2014). Another explanation could be that the uniform IBDs have an increased tendency to interact with each other at higher concentrations, preventing them from binding to the ice.

Gwak et al. have also been able to increase the thermal hysteresis evoked by an AFP by site-directed mutagenesis. The study was performed on an AFP from an Antarctic marine diatom, *Chaetoceros neogracile*. This *Cn*-AFP also has a flat IBD, though the motifs are not regular TxT, but lack uniformity. However, substituting a small glycine in the IBD with threonine or tyrosine, the hysteresis was increased from 1.2 °C to 1.7 and 1.9 °C, respectively. Surprisingly, the substitution with the large tyrosine led to the highest hysteresis. The reason behind this is believed to be

Table 14.2 AFP ice-binding motifs

AFP	Ice-binding motif							References
iwAFP-FL2	TxTx**V**xTxT	TxTxTxTxT	TxTxTx**R**xT	TxTxTxTxT				Lin et al. (2011)
RiAFP	Tx**H**x**K**xT	TxTxTxT	TxTxTxT	TxTxTxT	TxTxTxT	TxTxTxT		Kristiansen et al. (2011)
RmAFP1	Tx**H**x**K**xT	TxTxTxT	TxTxTxT	TxTxTxT	TxTxTxT	TxTxTx**S**		Kristiansen et al. (2012)
Tm4-9	**A**xT	TxT	TxT	TxT	TxT	TxT	**A**xT	Liou et al. (1999)
Cf501	**V**xT	TxT	**I**xT	**V**xT	TxT	TxT	TxT	Doucet et al. (2000)
MpAFP149	**A**xT	TxT	TxT	TxT	TxT	TxT	**A**xT	Qiu et al. (2010b)
DAFP1	**S**xT	**A**xT	TxT	TxT	TxT	**A**xT		Huang et al. (2002)
ApAFP752	**Q**xT	**E**xT	TxT	TxT	Tx**I**	Tx**I**		Mao et al. (2011)

The repetitive sequence, TxT, of insect AFPs contains deviations from the consensus sequence, marked with bold the tyrosine forms a hydrogen bond with a nearby aspartic acid, arranging the tyrosine ring almost parallel to the IBD (Gwak et al. 2014). The flatness of the IBD thus seems crucial for the degree of thermal hysteresis the antifreeze proteins can exert.

14.7 Creating Cold-Tolerant Organisms

Quite different investigations of the AFPs, that also entail genetic manipulation, involve the creation of cold- or freeze-tolerant organisms. In these studies, an AFP gene is transferred to an embryo of a species that does not naturally have AFPs. The grown individual from this embryo will then be able to express AFPs, and then, potentially, have acquired some degree of cold-hardiness. These experiments have been conducted on plants, animals, and unicellular organisms, using AFPs from different sources and with different potencies.

14.7.1 Plants

The goal of creating new cold-hardy plants is clear: to make freeze-sensitive crops more robust to brief temperature drops to below 0 °C that could otherwise weaken or kill the crops. The advances on this field are described in Sect. 7.10, and will only be described briefly here.

Though the main hypothesis is that the introduction of AFPs should make an organism more cold-hardy, the results are not unambiguous. For example, experiments on strains of the tobacco plant, *Nicotiana tabacum*, have shown both positive results (Wang et al. 2008) or no effect (Kenward et al. 1999). However, in experiments like these, there are many factors that can affect the outcome of the experiment, such as the type of AFPs used, how the transgenic organisms were constructed, the degree of expressed AFP, in which tissues etc. Furthermore, how the experiment to evaluate cold-hardiness is conducted (e.g. temperatures, duration, plant age) could also affect the result.

For example, the engineering of cold-hardy plants seems to be more likely to succeed when using AFPs from plants, rather than from fish or insects (Bredow and Walker 2017). The plant AFPs show low thermal hysteresis, and is more known to exert inhibition of recrystallisation, which could be the reasoning behind this tendency.

In general, the experiments show good results, and the potential of creating cold-hardy crops seems to be real, but it is likely that the solution is dependent on the crop to be modified, and the climate it needs to overcome (i.e. the magnitude of temperature drops).

14.7.2 Animals

A few different transgenic animal species have also been constructed and investigated. Bagis et al. have constructed mice expressing the *M. americanus* type III AFP, in all the tissues tested, including testis and ovarian tissues (Bagis et al. 2006). Their aim was not to create cold-hardy mice, but instead to improve the cold and freeze tolerance of the ovarian and testis tissue, to improve the survival during cryopreservation. In this way, the cryopreserved cells will express AFPs on their own, instead of adding the AFPs extracellular to the cryopreserving solutions, as is the case for most cryopreservation experiments involving AFPs (see Chap. 11). Their experiments showed that the AFPs protected the ovarian and testis tissue during hypothermic storage at 4 °C (Bagis et al. 2006), and resulted in continuing fertility of ovarian tissue after being vitrified at −196 °C (Bagis et al. 2008).

Heisig et al. have also studied the cold tolerance of transgenic mice, expressing the IAFGP from the tick *Ixodes scapularis*. After exposing skin fibroblast to 4 °C for 21 days they observed improved cell survival and increased in the relative cell count after cold treatment in the transgenic mice compared to the control. Likewise, it was also demonstrated that transgenic mice were more freeze tolerant. By submerging the tails of anesthetised mice in a −22 °C cold bath, and assess the following necrosis, it was observed that while only 11% of the control mice were free of frost bites during the follow-up period, this number was 60% for the transgenic mice, concluding a cryoprotective effect of the IAFGP in the transgenic mice (Heisig et al. 2015).

The fruit fly *Drosophila melanogaster* has also been subject of transgenic studies. As well as for the plants, the results are not unambiguous. Again, this could likely be linked to the different AFPs, construct and experimental setups that are used amongst the researchers.

Duncker et al. have constructed a transgenic *Drosophila* line, with the females producing Atlantic wolfish, *Anarhichas lupus*, type III AFP, while the males do not express the AFP. This was verified with measurements of thermal hysteresis. Both constructs express xanthine dehydrogenase, used as a marker gene to verify the transgenics. In cold exposure trials at 0 and −7 °C both sexes had a higher survival rate than the non-transgenic control, and the AFP could thus not be responsible for the effect. As xanthine dehydrogenase is part of the synthesis of uric acid—which confers protection against free radicals that are thought to rise after hypothermic exposure—this enzyme is thought to be the cause for the improved survival of the transgenic flies (Duncker et al. 1995).

Likewise, Tyshenko and Walker have constructed transgenic *Drosophila* lines expressing a *C. fumiferana* AFP. The different lines showed different degrees of thermal hysteresis in the hemolymph. The flies were exposed to 0 °C for 24–36 h and tested for survival. Their results showed no significant difference between the lines with the highest antifreeze activity and the control at any time point tested, demonstrating that AFPs does not confer cold tolerance in *Drosophila* (Tyshenko and Walker 2004).

Amongst experiments where an improvement of cold tolerance is seen is a study by Lin et al. Here, *Drosophila* lines were constructed to express *Dendroides canadensis* AFPs DAFP-1 and/or DAFP-4, with either one or two copies of the gene. The thermal hysteresis in the hemolymph of the different lines with AFPs spanned from 2 to 7 °C. The flies were subjected to cold exposure trials at 0 and 4 °C for various time periods. Firstly, they observed that the female flies had a greater ability to survive the cold trials, and therefore decided to perform the analyses of the sexes separately. At 0 °C the survival rate of the controls were much lower than the AFP-expressing lines after 30 h, and at 34 h the controls were dead, while a survivorship of 20–70% of the different AFP-expressing lines were observed. The same tendency was seen amongst males, though with generally lower surviving rates and a less clear difference between controls and transgenics, however still significant. At 4 °C, the flies generally lived longer, and again the AFP-expressing lines outlived the controls amongst females. The picture amongst males was not as clear, as the lines expressing DAFP-4 only, did not show better survivorship than the controls. In conclusion, the *D. canadensis* AFPs were effective choices for the promotion of cold tolerance of *Drosophila*, with the possibility of further improvement with expression of additional AFPs or increased expression levels (Lin et al. 2010).

Neelakanta et al. have also conducted experiments that show increased cold tolerance of *Drosophila*. In their studies, a transgenic line expressing *I. scapularis* AFGP was used. Firstly, they investigated the effect of cold exposure trials, and found that the time before half of the flies were killed was increased by 1 day in the AFGP-expressing line for both males and females. Interestingly, the males generally showed the best tolerance in this experiment, contrary to that of Lin et al. (2010). Secondly, hatching rates of *Drosophila* embryos after storage at −5 °C for 60–150 min were found to be two- to threefold higher for the AFGP-expressing line compared to the controls. Lastly, they showed that the expression of AFGP protected the flies from apoptotic cell death during cold trials (7 days at 4 °C) using the TUNEL assay, suggesting that the *I. scapularis* AFGP increases the cold tolerance of the flies, at least in part, by preventing the initiation of apoptosis (Neelakanta et al. 2012).

14.7.3 Unicellular Organisms

As unicellular organisms, most often bacteria, are utilised to express various recombinant AFPs, numerous of such constructs has been made. However, the cells are usually lysed in order to isolate the AFPs for studies, and are therefore not used for in vivo experiments. While the in vivo effects of AFPs are sometimes included in the experiments, the results are usually just included as a passing remark. A few studies with the main scope of testing the AFPs' cold-hardening effect in the unicellular organisms have been conducted, and will be described here.

In 1991, McKown and Warren created a *Saccharomyces cerevisiae* (baker's yeast) that expressed the type I AFP from *P. americanus*. The AFP was expressed as a chimeric protein with staphylococcal protein A (Spa). The presence of active AFP was confirmed with the splat cooling essay where inhibition of recrystallisation was observed. To test the freeze tolerance, the yeast cells (approx. 3×10^7 cells/ml) were cooled to −196 °C and held for 30 min. The cells were heated at three difference rates and spread out on agar plates to quantify cell survival. At the highest heating rate (64 °C/min) the AFP-expressing yeast cells showed more than twice as high survival rate compared to the control. At lower heating rates the positive effect of AFPs was not as profound, despite the fact that detrimental recrystallisation (inhibited by the AFP) is expected to be higher, the lower the heating rate. Overall, transgenic yeast showed higher survival after freezing, thus indicating that the AFPs did improve the freezing tolerance of *S. cerevisiae* (McKown and Warren 1991).

As the use of frozen dough for bread making has become popular, the interest for freeze-tolerant yeast has also increased. Panadero et al. have investigated two *S cerevisiae* strains expressing a type I AFP from *Myoxocephalus aenaeus* (grubby sculpin) both in regard to yeast freeze tolerance and frozen dough quality. The strains were designed to secrete the AFP from the cells. Their results showed increased yeast survival (up to 32-fold for one strain) during storage at −20 °C for 13 days compared to the control. They also observed an increased gas production capacity (important for dough leavening) after long-term (40 days) storage at −20 °C for the AFP-expressing yeast strains. However, this effect was not evident at short-term (hours) storage. Lastly they investigated the strains' ability to withstand osmotic stresses as this is a key point in the bread making process, but here no differences amongst the strains were observed. So though improving freeze tolerance, the AFP-expressing yeast strains did not obtain increased osmoprotection (Panadero et al. 2005).

In similar experiments, expression of AFP has also increased freeze tolerance of *Lactococcus lactis*, an important bacterium for manufacturing dairy products where it could have great potentials (Zhang et al. 2018), as well as in *E. coli*, where an increased osmotic tolerance also was observed (Meijer et al. 1996; Holmberg et al. 1994).

In the experiments on the transgenic organisms, it is often seen that AFPs have a positive effect on cold exposures above 0 °C, thus increasing the cold-hardiness of the organisms. This however, cannot be contributed the AFPs' abilities to bind to ice and lower the freezing point or inhibit ice recrystallisation. The protective effect is likely linked to a protection against both chronic and acute cold stress to membrane components, which is believed to be evoked by an interaction between AF(G)Ps and integral cell membranes (Hays et al. 1996; Rubinsky et al. 1990; Tablin et al. 1996; Tomczak et al. 2001, 2002).

14.8 Conclusions

The mutation studies that have been conducted on AFPs have led to useful insight into several subjects. Fluorescent fusion proteins unveil which planes of the ice the different AFPs bind to, substitution with obstructive amino acids gives strong indication of the location of the IBD and systematic substitution of the side chains of key amino acids have shed some light on the important features of the ice-binding mechanism. Also, some studies show that by transferring AFPs to cold-susceptible organisms, the trait of cold-hardiness is inherited as well, at least in part. Looking forward, the studies have also demonstrated that the activity or efficiency of the AFPs can be manipulated, opening up for the opportunity to improve the antifreeze activity. By changing the physicochemical properties on AFPs, e.g. the isoelectric point or temperature and pH stability, one can also tailor the AFPs to specific applications, making the proteins suitable for various commercial markets. For example, a study has shown that even a few extra amino acids at the ends of a type III AFP have increased the heat stability by 5 °C without affecting the activity (Li et al. 1991). Tailoring of the AFPs does seem feasible, as the proteins have shown to be receptive to various mutations and maintaining their tertiary structure. Even though the course of evolution has optimised the AFPs, it has only been in the aspect of the proteins' biological role, in a complex system of countless other proteins, cells and compartments of which undesirable interactions must be avoided. Thus, in a different environment, with no other proteins, a different pH or salinity, the AFPs needs to be optimised to evoke full antifreeze activity, which might be stronger than seen by the wild type.

So far, only the rational design strategy has been applied to create mutant AFPs. Though having high success rate, the process is time-consuming and relatively few mutants can be studied. The strategy of directed evolution would provide much more mutants, many of which would be rubbish, but also some that could turn out to be highly active and consist of mutations we ourselves would not have foreseen to have a positive effect. However, the main obstacle for this strategy is the lack of a suitable method for screening large libraries of mutants for their antifreeze activity. Development of an efficient screening method would thus open up for large-scale mutant screening and speed up the process of developing the AFPs of the future.

References

Baardsnes J, Kondejewski LH, Hodges RS, Chao H, Kay C, Davies PL (1999) New ice-binding face for type I antifreeze protein. FEBS Lett 463(1–2):87–91

Baardsnes J, Jelokhani-Niaraki M, Kondejewski LH, Kuiper MJ, Kay CM, Hodges RS, Davies PL (2001) Antifreeze protein from shorthorn sculpin: identification of the ice-binding surface. Protein Sci 10(12):2566–2576. https://doi.org/10.1110/ps.ps.26501

Baardsnes J, Kuiper MJ, Davies PL (2003) Antifreeze protein dimer: when two ice-binding faces are better than one. J Biol Chem 278(40):38942–38947. https://doi.org/10.1074/jbc.M306776200

Bagis H, Aktoprakligil D, Mercan HO, Yurdusev N, Turgut G, Sekmen S, Arat S, Cetin S (2006) Stable transmission and transcription of newfoundland ocean pout type III fish antifreeze protein (AFP) gene in transgenic mice and hypothermic storage of transgenic ovary and testis. Mol Reprod Dev 73(11):1404–1411. https://doi.org/10.1002/mrd.20601

Bagis H, Akkoc T, Tass A, Aktoprakligil D (2008) Cryogenic effect of antifreeze protein on transgenic mouse ovaries and the production of live offspring by orthotopic transplantation of cryopreserved mouse ovaries. Mol Reprod Dev 75(4):608–613. https://doi.org/10.1002/mrd.20799

Bang JK, Lee JH, Murugan RN, Lee SG, Do H, Koh HY, Shim HE, Kim HC, Kim HJ (2013) Antifreeze peptides and glycopeptides, and their derivatives: potential uses in biotechnology. Mar Drugs 11(6):2013–2041. https://doi.org/10.3390/md11062013

Basu K, Garnham CP, Nishimiya Y, Tsuda S, Braslavsky I, Davies P (2014) Determining the ice-binding planes of antifreeze proteins by fluorescence-based ice plane affinity. J Vis Exp 83: e51185. https://doi.org/10.3791/51185

Basu K, Wasserman SS, Jeronimo PS, Graham LA, Davies PL (2016) Intermediate activity of midge antifreeze protein is due to a tyrosine-rich ice-binding site and atypical ice plane affinity. FEBS J 283(8):1504–1515. https://doi.org/10.1111/febs.13687

Bredow M, Walker VK (2017) Ice-binding proteins in plants. Front Plant Sci 8:2153–2153. https://doi.org/10.3389/fpls.2017.02153

Celik Y, Drori R, Pertaya-Braun N, Altan A, Barton T, Bar-Dolev M, Groisman A, Davies PL, Braslavsky I (2013) Microfluidic experiments reveal that antifreeze proteins bound to ice crystals suffice to prevent their growth. Proc Natl Acad Sci U S A 110(4):1309–1314. https://doi.org/10.1073/pnas.1213603110

Chakrabartty A, Hew CL (1991) The effect of enhanced alpha-helicity on the activity of a winter flounder antifreeze polypeptide. Eur J Biochem 202(3):1057–1063

Chao H, Sönnichsen FD, DeLuca CI, Sykes BD, Davies PL (1994) Structure-function relationship in the globular type III antifreeze protein: identification of a cluster of surface residues required for binding to ice. Protein Sci 3(10):1760–1769. https://doi.org/10.1002/pro.5560031016

Chao H, Houston ME, Hodges RS, Kay CM, Sykes BD, Loewen MC, Davies PL, Sönnichsen FD (1997) A diminished role for hydrogen bonds in antifreeze protein binding to ice. Biochemistry 36(48):14652–14660. https://doi.org/10.1021/bi970817d

DeLuca CI, Chao H, Sönnichsen FD, Sykes BD, Davies PL (1996) Effect of type III antifreeze protein dilution and mutation on the growth inhibition of ice. Biophys J 71(5):2346–2355. https://doi.org/10.1016/S0006-3495(96)79476-6

DeLuca CI, Comley R, Davies PL (1998a) Antifreeze proteins bind independently to ice. Biophys J 74(3):1502–1508. https://doi.org/10.1016/S0006-3495(98)77862-2

DeLuca CI, Davies PL, Ye Q, Jia Z (1998b) The effects of steric mutations on the structure of type III antifreeze protein and its interaction with ice. J Mol Biol 275(3):515–525. https://doi.org/10.1006/jmbi.1997.1482

Devries AL, Lin Y (1977) Structure of a peptide antifreeze and mechanism of adsorption to ice. Biochim Biophys Acta 495(2):388–392

Doucet D, Tyshenko MG, Kuiper MJ, Graether SP, Sykes BD, Daugulis AJ, Davies PL, Walker VK (2000) Structure-function relationships in spruce budworm antifreeze protein revealed by isoform diversity. Eur J Biochem 267(19):6082–6088

Duncker BP, Chen CP, Davies PL, Walker VK (1995) Antifreeze protein does not confer cold tolerance to transgenic Drosophila melanogaster. Cryobiology 32(6):521–527. https://doi.org/10.1006/cryo.1995.1054

Ebbinghaus S, Meister K, Prigozhin MB, Devries AL, Havenith M, Dzubiella J, Gruebele M (2012) Functional importance of short-range binding and long-range solvent interactions in helical antifreeze peptides. Biophys J 103(2):L20–L22. https://doi.org/10.1016/j.bpj.2012.06.013

Ewart KV, Fletcher GL (1993) Herring antifreeze protein: primary structure and evidence for a C-type lectin evolutionary origin. Mol Mar Biol Biotechnol 2(1):20–27

Ewart KV, Rubinsky B, Fletcher GL (1992) Structural and functional similarity between fish antifreeze proteins and calcium-dependent lectins. Biochem Biophys Res Commun 185 (1):335–340

Ewart KV, Li Z, Yang DS, Fletcher GL, Hew CL (1998) The ice-binding site of Atlantic herring antifreeze protein corresponds to the carbohydrate-binding site of C-type lectins. Biochemistry 37(12):4080–4085. https://doi.org/10.1021/bi972503w

Friis DS, Kristiansen E, von Solms N, Ramløv H (2014) Antifreeze activity enhancement by site directed mutagenesis on an antifreeze protein from the beetle Rhagium mordax. FEBS Lett 588 (9):1767–1772. https://doi.org/10.1016/j.febslet.2014.03.032

Garnham CP, Natarajan A, Middleton AJ, Kuiper MJ, Braslavsky I, Davies PL (2010) Compound ice-binding site of an antifreeze protein revealed by mutagenesis and fluorescent tagging. Biochemistry 49(42):9063–9071. https://doi.org/10.1021/bi100516e

Graether SP, Kuiper MJ, Gagné SM, Walker VK, Jia Z, Sykes BD, Davies PL (2000) Beta-helix structure and ice-binding properties of a hyperactive antifreeze protein from an insect. Nature 406(6793):325–328. https://doi.org/10.1038/35018610

Gwak Y, Jung W, Lee Y, Kim JS, Kim CG, Ju JH, Song C, Hyun JK, Jin E (2014) An intracellular antifreeze protein from an Antarctic microalga that responds to various environmental stresses. FASEB J 28(11):4924–4935. https://doi.org/10.1096/fj.14-256388

Haymet AD, Ward LG, Harding MM, Knight CA (1998) Valine substituted winter flounder 'antifreeze': preservation of ice growth hysteresis. FEBS Lett 430(3):301–306

Haymet ADJ, Ward LG, Harding MM (1999) Winter flounder "antifreeze" proteins: synthesis and ice growth inhibition of analogues that probe the relative importance of hydrophobic and hydrogen-bonding interactions. J Am Chem Soc 121(5):941–948. https://doi.org/10.1021/ja9801341

Haymet AD, Ward LG, Harding MM (2001) Hydrophobic analogues of the winter flounder 'antifreeze' protein. *FEBS Lett* 491(3):285–288

Hays LM, Feeney RE, Crowe LM, Crowe JH, Oliver AE (1996) Antifreeze glycoproteins inhibit leakage from liposomes during thermotropic phase transitions. Proc Natl Acad Sci U S A 93 (13):6835–6840

Heisig M, Mattessich S, Rembisz A, Acar A, Shapiro M, Booth CJ, Neelakanta G, Fikrig E (2015) Frostbite protection in mice expressing an antifreeze glycoprotein. PLoS One 10(2):e0116562. https://doi.org/10.1371/journal.pone.0116562

Holmberg N, Lilius G, Bulow L (1994) Artificial antifreeze proteins can improve NaCl tolerance when expressed in E. coli. FEBS Lett 349(3):354–358

Huang T, Nicodemus J, Zarka DG, Thomashow MF, Wisniewski M, Duman JG (2002) Expression of an insect (Dendroides canadensis) antifreeze protein in Arabidopsis thaliana results in a decrease in plant freezing temperature. Plant Mol Biol 50(3):333–344

Jiang M, Ma J, Qiu LM (2011) Cryoprotective effect of an insect antifreeze protein MpAFP 698 and its mutants from the desert beetle Microdera punctipennis. Cryo Letters 32(5):436–446

Jorgensen WL, Tirado-Rives J (1988) The OPLS potential functions for proteins, energy minimizations for crystals of cyclic peptides and crambin. J Am Chem Soc 110(6):1657–1666. https://doi.org/10.1021/ja00214a001

Kenward KD, Brandle J, McPherson J, Davies PL (1999) Type II fish antifreeze protein accumulation in transgenic tobacco does not confer frost resistance. Transgenic Res 8(2):105–117. https://doi.org/10.1023/A:100888662

Kim H-E, Lee A-R, Lee Y-M, Jeong M, Park C-J, Lee J-H (2013) Hydrogen exchange study of winter flounder type I antifreeze protein. Bull Kor Chem Soc 34(10):3137–3140

Knight CA, Cheng CC, DeVries AL (1991) Adsorption of alpha-helical antifreeze peptides on specific ice crystal surface planes. Biophys J 59(2):409–418. https://doi.org/10.1016/S0006-3495(91)82234-2

Kondo H, Hanada Y, Sugimoto H, Hoshino T, Garnham CP, Davies PL, Tsuda S (2012) Ice-binding site of snow mold fungus antifreeze protein deviates from structural regularity and high conservation. Proc Natl Acad Sci U S A 109(24):9360–9365. https://doi.org/10.1073/pnas.1121607109

Kristiansen E, Ramløv H, Højrup P, Pedersen SA, Hagen L, Zachariassen KE (2011) Structural characteristics of a novel antifreeze protein from the longhorn beetle Rhagium inquisitor. Insect Biochem Mol Biol 41(2):109–117. https://doi.org/10.1016/j.ibmb.2010.11.002

Kristiansen E, Wilkens C, Vincents B, Friis D, Lorentzen AB, Jenssen H, Løbner-Olesen A, Ramløv H (2012) Hyperactive antifreeze proteins from longhorn beetles: some structural insights. J Insect Physiol 58(11):1502–1510. https://doi.org/10.1016/j.jinsphys.2012.09.004

Leinala EK, Davies PL, Doucet D, Tyshenko MG, Walker VK, Jia Z (2002) A beta-helical antifreeze protein isoform with increased activity. Structural and functional insights. J Biol Chem 277(36):33349–33352. https://doi.org/10.1074/jbc.M205575200

Li XM, Trinh KY, Hew CL (1991) Expression and characterization of an active and thermally more stable recombinant antifreeze polypeptide from ocean pout, Macrozoarces americanus, in Escherichia coli: improved expression by the modification of the secondary structure of the mRNA. Protein Eng 4(8):995–1002

Li Z, Lin Q, Yang DS, Ewart KV, Hew CL (2004) The role of Ca2+-coordinating residues of herring antifreeze protein in antifreeze activity. Biochemistry 43(46):14547–14554. https://doi.org/10.1021/bi048485h

Lin X, O'Tousa JE, Duman JG (2010) Expression of two self-enhancing antifreeze proteins from the beetle Dendroides canadensis in Drosophila melanogaster. J Insect Physiol 56(4):341–349. https://doi.org/10.1016/j.jinsphys.2009.11.005

Lin FH, Davies PL, Graham LA (2011) The Thr- and Ala-rich hyperactive antifreeze protein from inchworm folds as a flat silk-like β-helix. Biochemistry 50(21):4467–4478. https://doi.org/10.1021/bi2003108

Liou YC, Thibault P, Walker VK, Davies PL, Graham LA (1999) A complex family of highly heterogeneous and internally repetitive hyperactive antifreeze proteins from the beetle Tenebrio molitor. Biochemistry 38(35):11415–11424. https://doi.org/10.1021/bi990613s

Liu K, Jia Z, Chen G, Tung C, Liu R (2005) Systematic size study of an insect antifreeze protein and its interaction with ice. Biophys J 88(2):953–958. https://doi.org/10.1529/biophysj.104.051169

Loewen MC, Gronwald W, Sönnichsen FD, Sykes BD, Davies PL (1998) The ice-binding site of sea raven antifreeze protein is distinct from the carbohydrate-binding site of the homologous C-type lectin. Biochemistry 37(51):17745–17753

Loewen MC, Chao H, Houston ME, Baardsnes J, Hodges RS, Kay CM, Sykes BD, Sönnichsen FD, Davies PL (1999) Alternative roles for putative ice-binding residues in type I antifreeze protein. Biochemistry 38(15):4743–4749. https://doi.org/10.1021/bi982602p

Mao X, Liu Z, Ma J, Pang H, Zhang F (2011) Characterization of a novel β-helix antifreeze protein from the desert beetle Anatolica polita. Cryobiology 62(2):91–99. https://doi.org/10.1016/j.cryobiol.2011.01.001

Marshall CB, Daley ME, Graham LA, Sykes BD, Davies PL (2002) Identification of the ice-binding face of antifreeze protein from Tenebrio molitor. FEBS Lett 529(2-3):261–267

Marshall CB, Daley ME, Sykes BD, Davies PL (2004) Enhancing the activity of a beta-helical antifreeze protein by the engineered addition of coils. Biochemistry 43(37):11637–11646. https://doi.org/10.1021/bi0488909.

McKown RL, Warren GJ (1991) Enhanced survival of yeast expressing an antifreeze gene analogue after freezing. Cryobiology 28(5):474–482

Meijer PJ, Holmberg N, Grundstrom G, Bulow L (1996) Directed evolution of a type I antifreeze protein expressed in Escherichia coli with sodium chloride as selective pressure and its effect on antifreeze tolerance. Protein Eng 9(11):1051–1054

Middleton AJ, Brown AM, Davies PL, Walker VK (2009) Identification of the ice-binding face of a plant antifreeze protein. FEBS Lett 583(4):815–819. https://doi.org/10.1016/j.febslet.2009.01.035

Mok YF, Lin FH, Graham LA, Celik Y, Braslavsky I, Davies PL (2010) Structural basis for the superior activity of the large isoform of snow flea antifreeze protein. Biochemistry 49 (11):2593–2603. https://doi.org/10.1021/bi901929n

Neelakanta G, Hudson AM, Sultana H, Cooley L, Fikrig E (2012) Expression of Ixodes scapularis antifreeze glycoprotein enhances cold tolerance in Drosophila melanogaster. PLoS One 7(3): e33447. https://doi.org/10.1371/journal.pone.0033447

Nishimiya Y, Ohgiya S, Tsuda S (2003) Artificial multimers of the type III antifreeze protein. Effects on thermal hysteresis and ice crystal morphology. J Biol Chem 278(34):32307–32312. https://doi.org/10.1074/jbc.M304390200

Nishimiya Y, Kondo H, Takamichi M, Sugimoto H, Suzuki M, Miura A, Tsuda S (2008) Crystal structure and mutational analysis of Ca2+-independent type II antifreeze protein from longsnout poacher, Brachyopsis rostratus. J Mol Biol 382(3):734–746. https://doi.org/10.1016/j.jmb.2008.07.042

Panadero J, Randez-Gil F, Prieto JA (2005) Heterologous expression of type I antifreeze peptide GS-5 in baker's yeast increases freeze tolerance and provides enhanced gas production in frozen dough. J Agric Food Chem 53(26):9966–9970. https://doi.org/10.1021/jf0515577

Pertaya N, Marshall CB, DiPrinzio CL, Wilen L, Thomson ES, Wettlaufer JS, Davies PL, Braslavsky I (2007) Fluorescence microscopy evidence for quasi-permanent attachment of antifreeze proteins to ice surfaces. Biophys J 92(10):3663–3673. https://doi.org/10.1529/biophysj.106.096297

Pertaya N, Marshall CB, Celik Y, Davies PL, Braslavsky I (2008) Direct visualization of spruce budworm antifreeze protein interacting with ice crystals: basal plane affinity confers hyperactivity. Biophys J 95(1):333–341. https://doi.org/10.1529/biophysj.107.125328

Provencher SW, Glöckner J (1981) Estimation of globular protein secondary structure from circular dichroism. Biochemistry 20(1):33–37

Qiu LM, Ma J, Wang J, Zhang FC, Wang Y (2010a) Thermal stability properties of an antifreeze protein from the desert beetle Microdera punctipennis. Cryobiology 60(2):192–197. https://doi.org/10.1016/j.cryobiol.2009.10.014

Qiu L, Wang Y, Wang J, Zhang F, Ma J (2010b) Expression of biologically active recombinant antifreeze protein His-MpAFP149 from the desert beetle (Microdera punctipennis dzungarica) in Escherichia coli. Mol Biol Rep 37(4):1725–1732. https://doi.org/10.1007/s11033-009-9594-3

Rubinsky B, Arav A, Mattioli M, Devries AL (1990) The effect of antifreeze glycopeptides on membrane potential changes at hypothermic temperatures. Biochem Biophys Res Commun 173 (3):1369–1374

Sicheri F, Yang DS (1995) Ice-binding structure and mechanism of an antifreeze protein from winter flounder. Nature 375(6530):427–431. https://doi.org/10.1038/375427a0

Slaughter D, Fletcher GL, Ananthanarayanan VS, Hew CL (1981) Antifreeze proteins from the sea raven, Hemitripterus americanus. Further evidence for diversity among fish polypeptide antifreezes. J Biol Chem 256(4):2022–2026

Stevens CA, Drori R, Zalis S, Braslavsky I, Davies PL (2015) Dendrimer-linked antifreeze proteins have superior activity and thermal recovery. Bioconjug Chem 26(9):1908–1915. https://doi.org/10.1021/acs.bioconjchem.5b00290

Tablin F, Oliver AE, Walker NJ, Crowe LM, Crowe JH (1996) Membrane phase transition of intact human platelets: correlation with cold-induced activation. J Cell Physiol 168(2):305–313. https://doi.org/10.1002/(sici)1097-4652(199608)168:2<305::aid-jcp9>3.0.co;2-t

Tomczak MM, Hincha DK, Estrada SD, Feeney RE, Crowe JH (2001) Antifreeze proteins differentially affect model membranes during freezing. Biochim Biophys Acta 1511 (2):255–263

Tomczak MM, Hincha DK, Estrada SD, Wolkers WF, Crowe LM, Feeney RE, Tablin F, Crowe JH (2002) A mechanism for stabilization of membranes at low temperatures by an antifreeze protein. Biophys J 82(2):874–881. https://doi.org/10.1016/s0006-3495(02)75449-0

Tyshenko MG, Walker VK (2004) Hyperactive spruce budworm antifreeze protein expression in transgenic Drosophila does not confer cold shock tolerance. Cryobiology 49(1):28–36. https://doi.org/10.1016/j.cryobiol.2004.04.002

Wang X, DeVries AL, Cheng CH (1995) Antifreeze peptide heterogeneity in an antarctic eel pout includes an unusually large major variant comprised of two 7 kDa type III AFPs linked in tandem. Biochim Biophys Acta 1247(2):163–172

Wang Y, Qiu L, Dai C, Wang J, Luo J, Zhang F, Ma J (2008) Expression of insect (Microdera puntipennis dzungarica) antifreeze protein MpAFP149 confers the cold tolerance to transgenic tobacco. Plant Cell Rep 27(8):1349–1358. https://doi.org/10.1007/s00299-008-0562-5

Wen D, Laursen RA (1992) Structure-function relationships in an antifreeze polypeptide. The role of neutral, polar amino acids. J Biol Chem 267(20):14102–14108

Wen D, Laursen RA (1993a) A D-antifreeze polypeptide displays the same activity as its natural L-enantiomer. FEBS Lett 317(1-2):31–34

Wen D, Laursen RA (1993b) Structure-function relationships in an antifreeze polypeptide. The effect of added bulky groups on activity. J Biol Chem 268(22):16401–16405

Wen D, Laursen RA (1993c) Structure-function relationships in an antifreeze polypeptide. The role of charged amino acids. J Biol Chem 268(22):16396–16400

Zepeda S, Yokoyama E, Uda Y, Katagiri C, Furukawa Y (2008) Situ observation of antifreeze glycoprotein kinetics at the ice interface reveals a two-step reversible adsorption mechanism. Cryst Growth Des 8(10):3666–3672. https://doi.org/10.1021/cg800269w

Zhang W, Laursen RA (1998) Structure-function relationships in a type I antifreeze polypeptide. The role of threonine methyl and hydroxyl groups in antifreeze activity. J Biol Chem 273 (52):34806–34812

Zhang L, Jin Q, Luo J, Wu J, Wang S, Wang Z, Gong S, Zhang W, Lan X (2018) Intracellular expression of antifreeze peptides in food grade lactococcus lactis and evaluation of their cryoprotective activity. J Food Sci 83(5):1311–1320. https://doi.org/10.1111/1750-3841.14117

Part IV
Closure

Chapter 15
Summary and Future Directions

Hans Ramløv and Dennis Steven Friis

Despite the fact that much progress of understanding the antifreeze proteins, AFPs, (ice-binding proteins ad recrystallization inhibitor proteins) has been made during the last four decades, many questions are still unanswered and each time one question is answered, more seems to appear.

When studying the cold-tolerant organisms which synthesize AFPs it soon becomes clear that there is much more to be learned about the interplay between general physiology of the organisms and the physiology and natural history of their cold tolerance. AFPs are found both in circulation in the blood and in the hemolymph of various species, but they are also found in the skin and intestinal system of different fishes as well as intracellularly in a number of animals. Some isoforms of AFPs also only differ with respect to whether they are expressed with a signal peptide or not, which could indicate that they are either exported to the extracellular compartments of retained inside the cells. Thus it is not only the organism as a whole which is protected by AFPs, but specific compartments and structures which may be especially vulnerable or which are in contact with ice for very long periods are apparently specifically fortified with specific AFPs. In insects, for example, to a large extent it remains to be investigated where the different isoforms are found. As we have seen in, for example, the freeze-tolerant alpine cockroach *Celattoblatta quinquemaculata* AFPs are found only in association with the digestive tract. One can speculate that this is to avoid the growth of large amounts of ice in the intestine, which would draw water out of the tissues into the digestive tract leading to rupture of this as the ice crystals grow. However, this is pure speculation and experiments performed to investigate the role of AFPs in *C. quinquemaculata* are needed to

H. Ramløv (✉)
Department of Natural Sciences, Roskilde University, Roskilde, Denmark
e-mail: hr@ruc.dk

D. S. Friis
Copenhagen, Denmark

H. Ramløv, D. S. Friis (eds.), *Antifreeze Proteins Volume 2*,
https://doi.org/10.1007/978-3-030-41948-6_15

further our understanding of the role of the AFPs in the general cold tolerance physiology of this animal. Similar remarks can be made about many of the insects in which AFPs are found. In *Rhagium mordax* some of the isoforms isolated are lacking a signal peptide, but it still remains to be elucidated if this animal retains an intracellular complement of AFPs to avoid intracellular ice formation. Thus, careful dissection of specimens of this beetle to separate various tissues for isolating the AFP complement of the single tissues could be performed. This would still not prove that the purpose of intracellular AFPs is to avoid intracellular ice formation wherefore experiments with supercooling points of the single tissues could be performed.

In Antarctic fishes it has been shown that ice crystals are mainly found in the spleen, but a small number may be found in other tissues as well. However, as stated in the chapter by A.L. DeVries, at this stage it is not known how the ice crystals end up in the spleen, let alone where they are located and how they are voided from the spleen so that the spleen does not end up being totally infested with ice crystals. Experiments designed to solve such a problem are not without challenges; we are looking at ice crystals which will readily disappear at the slightest rise in temperature, the fish are found in the Antarctic and they have to be caught in the ocean and transported to the laboratory at cold temperature. The latter is due to the fact that it has so far not been possible to artificially induce ice crystals in fish which have been kept in aquaria above the melting temperature of the ice crystals. This makes studies of the ice crystals in the spleen difficult as the ice crystals can in no way be marked. However, isotope experiments with ice crystals made from water containing the oxygen radioisotope ^{15}O could (although the half-life is relatively short) possibly be performed once it has been discovered how the ice crystals enter the fish in nature. This would give a possibility to study the route of the ice crystals from the blood to the spleen and possibly into the cells. But there is no doubt that the technical challenges in such a study are significant. The occurrence and possible distribution of ice crystals in the tissues of north polar fish has not yet been determined. In these fishes the problem of the accumulation of ice crystals is less severe as the north polar seas often warm to higher temperatures than the melting point of the blood of the fishes living there during the summer. Thus, ice crystals which have possibly entered the blood and tissues may be melted once the sea temperature increases during summer. However, it is not known if the north polar fishes accumulate ice crystals during winter at all.

The interplay between AFPs and the immune response insects has just only been touched upon. The so far published results indicate that the AFPs can enhance the immune response and disrupt the formation of biofilms produced by pathogens in some insects. Both from an adaptive and from an application point of view more studies are of great interest. Thus AFPs in these animals may have a double role and in an evolutionary context this may be of great interest when the evolution of AFP genes in insects is going to be studied in detail.

Most of the studies investigating the interplay between AFPs and biological membranes (mostly model membranes) have been performed with fish AFPs and the conclusions are drawn in that ecological context. However, as a large number of cold-tolerant organisms synthesize AFPs, it is now of great interest to study the interaction between, for example, insects' AFPs and various model membrane systems. As seen above it is possible that AFPs have second roles in the cold tolerance physiology of the organisms which still are unexplored. Even if interactions between a model membrane system and a certain kind of AFP cannot be considered a second role of the AFP or maybe even not being adaptive to the intact organism the interaction may be of interest both to understand more about the nature of the AFP, membranes interactions with various substances, and in an application context.

It has been shown by various authors that the synthesis of AFPs is controlled both by temperature and light. It has also been shown that it is not necessarily enough to acclimatize an organism to low temperature and initiation of AFP synthesis via a simple drop in temperature or by changing the light cycle so it has longer darkness than light. However, more complex temperature regimes such as several cycles of decreased temperature and subsequent return to a higher temperature may be necessary to evoke the synthesis of cryoprotective substances some of which may be AFPs. These controlling factors are quite complex to study and only few studies have been performed and even fewer not only on the organismal level but also on the molecular level. Thus, future studies of the complex cues for initiation of AFP synthesis both on the organismal level and the molecular level are of great interest.

An observation that for many years has puzzled researchers of AFPs is the occurrence of various isoforms of the same AFP in one organism. For example, the antifreeze glycoproteins (AFGPs) in Antarctic notothenioid fishes occur in a number of sizes with rather large differences in antifreeze activity in the blood of these. It has been speculated that in concert they may have a synergistic effect so that the antifreeze activity is enhanced in solutions where the AFPs occur together. Some studies have not been able to show any enhancement or synergy whereas others have (Baardsnes et al. 2003; Nishimiya et al. 2005; Ramlov et al. 2005). In AFP type III it is well established that the rather passive SP isoforms, that are found in much higher abundance in AFP type III bearing fishes than the QAE isoforms, are enhanced by the QAE isoforms (Nishimiya et al. 2005; Takamichi et al. 2009). However, until now the exact mechanism of the synergistic effect resulting from the combination of one or more isoforms of AFPs has not been resolved. One explanation may be that there is a sequential binding of the isoforms to the ice surface due to differences in adsorption rates and in the strength of the binding to the ice. This would mean that the isoform with the highest adsorption rate binds first, slowing down the ice growth which facilitates the binding of the slower and more passive isoform. In a recent paper Berger et al. studied the combination effects on enhancement or reduction of antifreeze activity synergistic effects on a combination of AFPI, AFPIII-QAE, AFPIII-SP as well as AF(G)Ps isoforms from fish and plants using the nanoliter osmometer, microfluidics and fluorescence microscopy. The authors showed that in experiments with various AF(G)Ps even in experiments which included cross-organismal AFPs, there was no direct correlation between the structure of the protein or its biological origin in its effect of the antifreeze activity. The three dimensional structure of the AFP determines to which ice crystal plane it binds and that in turn determines the effect on the antifreeze activity in the mixture. From these experiments the authors were able to put forward a kinetic model that explains the synergistic or additive effects on the antifreeze activity in binary mixtures of isoforms. The model is limited to AFPs that cannot bind to basal planes and thus combinatory effects of insect AFPs are not explained by the model. In the proposed model the active isoform of AF(G)P show high adsorption rate as well as binding to the prism plane. The passive isoform binds to the pyramidal planes and at a slower rate. This way a so-called complex consisting of ice and the active isoform bound to the prism plane and the passive isoform bound to the pyramidal plane. Thus the inhibition of ice growth by the AF(G)Ps is a result of solution concentration (the passive form having the highest concentration) and their intrinsic adsorption rates (Berger et al. 2019). Future experiments on insect AFPs using a similar approach may indeed yield very interesting results concerning the interplay of various isoforms in the same organism.

It is beyond doubt that AFPs are adsorbing or binding (the denomination often rely on the context and on the researcher) to ice crystals and even to specific planes as exemplified in the paper by Berger et al. mentioned above. It can also for all practical purposes be stated that the binding by the AFPs to the ice surface is virtually irreversible as long as there is ice present. However, the description of the exact mechanism by which the AFPs are bound to the ice surface has so far mostly been a matter of macroscopic observations combined with theoretical models and molecular dynamic simulation models. Until recently the binding of AFPs to an ice surface has not been observed experimentally on the molecular level. Recently Meister et al. published results on the binding of the antifreeze RmAFP from the bark beetle *Rhagium mordax* to ice studied by surface-specific heterodyne-detected vibrational sum-frequency generation (HD-VSFG) spectroscopy and molecular dynamic simulations. Using this technique (HD-VSFG) the results show that unlike other proteins, RmAFP retains it hydrating water molecules upon binding to the ice surface. The hydrating water molecules are situated in a different hydrogen-bonding structure than that of the water molecules in the ice surface. This inhibits the insertion of water layers between the AFP and the ice surface. The ice-binding surface of the RmAFP consists of ridges and valleys and the authors propose that it is the hydrogen-bond structure of the water molecules in the valleys that inhibits the intercalation of ice layers in between the RmAFP molecules and the ice surface. Also, the proteins' hydration water molecules in the interfacial region are not associated with those in the ice crystal and the water domains between the ice-binding surface and the ice surface have a tetrahedrally coordinated arrangement (Meister et al. 2019).

A note on the studies of physical chemistry of AFPs may be on its place here. Rather few studies of this kind has been performed and to achieve a full understanding both of combinatory effects as mentioned above (for example, measuring and understanding adsorption rates) and of the mechanism by which the AFPs adsorb/ bind to the ice surface, we believe that an increased knowledge of the general physical chemistry of AFPs is one way to refine the various models having been put forward.

In the world of AFPs, we see many examples of convergent evolution and the proteins are found in a wide variety of organisms. It is thus possible that there are new groups of AFPs to be discovered and maybe even more potent ones than those we already know of. As we continuously get more insight into the mysteries of AFPs and the interactions with ice and water on a molecular plan, we will also understand more about what makes an AFP more or less potent, and eventually tailor AFPs by site-directed mutagenesis that can evoke the highest thermal hysteresis. But is this necessarily what we seek? If we need to exploit the qualities of AFPs, which many industries already speculate in, then the optimal goal is probably not the highest possible thermal hysteresis that is the objective. It is very likely that other properties such as stability, solubility, size, or maybe the specific ice plane the proteins bind to are of higher interest. Or maybe it is the recrystallization inhibition that is of interest and not the thermal hysteresis at all.

The ability to tailor AFPs to quite specific purposes is definitely there, as several experiments using site-directed mutagenesis have shown that the activity and overall structure are intact even after relatively large interventions such as inserting extra coils within the protein. Also, it has been possible to change single amino acids, changing the physicochemical properties of the proteins, without affecting antifreeze activity.

One of the industries interested in utilizing AFPs, and already does so to some extent, is the food industry. Here it is especially the ice recrystallization inhibition ability that has great interest, and some frozen foods already contain AFPs to improve the texture. Another aspect is prolongation of shelf-lives of frozen foods. However, it is often difficult to get the AFPs uniformly distributed into the food unless it is a homogenous product. A solution could be to construct transgenic organisms, expressing their own AFPs in their tissues, providing a "natural" ice recrystallization inhibition mechanism. However, it takes a lot of effort creating transgenic organisms, not to mention getting them cleared by authorities to ending up in the refrigerated counters.

Another field of application of AFPs is cryopreservation which in many ways is the application nature invented it for. The preservation could be of cells, tissue, or organs. However, it has shown to be extremely complex to exploit the cryopreserving effect of AFPs, and many studies have shown both positive and negative effects of adding the proteins to preserving solutions. There is a complex interaction between AFPs, additives, and the biological materiel that is to be preserved, and there a numerous parameters to take into account. There is no doubt about the potential, especially within organ preservations, where a prolongation of the shelf-lives could make the difference between life and death for terminal patients waiting for a donor organ.

There are many more potential industrial applications of the AFPs, e.g., inhibition of gas hydrates and surface treatments described in this book, and many more will probably arise as the understanding of the AFPs becomes more complete.

In this book we have tried to incorporate all of the state-of-the-art knowledge about AFPs/ice binding proteins at this time. It is our hope that the book has given the reader a background for going even more into detail with various areas of AFP research and that it also works as a large bibliography which can lead interested scholars to the right literature within the field. The book should hopefully also have led to the opening of the readers' imagination so the reader ponders over new ideas both to the solution of existing questions within the field but also to the posing of new questions leading to further insights and studies into the full picture of the role of the AFPs in the intact organism, the evolution, not only of AFPs, but also evolutionary mechanisms in general, the molecular function and its mechanism as well as to potential applications.

References

Baardsnes J, Kuiper MJ, Davies PL (2003) Antifreeze protein dimer: when two ice-binding faces are better than one. J Biol Chem 278:38942–38947

Berger T, Meister K, DeVries AL, Eves R, Davies PL, Drori R (2019) Synergy between antifreeze proteins is driven by complementary ice-binding. J Am Chem Soc 141:19144

Meister K, Moll CJ, Chakraborty S, Jana B, DeVries AL, Ramlov H, Bakker HJ (2019) Molecular structure of a hyperactive antifreeze protein adsorbed to ice. J Chem Phys 150:131101

Nishimiya Y, Sato R, Takamichi M, Miura A, Tsuda S (2005) Co-operative effect of the isoforms of type III antifreeze protein expressed in Notched-fin eelpout, Zoarces elongatus Kner. FEBS J 272:482–492

Ramlov H, DeVries AL, Wilson PW (2005) Antifreeze glycoproteins from the antartic fish Dissostichus mawsoni studied by differential scanning calorimetry (DSC) in combination with nanolitre osmometry. CryoLetters 26:73–84

Takamichi M, Nishimiya Y, Miura A, Tsuda S (2009) Fully active QAE isoform confers thermal hysteresis activity on a defective SP isoform of type III antifreeze protein. FEBS J 276:1471–1479

The manufacturer's authorised representative in the EU is Springer Nature Customer Service Centre GmbH, Europaplatz 3, 69115 Heidelberg, Germany. If you have any concerns regarding our products, please contact ProductSafety@springernature.com

Printed and bound by CPI Group (UK) Ltd, Croydon, CR0 4YY
30/09/2026
02225240-0008